INTRODUCTION TO ENVIRONMENTAL SOIL PHYSICS

INTRODUCTION TO ENVIRONMENTAL SOIL PHYSICS

Daniel Hillel

Professor Emeritus
Plant, Soil, and Environmental Sciences
University of Massachusetts

ELSEVIER
ACADEMIC
PRESS

Amsterdam Boston Heidelberg London New York Oxford Paris
San Diego San Francisco Singapore Sydney Tokyo

This book is printed on acid-free paper. ♾

Academic Press
An imprint of Elsevier
525 B Street, Suite 1900, San Diego, California 92101-4495, USA
http://www.academicpress.com

Academic Press
An imprint of Elsevier
84 Theobald's Road, London WC1X 8RR, UK
http://www.academicpress.com

Academic Press
An imprint of Elsevier
200 Wheeler Road, Burlington, Massachusetts 01803, USA
www.academicpressbooks.com

Library of Congress Cataloging-in-Publication Data: A catalog record for this book is available from the Library of Congress.

ISBN-13: 978-0-12-348655-4
ISBN-10: 0-12-348655-6

Printed and bound by CPI Group (UK) Ltd, Croydon, CR0 4YY

Transferred to Digital Print 2011

I dedicate this edition
to my beloved daughter Sari
whose young life sparkled
with kindness and wisdom and joy

CONTENTS

LIST OF TEXT BOXES

To see a world in a grain of sand
and heaven in a wild flower
Hold infinity in the palm of your hand
and eternity in an hour.
William Blake (1757–1827)
Songs of Innocence and Experience

PREFACE

What is special about this fragmented and loose outer layer of the earth's continental surface, which we call "soil"? What role does it fulfill in generating and sustaining life on earth, and why do we need to study it?

Considering the height of the atmosphere, the thickness of the earth's rock mantle, and the depth of the ocean, we note that the soil is an amazingly thin body — typically not much more than one meter thick and often less than that. Yet it is the crucible of terrestrial life, within which biological productivity is generated and sustained. It acts like a composite living entity, being home to a community of innumerable microscopic and macroscopic plants and animals. A mere fistful of soil typically contains millions of microorganisms, which perform the most vital functions of biochemistry. Another amazing attribute of the soil is its spongelike porosity and its enormous internal surface area. That same fistful of soil may actually consist of several acres of active surface upon which physicochemical processes take place continuously.

Realizing humanity's dependence on the soil, ancient peoples, who lived intimately with nature, actually revered the soil. It was their source of livelihood, as well as the material with which they built homes and that they learned to shape, heat, and fuse into household vessels and writing tablets (ceramic being the first synthetic material in the history of technology). In the Bible, the name assigned to the first human being was Adam, derived from the Hebrew word *adama*, meaning "soil." The name given to his mate was Hava (Eve, in transliteration), meaning "living" or "life-giving." Together, therefore, Adam and Eve signified quite literally "soil and life."

The same powerful metaphor is echoed in the Latin name for the human species — *Homo* — derived from *humus*, the material of the soil. Hence, the adjective *human* also implies "of the soil." Other ancient cultures evoked equally powerful associations. To the ancient Greeks, the earth was a manifestation of Gaea, the maternal goddess who, impregnated by Uranus (god of the sky), gave birth to all the disparate gods of the Greek pantheon.

Our civilization depends on the soil more crucially than ever, because our numbers have grown while available soil resources have diminished and deteriorated. Paradoxically, however, even as our dependence on the soil has increased, most of us have become physically and emotionally detached from it. The majority of the people in the so-called "developed" countries spend their lives in the artificial environment of a city, insulated from direct exposure to nature. Many children now assume as a matter of course that food originates in supermarkets.

Detachment has bred ignorance, and out of ignorance has come the arrogant delusion that our civilization has risen above nature and has set itself free of its constraints. Agriculture and food security, erosion and salination, degradation of natural ecosystems, depletion and pollution of surface waters and aquifers, and decimation of biodiversity — all these processes, which involve the soil directly or indirectly, have become mere abstractions to most people. The very language we use betrays disdain for that common material underfoot, often referred to as "dirt." Some fastidious parents prohibit their children from playing in the mud and rush to wash their "soiled" hands when the children nonetheless obey an innate instinct to do so. Thus is devalued and treated as unclean what is in fact the terrestrial realm's principal medium of purification, wherein wastes are decomposed and nature's productivity is continually rejuvenated.

Scientists who observe the soil discern a seething foundry in which matter and energy are in constant flux. Radiant energy from the sun streams onto the field and cascades through the soil and the plants growing in it. Heat is exchanged, rainwater percolates in the intricate passages of the soil, plant roots suck up that water and transmit it to their leaves, which transpire it back to the atmosphere. The leaves absorb carbon dioxide from the air and synthesize it with soil-derived water to form the primary compounds of life: carbohydrates, fats, proteins, and numerous other compounds (many of which provide medicinal as well as nutritional value). Oxygen emitted by the leaves makes the air breathable for animals, which feed on and in turn fertilize the plants.

The soil is thus a self-regulating biophysical factory, utilizing its own materials, water, and solar energy. It also determines the fate of rainfall and snowfall reaching the ground surface — whether the water thus received will flow over the land as runoff or seep downward to the subterranean reservoir called groundwater, which in turn maintains the steady flow of springs and streams. With its finite capacity to absorb and store moisture, the soil regulates all of these phenomena. Without the soil as a buffer, rain falling over the continents would run off immediately, producing violent floods rather than sustained stream flow.

The soil naturally acts as a living filter in which pathogens and toxins that might otherwise accumulate to foul the terrestrial environment are rendered harmless and transmuted into nutrients. Since time immemorial, humans and other animals have been dying of all manner of diseases and have then been buried in the soil, yet no major disease is transmitted by it. The term *antibiotic* was coined by soil microbiologists, who, as a consequence of their studies of soil bacteria and actinomycetes, discovered streptomycin (an important cure for tuberculosis and other infections). Ion exchange, a useful process of water purification, was also discovered by soil scientists studying the passage of solutes through beds of clay.

However unique in form and function, the soil is not an isolated body. It is, rather, a central link in the chain of interconnected domains comprising the terrestrial environment. The soil interacts with both the overlying atmosphere and the underlying strata, as well as with surface and underground bodies of water. Especially important is the interrelation between the soil and the climate. In addition to its function of regulating the cycle of water, soil regulates energy exchange and temperature. When virgin land is cleared of vegetation and turned into a cultivated field, the native biomass above the ground is often burned and the organic matter within the soil tends to decompose rapidly. These processes release carbon dioxide into the atmosphere, thus contributing to the earth's greenhouse effect and to global warming. On the other hand, the opposite action of soil enrichment with organic matter, such as can be achieved by means of reforestation and conservation farming, may help to absorb carbon dioxide from the atmosphere. To an extent, the soil's capacity to absorb and sequester carbon can thus help to mitigate the atmosphere's so-called greenhouse effect.

It takes nature thousands of years to create life-giving soil out of sterile bedrock. It takes but a few decades for unknowing or uncaring humans to destroy that wondrous work of nature. It is for us who do care for future generations to treat the soil with respect and humility, another word derived from *humus*. In the Book of Genesis, humans are said to have been placed in the Garden of Eden for a purpose, "to serve and preserve it." There is a profound truth in that perception. The earth and its soil can be a veritable Garden of Eden, but only if we do not despoil it and thereby banish ourselves from a life of harmony within it.

"To the wise man, the whole world's a soil," wrote Ben Johnson (1573–1637). His thought was echoed by William Butler Yeats (1865–1939): "All that we did, all that we said or sang, must come from contact with the soil."

This book is an abridged and updated version of my earlier book, published in 1998 under the title "Environmental Soil Physics." As such, this version is intended to serve as a basic text for introductory undergraduate courses in soil physics for students in the environmental, agricultural, and engineering sciences. The book is also meant to appeal to a broad range of students and professionals, as well as educated lay readers outside those formal categories, who may wish to acquire a fundamental understanding of the principles and processes governing the ways the soil functions in natural and in managed ecosystems. Those who seek a more detailed treatment of the various topics introduced herewith are invited to consult the larger book (1998), as well as the numerous references listed in the text and the Bibliography.

A textbook on so vital a subject ought by right to capture and convey the special sense of wonderment and excitement that impels the scientist's quest to comprehend the workings of nature, and hence should give some pleasure in the reading. It is my hope that this book might indeed be pondered, not merely studied, that its readers might find within it a few insights as well as facts, and that it will deepen their understandings as well as broaden their knowledge.

Acknowledgments

I am grateful to Dr. Charles Crumly, Editor-in-Chief of Applied Sciences at Academic Press/Elsevier, for initiating and encouraging the writing of this book, and to Ms. Angela Dooley, the Production Editor, for her empathy and competence in nursing the manuscript to publication.

Daniel Hillel
May, 2003

Part I

BASIC RELATIONSHIPS

1. SOIL PHYSICS AND SOIL PHYSICAL CHARACTERISTICS

SOIL SCIENCE

To begin, we define *soil* as the weathered and fragmented outer layer of the earth's terrestrial surface. It is formed initially through disintegration, decomposition, and recomposition of mineral material contained in exposed rocks by physical, chemical, and biological processes. The material thus modified is further conditioned by the activity and accumulated residues of numerous microscopic and macroscopic organisms (plants and animals). In a series of processes that may require hundreds or even thousands of years and that is called *soil genesis*, the loose debris of rock fragments is transmuted into a more or less stable, internally ordered, actively functioning natural body. Ultimately, this culminates in the formation of a characteristic *soil profile*, which resembles a layer cake. We can visualize the soil profile as a composite living body, in the same way that we think of the human body as a distinct organism, even though in reality it is an ensemble of numerous interdependent and symbiotically coordinated groups of organelles, cells, organs, and colonies of myriad organisms.

Soil science is the study of the soil in all its ramified manifestations and facets: as a central link in the biosphere, as a medium for the production of agricultural commodities, and as a raw material for industry and construction. As such, it shares interests with geology, sedimentology, terrestrial ecology, and geobotany as well as with such applied sciences as agronomy and engineering. Because of its varied interests and concerns, soil science itself is commonly divided into several subdivisions, including pedology (soil formation and classification), soil chemistry, soil mineralogy, soil biology, soil fertility, and soil mechanics. However, such distinctions are often arbitrary, because in fact all of the environmental sciences are inextricably interconnected.

SOIL PHYSICS

Soil physics is one of the major subdivisions of soil science. It seeks to define, measure, and predict the physical properties and behavior of the soil, both in its natural state and under the influence of human activity. As physics deals in general with the forms and interactions of matter and energy, so soil physics deals specifically with the state and movement of matter and with the fluxes and transformations of energy in the soil. On the one hand, the fundamental study of soil physics aims at understanding the mechanisms governing such processes as terrestrial energy exchange, the cycles of water and of transportable materials, and the growth of plants in the field. On the other hand, the practical application of soil physics aims at the proper management of the soil by means of cultivation, irrigation, drainage, aeration, improvement of soil structure, control of infiltration and evaporation, regulation of soil temperature, and prevention of erosion.

Soil physics is thus both a basic and an applied science, with a very wide range of interests. The study of soil science in general and of soil physics in particular is driven not only by the innate curiosity that is our species' main creative impulse, but also by urgent necessity. The intensifying pressure of population and development has diminished the soil resources of our small planet and has led to their unsustainable use and degradation in too many parts of the world.

Since the soil is not an isolated medium but is in constant dynamic interaction with the larger environment, soil physics is an aspect of the more encompassing field of environmental physics (sometimes called biospheric physics) and of the overall science of geophysics.

The early soil physicists were interested primarily in the engineering and the agricultural aspects of their discipline, hence their research focused on the soil as a material for construction or as a medium for the production of crops. Recent decades have witnessed an increasing emphasis on the environmental aspects and applications of soil physics. Consequently, research in soil physics has expanded its scope to include phenomena related to natural ecosystems and to processes affecting the quality of the environment. Processes occurring in the soil are now seen to affect the entire terrestrial environment, including local and regional climates, the natural food chain, biodiversity, and the fate of the voluminous waste products of our civilization (among which are many pathogenic and toxic agents).

Increasingly, the main concern of soil physics has shifted from the laboratory to the field and from a restricted one-dimensional view to an expansive three-dimensional view interfacing with the domains of sister disciplines such as meteorology and climatology, hydrology, ecology, and geochemistry. The larger domain of soil physics now encompasses greater complexity and variability in space and time, the treatment of which requires reliance on stochastic as well as deterministic methods. Consequently, the science is becoming ever more interesting and relevant.

The task of soil physics is made difficult by the enormous and baffling intricacy of a medium containing myriad mineral and organic components, all irregularly fragmented and variously associated in a geometric pattern that is so complex and labile as to challenge our imagination and descriptive powers. Some of the solid material consists of crystalline particles, while some is made up of amorphous gels that may coat the crystals and modify their behavior. The solid phase in the soil interacts with the fluids, water, and air that permeate soil pores. The entire soil is hardly ever in equilibrium as it alternately wets and dries, swells and shrinks, disperses and flocculates, hardens and softens, warms and cools, freezes and thaws, compacts and cracks, absorbs and emits gases, adsorbs and releases exchangeable ions, dissolves and precipitates salts, becomes acidic or alkaline, and exhibits aerobic or anaerobic conditions leading to chemical oxidation or reduction.

THE SOIL PROFILE

The most obvious part of any soil is its surface zone. Through it, matter and energy are transported between the soil and the atmosphere. The surface may be smooth or pitted, granular or crusted and cracked, friable or hard, level or sloping, vegetated or fallow, mulched or exposed. Such conditions affect the processes of radiation and heat exchange, water and solute movement, and gaseous diffusion.

Important though the surface is, however, it does not necessarily portray the character of the soil as a whole. To describe the latter, we must examine the soil in depth. We can do this by digging a trench and sectioning the soil from the surface downward, thus revealing what is commonly termed the *soil profile*.

The soil profile typically consists of a succession of more-or-less distinct strata. These may result from the pattern of deposition, or sedimentation, as can be observed in wind-deposited (*aeolian*) soils and particularly in water-deposited (*alluvial*) soils. If, however, the strata form in place by internal soil-forming (*pedogenic*) processes, they are called *horizons*.

The top layer, or *A horizon*, is the zone of major biological activity and is therefore generally enriched with organic matter and darker in color than the underlying soil. Here, plants and animals and their residues interact with an enormously diverse and labile multitude of microorganisms, such as bacteria, protozoa, and fungi, millions of which can be found in a mere handful of topsoil. In addition, there are usually varied forms of macroorganisms (including earthworms, arthropods, and rodents) that burrow in the soil. The A horizon is generally the most fertile zone of the soil, but it is also the zone most vulnerable to erosion by water and wind (especially if it is denuded of vegetative cover or its protective residues).

Underneath the A horizon is the *B horizon*, where some of the materials (e.g., clay or carbonates) that are leached from the A horizon by percolating water tend to accumulate. The B horizon is often thicker than the A horizon. The pressure of the overlying soil tends to reduce the porosity of the deeper layers. In some cases, an overly dense or indurated B horizon may inhibit gas exchange, water drainage, and root penetration.

Underlying the B horizon is the C *horizon*, which is the soil's parent material. In a soil formed of bedrock in situ (called a *residual soil*), the C horizon consists of the weathered and fragmented rock material. In other cases, the C horizon may consist of alluvial, aeolian, or glacial deposits.

The character of the profile depends primarily on the climate that prevailed during the process of soil formation. It also depends on the parent material, the vegetation, the topography, and time. These five variables are known as the *factors of soil formation* (Fanning and Fanning, 1989). Mature soils are such that have been subjected to those factors for a sufficient length of time so that full profile development has taken place. The A, B, C sequence of horizons is clearly recognizable in some cases, as, for example, in a typical zonal soil (i.e., a soil associated with a distinct climatic zone), such as a *podzol* (also known as a *spodosol*). In other cases, no clearly developed B horizon is discernible, and the soil is then characterized by an A,C profile. In a recent alluvium, hardly any profile differentiation is apparent.

The typical development of a soil and its profile, called *pedogenesis* (Buol et al., 2003), can be summarized: It begins with the physical disintegration of an exposed rock formation, which provides the soil's parent material. Gradually, the loosened material is colonized by living organisms. The consequent accumulation of organic residues at and below the surface brings about the development of a discernible A horizon. That horizon may acquire an aggregated structure, stabilized to some degree by gluelike components of the organic matter complex (known as *humus*) resulting from the decomposition of plant and animals residues. Continued weathering (decomposition and recomposition) of minerals may bring about the formation of clay. Some of the clay thus formed tends to migrate downward, along with other transportable materials (such as soluble salts), and to accumulate in an intermediate zone (the B horizon) between the surface zone of major biological activity and the deeper parent material of the so-called C horizon.

Important aspects of soil formation and profile development are the twin processes of *eluviation* and *illuviation* ("washing out" and "washing in," respectively), wherein clay and other substances emigrate from the overlying eluvial A horizon and accumulate in the underlying illuvial B horizon. The two horizons come to differ substantially in composition and structure.

Throughout these processes, the profile as a whole deepens as the upper part of the C horizon is gradually transformed, until eventually a quasi-stable condition is approached in which the counterprocesses of soil formation and soil erosion are more or less in balance. In the natural state, the A horizon may have a thickness of 0.1–0.5 meters. When stripped of vegetative cover and pulverized or compacted by tillage or traffic, this horizon may lose half or more of its original thickness within a few decades.

In arid regions, salts such as calcium sulfate and calcium carbonate, dissolved from the upper part of the soil, may precipitate at some depth to form a cemented *pan* (sometimes called *caliche*, from the Spanish word for "lime"). Erosion of the A horizon may bring the B horizon to the surface. In extreme cases, both the A and B horizons may be scoured off by natural or human-induced erosion so that the C horizon becomes exposed and a new cycle of soil formation may then begin. In other cases, a mature soil may be covered by a new layer of sediments (alluvial or aeolian) so that a new soil may form over a "buried" old soil. Where episodes of deposition occur repeatedly over a long period of time, a sequence of soils may be formed in succession, thus recording the pedological history — called the *paleopedology* — of the region (including evidence of the climate and vegetation that had prevailed at the time of each profile's formation).

Numerous variations of the processes described are possible, depending on local conditions. The characteristic depth of the soil, for instance, varies from location to location. Valley soils are typically deeper than hillslope soils, and the depth of the latter depends on slope steepness. In places, the depth of the profile can hardly be ascertained, because the soil blends into its parent material without any distinct boundary. However, the zone of

BOX 1.1 Soil Physics and the Environment

As human populations have grown and living standards have risen, the requirements for agricultural products have increased enormously. More land has been brought under cultivation, including marginal land that is particularly vulnerable to degradation by such processes as erosion, depletion of organic matter and nutrients, pollution, waterlogging, and salination. Other forms of land use — towns, roads, factories, airports, feedlots, waste disposal sites, and recreational facilities — usurp ever more land.

Consequently, the domain remaining for natural ecosystems has shrunk and been divided into smaller enclaves, to the detriment of numerous species. The mutual checks and balances that have long sustained the rich diversity of life on earth are now threatened by the human appetite for resources and the wanton way they are used and their waste products discarded. The task therefore is to supply human needs in ways that are sustainable locally and do not damage the larger environment.

Two alternative approaches have been proposed to prevent further destruction of the remaining natural ecosystems and to relieve pressure on fragile marginal lands. One way is to restrict human activities to choice areas, where production can be intensified. This calls for optimizing all production factors so as to achieve maximum efficiency in the utilization of soil, water, energy, and other necessary inputs (e.g., nutrients and pest control measures). The problems are that soil processes are difficult to control, and, because the soil is an open system in constant interaction with its surroundings, the time-delayed and space-removed consequences of soil processes are difficult to predict. Full control of agricultural production can ultimately be achieved only in enclosed spaces such as greenhouses or in confined fields.

Another approach is to devise more naturalistic modes of production that are compatible with the environment and do not require the drastic isolation of production from neighboring ecosystems. This agro-ecosystem approach is exemplified by the trends toward organic farming and agroforestry.

Either way, the physical attributes and processes of the soil are of prime importance. Physical factors strongly affect whether the soil is to be cool or warm, anaerobic or aerobic, wet or dry, compact or highly porous, hard or friable, dispersed or aggregated, impervious or permeable, eroded or conserved, saline or salt free, leached or nutrient rich. All these, in turn, determine whether the soil can be a favorable or unfavorable medium for various types of plants and other living organisms as well as for alternative modes of production; in short, whether the soil can be managed productively and sustainably while neutralizing — rather than transmitting — environmental pollutants.

biological activity seldom extends below 2–3 meters and in many cases is shallower than 1 meter.

SOILS OF DIFFERENT REGIONS

Each climatic zone exhibits a characteristic group of soils. In the humid tropics, there is a tendency to dissolve and leach away the silica and to accumulate iron and aluminum oxides. As a result, the soils are typically colored red, the hue of oxidized iron. Chunks or blocks excavated from such soils and dried in the sun may harden to form bricks; hence these soils are called *laterites*, from the Latin word *later*, meaning "brick." On the other hand, soils of the humid cool regions often exhibit an A horizon consisting of a thin surface layer darkened by organic matter and underlain by a bleached, ashlike layer; in turn, this overlies a clay-enriched B horizon. These soil were called *podzols* by the early Russian pedologists (from the Russian words *pod* = "ground" and *zola* = "ash") and are known as *spodosols* in the current American system. Soils that are poorly drained ("waterlogged") may exhibit conditions of chemical reduction in the profile, indicated by streaks of discoloration ("mottling").

In contrast with the soils of humid regions, from which nearly all readily soluble salts have been leached, the soils of arid regions tend to precipitate the moderately soluble salts of calcium and magnesium. (Especially prevalent are accumulations of calcium carbonate [lime] and calcium sulfate [gypsum].) Under certain conditions, arid-zone soils may even accumulate the more readily soluble salts of sodium (NaCl and Na_2CO_3) and of potassium. Such soils are prone to excessive salinity, in extreme cases of which they become practically sterile. Irrigated soils in poorly drained river valleys of arid regions are particularly liable to undergo the process of salination. When such soils are leached of excess salts, they must be treated with soil amendments (such as gypsum) to replace the exchangeable sodium ions with calcium, lest the sodium ion cause dispersion of the clay and the breakdown of soil structure.

An outstanding soil formed naturally in some intermediate semihumid to semiarid regions (e.g., Ukraine, Argentina, and the prairie states of North America — the so-called "corn belt" of the United States) is the soil classically called *chernozem* (Russian for "black earth") with its unusually thick, humus-rich, fertile A horizon.

Whereas Russian pedologists were the first to develop a universal classification system of soils over a century ago, other schools of pedology have since offered alternative taxonomies, claimed to be more detailed and comprehensive. Notable among these is the systems offered by the U.N. Food and Agriculture Organization and the one developed by the U.S. Soil Survey. Each system recognizes hundreds of soil types and their variants.

By way of illustration, we present a hypothetical soil profile in Fig. 1.1. This is not a "typical" soil, for among the myriad of differing soil types recognized by pedologists no single type can be considered typical. Our figure is only meant to suggest the sort of contrasts in appearance and structure among different horizons that may be encountered in a soil profile. Pedologists classify soils by their mode of formation (genesis) and recognizable properties (see Fig. 1.2).

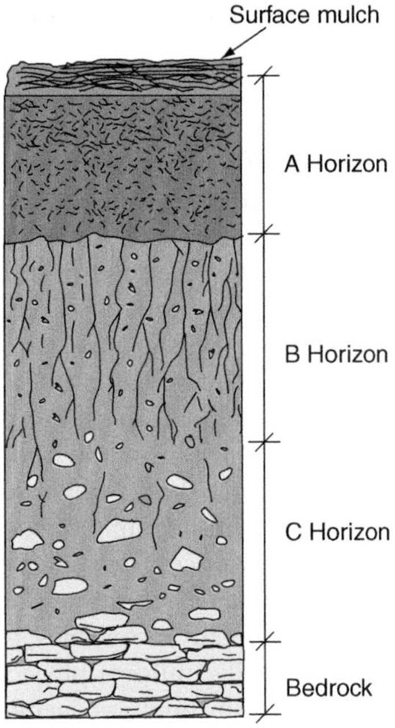

Fig. 1.1. Schematic representation of a hypothetical soil profile. The A horizon is shown with an aggregated crumblike structure, the B horizon with columnar structure, and the C horizon with incompletely weathered rock fragments.

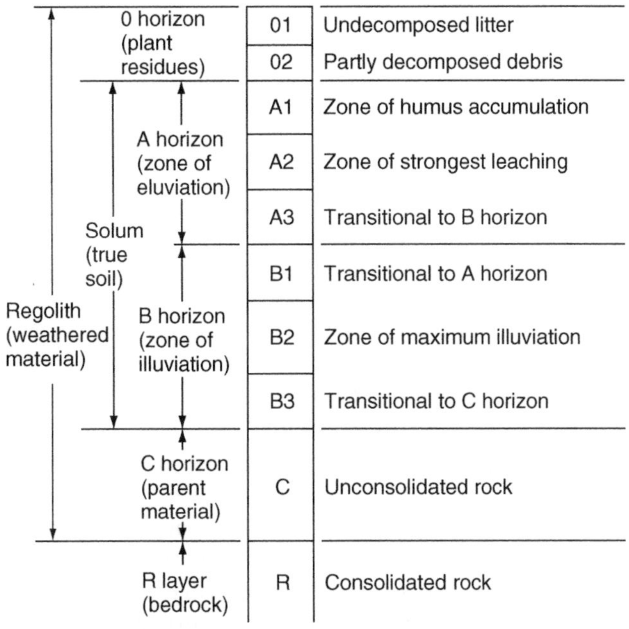

Fig. 1.2. Descriptive terminology for soil profile horizones.

SOIL AS A DISPERSE THREE-PHASE SYSTEM

The term *system* refers to a group of interacting, interrelated, or interdependent elements constituting an integrated entity. Since all of nature is in fact an integrated entity, defining any domain within nature to be a "system" is an admittedly arbitrary exercise. We choose to do so for reasons of convenience, since our own limitations prevent us from dealing with the entire complexity of nature all at once. However, the part of nature on which we may wish to focus our attention at any moment is necessarily a subsystem inside a larger system, with which our system interacts continuously. The modes of interaction generally include transfers or exchanges of materials and of energy.

Systems may vary widely in size, shape, and degree of complexity. They may consist of one or more substances and of one or more phases. The simplest system is one that is comprised of a single substance that has the same physical properties throughout. An example of such a system is a body of water consisting entirely of uniform ice. Such a system is homogeneous. A system comprised of a single chemical compound can also be heterogeneous if that substance exhibits different properties in different regions of the system. A region that is internally uniform physically is called a *phase*. A mixture of ice and liquid water, for instance, is chemically uniform but physically heterogeneous, because it includes two phases. The three ordinary phases in nature are the solid, liquid, and gaseous phases.

A system containing several substances may also be monophasic. A solution of salt and water, for example, is a homogeneous liquid. A system of several substances can also be heterogeneous. In a heterogeneous system the properties may differ not only between one phase and another, but also between the internal parts of each phase and the interfaces over which the phase comes into contact with one another. Interfaces exhibit specific phenomena resulting from the interaction of the phases. The importance of such phenomena (including adsorption, surface tension, and friction) in determining the behavior of the system as a whole depends on the magnitude of the interfacial area per unit volume of the system.

Systems in which at least one of the phases is subdivided into numerous minute particles, which together present a very large interfacial area per unit volume, are commonly called *disperse systems*. Colloidal sols (including aerosols), gels, emulsions, and aerosols are examples of disperse systems.

The soil is a heterogeneous, polyphasic, particulate, disperse, and porous system, with a large interfacial area per unit volume. (A handful of clay, for instance, may have an internal surface area of several hectares, a hectare being equal to 2.5 acres!). The disperse nature of soil and its consequent interfacial activity give rise to such phenomena as adsorption of water and chemicals, capillarity, ion exchange, swelling and shrinking, dispersion and flocculation.

The three phases of ordinary nature are represented in the soil as follows: the solid phase forms the *soil matrix*; the liquid phase is the water in the soil, which always contains dissolved substances, so it should properly be called the *soil solution*; and the gaseous phase is the *soil atmosphere*. The solid matrix of the soil consists of particles that vary in chemical and mineralogical composition as well as in size, shape, and orientation. It also contains amorphous substances, particularly organic matter, which is attached to the mineral grains

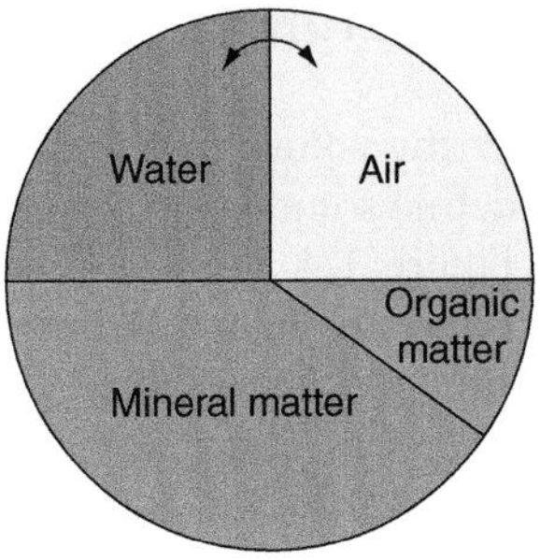

Fig. 1.3. Schematic composition (by volume) of a medium-textured soil at a condition considered optimal for plant growth. Note that the solid matter constitutes 50% and the pore space 50% of the soil volume, with the latter divided equally between water and air. the arrows indicate that water and air are related so that an increase in one is associated with a decrease in the other.

and may bind them together in assemblages called *aggregates*. The organization of the solid components of the soil determines the geometric characteristics of the pore spaces in which water and air are transmitted and retained. Finally, soil water and soil air vary in composition, both in time and in space.

The relative proportions of the three phases in the soil are not fixed but vary continuously, depending on such variables as weather, vegetation, and management. Figure 1.3 presents the hypothetical volume composition of a medium-textured soil at a condition considered approximately optimal for plant growth.

BOX 1.2 The Concept of Representative Elementary Volume

Some soil properties (for example, temperature) can be measured at a point, whereas other properties are volume dependent. Suppose we wish to measure some volume-dependent soil property, such as porosity. If our sample is very small, say, the size of a single particle or pore, the measured porosity may vary between zero and 100 percent, depending on the point at which we make our measurement (whether at a particle or at a pore). If we measure the porosity repeatedly at several adjacent points, the results will fluctuate widely. However, if we increase the scale or volume of the each sample so as to encompass within it both particles and pores, the fluctuations among repeated measurements at adjacent locations will diminish. If we keep enlarging the sample progressively, we will eventually obtain a consistent measurement of the soil's average porosity. The minimal volume of sample needed to obtain a consistent value of a measured parameter has been called the *representative elementary volume* (REV) (Bear, 1969). Obviously, the REV becomes larger in soils that are strongly aggregated (as well as in soils that are cracked or otherwise heterogeneous) than in more uniform soils.

The problem with the REV concept is that different parameters may exhibit different spatial or temporal patterns, so the REV for one parameter or property may differ from those for other parameters. That is to say, each property may have its own characteristic scale. Even more serious may be the failure of the REV concept in the case of "structured" fields, i.e., in fields that vary systematically in one direction or another. In such fields, increasing the size of the sample measured may not produce a consistent value at all.

VOLUME AND MASS RELATIONSHIPS OF SOIL CONSTITUENTS

Let us consider the volume and mass relationships among the three phases of the soil, and define some basic parameters that can help to characterize the soil physically. Figure 1.4 is a schematic depiction of a hypothetical soil in which the three phases have been separated and stacked one atop the other for the purpose of showing their relative volumes and masses.

In the figure, the masses of the phases are indicated on the right-hand side: the mass of air M_a, which is negligible compared to the masses of solids and water; the mass of water M_w; the mass of solids M_s; and the total mass M_t. (These masses can also be represented in terms of their weights, being the product of each mass and gravitational acceleration.) The volumes of the same components are indicated on the left-hand side of the diagram: volume of air V_a, volume of water V_w, volume of pores $V_f = V_a + V_w$, volume of solids V_s, and the total volume of the representative sample V_t.

On the basis of this diagram, we can now define terms that are generally used to express the quantitative interrelations of the three primary soil phases.

Density of Solids (Mean Particle Density) ρ_s

$$\rho_s = M_s / V_s \tag{1.1}$$

In most mineral soils, the mean mass per unit volume of solids is about 2600–2700 kg/m^3. This is close to the density of quartz, which is generally the most prevalent mineral in the coarsest fraction of the soil. Some of the minerals composing the finest fraction of the soil have a similar density. However, the presence of iron oxides and of various other "heavy" minerals (generally defined as those having a density exceeding 2900 kg/m^3) increases the average value of ρ_s, whereas the presence of low-density organic matter generally

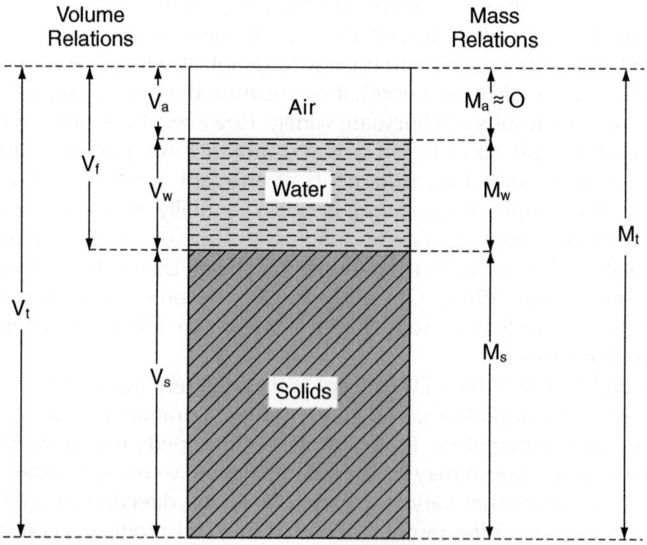

Fig. 1.4. Schematic diagram of the soil as a three-phase system.

lowers the mean density of the solid phase. Sometimes the density is expressed in terms of the *specific gravity*, σ_g, which is the ratio of the density of any material to that of water at 4°C and at atmospheric pressure. The latter density is about 1000 kg/m^3, so the specific gravity of the solid phase in a typical mineral soil is about 2.65, a value that is numerically (though not dimensionally) equal to the density expressed in the cgs. system of units (g/cm^3).

Dry Bulk Density ρ_b

$$\rho_b = M_s / V_t = M_s / (V_s + V_a + V_w) \tag{1.2}$$

The dry bulk density expresses the ratio of the mass of solids to the total soil volume (solids and pores together). Obviously, ρ_b is always smaller than ρ_s. If the pores constitute half the volume, then ρ_b is half of ρ_s, namely, about 1300–1350 kg/m^3, equivalent to a bulk specific gravity (i.e., the ratio of the soil's bulk density to the density of water at standard conditions) of 1.3–1.35. The bulk specific gravity of sandy soils with a relatively low volume of pores may be as high as 1.6, whereas that of aggregated loams and clay soils may be below 1.2. Whereas the mean particle density is typically constant, the bulk density is highly labile. It is affected by the structure of the soil, that is, its looseness or degree of compaction, as well as by its swelling and shrinkage characteristics. The latter depend both on clay content and water content. Even in extremely compacted soils, however, the bulk density remains appreciably lower than the density of the solid matter, since the particles can never interlock perfectly. However much the pore space can be reduced by compaction, it can never be eliminated.

Total (Wet) Bulk Density ρ_t

$$\rho_t = M_t / V_t = (M_s + M_w)/(V_s + V_a + V_w) \tag{1.3}$$

This is an expression of the total mass of a moist soil per unit volume. As such, this parameter depends more strongly than does the dry bulk density on soil wetness or water content.

Dry Specific Volume v_b

$$v_b = V_t / M_s = 1/\rho_b \tag{1.4}$$

The volume of a unit mass of a dry soil (the reciprocal of the dry bulk density) serves as another useful index of the degree of looseness or compaction of a soil body.

Porosity *f*

$$f = V_f / V_t = (V_a + V_w)/(V_s + V_a + V_w) \tag{1.5}$$

Porosity is an index of the relative pore space in a soil. Its value generally ranges from 0.3 to 0.6 (30–60%). Coarse-textured soils tend to be less porous than fine-textured soils, though the mean size of individual pores is greater in the former. In clayey soils, the porosity is highly variable because the soil alternately swells, shrinks, aggregates, disperses, compacts, and cracks. As generally defined, the term *porosity* refers to the volume fraction of pores, and this value should be equal, on average, to the areal porosity (the fraction of pores in a representative cross-sectional area) as well as to the average lineal porosity (the fractional length of pores along a straight line passing through the soil in any direction). However, the total porosity reveals nothing about the sizes or shapes of the various pores in the soil.

Void Ratio *e*

$$e = V_f/V_s = (V_a + V_w)/(V_t - V_f) \tag{1.6}$$

The void ratio is also an index of the fractional pore space, but it relates that space to the volume of solids rather than to the total volume of the soil. As such, it ranges between 0.3 and 2. The advantage of this index over the preceding one is that in the case of e any change in pore volume affects only the numerator of the defining equation, whereas in the case of f such a change affects both the numerator and the denominator. Void ratio is the index generally preferred by soil engineers, while porosity is more frequently used by agronomists.

Sample Problem

Prove the following relation of porosity to particle density and bulk density:

$$f = (\rho_s - \rho_b)/\rho_s = 1 - \rho_b/\rho_s$$

Substituting the definitions of f, ρ_s, and ρ_b, we can rewrite the equation as

$$V_f/V_t = 1 - (M_s/V_t)/(M_s/V_s)$$

Simplifying the right-hand side, we obtain

$$V_f/V_t = 1 - (V_s/V_t) = (V_t - V_s)/V_t$$

But since $V_t - V_s = V_f$, we have

$$V_f/V_t = V_f/V_t$$

Soil Wetness (Water Content)

The water content of a soil can be expressed in various ways: relative to the mass of solids, or to the total mass, or to the volume of solids, or to the total volume, or to the volume of pores. The various indexes are defined as follows.

Mass Wetness *w*

$$w = M_w/M_s \tag{1.7}$$

This is the mass of water relative to the mass of dry soil particles. The standard definition of *dry soil* refers to a mass of soil dried to equilibrium (in practice, over a 24-hour period) in an oven at 105°C, though a clay soil may still contain an appreciable amount of water at that state. Mass wetness is sometimes expressed as a decimal fraction but more often as a percentage. A sample of soil dried in "ordinary" air at ambient temperature (rather than in an oven) will generally retain several percent more water than if dried in the oven. Similarly, an oven-dry soil exposed to "ordinary" air will gradually gain appreciable moisture. This phenomenon results from the tendency of the soil's clay fraction to adsorb moisture from the air, a property known as *hygroscopicity*. The amount thus adsorbed depends on the type and content of clay in the soil as well as on the humidity of the ambient atmosphere. The water content at saturation (when all pores are filled with water) is also higher in clayey than in sandy soils. In different soils, w can range between 25% and 60%, depending on bulk density. In the special case of organic soils, such as peat or muck soils, the saturation water content on the mass basis may exceed 100%.

Volume Wetness θ

$$\theta = V_w/V_t = V_w/(V_s + V_f) \tag{1.8}$$

The volume wetness (often termed *volumetric water content*) is generally computed as a percentage of the total soil volume. At saturation, therefore, it is equal to the porosity. In sandy soils, θ at saturation is on the order of 40%; in medium-textured soils it is approximately 50%; and in clayey soils it can approach 60%. In the last, in fact, the volume of water at saturation can exceed the porosity of the dry soil, since clayey soils swell upon wetting. The use of rather than θ to express water content is often more convenient because it is more directly applicable to the computation of fluxes and water volumes added to soil by rain or irrigation and to quantities extracted from the soil by evaporation and transpiration. The volume ratio is also equivalent to the depth ratio of soil water, that is, the depth of water per unit depth of soil.

Sample Problem

Prove the following relation between volume wetness, mass wetness, bulk density, and water density ($\rho_w = M_w/V_w$):

$$\theta = w\rho_b/\rho_w$$

Again, we start by substituting the respective definitions of w_v, w_m, ρ_b, and ρ_w:

$$V_w/V_t = [(M_w/M_s)(M_s/V_t)]/(M_w/V_w)$$

Rearranging the right-hand side, we get

$$V_w/V_t = (V_w M_w M_s)/(M_w M_s V_t) = V_w/V_t$$

Water Volume Ratio v_w

$$v_w = V_w/V_s \tag{1.9}$$

For swelling soils, in which porosity changes markedly with wetness, it may be preferable to refer the volume of water present in a sample to the invariant volume of particles rather than to total volume. At saturation, v_w is equal to the void ratio e.

Degree of Saturation s

$$s = V_w / V_f = V_w / (V_a + V_w) \qquad (1.10)$$

This index expresses the water volume present in the soil relative to the pore volume. Index s ranges from zero in a completely dry soil to unity (100%) in a saturated soil. Complete saturation, however, is hardly ever attainable in field conditions, since some air is nearly always present. In a relatively dry soil the air phase occupies a continuous space, whereas in a very wet soil air may be occluded or encapsulated in the form of discontinuous bubbles.

Air-Filled Porosity (Fractional Air Content) f_a

$$f_a = V_a / V_t = V_a / (V_s + V_a + V_w) \qquad (1.11)$$

This is a measure of the relative content of air in the soil and as such is an important criterion of soil aeration. It is related negatively to the degree of saturation s (i.e., $f_a = f - s$). The relative volume of air in the soil may also be expressed as a fraction, a, of the pore volume (rather than of the total soil volume). Thus,

$$a = V_a / V_f = V_a / (V_a + V_w) \qquad (1.12)$$

Sample Problem

A sample of moist soil with a wet mass of 1.0 kg and a volume of 0.64 liters (6.4×10^{-4} m^3) was dried in the oven and found to have a dry mass of 0.8 kg. Assuming the typical value of particle density for a mineral soil (2650 kg/m^3), calculate the bulk density ρ_b, porosity f, void ration e, mass wetness w_m, volume wetness θ, water volume ratio v_w, degree of saturation s, and air-filled porosity f_a.

Bulk density: $\rho_b = M_s/V_t = 0.8\ \text{kg}/6.4 \times 10^{-4}\ \text{m}^3 = 1250\ \text{kg/m}^3$

Porosity: $f = 1 - \rho_b/\rho_s = 1 - 1250/2650 = 1 - 0.472 = 0.528$

Alternatively, $f = V_f/V_t = (V_t - V_s)/V_t$

and since $V_s = M_s/\rho_s = 0.8\ \text{kg}/2650\ \text{kg/m}^3 = 3.02 \times 10^{-4}\ \text{m}^3$

hence $f = (6.4 - 3.02) \times 10^{-4}\ \text{m}^3/6.4 \times 10^{-4}\ \text{m}^3 = 0.528 = 52.8\%$

Void ratio: $e = V_f/V_s = (V_t - V_s)/V_s = (6.4 - 3.02) \times 10^{-4}\ \text{m}^3/3.02 \times 10^{-4}\ \text{m}^3 = 1.12$

Mass wetness: $w = M_w/M_s = (M_t - M_s)/M_s = (1.0 - 0.8)\ \text{kg}/0.8\ \text{kg} = 0.25 = 25\%$

Volume wetness: $\theta = V_w/V_t = 2.0 \times 10^{-4}\ \text{m}^3/6.4 \times 10^{-4}\ \text{m}^3 = 0.3125 = 31.25\%$

(*Note*: $V_w = M_w/\rho_w$, wherein $\rho_w \approx 1000\ \text{kg/m}^3$)

Alternatively, $\theta = w\rho_b/\rho_w = 0.25(1250\ \text{kg/m}^3/1000\ \text{kg/m}^3) = 0.3125$

Water volume ratio: $v_w = V_w/V_s = 2.0 \times 10^{-4}\ \text{m}^3/3.02 \times 10^{-4}\ \text{m}^3 = 0.662$

Degree of saturation: $s = V_w/(V_t - V_s) = 2.0 \times 10^{-4}\ \text{m}^3/(6.4 - 3.02) \times 10^{-4}\ \text{m}^3 = 0.592$

Air-filled porosity: $f_a = V_a/V_t = (6.4 - 2.0 - 3.02) \times 10^{-4}\ \text{m}^3/6.4 \times 10^{-4}\ \text{m}^3 = 0.216$

Sample Problem

What is the equivalent depth of water contained in a soil profile 1 m deep if the mass wetness of the upper 0.4 m is 15% and that of the lower 0.6 m is 25%? Assume a bulk density of 1200 kg/m^3 in the upper layer and 1400 in the lower layer. How much water does the soil contain, in cubic meters per hectare of land?

Recall that $\theta = w(\rho_b/\rho_w)$, where $\rho_w = 1000$ kg/m^3

Volume wetness in the upper layer: $\theta_1 = 0.15(1200/1000) = 0.18$

Equivalent depth in upper 0.4 m = 0.18 × 0.4 m = 0.072 m = 72 mm

Volume wetness in lower layer: $\theta_2 = 0.25(1400/1000) = 0.35$

Equivalent depth in lower 0.6 m = 0.35 × 0.6 m = 0.21 m = 210 mm

Total depth of water in 1-m profile = 0.072 m + 0.210 m = 0.282 m

Volume of water contained in 1-m profile per hectare
(1 ha = 10^4 m^2) = 0.282 m × 1000 m^2 = 2820 m^3

2. WATER PROPERTIES IN RELATION TO POROUS MEDIA

THE FLUID OF LIFE

Our planet is the planet of life primarily because it is blessed with the precise ranges of temperature and pressure that make possible the existence in the liquid state of a singular substance called water. So ubiquitous is water on our globe, covering nearly three-quarters of its surface, that the entire planet really should be called "water" rather than "earth." However, as Coleridge's Ancient Mariner complained, most of the water everywhere is unfit to drink. Less than 3% of the water on earth is "fresh" (i.e., nonsaline), and that amount is unevenly distributed. Humid regions are endowed with an abundance of it, even with a surfeit, so that often the problem is how to dispose of excess water. Arid and semiarid regions, on the other hand, are afflicted with a chronic shortage of it.

Life as we know it began in an aquatic medium, and water is still the principal constituent of living organisms. It is, literally, the essence of life. As Vladimir Vernadsky wrote a century ago: "Life is animated water." Though we appear to be solid, we are really liquid bodies, similar to gelatin, which also seems solid but is in fact largely water, made consistent by the presence of organic material. The analogous material in our bodies is protoplasm. The body of a newborn infant contains nearly 90% water by mass, and even adults contain over 65% water. Growing herbaceous plants typically contain over 90% water. Far from being a bland, inert liquid, water is a highly reactive substance, a solvent, and a transporter of numerous substances.

The importance of water was recognized early in history, yet little was known about its real nature. In the Middle Ages, people believed that fresh water emanated magically from the bowels of the earth. They could not imagine that all the water flowing in innumerable springs and mighty rivers (such as the Nile, which appeared to the ancient Egyptians to come out of the driest desert!) could possibly result from so seemingly feeble a source as rain and snow. The first to conjecture this was Leonardo da Vinci, but only in the latter part of the 17th century did the English astronomer Edmond Halley and, separately, the Frenchman Claude Perrault, prove the principle by calculation and measurement. Water was long thought to be a single element, until early in the 18th century, when it was found to consist of hydrogen and oxygen in combination.

Notwithstanding its ubiquity, water remains something of an enigma, possessing unusual and anomalous attributes. Perhaps the first anomaly is that, being composed of two gases and having relatively low molecular weight, water is a liquid and not a gas at normal temperatures. (Its sister compound, hydrogen sulfide, H_2S, has a boiling point temperature of –60.7°C.) Compared to other common liquids, water has unusually high melting and boiling points, heats of fusion and vaporization, specific heat, dielectric constant, viscosity, and surface tension.

MOLECULAR STRUCTURE

One cubic meter of liquid water at 20°C contains some 3.4×10^{28} (34 billion billion billion) molecules, the diameter of which is about 3×10^{-10} meter (3×10^{-4} micrometers (μm), or 3 angstrom units). The chemical formula of water is H_2O, which signifies that each molecule consists of two atoms of hydrogen and one of oxygen. There are three isotopes of hydrogen (^1H, ^2H, ^3H) and three of oxygen (^{16}O, ^{17}O, ^{18}O), which can form 18 combinations. However, all isotopes but ^1H and ^{16}O are quite rare.

The hydrogen atom consists of a positively charged proton and a negatively charged electron. The oxygen atom consists of a nucleus having a positive charge of eight protons, surrounded by eight electrons, of which six are in the outer shell. Since the outer electron shell of hydrogen lacks one electron and that of oxygen lacks two electrons, one atom of oxygen can combine with two atoms of hydrogen in an electron-sharing molecule.

The strong intermolecular forces in liquid water are caused by the electrical polarity of the water molecule, which in turn is a consequence of the arrangement of electrons in its oxygen and hydrogen atoms (Fig. 2.1). The oxygen atom shares a pair of electrons with each of the two hydrogen atoms, through overlap of the 1s orbitals of the hydrogen atoms with two hybridized sp^3 orbitals of the oxygen atom. The H–O–H bond in water is not linear but bent at an angle of 104.5°:

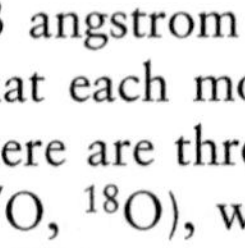

That angle deviates slightly from a perfectly tetrahedral arrangement of the oxygen atom's four possible sp^3 orbitals, which would have an angle of 109.5°. The mean H–O interatomic distance is 9.65×10^{-5} μm. The arrangement of

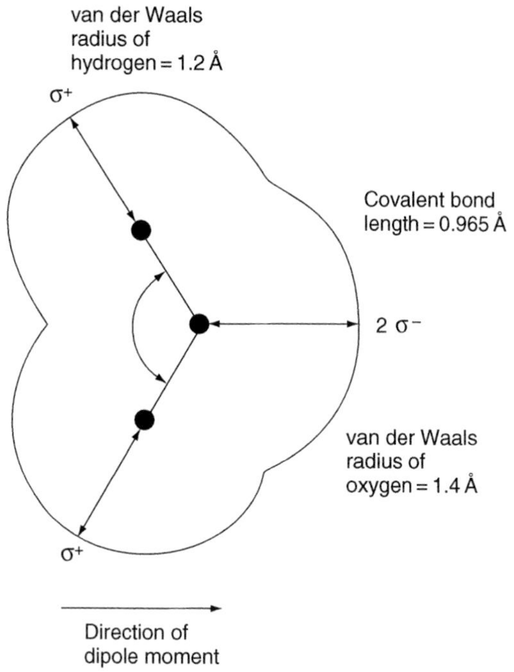

Fig. 2.1. Model of a water molecule. The curved lines represent the borders at which van der Waals attractions are counterbalanced by repulsive forces.

electrons in the molecule gives it electrical asymmetry. The electronegative oxygen atom attracts the electrons of the hydrogen atoms, leaving the hydrogen nuclei bare, so each of the two hydrogen atoms has a local partial positive charge. The oxygen atom, in turn, has a partial negative charge, located in the zone of the unshared orbitals. Though water molecules have no net charge, they form electrical dipoles.

HYDROGEN BONDING

Every hydrogen proton, though attached primarily to a particular molecule, is also attracted to the oxygen of a neighboring molecule, with which it forms a secondary link known as a *hydrogen bond.* Though the intermolecular link resulting from dipole attraction is not as strong as the primary link of the hydrogen to the oxygen of its own molecule, water can be regarded as a polymer of hydrogen-bonded molecules. This structure is most complete in ice crystals, in which each molecule is linked to four neighbors via four hydrogen bonds, thus forming a hexagonal lattice with a rather open structure (Fig. 2.2). When ice melts, this structure breaks partially, so additional molecules can enter the intermolecular spaces and each molecule can have more than four near neighbors. For this reason, liquid water can be more dense than ice at the same temperature, and thus lakes and ponds develop a surface ice sheet in winter rather than freeze solid from bottom to top as they would if ice were denser than liquid water.

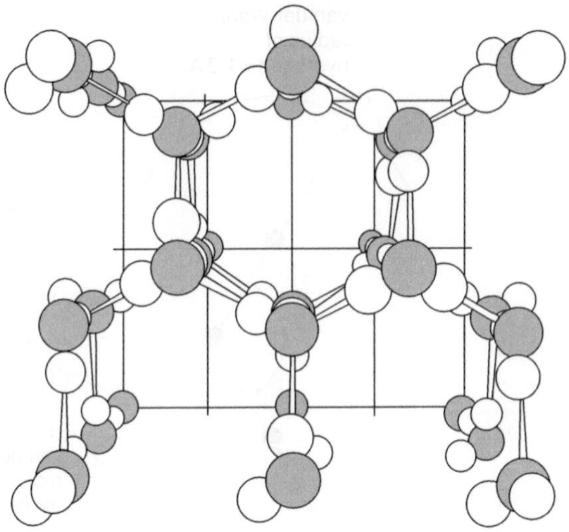

Fig. 2.2. Schematic structure of an ice crystal. The oxygen atoms are shown in gray and the hydrogen atoms in white. The pegs linking adjacent molecules represent hydrogen bonds.

STATES OF WATER

In the gaseous state, water molecules are largely independent of one another and occur mostly as monomers, designated $(H_20)_1$. Occasionally, colliding molecules fuse to form dimers $(H_20)_2$ or trimers $(H_20)_3$, but such combinations are rare. In the solid state, at the other extreme, a rigidly structured lattice forms with a tetrahedral configuration (Fig. 2.2) that can be depicted schematically as sheets of puckered hexagonal rings (Fig. 2.3). Nine alternative ice forms can occur when water freezes, depending on prevailing temperature and pressure. Figure 2.3 pertains to ice 1, the familiar form, which occurs and is stable at ordinary atmospheric pressure.

The orderly structure of ice does not totally disappear in the liquid state. The polarity and hydrogen bonds continue to bind water molecules together,

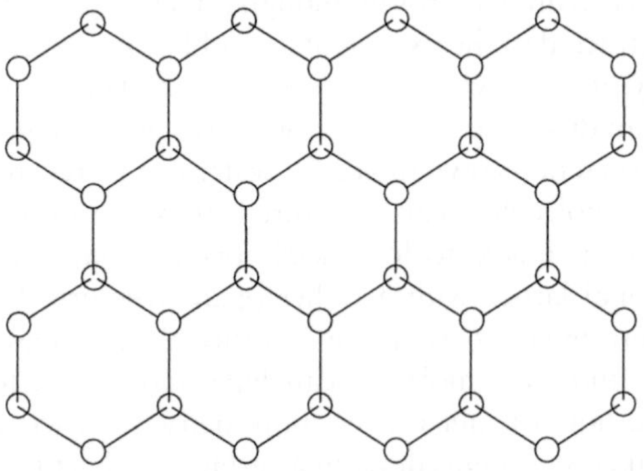

Fig. 2.3. The crystalline structure of ice.

though the structural forms that develop in the liquid state are much more flexible and transient than in the rigidly structured solid state. Hydrogen bonds in liquid water form an extensive three-dimensional network, the detailed features of which appear to be short-lived. According to the "flickering cluster" model, for instance, the molecules of liquid water associate and dissociate repeatedly in transitory or flickering polymer groups, designated $(H_2O)_n$, having a quasi-crystalline internal structure. These microcrystals, as it were, form and melt so rapidly and randomly that, on a macroscopic scale, water appears to behave as a homogeneous liquid.

In transition from solid to liquid and from liquid to gas, hydrogen bonds must be broken (while in freezing and condensation they are reestablished). Hence relatively high temperatures and energies are required to achieve these transitions. To thaw 1 kg of ice, 3.35×10^5 J (80 cal/g) must be supplied. Conversely, the same energy (the *latent heat of fusion*) is released in freezing.

At the boiling point (100°C at atmospheric pressure), water passes from the liquid to the gaseous state and in so doing absorbs 2.26×10^6 J/kg (540 cal/g). This amount of heat is known as the *latent heat of vaporization*. Water can be vaporized at temperatures below 100°C, but such vaporization requires greater heat. At 30°C, the latent heat is about 2.43×10^6 J/kg (580 cal/g). *Sublimation* is the direct transition from the solid state to vapor, and the heat absorbed by it is equal to the sum of the latent heats of fusion and of vaporization.

OSMOTIC PRESSURE

Owing to the constant thermal motion of all molecules in a fluid (above a temperature of absolute zero), solute species spread throughout the solution in a spontaneous tendency toward a state of equal concentration throughout. This migration of solutes in response to spatial differences in concentration is called *diffusion*.

If a physical barrier is interposed across the path of diffusion, and if that barrier is permeable to molecules of the solvent but not to those of the solute, the former will diffuse through the barrier in a process called *osmosis* (from the Greek *osmos*, meaning "push"). As in the case of unhindered diffusion, this process tends toward a state of uniform concentration, even across the barrier. Barriers permeable to one substance in a solution but not to another are called *selective* or *semipermeable membranes*. Membranes surrounding cells in living organisms, for example, exhibit selective permeability to water while restricting the diffusion of solutes between the cells' interior and their exterior environment. Water molecules cross the membrane in both directions, but the net flow of water is from the more dilute solution to the more concentrated.

Figure 2.4a is a schematic representation of a pure solvent separated from a solution by a semipermeable membrane. Solvent will pass through the membrane and enter the solution compartment, driving the solution level up the left-hand tube until the hydrostatic pressure of the column of dilute solution on the left is sufficient to counter the diffusion pressure of

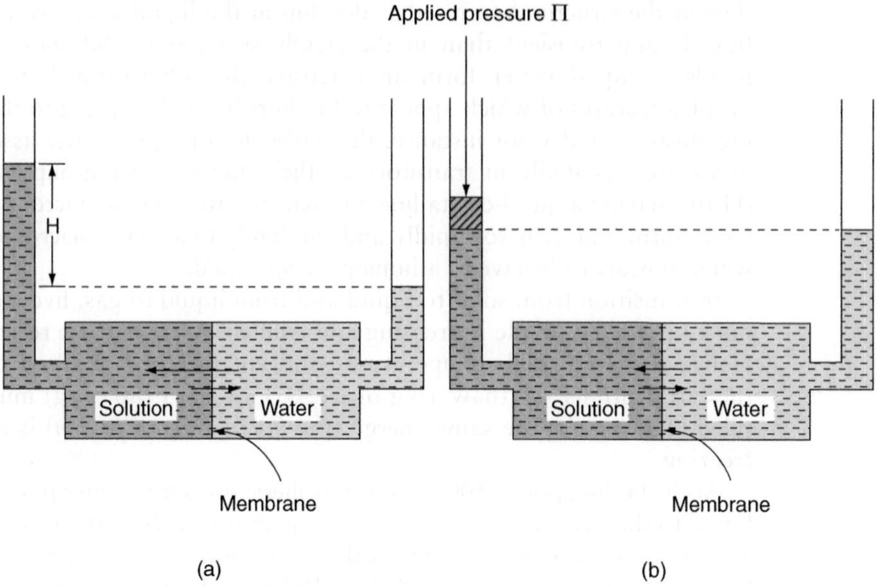

Fig. 2.4. Osmosis and osmotic pressure: (a) In osmosis, the flow of water molecules through the membrane into the solutions is at first greater than the reverse flow from the solution into the water compartment. The hydrostatic pressure due to the column of expanded solution increases the rate of water flow from the solution to the water compartment until, at equilibrium, the opposite flows are equal. (b) The osmotic pressure of the solution is equal to the hydrostatic pressure Π that must be applied to the solution to equalize the rate of flow to and from the solution and produce a net flow of zero.

the solvent molecules drawn into the solution through the membrane. The hydrostatic pressure at equilibrium, when solvent molecules are crossing the membranes in both directions at equal rates, is the osmotic pressure of the solution.

In dilute solutions, the osmotic pressure is proportional to the concentration of the solution and to its temperature, according to the following equation:

$$\Pi = MRT \tag{2.1}$$

Here Π is the osmotic pressure in atmospheres (to be multiplied by 0.101 to obtain megapascal units), M is the total molar concentration of the solute (whether molecules or dissociated ions), T is the temperature in degrees Kelvin, and R is the gas constant (0.08205 L atm/deg mole). The osmotic pressure increase with temperature is associated with the corresponding increase of the molecular diffusivity (self-diffusion coefficient) of water, D_w. According to the Einstein–Stokes equation,

$$D_w = kT/6\pi\eta r$$

where $k = R/N$, the Boltzmann constant (1.38×10^{-23} J/K), r is the rotation radius of the molecule ($\sim 1.5 \times 10^{-4}$), η is the viscosity, and N is Avagadro's number.

Sample Problem

Calculate the osmotic pressure of a 0.01 *M* solution of sodium chloride at 20°C, assuming complete dissociation into Na^+ and Cl^- ions.

$$\Pi = MRT = (2 \times 0.01 \text{ mol/L})(0.08205 \text{ L atm/mole K})(293 \text{ K}) = 0.48 \text{ atm} = 48.5 \text{ kPa}$$

SOLUBILITY OF GASES

The concentration of dissolved gases in water in equilibrium with a gaseous phase generally increases with pressure and decreases with temperature. According to Henry's law, the mass concentration of a dissolved gas c_m is proportional to the partial pressure of the gas p_i in the ambient atmosphere:

$$c_m = s_c p_i / p_0 \tag{2.2}$$

where s_c is the solubility coefficient of the particular gas in water and p_0 is the total pressure of the atmosphere. The volume concentration is similarly proportional:

$$c_v = s_v p_i / p_0 \tag{2.3}$$

where s_v is the solubility coefficient expressed in terms of volume ratios (i.e., c_v is the volume of dissolved gas relative to the volume of water). Both s_c and s_v are determined experimentally. If the gas does not react chemically with the liquid, these properties should remain constant over a range of pressures, especially at low partial pressures of the dissolved gases. Solubility is, however, strongly influenced by temperature. Table 2.1 gives the s_v values of several atmospheric gases at various temperatures.

The solubilities of various gases (particularly oxygen) in varying conditions strongly influence such vital soil processes as oxidation and reduction, as well as respiration by roots and microorganisms.

TABLE 2.1 Solubility Coefficients of Gases in Water

Temperature (°C)	Nitrogen (N_2)	Oxygen (O_2)	Carbon dioxide (CO_2)	Air (without CO_2)
0	0.0235	0.0489	1.713	0.0292
10	0.0186	0.0380	1.194	0.0228
20	0.0154	0.0310	0.878	0.0187
30	0.0134	0.0261	0.665	0.0156
40	0.0118	0.0231	0.530	—

Sample Problem

A liter of water at 25°C dissolves 0.0283 L of oxygen when the pressure of oxygen in equilibrium with the solution is 1 atm (101 kPa). From this we can find the proportionality constant s_v in Henry's law with $P_{O_2} = 1$ atm, as follows:

Restating Henry's law: $c_v = s_v(P_{O_2}/P_{total})$

(where c_v is the volume of gas dissolved at equilibrium with a partial pressure P_{O_2}/P_{total}) we can write

$$s_v = c_v/(P_{O_2}/P_{total}) = 0.0283 \text{ L/L atm}$$

With the prevailing pressure of oxygen in normal dry air equal to 159 mm Hg,

$$c_v = s_v(P_{O_2}/P_{total}) = (0.0283)(159/760) = 0.00592 \text{ L/L } H_2O$$
$$= 5.92 \text{ mL/L}$$

ADSORPTION OF WATER ON SOLID SURFACES

Adsorption is an interfacial phenomenon resulting from the differential forces of attraction or repulsion occurring among molecules or ions of different phases at their exposed surfaces. As a result of cohesive and adhesive forces coming into play, the zones of contact among phases may exhibit a concentration or a density of material different from that inside the phases themselves. As different phases come in contact, various types of adsorption can occur: adsorption of gases on solids, of gases on liquid surfaces, and of liquids on solids.

The interfacial forces of attraction or repulsion may themselves be of different types, including electrostatic or ionic (coulombic) forces, intermolecular forces such as van der Waals and London forces, and short-range repulsive (Born) forces. The adsorption of water upon solid surfaces is generally of an electrostatic nature. The polar water molecules attach to the charged faces of the solids and to the ions adsorbed on them. This adsorption of water is the mechanism causing the strong retention of water by clay at high suctions.

The interaction of the charges of the solid with the polar water molecules may impart to the adsorbed water a distinct structure in which the water dipoles assume an orientation dictated by the charge sites on the solids. This adsorbed "phase" may have mechanical properties of strength and viscosity that differ from those of ordinary ("free") liquid water at the same temperature. The adsorption of water on clay surfaces is an exothermic process, resulting in the liberation of an amount of heat known as the *heat of wetting*.

VAPOR PRESSURE

According to the kinetic theory, molecules in a liquid are in constant motion due to their thermal energy. These molecules collide frequently, and occasionally one or another at the surface absorbs sufficient momentum to leap out of the liquid and into the atmosphere above it. Such a molecule, by virtue of its kinetic energy, thus changes from the liquid to the gaseous phase. This kinetic energy is then lost in overcoming the potential energy of intermolecular attraction while escaping from the liquid. At the same time, some of the randomly moving molecules in the gaseous phase may strike the surface of the liquid and be absorbed in it.

The relative rates of these two directions of movement depend upon the concentration of vapor in the atmosphere. An atmosphere that is at equilibrium with a body of pure water at standard atmospheric pressure is considered to be saturated with water vapor, and the partial pressure of vapor in such an atmosphere is called the *saturation vapor pressure*. The vapor pressure at equilibrium with any body of water depends on its physical condition (pressure and temperature) and on its chemical state (affected by the nature and concentration of solutes) but is independent of the absolute or relative quantity of liquid or gas in the system.

The saturation vapor pressure rises with temperature. As the kinetic energy of the molecules in the liquid increases, so does the evaporation rate. Consequently, a higher concentration of vapor in the atmosphere is required for the rate of return to the liquid to match the rate of escape from it. A liquid arrives at its boiling point when the vapor pressure becomes equal to the atmospheric pressure. If the temperature range is not too wide, the dependence of saturation vapor pressure on temperature is expressible by the following equation (see Table 2.2):

$$\ln p_0 = a - b/T \tag{2.4}$$

where $\ln p_0$ is the logarithm to the base e of the saturation vapor pressure p_0, T is the absolute temperature, and a and b are constants.

TABLE 2.2 Physical Properties of Water Vapor

Temperature (°C)	Saturation vapor pressure (torr): Over liquid	Saturation vapor pressure (torr): Over ice	Vapor density in saturated air (kg/m³): Over liquid	Vapor density in saturated air (kg/m³): Over ice	Diffusion coefficient (m²/sec)
−10	2.15	1.95	2.36×10^{-3}	2.14×10^{-3}	0.211×10^{-4}
−5	3.16	3.01	3.41×10^{-3}	3.25×10^{-3}	—
0	4.58	4.58	4.85×10^{-3}	4.85×10^{-3}	0.226×10^{-4}
5	6.53	—	6.80×10^{-3}	—	
10	9.20	—	9.40×10^{-3}	—	0.241×10^{-4}
15	12.78	—	12.85×10^{-3}	—	
20	17.52	—	17.30×10^{-3}	—	0.257×10^{-4}
25	23.75	—	23.05×10^{-3}	—	
30	31.82	—	30.38×10^{-3}	—	0.273×10^{-4}
35	42.20	—	39.63×10^{-3}	—	
40	55.30	—	51.1×10^{-3}	—	0.289×10^{-4}
45	71.90	—	65.6×10^{-3}	—	
50	92.50	—	83.2×10^{-3}	—	

The vapor pressure also depends also on the hydrostatic pressure of the liquid water. At equilibrium with drops of water (which have a hydrostatic pressure greater than atmospheric), the vapor pressure is greater than in a state of equilibrium with free water (which has a flat interface with the atmosphere). On the other hand, in equilibrium with adsorbed or capillary water under a hydrostatic pressure smaller than atmospheric, the vapor pressure is smaller than in equilibrium with free water. The curvature of drops is considered to be positive, because these drops are convex toward the atmosphere, whereas the curvature of capillary water menisci is considered negative, because they are concave toward the atmosphere.

For water in capillaries with a concave air–water interface, the Kelvin equation applies:

$$-(\mu_l - \mu_l^\circ) = \mathrm{RT} \ln (p_l^\circ/p_l) = 2\gamma v_1 \cos \alpha/\mathrm{r_c}$$

in which $(\mu_l - \mu_l^\circ)$ is the change in potential of the water due to the curvature of the air–water interface, γ is the surface tension of water, α is the contact angle, v_1 is the partial molar volume of water, and r_c is the radius of the capillary.

Water present in the soil invariably contains solutes, generally electrolytic salts. The composition and concentration of the soil solution affect soil behavior. In humid regions the soil solution may have a concentration of only a few parts per million, but in arid regions it may be as high as several percent. The vapor pressure of electrolytic solutions is less than that of pure water:

$$v_1\Pi_o = RT \ln (p_l^\circ/p_l) = \mu_l - \mu_l^\circ$$

where Π_o is the osmotic pressure of the solution, μ_1° and p_1° are the chemical potential and vapor pressure of the pure liquid, and μ_l and p_l are the same for the solution. The soil solution has a lower vapor pressure than pure water, even when the soil is saturated. In unsaturated soil, capillary and adsorptive effects further lower the potential and hence also the vapor pressure.

SURFACE TENSION

Surface tension is a phenomenon occurring typically, though not exclusively, at the interface of a liquid and a gas. The liquid behaves as if it were covered by an elastic membrane in a constant state of tension that has the effect of causing the surface to contract. To be sure, no such membrane exists, yet the analogy is a useful one if not taken too literally.

If we draw an arbitrary line of length L on a liquid surface, there will be a force F pulling the surface to the right of the line and an equal force pulling the surface leftward. The ratio F/L is the surface tension, and its dimensions are those of force per unit length. The same phenomenon can also be described in terms of energy. Increasing the surface area of a liquid requires work, which remains stored as potential energy in the enlarged surface, just as energy can be stored in a stretched spring. That potential energy can perform work if the enlarged surface is allowed to contract again. Energy per unit area has the same dimensions as force per unit length.

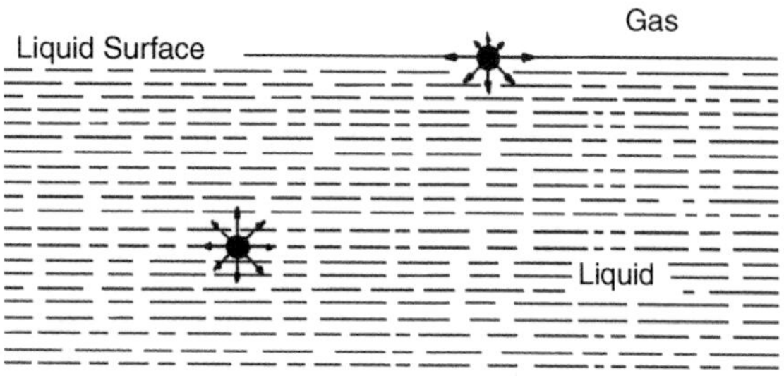

Fig. 2.5. Cohesive forces acting on a molecule inside the liquid and at its surface.

An explanation for the occurrence of surface tension is given in Fig. 2.5. A molecule inside the liquid is attracted to neighboring molecules in all directions equally by cohesive forces, while a molecule at the surface of the liquid is attracted into the relatively dense liquid phase by a net force greater than that attracting it toward the rarified gaseous phase. This unbalanced force draws the surface molecules inward into the liquid and results in the tendency for the surface to contract. This is why drops of a liquid in air (as well as bubbles of air in a liquid) assume the shape of a sphere, which is a body of minimal surface exposure relative to its volume.

Different liquids exhibit different surface tension values, as in the following list:

Water, 7.27×10^{-2} N/m (72.7 dyn/cm) at 20°C

Ethyl ether, 1.7×10^{-2} N/m (17 dyn/cm)

Ethyl alcohol, 2.2×10^{-2} N/m (22 dyn/cm)

Benzene, 2.9×10^{-2} N/m (29 dyn/cm)

Mercury, 0.43 N/m (430 dyn/cm)

CURVATURE OF WATER SURFACES AND HYDROSTATIC PRESSURE

Wherever an interface between fluids (say, between water and air) is not planar but curved, the resolution of forces due to surface tension creates a pressure differential across that interface. For a spherical interface (as in the case of a bubble of air immersed in a body of water), the pressure difference is proportional to the surface tension and inversely proportional to the curvature:

$$\Delta P = 2\gamma / R \tag{2.5}$$

Thus, the smaller the bubble is, the greater is its pressure.

If the bubble is not spherical, then instead of Eq. (2.5) we obtain

$$\Delta P = \gamma(1/R_1 + 1/R_2) \tag{2.6}$$

where R_1 and R_2 are the principal radii of curvature for any given point on the interface. Equation (2.6) reduces to (2.5) whenever $R_1 = R_2$.

Sample Problem

Calculate the hydrostatic pressure of water in raindrops of the following diameters: (a) 5; (b) 1; (c) 0.2 mm. Assume 4°C. What are the corresponding pressures at 25°C?

The relevant equation is: $\Delta P = 2\gamma/R$. At 4°C the surface tension γ is 7.5×10^{-2} N/m (75.0 dyn/cm).

(a) $\Delta P = (2 \times 7.5 \times 10^{-2}\ \text{N/m})/(2.5 \times 10^{-3}\ \text{m}) = 60\ \text{Pa} = 0.6\ \text{mbar}$
(b) $\Delta P = (2 \times 7.5 \times 10^{-2}\ \text{N/m})/(0.5 \times 10^{-3}\ \text{m}) = 300\ \text{Pa} = 3.0\ \text{mbar}$
(c) $\Delta P = (2 \times 7.5 \times 10^{-2}\ \text{N/m})/(0.1 \times 10^{-3}\ \text{m}) = 1500\ \text{Pa} = 15.0\ \text{mbar}$

At 25°C the surface tension of water is 7.19×10^{-2} N/m (71.9 dyn/cm).

(a) $\Delta P = (2 \times 7.19 \times 10^{-2}\ \text{N/m})/(2.5 \times 10^{-3}\ \text{m}) = 57.5\ \text{Pa} = 0.575\ \text{mbar}$
(b) $\Delta P = (2 \times 7.19 \times 10^{-2}\ \text{N/m})/(0.5 \times 10^{-3}\ \text{m}) = 287.6\ \text{Pa} = 2.876\ \text{mbar}$
(c) $\Delta P = (2 \times 7.19 \times 10^{-2}\ \text{N/m})/(0.1 \times 10^{-3}\ \text{m}) = 1438\ \text{Pa} = 14.38\ \text{mbar}$

We note that the hydrostatic pressure in drops is greatly influenced by drop size but not much by temperature in the range considered.

CONTACT ANGLE OF WATER ON SOLID SURFACES

If we place a drop of liquid on a dry solid surface, the liquid will usually displace the gas that covered the surface of the solid and it will spread over that surface to a certain extent. Where its spreading ceases and the edge of the drop comes to rest, it will form a typical angle with the surface of the solid. This angle, termed *contact angle*, is illustrated in Fig. 2.6.

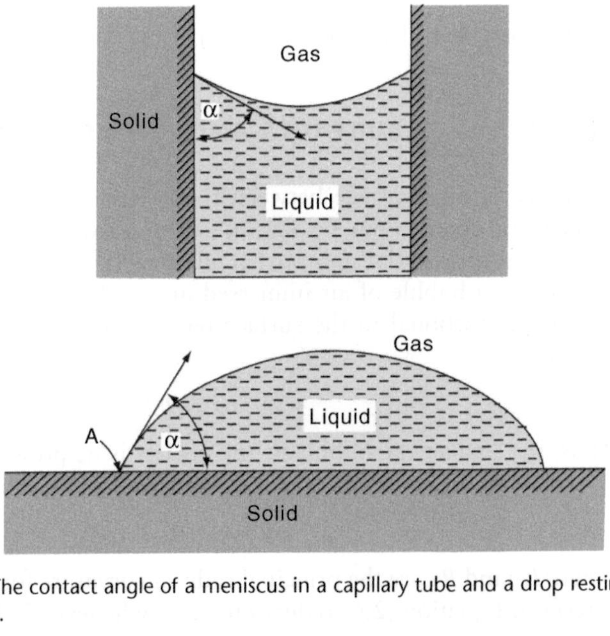

Fig. 2.6. The contact angle of a meniscus in a capillary tube and a drop resting on the surface of a solid.

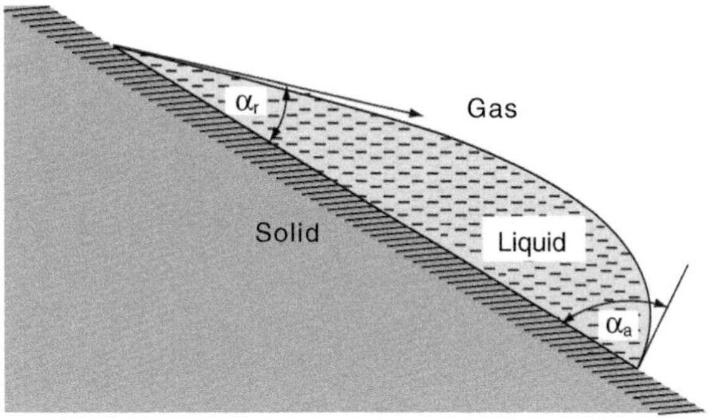

Fig. 2.7. Hypothetical representation of a drop resting on an inclined surface. The contact angle α_3 at the advancing edge of the drop is shown to be greater than the corresponding angle α_r at the receding edge.

We now consider what factors determine the magnitude of the angle α. We can expect that angle to be acute if the adhesive affinity between the solid and liquid is strong relative to the cohesive forces inside the liquid itself and to the affinity between the gas and the solid. We can then say that the liquid "wets" the solid. A contact angle of zero would mean the complete flattening of the drop and the perfect wetting of the solid surface by the liquid. On the other hand, a contact angle of 180° would imply a complete nonwetting or rejection of the liquid by the gas-covered solid. In that case the drop would retain its spherical shape without spreading over the surface at all (assuming no gravity effect). Surfaces on which water exhibits an obtuse contact angle are called *water repellent* or *hydrophobic* (Greek: "water-hating").

The contact angle of a given liquid on any particular solid is generally characteristic of their interaction under given physical conditions. This angle, however, may be different in the case of a liquid that is advancing over the surface than in the case of the same liquid receding over the surface. This phenomenon, where it occurs, is called *contact angle hysteresis*. The wetting angle of pure water on clean and smooth mineral surfaces is generally zero, but where the surface is rough or coated with adsorbed surfactants of a hydrophobic nature, the contact angle, and especially the wetting angle, can be considerably greater than zero and may even exceed 90°. This phenomenon is illustrated in given Fig. 2.7.

THE PHENOMENON OF CAPILLARITY

A capillary tube dipped in a body of free water will form a meniscus as the result of the contact angle of water with the walls of the tube. The curvature of this meniscus will be greater (i.e., the radius of curvature will be smaller) the narrower the tube. The occurrence of curvature causes a pressure difference to develop across the liquid–gas interface.

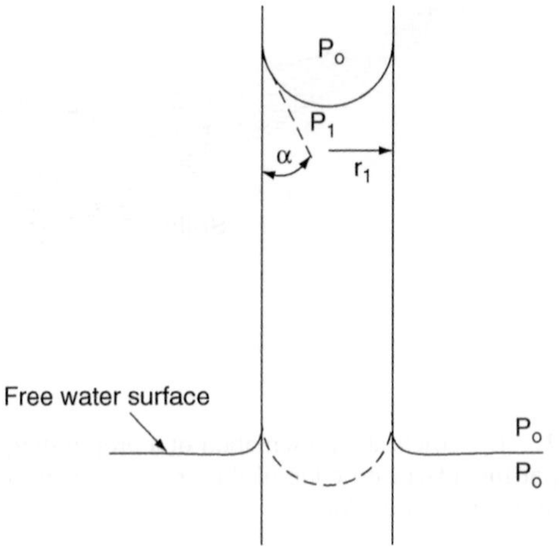

Fig. 2.8. Capillary rise.

A liquid with an acute contact angle (e.g., water on glass) will form a concave meniscus, and therefore the liquid pressure under the meniscus P_1 will be smaller than the atmospheric pressure P_0 (Fig. 2.8). Hence, water inside the tube will be driven up the tube from its initial location (shown as a dashed curve in Fig. 2.8) by the greater pressure of the free water (i.e., water at atmospheric pressure, under a horizontal air–water interface) at the same level.

The upward movement will stop when the pressure difference between the water inside the tube and the water under the flat surface outside the tube is countered by the hydrostatic pressure exerted by the water column in the capillary tube. If the capillary tube is cylindrical and if the contact angle of the liquid on the walls of the tube is zero, the meniscus will be a hemisphere (and in a two-dimensional drawing can be represented as a semicircle) with its radius of curvature equal to the radius of the capillary tube. If, however, the liquid contacts the tube at an angle greater than zero but smaller than 90°, then the diameter of the tube ($2r$) is the length of a chord cutting a section of a circle with an angle of $\pi - 2\alpha$, as shown in Fig. 2.9. Thus,

$$R = r/\cos\alpha \tag{2.7}$$

Here R is the radius of curvature of the meniscus, r is the radius of the capillary, and α is the contact angle.

The pressure difference ΔP between the capillary water (under the meniscus) and the atmosphere, therefore, is

$$\Delta P = (2\gamma\cos\alpha)/r \tag{2.8}$$

Recalling that hydrostatic pressure is proportional to the depth d below the free water surface (i.e., $P = \rho g d$, where ρ is liquid density and g is gravitational acceleration), we can infer that the hydrostatic tension (negative pressure) in

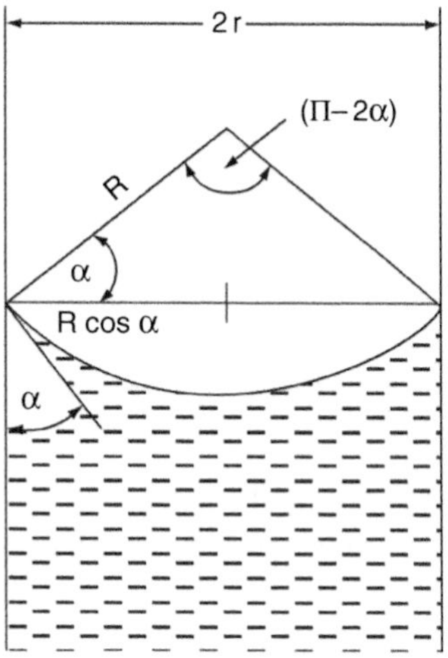

Fig. 2.9. The geometric relationship of the radius of curvature R to the radius of the capillary r and the contact angle α.

a capillary tube is proportional to the height h above the free water surface. Hence the height of capillary rise is

$$h_c = (2\gamma \cos \alpha)/g(\rho_l - \rho_g)r \qquad (2.9)$$

where ρ_g is the density of the gas (generally neglected), ρ_l is the density of the liquid, g is the acceleration of gravity, r is the capillary radius, α is the contact angle, and γ is the surface tension.

When the liquid surface is concave, the center of curvature lies outside the liquid and the curvature, by convention, is regarded as negative. Thus, for a concave meniscus such as that of water in a clean glass capillary, ΔP is negative with reference to the atmosphere, indicating a capillary pressure deficit, or subpressure, commonly called *tension*. On the other hand, in a capillary tube that forms a convex meniscus (such as in the case of mercury in glass or of water in a waxy or otherwise water-repellent tube), ΔP would be positive, and hence capillary depression, rather than capillary rise, will result.

Sample Problems

Calculate the equilibrium capillary rise of water and mercury at 20°C in glass cylindrical capillary tubes of the following diameters: (a) 2 mm; (b) 0.5 mm; (c) 0.1 mm. Disregard the density of atmosphere.

$$\Delta h = (2\gamma \cos \alpha)/ (\rho g r)$$

For water: $\gamma = 7.27 \times 10^{-2}$ kg/sec^2 (= N/m); $\alpha = 0$; $\rho = 998$ kg/m^3

(a) $r = 10^{-3}$ m:

$$\Delta h = (2 \times 7.27 \times 10^{-2} \text{ kg/sec}^2)/(998 \text{ kg/m}^3 \times 9.81 \text{ m/sec}^2 \times 10^{-3}\text{m})$$
$$= 1.48 \times 10^{-2} \text{ m} = 1.48 \text{ cm}$$

(b) $r = 2.5 \times 10^{-4}$ m:

$\Delta h = (2 \times 7.27 \times 10^{-2}$ kg/sec^2)/(998 kg/m$^3 \times 9.81$ m/sec$^2 \times 2.5 \times 10^{-4}$ m
$= 5.92 \times 10^{-2}$ m $= 5.92$ cm

(c) $r = 5.0 \times 10^{-5}$ m:

$\Delta h = (2 \times 7.27 \times 10^{-2}$ kg/sec^2)/(998 kg/m$^3 \times 9.81$ m/sec$^2 \times 5.0 \times 10^{-5}$ m
$= 29.6 \times 10^{-2}$ m $= 29.6$ cm

For mercury: $\gamma = 0.43$ kg/sec^2; $\alpha = 180°$; $\rho = 13{,}600$ kg/m^3

(a) $r = 10^{-3}$ m:

$\Delta h = [2 \times 0.43$ kg/sec$^2 \times (-1)]/(13{,}600$ kg/m$^3 \times 9.81$ m/sec$^2 \times 10^{-3}$ m)
$= -0.64 \times 10^{-2}$ m $= -0.64$ cm (capillary depression)

(b) $r = 2.5 \times 10^{-4}$ m:

$\Delta h = [2 \times 0.43$ kg/sec$^2 \times (-1)]/(13{,}600$ kg/m$^3 \times 9.81$ m/sec$^2 \times 2.5 \times 10^{-4}$ m)
$= -2.58 \times 10^{-2}$ m $= -2.58$ cm

(c) $r = 5.0 \times 10^{-5}$ m:

$\Delta h = [2 \times 0.43$ kg/sec$^2 \times (-1)]/(13{,}600$ kg/m$^3 \times 9.81$ m/sec$^2 \times 5.0 \times 10^{-5}$ m
$= -12.9 \times 10^{-2}$ m $= -12.9$ cm

DYNAMIC AND KINEMATIC VISCOSITY

When a fluid is moved in shear (i.e., when adjacent layers of fluid are made to slide over each other), the force required is proportional to the velocity of shear. The proportionality factor is called the *viscosity*. As such, it is the property of the fluid to resist the rate of shearing and can be visualized as an internal friction. The coefficient of viscosity η is defined as the force per unit area necessary to maintain a velocity difference of 1 m/sec between two parallel layers of fluid that are 1 m apart. The viscosity equation is

$$\tau = F_s/A = \eta du/dx \tag{2.10}$$

where τ is the shearing stress, consisting of a force F_s acting on an area A; η [dimensions: mass/ (length × time)] is the *coefficient of dynamic viscosity*; and du/dx is the velocity gradient perpendicular to the stressed area A.

The ratio of the dynamic viscosity of a fluid to its density is called the *kinematic viscosity*, designated ν. It expresses the shearing-rate resistance of a fluid independent of the density. While the dynamic viscosity of water exceeds that of air by a factor of about 50 (at room temperature), its kinematic viscosity is actually lower.

Liquids of lower viscosity flow more readily and are said to possess greater *fluidity* (the reciprocal of viscosity). The viscosity of water diminishes by over 2% per 1°C rise in temperature, and thus it decreases by more than half as the temperature increases from 5 to 35°C. The viscosity is also affected by the type and concentration of solutes present.

BOX 2.1 The Shape of a Raindrop

What is the shape of a free-falling raindrop? Most people answer that question unhesitatingly: A raindrop is shaped like a teardrop, flattened at the bottom and tapering to a point at the top. That nugget of conventional wisdom is so embedded in our collective consciousness that the "standard" teardrop is used not only in cartoons but also in respectable scientific publications. Irrigation companies emblazon it on their brochures as a logo, and so do agencies of the United Nations Organization dealing with the supply and utilization of water by the world's thirsty population.

Alas, the vertically elongated, top-pointed teardrop falling through the air is a physical impossibility. Although a drop may have that shape at the instant of detachment from the tip of a wire or a capillary, it will naturally "ball up" to form a sphere once it is suspended in air. Raindrops, in any case, do not form at the tip of a wire, but are condensed from vapor as small droplets, which later grow by coalescence. They are spherical from the start. Then, as they fall vertically through the air, other effects modify their shape somewhat.

Air flowing around the falling drop has a greater velocity near the horizontal "equator" of the drop than near its top and bottom "poles." According to Bernoulli's law, relating the variation of speed to the variation of pressure along streamlines ($p + \rho q^2/2 = \text{constant}$, where p is the pressure, q is the velocity, and ρ is the density of the fluid), the air around the sides of the drop will acquire a pressure smaller than the air above and below the drop, hence the drop will tend to compress vertically and will come to resemble an ellipsoid (with its horizontal axis longer than its vertical axis). If the drop were to continue accelerating, it might eventually come to resemble a pancake. However, Stokes' law enters the picture, with air resistance countering the acceleration and establishing a terminal velocity (which depends on drop size). Turbulence at its wake may cause the drop to resemble a hamburger bun, and random air currents might cause the drop to flutter, but its shape remains more or less the same. And so we can bid farewell to sad tears, as myriad happy raindrops hit the ground surface with their flattened faces and are absorbed into the soil, providing vital moisture to the roots of growing plants.

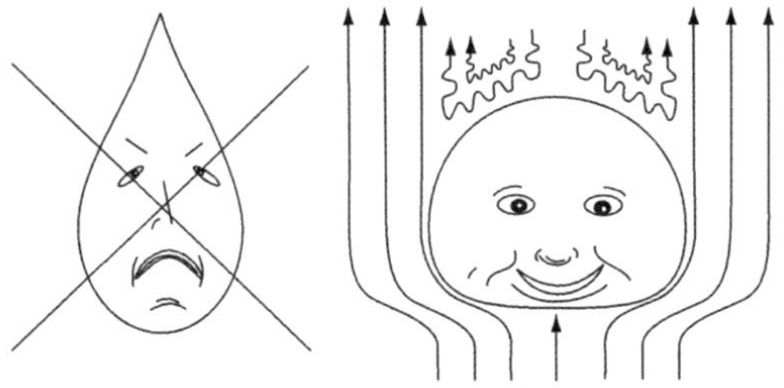

Fig. B2.1. Conventional and real raindrops.

Part II

THE SOLID PHASE

3. PARTICLE SIZES, SHAPES, AND SPECIFIC SURFACE

SOIL TEXTURE

Having defined the soil as a three-phase system, we now focus attention on the solid phase, which is the permanent component of the soil. We can readily imagine a soil without air or without water and, in a vacuum, without both (as is the case with the "soil" found on the moon), but there could hardly be a soil without the solid phase. The soil's solid phase consists of mineral particles having various shapes and sizes, as well as of amorphous compounds such as organic matter and hydrated iron oxides, which are generally attached to the particles. Because the content of the amorphous material is generally (though not invariably) small, we can in most cases represent the solid phase as consisting mainly of discrete particles. The largest soil particles are often visible to the naked eye, whereas the smallest are colloidal and can be observed only with the aid of an electron microscope.

In general, it is possible to classify or group soil particles according to their sizes and to characterize the soil as a whole in terms of the relative proportions of those size groups. The groups may differ from one another in mineral composition as well as in particle size. It is these attributes of the soil solid phase, particle size and mineral composition, that largely determine the nature and behavior of the soil: its internal geometry and porosity, its interactions with fluids and solutes, as well as its compressibility, strength, and thermal regime.

The term *soil texture* refers to the range of particle sizes in a soil, that is, whether a given soil contains a wide or relatively narrow array of particle sizes and whether the particles are mainly large, small, or of intermediate size. The term thus carries both qualitative and quantitative connotations. Qualitatively, it represents the "feel" of the soil material, whether coarse and gritty or fine and smooth to the touch. An experienced soil classifier can feel, by rubbing the moistened soil with his fingers, whether it is coarse textured or fine textured, and can even assess semiquantitatively to which of the textural "classes" the soil might belong.

In a more rigorously quantitative sense, however, the term *soil texture* denotes the precisely measured distribution of particle sizes and the proportions of the various size ranges of particles composing a given soil. As such, soil texture is an intrinsic attribute of the soil and the one most often used to characterize its physical makeup.

TEXTURAL FRACTIONS

The traditional method of characterizing particle sizes in soils is to divide the array of particle diameters into three conveniently separable size ranges known as *textural fractions* or *separates*, namely, *sand*, *silt*, and *clay*. The actual procedure of separating out these fractions and of measuring their proportions is called *mechanical analysis*, for which standard techniques have been devised. The results of this analysis yield the *mechanical composition* of the soil, a term often used interchangeably with *soil texture*.

Unfortunately, there is as yet no universally accepted scheme for classifying particle sizes. For instance, the classification standardized in America by the U.S. Department of Agriculture differs from that of the International Soil Science Society (ISSS), as well as from those promulgated by the American Society for Testing Materials (ASTM), the Massachusetts Institute of Technology (MIT), and various national institutes abroad. The classification followed by soil engineers often differs from that of agricultural soil scientists. The same terms are used to designate differing size ranges, an inconsistency that can be confusing indeed. Several of the often-used particle size classification schemes are compared in Fig. 3.1.

An essential criterion for determining soil texture is the upper limit of particle size that is to be included in the definition of *soil material*. Some soils contain large rocks that obviously do not behave like soil, but, if numerous, might affect the behavior of the soil in bulk. The conventional definition of soil material includes particles smaller than 2 mm in diameter. Larger particles are generally referred to as *gravel*, and still larger rock fragments, several centimeters in diameter, are variously called *stones*, *cobbles*, or — if very large — *boulders*.

The largest particles that are generally recognized as soil material are designated *sand*, defined as particles ranging in diameter from 2000 μm (2 mm) to 50 μm (USDA classification) or to 20 μm (ISSS classification). The sand fraction is often further divided into subfractions such as coarse, medium, and fine sand. Sand grains usually consist of quartz but may also be fragments of feldspar, mica, and, occasionally, heavy minerals such as zircon, tourmaline, and hornblende, though the latter are rather rare. In most cases, sand grains have more or less uniform dimensions and can be represented as spherical, though they are not necessarily smooth and may in fact have quite jagged surfaces (Fig. 3.2). That, together with their hardness, accounts for their abrasiveness.

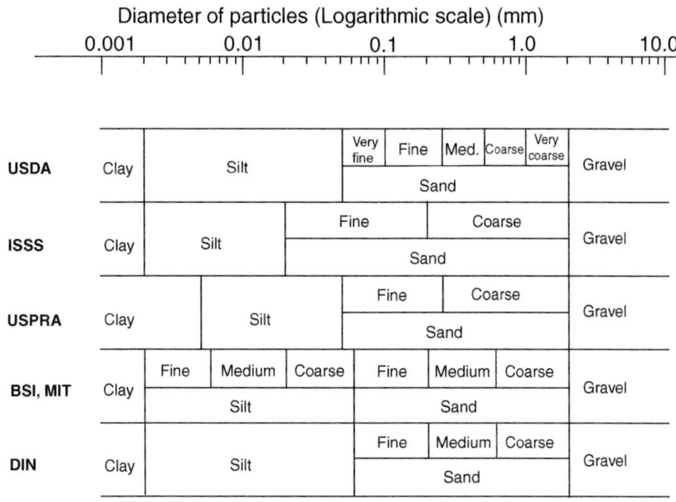

Fig. 3.1. Several conventional schemes for the classification of soil fractions according to particle diameter ranges; U.S. Department of Agriculture (USDA); International Soil Science Society (ISSS); U.S. Public Roads Administration (USPRA); British Standards Institute (BSI); Massachusetts Institute of Technology (MIT); German Standards (DIN).

The next fraction is silt, which consists of particles intermediate in size between sand and clay. Mineralogically and physically, silt particles generally resemble sand particles. However, since the silt are smaller, particles have a greater surface area per unit mass, and are often coated with strongly adherent clay, silt may exhibit, to a limited degree, some of the physicochemical characteristics generally attributed to clay.

The clay fraction, with particles ranging from 2 μm downwards, is the colloidal fraction. Clay particles are characteristically platelike or needlelike in shape and generally belong to a group of minerals known as the *aluminosilicates*. These are secondary minerals, formed in the soil itself in the course of its evolution from the primary minerals that were contained in the original rock. In some cases, however, the clay fraction may include particles (such as iron oxide and calcium carbonate) that do not belong to the aluminosilicate clay mineral category.

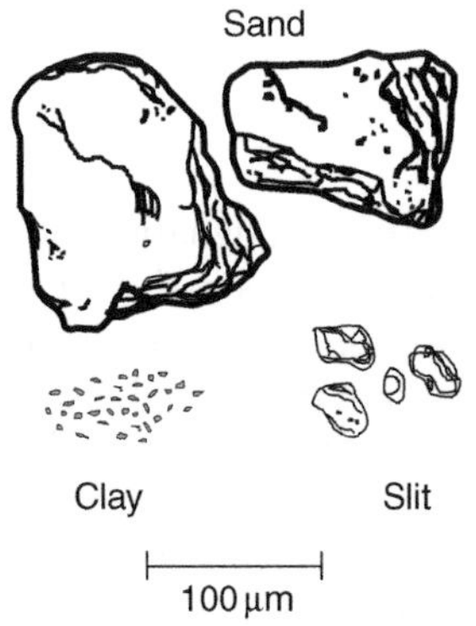

Fig. 3.2. A visual representation of the comparative sizes of sand, silt, and clay particles.

BOX 3.1 The Life Story of Sand

Sand is ubiquitous. We associate sand most commonly with beaches and deserts, but in fact it is an important constituent of nearly all soils. Where does sand come from?

Most sand originates in outcropping continental rocks, especially in mountainous areas that are exposed to weathering. Mechanical breakdown first dislodges rock fragments. Chemical assault follows mechanical disintegration in dissolving and recombining the more reactive constituent minerals, such as micas and feldspars, while the less reactive (more resistant) minerals — primarily quartz — remain as individual grains.

Sand grains are then carried downslope by surface runoff and are generally delivered into a stream. There they roll and bounce along the bottom of the streambed, resting temporarily in stagnant pools and then resuming their sporadic motion when a new flood arrives. Some of the river sand eventually reaches the seashore, though the journey may take hundreds of years. In the course of this journey, the grains are smoothed and polished by friction. Where masses of sand are deposited and dried, the winds take over, further sorting and redistributing the sand. The grains are propelled in spurts, alternately rising a few centimeters above the ground surface and bumping against other grains as they fall back to the surface at an oblique angle.

Because of its far greater surface area per unit mass and its resulting physicochemical activity, clay is the fraction with the most influence on soil behavior. Clay particles adsorb water and hydrate, thereby causing the soil to swell upon wetting and to shrink upon drying. Clay particles typically carry a negative electrostatic charge and, when hydrated, form an electrostatic double layer with exchangeable ions in the surrounding solution. Another expression of surface activity is the heat that evolves when dry clay is wetted, called the *heat of wetting*. A body of clay will typically exhibit plastic behavior and become sticky when moist and then cake up and crack to form cemented hard fragments when desiccated.

The relatively inert sand and silt fractions can be called the soil "skeleton," while the clay, by analogy, can be thought of as the "flesh" of the soil. Together, all three fractions of the solid phase, as they are combined in various configurations, constitute the *matrix* of the soil.

TEXTURAL CLASSES

The overall textural designation of a soil, called *textural class*, is conventionally based on the mass ratios of the three fractions. Soils with different proportions of sand, silt, and clay are assigned to different classes, as shown in the triangular diagram of Fig. 3.3. To illustrate the use of the textural triangle, let us assume that a soil is composed of 50% sand, 20% silt, and 30% clay. Note that the lower left apex of the triangle represents 100% sand and the right side of the triangle represents 0% sand. Now find the point of 50% sand on the bottom edge of the triangle and follow the diagonally leftward line rising from that point and parallel to the zero line for sand. Next, identify the 20%

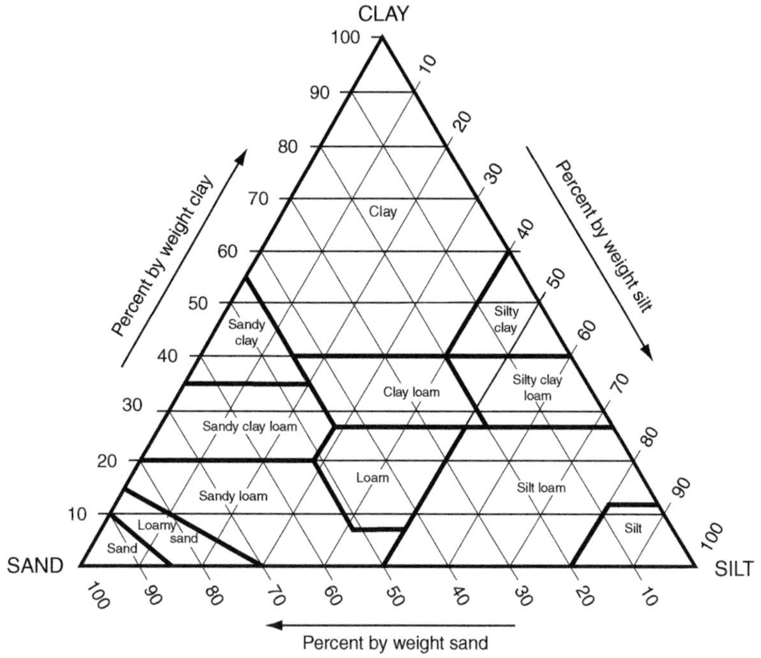

Fig. 3.3. Textural triangle, showing the percentages of clay (below 0.002 mm), silt (0.002–0.05 mm), and sand (0.05–2.0 mm) in the conventional soil textural classes.

line for silt, which is parallel to the zero line for silt, namely, the left edge of the triangle. Where the two lines intersect, they meet the 30% line for clay. The soil in this example happens to fit the category of "sandy clay loam."

Note that a class of soils called *loam* occupies a central location in the textural triangle. It refers to a soil that contains a "balanced" mixture of coarse and fine particles, so its properties are intermediate among those of a sand, a silt, and a clay. As such, a loam is often considered to be the optimal soil for crop growth. Its capacity to retain water and nutrients is better than that of sand, while its drainage, aeration, and tillage properties are more favorable than those of clay. There are, however, exceptions to this generalization. Under different environmental conditions and for different plant species, a sand or a clay may be more suitable than a loam.

Sample Problem

Determine the textural class designations for soils with the distributions of particle sizes shown in Table 3.1.

Using the USDA classification (Fig. 3.1) and the textural triangle (Fig. 3.3), we obtain the following textural classes:

(a) % sand = 40, % silt = 45, % clay = 15; soil class: loam.
(b) % sand = 25, % silt = 60, % clay = 15; soil class: silt-loam.

(c) % sand = 20, % silt = 30, % clay = 50; soil class: clay.
(d) % sand = 60, % silt = 30, % clay = 10; soil class: sandy loam.

TABLE 3.1 Particle Size Distributions for Sample Problem 1

<0.0002 (%)	0.0002–0.002 (%)	0.002–0.01 (%)	0.01–0.05 (%)	0.05–0.25 (%)	0.25–2.0 (mm) (%)
(a) 5	10	20	25	20	20
(b) 6	9	30	30	15	10
(c) 10	30	30	10	10	10
(d) 4	6	10	20	30	30

PARTICLE SIZE DISTRIBUTION

Any attempt to divide into discrete fractions what is usually a continuous array of particle sizes is necessarily arbitrary, and the further classification of soils into distinct textural classes is doubly so. Although this approach is widely followed and in some ways useful, greater information is to be gained from measuring and displaying the complete array and distribution of particle sizes. Figure 3.4 presents typical *particle size distribution curves*. The ordinate of the graph indicates the percentage of soil particles with diameters smaller than the diameter denoted in the abscissa. The latter is drawn on a logarithmic scale to encompass several orders of magnitude of particle diameters while

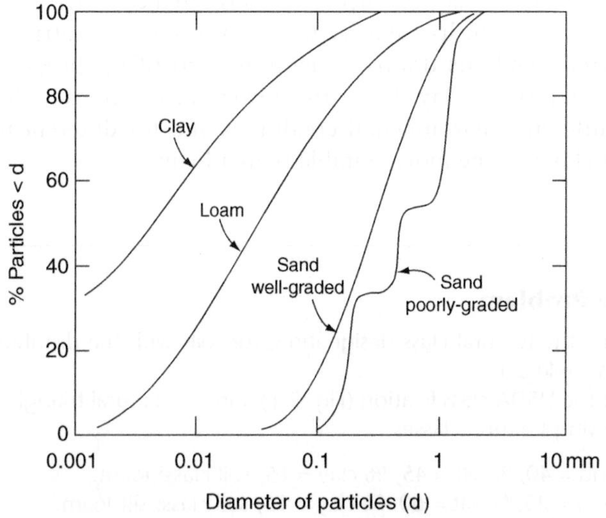

Fig. 3.4. Particle size distribution curves for various types of soil material (schematic).

allowing sufficient space for representing the fine particles. Note that this graph gives an integral, or cumulative, representation. In practice, the particle size distribution curve is constructed by connecting a series of *n* points, each expressing the cumulative mass fraction of particles finer than each of the *n* diameters measured ($F_1, F_2, \ldots, F_i, \ldots, F_n$). Thus,

$$F_i = (M_s - \Sigma M_i)/M_s \qquad (3.1)$$

in which *Ms* is the total mass of the soil sample analyzed and ΣM_i is the cumulative mass of particles finer than the *i*th diameter measured.

The information obtainable from this representation of particle size distribution includes the diameter of the largest grains in the assemblage and the grading pattern, that is, whether the soil is composed of distinct groups of particles each of uniform size or whether it consists of a more or less uniform distribution of particle sizes. Soils having a preponderance of particles of one or several distinct sizes, indicating a steplike distribution curve, are called *poorly graded.* Soils with a flattened and smooth distribution curve (without apparent discontinuities) are called *well graded.*

This aspect of the particle size distribution can be expressed in terms of the so-called *uniformity index*, I_u, defined as the ratio of the diameter d_{60}, which includes 60% of the particles, to the smaller diameter d_{10}, which includes 10% of the particles (as shown in Fig. 3.4). This index, also called the *uniformity coefficient*, is used mostly with coarse-grained soils. For a soil material consisting entirely of equal-sized particles, if such were to exist, I_u would be unity. Some sand deposits may have uniformity indexes smaller than 10. Some well-graded soils, on the other hand, have I_u values greater than 1000.

The particle size distribution curve can also be differentiated graphically to yield a frequency distribution curve for grain sizes, with a peak indicating the most prevalent grain size. Attempts have been made to correlate this index, as well as the harmonic mean diameter of grains, with various soil properties, such as permeability (see Chapter 7).

MECHANICAL ANALYSIS

Mechanical analysis is the procedure for determining the particle size distribution of soil samples (Gee and Or, 2002). The first step in this procedure is to disperse the soil in an aqueous suspension. The primary soil particles, often naturally aggregated, must be separated and made discrete by removal of cementing agents (such as organic matter, calcium carbonate, and iron oxides) and by deflocculating the clay. Removal of organic matter is usually achieved by oxidation with hydrogen peroxide, and calcium carbonate can be dissolved by addition of hydrochloric acid. Deflocculation is carried out by means of a chemical dispersing agent (such as sodium metaphosphate) and by mechanical agitation (shaking, stirring, or ultrasonic vibration). The function of the dispersing agent is to replace the cations adsorbed to the clay, particularly divalent or trivalent cations, with sodium. This has the effect of increasing the hydration of the clay micelles, thus causing them to repel each other rather than coalesce, as they do in the flocculated state. Failure to disperse the soil completely will result in flocs of clay or aggregates settling as if they were silt-sized or sand-sized primary particles, thus biasing the results of mechanical analysis to indicate an apparent content of clay lower than the real value.

Actual separation of particles into size groups can be carried out by passing the suspension through graded sieves, down to a particle diameter of approximately 0.05 mm. To separate and classify still finer particles, the method of sedimentation is usually used, based on measuring the relative settling velocity of particles of various sizes in an aqueous suspension.

According to *Stokes' law*, the velocity of a spherical particle settling under the influence of gravity in a fluid of a given density and viscosity is proportional to the square of the particle's radius. We shall now derive this law, inasmuch as it governs the method of sedimentation analysis.

A particle falling in a vacuum will encounter no resistance as it is accelerated by gravity, so its velocity will increase as it falls. However, a particle falling in a fluid will encounter frictional resistance proportional to the product of its radius and velocity and to the viscosity of the fluid.

The resisting force due to friction F_r was shown by George Stokes in 1851 to be

$$F_r = 6\pi\eta ru \tag{3.2}$$

where η is the viscosity of the fluid, r is the radius of the particle, and u is its velocity. As the particle begins its fall, its velocity increases due to the acceleration of gravity. Eventually, the increasing resistance force equals the constant downward gravity force. Henceforth the particle continues to descend by inertia at a constant velocity known as the *terminal velocity* u_t.

The downward force due to gravity F_g is

$$F_g = (4/3)\pi r^3(\rho_s - \rho_f)g \tag{3.3}$$

Here $(4/3)\pi r^3$ is volume of the spherical particle, ρ_s is its density, ρ_f is the density of the fluid, and g is the acceleration of gravity. Subtracting fluid density from particle density accounts for the buoyancy that reduces the effective weight of particles in suspension (Archimedes' law).

Setting the two forces equal, we obtain Stokes' law:

$$u_t = (2/9)(r^2g/\eta)(\rho_s - \rho_f) = (d^2g/18\eta)(\rho_s - \rho_f) \tag{3.4}$$

where d is the diameter of the particle. Assuming that the terminal velocity is attained almost instantly, we can obtain the time t needed for the particle to fall through a vertical distance h:

$$t = 18h\eta/d^2g(\rho_s - \rho_f)$$

Rearranging and solving for the particle diameter gives

$$d = [18h\eta/tg(\rho_s - \rho_f)]^{1/2} \tag{3.5}$$

Since all terms on the right-hand side of the equation, except t, are constants, we can combine them into a single constant A and write

$$d = A/t^{1/2} \qquad \text{or} \qquad t = B/d^2$$

where $B = A^2$.

One way of measuring particle size distribution is to use a pipette to draw samples of known volume from a given depth in the suspension at regular times during sedimentation. An alternative method is to use a *hydrometer* to monitor the density of the suspension at a given depth as a function of time. The density of the suspension diminishes as the largest particles and then progressively smaller ones settle out of the region of the suspension being monitored.

The use of Stokes' law to measure particle sizes is based on certain simplifying assumptions:

1. The particles are large enough to be unaffected by random motion of the fluid molecules.
2. The particles are rigid, spherical, and smooth.
3. All particles have the same density.
4. The suspension is dilute enough so particles settle independent.
5. Fluid flow around the particles is laminar (slow enough to avoid onset of turbulence).

In fact, we know that soil particles, though mostly rigid, are neither spherical nor smooth, and some may be platelike. Hence, the diameter calculated from the settlement velocity does not necessarily correspond to the actual dimensions of the particle. Rather, we should speak of an *effective* or *equivalent* settling diameter.

The results of a mechanical analysis based on sieving may differ from those of a sedimentation process for the same soil material. Moreover, soil particles are not all of the same density. Most silicates have specific gravity values of 2.6–2.7, whereas iron oxides and other heavy minerals may be twice as dense. For all these reasons, the mechanical analysis of soils yields only approximate results. Its greatest shortcoming, however, is that it only gives the *amount* of clay but not the *type* of clay, which can greatly affect soil behavior.

Sample Problem

Using Stokes' law, calculate the time needed for all sand particles (diameter >50 μm) to settle out of a depth of 0.2 m in an aqueous suspension at 30°C. How long would it take for all silt particles to settle out? How long for "coarse" clay (>1 μm)?

We use Eq. (3.5):

$$t = 18h\eta/d^2g(\rho_s - \rho_f).$$

Substituting the appropriate values for depth h (0.20 m), viscosity η (0.0008 kg/m sec), particle diameter d (50 μm, 2 μm, and 1 μm for the lower limits of sand, silt, and coarse clay, respectively), gravitational acceleration g (9.81 m/sec^2), average particle density ρs (2.65 × 10 km/m^3), and water density (10^3 kg/m^3), we obtain the following answers.

(a) For all sand to settle out, leaving only silt and clay in suspension:

$$t = \frac{18 \times 0.2 \times (8 \times 10^{-4})}{(50 \times 10^{-6})^2 \times 9.81 \times (2.65 - 1.0) \times 10^3} = 71 \text{ sec}$$

(b) For all silt to settle out, leaving only clay in suspension:

$$t = \frac{18 \times 0.2 \times (8 \times 10^{-4})}{(2 \times 10^{-6})^2 \times 9.81 \times 1.65 \times 10^3} = 44,500 \text{ sec} = 12.36 \text{ hr}$$

(c) For all coarse clay to settle out, leaving only fine clay in suspension:

$$t = \frac{18 \times 0.2 \times (8 \times 10^{-4})}{(1 \times 10^{-6})^2 \times 9.81 \times 1.65 \times 10^3} = 178,800 \text{ sec} = 49.44 \text{ hr}$$

PARTICLE SHAPES

Particle shapes, as well as particle sizes, affect the properties and behavior of the soil in bulk. Irregularly shaped particles expose a larger surface per unit mass than do regularly shaped particles. Whether particles are smooth or jagged has a bearing on the tendency of particles to fit closely together or interlock and, hence, on the resulting bulk density, the geometric configuration of the pore space, the interparticle bonding, and the resistance of the soil to deformation. This resistance determines the reaction of soil bodies to forces tending to cause compression as well as shearing. The jaggedness (as well as hardness) of particles also affects the frictional abrasiveness of soils toward traversing vehicles and tillage tools.

The techniques employed to examine particle shapes include direct observation via a microscope, image analysis, and the measurement of viscosity (Egashira and Matsumoto, 1981) and the light-scattering properties of suspensions

For particles that can be assigned the dimensions of length (L), breadth (B), and thickness (T), Heywood (1947) first defined the following shape parameters:

$$\text{Flatness ratio} = B/T \tag{3.6}$$

$$\text{Elongation ratio} = L/B \tag{3.7}$$

Other, dimensionless parameters that describe particle shapes are (Skopp, 2002):

$$\text{Sphericity} = A_s/A_a \tag{3.8}$$

$$\text{Circularity} = C_a/P_a \tag{3.9}$$

$$\text{Rugosity} = P_a/P_c \tag{3.10}$$

where A_a and A_s are, respectively, the surface area of an actual particle and the surface area of an "equivalent" sphere (i.e., a sphere that has the same apparent settling or sieve-passing properties as the particle C_a and P_a are, respectively, the circumference of a circle with the same area as the cross-sectional area of a particle and the particle's actual perimeter P_a; and P_c is the circumference of a circle circumscribing the particle.

These parameters can be used to describe the shapes of idealized particles and to model the characteristics of assemblages of such particles. However, the same criteria are of limited use in characterizing soils in which the particles are too irregular and variable to obey any simplistic criterion. Attempts have nonetheless been made to characterize the shapes of soil particles using empirically determined dimensionless shape factors.

SPECIFIC SURFACE

The specific surface of soil material can be defined as the surface area A_s of particles per unit mass (a_m) or per unit volume of particles (a_v) or per unit bulk volume of soil (a_b):

$$a_m = A_s/M_s \tag{3.11}$$

$$a_v = A_s/V_s \tag{3.12}$$

$$a_b = A_s/V_t \tag{3.13}$$

where M_s is the mass of particles of volume V_s contained in a bulk volume V_t of soil.

Specific surface has traditionally been expressed in terms of square meters per gram (a_m) or square meters per cubic centimeter (a_v). To convert from m^2/gm to m^2/kg, one need only multiply by 10^3. To convert from m^2/cm^3 to m^2/m^3, one needs to multiply by 10^6.

Specific surface obviously depends on the sizes of the soil particles. It also depends on their shapes. Flattened or elongated particles expose greater surface per unit mass or volume than do equidimensional (cubical or spherical) particles. Since clay particles are generally platy, they contribute more to the overall specific surface of a soil than is indicated by their small size alone. In addition to their external surfaces, some clay particles exhibit internal surface areas, such as are exposed when the accordion-like crystal lattice of montmorillonite expands on imbibing water. While the specific surface of sand may be no more than 1 or 2 square meters per gram, that of clay may be a hundred times higher. In fact, it is the clay fraction — its content and mineral composition — that largely determines the specific surface of a soil.

MEASURING SPECIFIC SURFACE BY ADSORPTION

The usual procedure for determining surface area is to measure the amount of gas or liquid needed to form a *monomolecular layer* over the entire surface in a process of adsorption (Pennell, 2002). The standard method is to use an inert gas such as nitrogen. Water vapor and organic liquids (e.g., glycerol and ethylene glycol) are also used. The adsorption phenomenon was described by de Boer (1953). At low gas pressures, the amount of a gas adsorbed per unit area, σ_a, is related to the gas pressure P, the temperature T, and the heat of adsorption Q by the equation

$$\sigma_a = k_i P \exp(Q_a/RT) \tag{3.14}$$

where R is the gas constant and k_i is also a constant. Thus, the amount of adsorption increases with pressure but decreases with temperature.

The equation of Langmuir (1918) relates the gas pressure P to the volume of gas adsorbed per unit mass of adsorbent m_a at constant temperature:

$$P/m_a = 1/kv_1 + P/v_1 \qquad (3.15)$$

where k is a constant and v_1 is the volume of adsorbed gas forming a monomolecular layer over the adsorbent. Parameter v_1 is the slope of the curve of P/m_a versus P. The specific surface of the adsorbent can be calculated by determining the number of molecules in v_1 and multiplying this by the cross-sectional area of these molecules. Langmuir's equation assumes that only one layer of molecules is adsorbed and that the heat of adsorption is uniform during the process.

Brunauer, Emmett, and Teller (1938) derived what has come to be known as the BET equation, based on multilayer adsorption theory:

$$P/v(P_0 - P) = (1/v_1C) + (C - 1)P/v_1CP_0 \qquad (3.16)$$

where v is the volume of gas adsorbed at pressure P, v_1 is the volume of a single molecular layer over the surface of the adsorbent, P_0 is the gas pressure required for monolayer saturation at ambient temperature, and C is a constant for the particular gas, adsorbent, and temperature. The volume v_1 can be obtained from the BET theory by plotting $P/v(P_0 - P)$ versus P/P_0. The density of the adsorbed gas is usually assumed to be that of the liquefied or solidified gas.

Polar adsorbents (such as water) may not obey the BET and Langmuir equations (which are similar at low pressures), because their molecules or ions tend to cluster at charged sites rather than spread evenly over the surface of the adsorbent. Techniques for measuring the specific surface of soils by using various adsorbed materials were described by Carter et al. (1986).

ESTIMATING SPECIFIC SURFACE BY CALCULATION

An estimation of the specific surface area can also be made by calculation based on the sizes, shapes, and relative abundance of different types of particles in the soil.

For a sphere of diameter d, the ratio of surface to volume is

$$a_v = (\pi d^2)/(\pi d^3/6) = 6/d \qquad (3.17)$$

and the ratio of surface to mass is

$$a_m = a_v/\rho_s = 6/\rho_s d \qquad (3.18)$$

For particles having a density ρ_s of about 2.65 g/cm^3 (2650 kg/m^3), Equation (3.18) can be approximated by

$$a_m \approx 2.3/d \qquad (3.19)$$

Now suppose the particles are not spherical but rectangular: For a cube of edge L, the ratio of surface to volume is

$$a_v = 6L^2/L^3 = 6/L \qquad (3.20)$$

and the ratio of surface to mass is

$$a_m = 6/\rho_s L \qquad (3.21)$$

Thus, the expressions for particles of nearly equal dimensions, such as most sand and silt grains, are similar, and knowledge of the particle size distribution can allow us to calculate the approximate specific surface by the summation equation:

$$a_m = (6/\rho_s)\Sigma\, c_i(d_i^2/d_i^3) = (6/\rho_s)\Sigma\,(c_i/d_i) \quad (3.22)$$

in which c_i is the mass fraction of particles of average diameter d_i.

Now let's consider a platy particle. For the sake of simplicity, assume that a hypothetical plate is square shaped, with sides L and thickness z. The surface-to-volume ratio is

$$a_v = (2L^2 + 4Lz)/L^2z \quad (3.23)$$

and the surface-to-mass ratio is

$$a_m = 2(L + 2z)/\rho_s Lz \quad (3.24)$$

If the platelet is very thin, so that its thickness z is negligible compared to principal dimension L, and if $\rho_s = 2.65$ g/cm^3 (2650 kg/m^3), then

$$a_m \approx 2/\rho_s z \approx 0.75/z \text{ cm}^2/\text{g} \quad (3.25)$$

Thus, the specific surface of a clay can be estimated if the thickness of its platelets is known. The platelet thickness of dispersed montmorillonite is about 10 Å (or 10^{-7} cm = 10^{-9} m). Hence $a_m = 0.75/10^{-7}$ cm^2/g, or 750 m^2/g, which compares closely with the measured value of about 800 m^2/g. The average platelet thickness for illite clay is about 50 Å (or 5×10^{-11} m), and for kaolinite clays it is a few hundred angstroms.

Sample Problem

Estimate the approximate specific surface (m^2/g) of a soil composed of 10% coarse sand (average diameter 1 mm), 20% fine sand (average diameter 0.1 mm), 30% silt (average diameter 0.02 mm), 20% kaolinite clay (average platelet thickness 4×10^{-8} m), 10% illite clay (average thickness 0.50×10^{-8} m), and 10% montmorillonite (average thickness 10^{-9} m).

For the sand and silt fractions, we use Eq. (3.17):

$$a_m = (6/2.65)[(0.1/0.1) + (0.2/0.01) + (0.3/0.002)] = 0.0387 \text{ m}^2/\text{g} = 38.7 \text{ m}^2/\text{kg}$$

For the clay fraction, we use Eq. (3.22) in summation form to include the partial specific surface values for kaolinite, illite, and montmorillonite, respectively:

$$\begin{aligned} a_m &= (0.2 \times 0.75)/(400 \times 10^{-8}) + 0.1 \times 0.75/(50 \times 10^{-8}) + 0.1 \times 0.75/(10 \times 10^{-8}) \\ &= 3.78 \text{ m}^2/\text{g (kaol.)} + 15.09 \text{ m}^2/\text{g (ill.)} + 75.45 \text{ m}^2/\text{g (mont.)} \\ &= 94.32 \text{ m}^2/\text{g} = 94{,}320 \text{ m}^2/\text{kg} \end{aligned}$$

Total for the soil = 0.0387 + 94.32 = 94.36 m^2/g = 94,360 m^2/kg

Note: The clay fraction, only 40% of the soil mass, accounts for 99.96% of the specific surface. Montmorillonite alone (10% of the mass) accounts for nearly 80% of this soil's specific surface.

4. CLAY, THE COLLOIDAL COMPONENT

DEFINITIONS OF CLAY

The term *clay* carries several connotations, which are not necessarily mutually consistent. In daily language, it suggests a soil that tends to retain water and to become soft and sticky when wet. In the context of soil texture, it designates a range of particle sizes (namely, particles smaller than 2 micrometers) or a soil material with a high fractional content of this particle-size range. (Much of the clay therefore fits within the physicochemical category of *colloids*, which are finely divided materials that tend to disperse within gases and liquids and exhibit a very high degree of surface activity.) Finally, in the mineralogical sense, clay refers to a particular group of minerals, many of which occur in the clay fraction of the soil. That fraction thus differs from sand and silt not only in the range of particle sizes but also in mineralogical composition. Sand and silt consist mainly of weathering-resistant *primary minerals*, that is, minerals present in the original rock from which the soil was formed. Clay, however, includes *secondary minerals* formed in the soil itself by decomposition of the primary minerals and their recomposition into new ones.

The various *clay minerals* differ from one another in prevalence and properties and in the way they affect soil behavior. Rarely do any of these minerals occur in homogeneous deposits; in the soil they generally appear in mixtures, the specific composition of which depends in each case on the combination of conditions that governed soil formation. To understand why and how the clay fraction serves as the active constituent of soils, we must consider the structure and function of clay minerals.

STRUCTURE OF CLAY MINERALS

The forerunners of modern soil science assumed at first that clay consists of particles that are essentially similar to those of sand and silt, differing from them only in size. Later, they began to notice that when clays were dried from aqueous suspensions they tended to form thin flakes and also that moist clay could be skimmed and polished to form a smooth and shiny surface. These observations suggested that clay particles may be flat and capable of being oriented in different ways. But it was not until the advent of X-ray diffraction (described by Whittig and Allardice, 1986), of differential thermal analysis (Tan et al., 1986), and electron microscopy that the crystalline nature of clay minerals was proven and their structures were described.

The most prevalent minerals in the clay fraction of temperate region soils are silicate clays, whereas in tropical regions hydrated oxides of iron and aluminum are more prevalent. The typical aluminosilicate clay minerals appear as laminated microcrystals, composed mainly of two basic structural units: a tetrahedron of four oxygen atoms surrounding a central cation, usually Si^{4+}, and an octahedron of six oxygen atoms or hydroxyls surrounding a somewhat larger cation of lesser valency, usually Al^{3+} or Mg^{2+}. These basic building blocks are illustrated in Fig. 4.1.

The tetrahedra are joined together at their basal corners by means of shared oxygen atoms, in an hexagonal network that forms a flat sheet only 0.493 nanometers (4.93 Å) thick. This is illustrated in Fig. 4.2. The octahedra are similarly joined along their edges to form a triangular array as shown in Fig. 4.3. These sheets are about 0.505 nm (5.05 Å) thick.

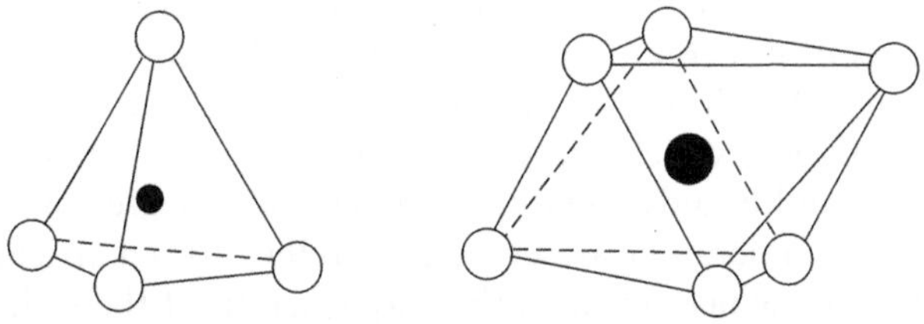

Fig. 4.1. The basic structural units of aluminosilicate clay minerals: a tetrahedron of oxygen atoms surrounding a silicon ion (left), and an octahedron of oxygens or hydroxyls enclosing an aluminum ion (right).

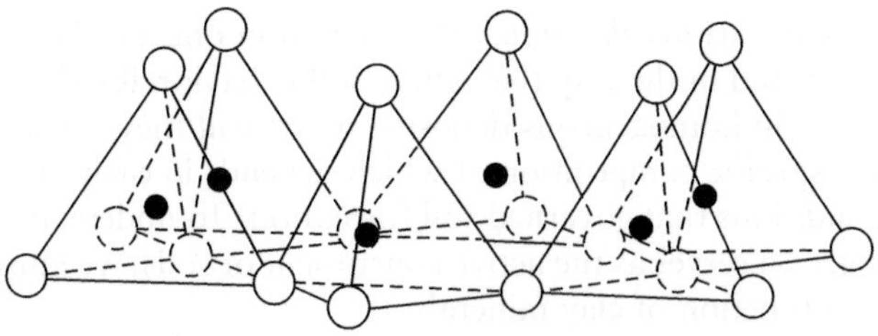

Fig. 4.2. Hexagonal network of tetrahedra forming a silica sheet.

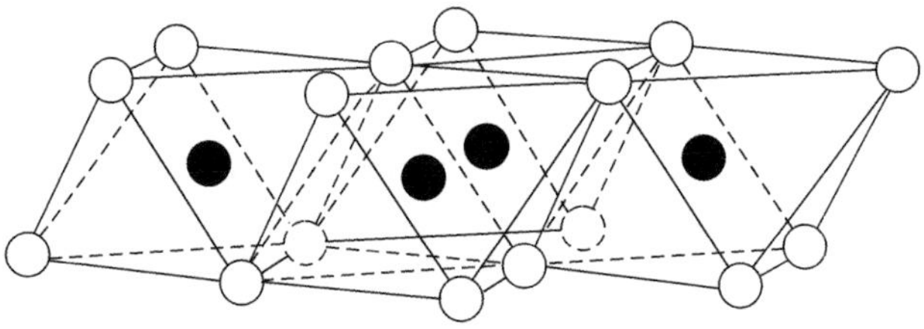

Fig. 4.3. Structural network of octahedra forming an alumina sheet.

The layered aluminosilicate clay minerals are of two main types, depending on the ratios of tetrahedral to octahedral sheets, whether 1:1 or 2:1 (2:2 minerals are also recognized). In 1:1 minerals like kaolinite, an octahedral sheet is attached by the sharing of oxygens to a single tetrahedral sheet. In 2:1 minerals like montmorillonite, it is attached in the same way to two tetrahedral sheets, one on each side. This is shown in Fig. 4.4. A clay particle is composed of multiply stacked composite layers (or unit cells) of this sort, called *lamellae*.

The structure described is an idealized one. Typically, some substitutions of ions of approximately equal radii, called *isomorphous replacements*, take place during crystallization. In the tetrahedral sheets, Al^{3+} may take the place of Si^{4+}, whereas in the octahedral layer Mg^{2+} may occasionally substitute for Al^{3+}. Consequently, internally unbalanced negative charges occur at different sites in the lamellae. Another source of unbalanced charge in clay crystals is the incomplete charge neutralization of terminal ions on lattice edges.

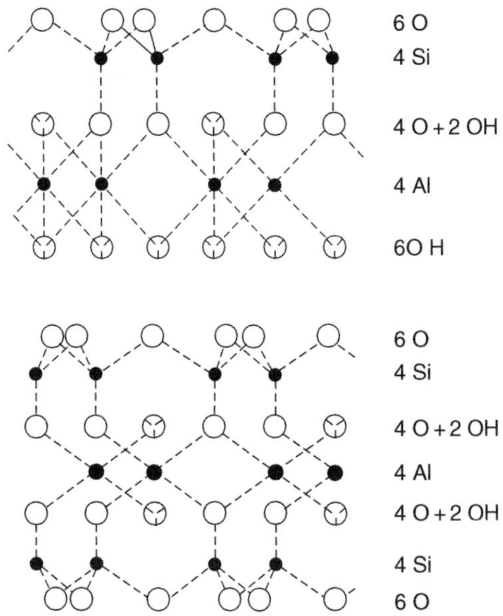

Fig. 4.4. Schematic representation of the structure of aluminosilicate minerals: kaolinite (top) and montmorillonite (bottom).

These unbalanced charges must be compensated externally by the adsorption of ions (mostly cations) from the solution surrounding the clay when it is wetted. These ions tend to concentrate near the external surfaces of the clay particle and occasionally penetrate into the interlamellar spaces. The adsorbed cations, including Na^+, K^+, H^+, Mg^{2+}, Ca^{2+}, and Al^{3+}, are not an integral part of the lattice structure and can be replaced, or exchanged, by other cations in the soil solution. The *cation exchange* phenomenon is of great importance in the soil, because it affects the retention and release of nutrients and other salts as well as the flocculation–dispersion processes of soil colloids.

PRINCIPAL CLAY MINERALS

Clay minerals are usually classified into two main groups, structured and amorphous. The structured clays are subclassified according to their internal crystal structure (i.e., the layered arrangement of tetrahedra and octahedra sheets) into two principal types, 1:1 and 2:1 minerals. The 2:1 clay minerals are further divided into expanding and nonexpanding types. Finally, each of these types includes a number of specific minerals, which can be identified on the basis of X-ray, electron microscope, or thermal analysis techniques.

The most common mineral of the 1:1 type is *kaolinite*. Other minerals in the same group are *halloysite* and *dickite*. The basic layer in the crystal structure is a pair of silica-alumina sheets, and these are stacked in alternating fashion and held together by hydrogen bonding in a rigid, multilayered lattice that often forms an hexagonal platelet. Since water and ions cannot enter between the basic layers, these cannot ordinarily be cleaved or separated. Moreover, since only the outer faces and edges of the platelets are exposed, kaolinite has a relatively low specific surface. Kaolinite crystals generally range in planar diameter from 0.1 to 2 μm, with a variable thickness in the range of about 0.02–0.05 μm. Owing to its relatively large particles and low specific surface, kaolinite exhibits less plasticity, cohesion, and swelling than most other clay minerals. The unit layer formula of kaolinite is $Al_4Si_4O_{10}(OH)_8$.

At the opposite end of the spectrum of aluminosilicate clay minerals is *montmorillonite*, a 2:1 mineral of the expanding type, which also includes *vermiculite* and *beidellite*. The lamellae of montmorillonite are stacked in loose assemblages called *tactoids*. Water and ions are drawn into the cleavage planes between the lamellae, and as the crystal expands like an accordion, it can readily be separated into flakelike thinner units and, ultimately, into individual lamellae, which are only 1 nm thick. As the montmorillonite crystals expand, their internal surfaces as well as external ones come into play, thus increasing the effective specific surface severalfold. Because of its tendency to expand and to disperse, montmorillonite exhibits pronounced swelling–shrinking behavior as well as high plasticity and cohesion. On drying, montmorillonitic soils, especially if dispersed, tend to crack and form hard clods. When tactoids are heated to several hundred degrees they tend to close irreversibly so that only their external areas act as adsorbing surfaces. The unit layer formula of montmorillonite is $Al_{3.5}Mg_{0.5}Si_8O_{20}(OH)_4$.

A clay mineral intermediate in properties between kaolinite and montmorillonite is *illite*. It belongs to a group of clay minerals called *hydrous micas*, which have a 2:1 silica:alumina ratio but are of the nonexpanding type. Isomorphous substitution of aluminum ions for silicon ions in the tetrahedral sheets (rather than substitutions of Mg^{2+} for Al^{3+} in the octahedral sheets, as is the case in montmorillonite), to the extent of about 15%, accounts for the relatively high density of negative charges in these sheets. This, in turn, attracts potassium ions and "fixes" them tightly between adjacent lamellae. As a result, the layers are bound together, so their separation and, hence, expansion of the entire lattice are effectively prevented. The unit layer formula of illite is $Al_4Si_7AlO_{20}(OH)_4K_{0.8}$, with the potassium occurring between the crystal units.

An example of a 2:2-type mineral is *chlorite*, in which magnesium rather than aluminum ions predominate in the octahedral sheets, which are in combination with the tetrahedral silica sheets. Its unit layer formula is $Mg_6Si_6Al_2O_{20}(OH)_4$, with $Mg_6(OH)_{12}$ occurring between the layers. In behavior, chlorite resembles illite. An additional group of silicate clays, in which the lattice is continuous in only one dimension, is known as *attapulgite* or *palygorskite*. The particles of this group are needlelike or tubelike, with microcavities providing internal surfaces.

Frequently, the various clay minerals occur not separately but in a complex mixture. At times, even the internal structure is mixed or interstratified, giving rise to composite minerals, which are somewhat loosely termed *bravaisite* (illite-montmorillonite, chlorite-illite, vermiculite-chlorite, etc.).

The clay fraction may also contain appreciable quantities of noncrystalline (amorphous) mineral colloids. *Allophanes*, for instance, are random combinations of poorly structured silica and alumina components expressible in the general formula $Al_2O_3{\cdot}2SiO_2{\cdot}H_2O$. The actual mole ratio of alumina to silica ranges in this group from 0.5 to 2.0. Phosphorus and iron oxides are frequently present in allophane. Notwithstanding its variable composition, allophane is sufficiently distinctive to be identified as a clay mineral. In behavior, this amorphous clay is similar to the crystalline clays with respect to adsorption, ion exchange, and plasticity.

Still another constituent of the clay traction is a group of hydrous oxides of iron and aluminum known as *sesquioxides*, which are prevalent mainly in the soils of tropical and subtropical regions and are responsible for the predominantly reddish or yellowish hue of these soils. Their composition can be formulated as $Fe_2O_3{\cdot}nH_2O$ and $Al_2O_3{\cdot}nH_2O$, in which the hydration ratio n is variable. Limonite and goethite are typical hydrated iron oxides, and gibbsite is a frequently encountered aluminum oxide. The sesquioxide clays may be partly crystallized but are often amorphous. Their electrostatic, adsorptive, and plasticity properties are less distinct than those of most of the silicate clays. Frequently, these oxides serve as cementing agents in the stabilization of soil aggregates, particularly in subtropical and tropical regions. Table 4.1 summarizes the properties of a few of the silicate clay minerals.

TABLE 4.1 Typical Properties of Selected Clay Minerals (Approximate Values)

Properties	Clay mineral				
	Kaolinite	Illite	Montmorillonite	Chlorite	Allophane
Planar Diameter (μm)	0.1–4	0.1–2	0.01–1	0.1–2	
Basic layer thickness (Å)	7.2	10	10	14	
Particle thickness (Å)	500	50–300	10–100	100–1000	
Specific surface (m^2/g)	5–20	80–120	700–800	80	
Cation exchange capacity (mEq/100 g)	3–15	15–40	80–100	20–40	40–70
Area per charge ($Å^2$)	25	50	100	50	120

THE ELECTROSTATIC DOUBLE LAYER

When a colloidal particle is more or less dry, the neutralizing counterions are attached to its surface. Upon wetting, however, some of the ions dissociate from the surface and enter into solution. A hydrated colloidal particle of clay or humus thus forms a *micelle*, in which the adsorbed ions are spatially separated, to a greater or lesser degree, from the negatively charged particle (Fig. 4.5). Together, the particle surface, acting as a multiple anion, and the "swarm" of cations hovering about it form an *electrostatic double layer.*

The assemblage of adsorbed cations can be regarded as consisting of (a) a layer more or less fixed in position at the immediate proximity of the particle surface and known as the *Stern layer* and (b) a diffuse cloud of cations extending some distance away from the particle surface and gradually diminishing in concentration. This distribution is illustrated in Fig. 4.6. It results from an equilibrium between two opposing tendencies: (1) the electrostatic

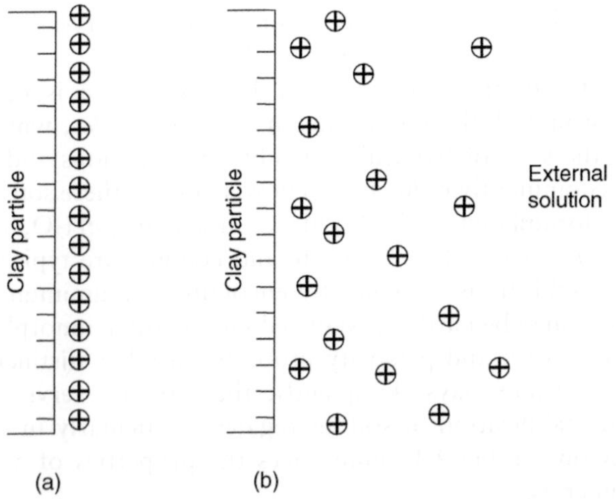

Fig. 4.5. Formation of a diffuse double layer in a dry (a) vs. hydrated (b) micelle.

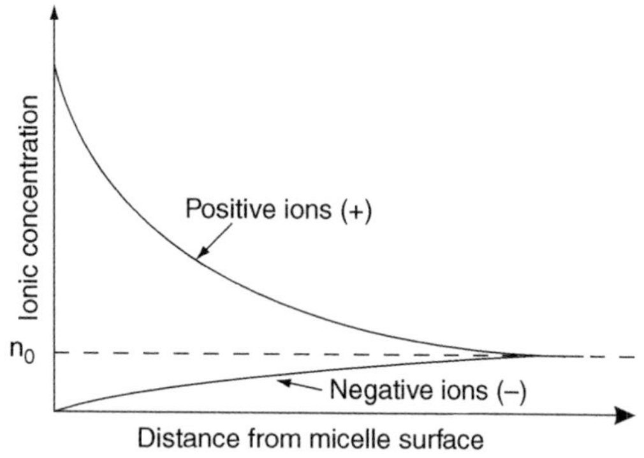

Fig. 4.6. The distribution of positive and negative ions in solution with distance from the surface of a clay micelle bearing a net negative charge. Here n_0 is the ionic concentration in the bulk solution outside the electrical double layer.

(coulombic) attraction of the negatively charged surface for the positively charged ions, which tends to pull the cations inward so as to attain the minimum potential energy level; and (2) the kinetic (Brownian) motion of the liquid molecules, inducing the outward diffusion of the adsorbed cations in a tendency toward equalizing the concentration throughout the solution phase, thus maximizing entropy. The equilibrium distribution of cations is such as to minimize the free energy of the system as a whole. The actual concentration of cations inside the double layer can be 100 or even 1000 times greater than in the external, or intermicellar, solution (i.e., outside the range of influence of the particle's electrostatic force field).

As cations are adsorbed positively by the colloidal particle, anions are generally repelled, or adsorbed negatively, and thus are relegated from the micellar to the intermicellar region of the solution surrounding the clay particles. In some special cases, however, anions may also be attracted to specific sites on colloidal surfaces, but this phenomenon of *anion adsorption* is much less prevalent than *cation adsorption* in soils.

Theoretical treatments of the electrostatic double layer have generally been based on the *Gouy–Chapman theory*, in which the negative charges are assumed to be constant and to be distributed uniformly over the particle surfaces (although in actual fact they may originate within the crystal lattice). The strength of the surface charge is proportional to the charge density (i.e., the number of charged sites per unit area).

The effective thickness of the diffuse double layer can be estimated by means of the following equation:

$$z = (1/ev)\ [(\varepsilon kT)/(8\pi n_0)]^{0.5} \tag{4.1}$$

in which z is the characteristic length, or extent, of the double layer, defined as the distance from the clay surface to where the ionic concentration is very nearly that of the external (intermicellar) solution; e is the elementary charge of an electron (4.77×10^{-10} esu); ε is the dielectric constant; k is the Boltzmann constant ($k = R/N$, where R is the gas constant and N is Avogadro's number); v is the valence of the ions in solution; n_0 is the concentration of the ions in the bulk solution; and T is the temperature in Kelvin.

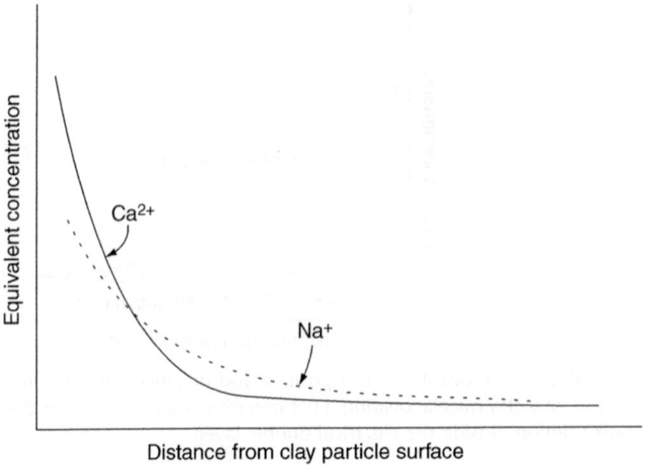

Fig. 4.7. Comparison of the distribution of a monovalent cation (Na^+) with a divalent cation (Ca^{2+}) in the diffuse double layer. Thanks to its greater charge, the divalent cation is attracted more strongly and held more closely to the particle surface. The area under each curve, signifying milliequivalents of cations per unit surface, is equal.

As the equation indicates, the diffuse double layer is compressed as the valence of the ions in solution is increased. For instance, if a solution of monovalent cations is replaced by a solution of divalent cations, the double layer becomes only half as thick (Fig. 4.7). The double-layer thickness is also affected by the solution's concentration (Fig. 4.8), because z of Eq. (4.1) is inversely proportional to the square root of n. Thus a tenfold increase of concentration will reduce the double layer to $1/10^{0.5}$, that is, to about a third of its previous thickness.

The foregoing considerations do not account for interparticle interaction, that is, where the diffuse ionic clouds of neighboring micelles intermingle. In

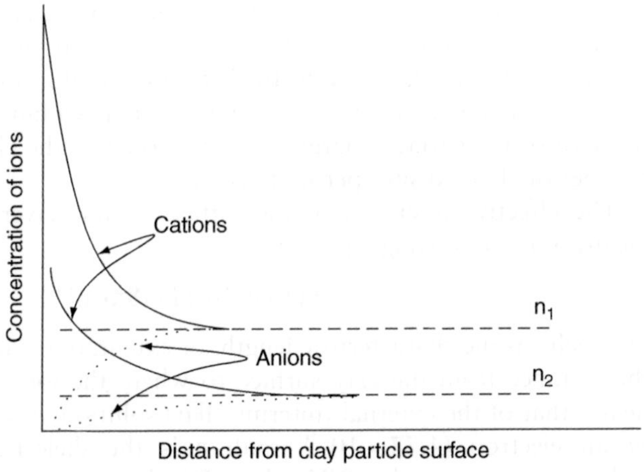

Fig. 4.8. Influence of ambient solution concentration (n_1, n_2) on thickness of the diffuse double layer. The solid curves represent cations and the dotted curves anions. Higher concentration of the external solution is seen to compress the double layer.

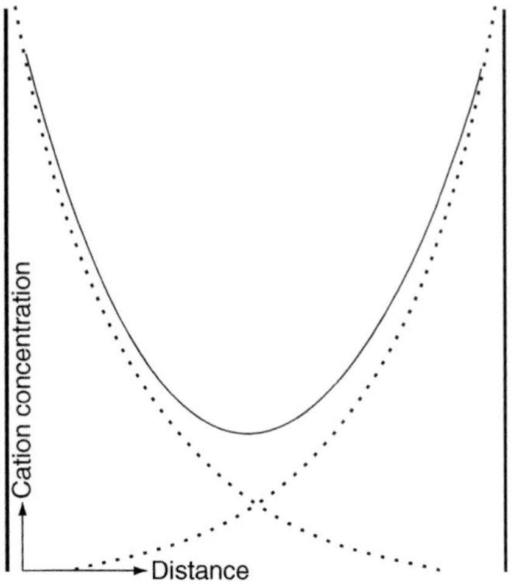

Fig. 4.9. Schematic illustration of the increased concentration of cations in the double layer due to interaction of adjacent particles. The dotted curves indicate the hypothetical distribution of cations for each of the particles if it were suspended alone in the same ambient solution.

this case, the concentration in the median plane, rather than in the external solution, is referred to (Langmuir, 1938):

$$n_c = \pi^2/[v^2 B(d + a)^2 \times 10^{-16}] \quad (4.2)$$

Here n_c is cation concentration at the median plane (moles/liter), v is the valence of the exchangeable cations, d is the distance from either plate to the median plane (angstroms), a is a correction factor (1–4 Å), and B is a constant related to the temperature and dielectric constant (10^{15} cm/mmol). The distribution of cations between two charged plates is illustrated in Fig. 4.9.

ION EXCHANGE

The cations in the double layer can be replaced or exchanged by other cations introduced into the solution. Under chemically neutral conditions, the total number of exchangeable cation charges, expressed in terms of chemical equivalents per unit mass of soil particles, is nearly constant and independent of the species of cations present. It is thus considered to be an intrinsic property of the soil material, generally called the *cation exchange capacity* (or the *cation adsorption capacity*), and has traditionally been reported in terms of milliequivalents of cations per 100 grams of soil. [The term *equivalent* is defined in chemistry as one gram atomic weight (mole) of hydrogen or the mass of any ion that can combine with or replace it in chemical reactions.] Thus expressed, the cation exchange capacity ranges from nil in sands to 100 meq/100 g, or even more, in clays and in organic soils.

The cation exchange capacity depends not only on clay content but also on clay type (i.e., on specific surface and charge density), as indicated in

Table 4.1. In the top layer of soils rich in organic matter, humus can account for as much as 1/4 or 1/3 of the exchange capacity. The cation exchange phenomenon affects the movement and retention of ions in the soil, which can be important in plant nutrition and in environmental processes involving the transport of pollutants. Cation exchange also affects the flocculation–dispersion processes of soil colloids and hence the development and degradation of soil structure. The total number of electrostatic charges of clay particles divided by the total exposed surface area is called the *surface charge density*.

Because of differences in their valences, radii, and hydration properties, different cations are adsorbed with varying degrees of tenacity or preference and hence are either more readily or less readily exchangeable. In general, the smaller the ionic radius and the greater the valence, the more closely and strongly is the ion adsorbed. However, the greater the ion's hydration, the farther it is from the adsorbing surface and the weaker its adsorption. Sodium ion, for example, has an atomic radius of only 9.8×10^{-11} m when in the "naked" state, but it tends to be strongly hydrated and its effective radius increases eightfold when it is enveloped by water molecules. Monovalent cations, attracted by only a single charge, can be replaced more easily than divalent or trivalent cations. The order of preference in exchange reactions is as follows:

$$Al^{3+} > Ca^{2+} > Mg^{2+} > NH^{+} > K^{+} > H^{+} > Na^{+} > Li^{+} \tag{4.3}$$

An example of an exchange reaction is the following:

$$Na_2\,[\text{clay}] + Ca^{2+} \leftrightarrow Ca\,[\text{clay}] + 2Na^{+} \tag{4.4}$$

In nature, soil material is seldom adsorbed homogeneously with a single ionic species. Typically, the exchange capacity is taken up by several cations in varying proportions, all of which together constitute the so-called *exchange complex*. A typically heterogeneous exchange complex can be represented in the following way (with subscripts a, b, c, d, etc., indicating equivalent fractions of the cation exchange capacity, adding up to 100%):

$$\begin{array}{c|c|} Al_a & \\ Ca_b & \\ Mg_c & \text{Clay} \\ Na_d & \\ \cdot & \\ \cdot & \\ \cdot & \end{array}$$

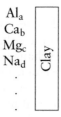

In arid regions, calcium, magnesium, and sometimes sodium tend to predominate in the exchange complex. In humid regions, where soils are highly leached and often acid, hydrogen and aluminum ions, the latter often released from the clay crystal lattice under low pH conditions, play an important role. The presence of exchangeable hydrogen is usually referred to as *base unsaturation* of the exchange complex, and its measure in equivalent terms for the soil mass of an agricultural field is taken to be an indication of the amount of lime needed to neutralize the soil's *reserve acidity*.

The composition of the exchange complex depends on the concentration and ionic composition of the ambient solution. This is expressed in the *Gapon equation*:

$$A_e/B_e = c([A_s]^{1/a}/[B_s]^{1/b}) \quad (4.5)$$

in which A and B are fractional concentrations of cations with valences of a and b, respectively, and subscripts e and s refer to concentrations in the exchange complex and in the ambient solution, respectively. The coefficient c depends on the nature of the charged surface and the nature of the adsorbed cation. This equation indicates that the adsorption mechanism favors cations of higher valence and that this preference increases as the solution becomes increasingly dilute.

For the important case of calcium–sodium exchange, the *selectivity coefficient* (Shainberg, 1973) is given by

$$\sigma_{Ca-Na} = n_0([Ca_e][Na_s]^2/[Na_e]^2[Ca_s]) \quad (4.6)$$

Here, subscript e denotes the ionic equivalent fractions of Ca and Na in the exchange phase, subscript s denotes the mole fractions of these ions in the solution phase, and n_0 is the total molar concentration of the solution. The selectivity coefficient σ for a range of exchangeable sodium fractions has been reported to be about 4 in typical soils in Israel (Levy and Hillel, 1968).

For ordinarily encountered solution concentrations in nonsaline soils, divalent cations predominate in the exchange complex if present in appreciable concentrations. In a mixed calcium–sodium system most of the calcium ions are usually adsorbed tightly in the Stern layer, whereas the sodium ions are relegated to the diffuse region of the double layer.

Cation exchange reactions are rapid and reversible; so the composition of the exchange complex responds to frequent changes in the composition and concentration of the soil solution. The composition of the soil's exchange complex in turn governs the soil's pH as well as swelling and flocculation–dispersion tendencies.

In addition to the *structural charges* due to isomorphous substitutions within the crystal lattice of the clay, there are *surface charges* due to the imbalance of proton and hydroxyl charges on the exposed peripheries or edges (as opposed to the faces) of the clay particles, where the lattice bonds are broken (Sposito, 1984). The former charges are permanent and are generally unaffected by the ambient solution surrounding the particles, whereas the latter charges are affected by the pH of the ambient solution. The structural charges are generally negative, but the surface charges may become positive at low values of pH. Consequently, pH-sensitive clay minerals may display a capacity for anion adsorption. This has been noticed specifically in kaolinite. Anion adsorption is selective. Among the common anions, silicate and phosphate ions are more strongly adsorbed than are sulfate, nitrate, or chloride ions. Anion adsorption can be important in the retention of phosphate in soils. Various nonionic substances (including organic compounds) can also be adsorbed by clay, generally by means of hydrogen and van der Waals bonds. Organic matter (humus) can also exhibit anion exchange, particularly at low pH values.

BOX 4.1 The Uses of Clay

Some eight millennia ago, humans discovered the use of clay for pottery making. In Mesopotamia, an important use of clay was for writing. Clay tablets were fashioned and then while still moist imprinted with a sharp reed stylus. If then fired, such tablets could become permanent records of political and commercial deeds, legal codes, myths, and poetry.

The value of clay for ceramics arises from its plasticity and its tendency to harden when heated. Hence it can be shaped at will and the desired shape can then be made durable, strong, thermally and electrically resistant, and either pervious or impervious (depending on the type of clay and on the temperature to which it is heated). When the temperature is high enough, the clay particles tend to fuse so the shaped object becomes practically impervious. Wall tiles, ceramic pipes, and stoneware pots are among the objects made in this way. Sanitary faiences are made with kaolinite fired at very high temperatures. When the kaolinite is quite pure and fusing is complete, a translucent glass is obtained, called *porcelain,* which is used for the manufacture of fine chinaware as well as electric insulators.

An important use of clay is in construction. Even today, an estimated third of all humans live in houses made of clay. Clay mixed with sand or straw is the plasterlike material called *daub.* Packed into molds and then dried, it is made into mudbricks. When fired, it is hardened into stonelike bricks. Cement is a mixture of limestone and clay (about 25%) fired at a temperature of 1400°C.

The fact that clay can be made into impermeable pastes has been utilized since antiquity. Cisterns and channels were coated with clay so that they would hold and convey water with minimum leakage. Porous soil can be made impermeable by the injection of mixtures of water and clay. Injection of clayey mud is a common practice in drilling for oil. Pumped into a drill pipe, the mud cools the drill bit and rises along the pipe, carrying away the drill cuttings.

The adsorptive capacities of clay are important in agriculture and in industry. Adsorbed ions are loosely bound and therefore are easily exchanged with hydrogen from the roots of plants. Hence the clay minerals act to provide a source for the nutrient mineral required by plants. The ability of smectite clays to form stable colloidal suspensions is useful in the preparation of polishes, cosmetics, and pesticides, among many products. Kaolin, or white clay, serves widely for coating paper. (The paper on which this book is printed may contain over 20% kaolin.) Without it the paper would be much less opaque and would absorb the ink so that the lines of illustrations would become blurred.

HYDRATION AND SWELLING

Clay particles are hardly ever completely dry. Even after oven-drying at 105°C for 24 hr, which is the standard for drying soil material, clay particles still retain appreciable amounts of adsorbed water, as shown in Fig. 4.10. The affinity of clay surfaces for water is demonstrated by the hygroscopic nature of clay soils, that is, their ability to sorb and condense water vapor from the air. So-called "air-dry" soil commonly contains several percent water, the exact percentage depending of course on the kind and quantity of clay as well as on the humidity of the air.

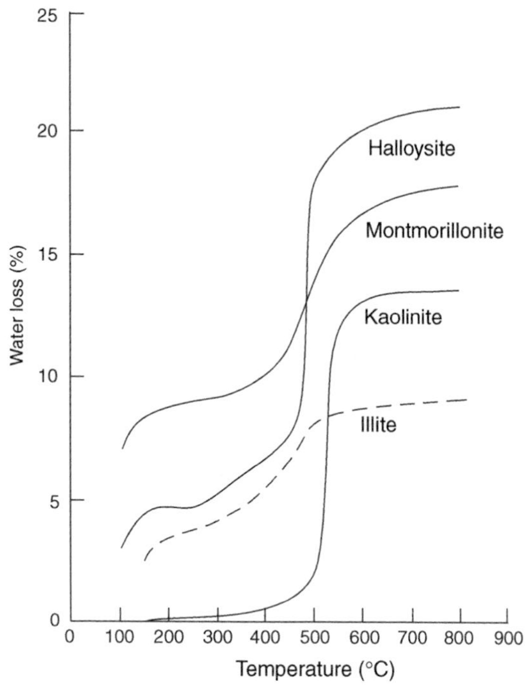

Fig. 4.10. Dehydration curves of clay minerals. (After Marshall, C. E., 1964)

In the oven-dry state, the water associated with clay is so tightly held that it can be considered a part of the clay itself. When water is added to initially dry clay, the water films surrounding each particle thicken and the water is then more loosely held. The entire physical behavior of a clay-containing soil mass (including strength, consistency, plasticity, and the conduction of water and heat) is strongly influenced by the degree of hydration.

Water is attached to clay surfaces by several mechanisms, including the electrostatic attraction of the dipolar, oriented water molecules to charged sites as well as their hydrogen bonding to exposed oxygen atoms on the clay crystal. Still another mechanism of hydration arises from the presence of adsorbed cations. Because the cations associated with the clay also tend to hydrate, they contribute to the overall hydration of the clay system. Quantitatively, this effect depends on the type of cations present and on the cation adsorption capacity of the clay.

The strength of clay–water adsorption is clearly greatest for the first layer of water molecules. The second layer is attached to the first by hydrogen bonding, and the third to the second, and so forth, so the influence of the attractive force field of the clay surface diminishes rapidly with distance, to become vanishingly small beyond a few molecular layers. The effective thickness and the physical properties of the adsorbed water in the vicinity of the clay surface have been a subject of controversy, with some investigators contending that adsorbed water is quasi-crystalline and hence differs significantly from bulk water in viscosity, diffusivity to ions, dielectric constant, and density. Others claim that the difference in properties between adsorbed water and capillary water in the soil is scarcely detectable and is in any case inconsequential.

Sample Problem

The diameter of a water molecule is about 3×10^{-8} cm (3×10^{-10} m). One cubic centimeter of water at standard temperature and pressure has a mass of about 1 gram and contains about 3.4×10^{22} molecules. The mass per molecule M_m is therefore about 3×10^{-23} g.

Calculate the mass of water adsorbed by 100 g of montmorillonite, illite, kaolinite, and fine quartz sand if the specific surface areas a_s are 800, 100, 10, and 1 m^2/g, respectively. Assume monomolecular adsorption.

For montmorillonite:

Surface area of sample: $A_s = ma_s = 100 \text{ g} \times 800 \text{ m}^2/\text{g} = 8 \times 10^4 \text{ m}^2 = 8 \times 10^8 \text{ cm}^2$

Area of one molecule $A_m \approx (3 \times 10^{-8} \text{ cm})^2 \approx 9 \times 10^{-16} \text{ cm}^2$

Number of molecules adsorbed: $n = A_s/A_m = 8 \times 10^8 \text{ cm}^2 / 9 \times 10^{-16} \text{ cm}^2 \approx 9 \times 10^{23}$

Mass of water adsorbed: $M_w = nM_m \approx (9 \times 10^{23})(3 \times 10^{-23} \text{ g}) = 27 \text{ g}$

For illite: $M_w \approx 3.375$ g; for kaolinite ≈ 0.338 g; for fine quartz sand ≈ 0.034 g.

When a confined body of clay sorbs water, *swelling pressures* (defined as the pressure that must be applied to a body of soil that is allowed to imbibe water, to prevent it from expanding) develop, related to the osmotic pressure difference between the adsorbed water and the external solution. In a partially hydrated micelle, the thickness of the enveloping water is less than the potential thickness of the fully expanded diffuse double layer. The double layer, thus truncated, will tend to expand to its full potential thickness and dilution by the osmotic absorption of additional water, if available. As each micelle expands, its swarm of positively charged cations repels those of the adjacent micelles. Thus the micelles tend to push each other apart. Though this causes the system to swell and increase its overall porosity, it may have the internal effect of closing the soil's larger pores, thus reducing its permeability.

We have already shown that the concentration of ions between associated clay micelles is greater than in the external solution. The actual concentration difference depends on the interparticle distance (i.e., on the degree of hydration of the clay) and on the potential extent of the diffuse double layer (which, in turn, depends upon the valences and concentrations of the adsorbed cations). The osmotic attraction of a clay assemblage for "external" water may be roughly twice as high with monovalent than with divalent cations, since there are normally twice as many of the former than of the latter. Hence swelling is greatest with monovalent cations, such as sodium, and with distilled water as the external solution. With calcium as the dominant cation in the exchange complex, swelling is reduced. A similar restraining effect is caused at low pH values by the presence of trivalent aluminum. High salinity

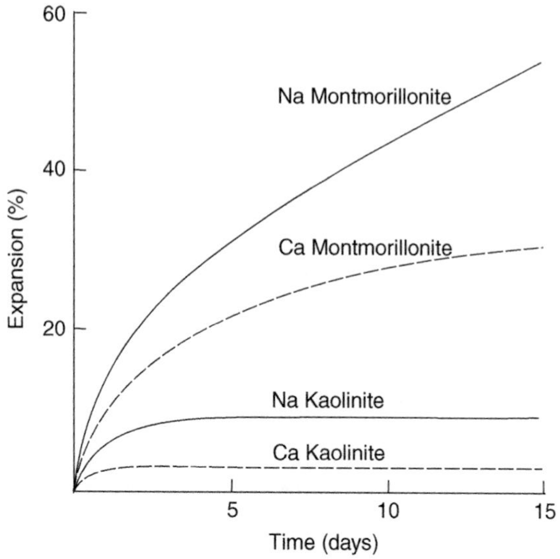

Fig. 4.11. Volume changes of montmorillonite and kaolinite clays during hydration.

of the soil solution will also suppress swelling. However, if a saline soil in which sodium salts predominate is leached of excess salts with freshwater without concurrent addition of calcium (e.g., in the form of gypsum), strong swelling may result from the predominance of sodium ions in the adsorbed phase.

The time-dependent volume increase of clays and clay–sand mixtures in the process of hydration is illustrated in Fig. 4.11. This time dependence is due to the low permeability of clay systems. The eventual swelling is seen to depend on the amount and nature of the clay present. In general, swelling increases with increasing specific surface area. It is also affected by the arrangement or orientation of soil particles and by the possible occurrence of interparticle cementation by such materials as iron or aluminum oxides, carbonates, and humus, which may constrain the expansion of the soil matrix.

When a hydrated body of clay is dried, a process opposite to swelling occurs, namely, shrinkage. In the field, the shrinkage that typically begins at the surface often causes the formation of numerous cracks, which break the soil mass into fragments of various sizes, from small aggregates to large blocks. An extreme example of this can be seen in *vertisols*, which are soils rich in expansive clay (e.g., montmorillonite). When subject to alternating wetting and drying, as in a semiarid region, such soils tend to heave and then to settle and to form wide, deep cracks and slanted sheer planes extending deep into the soil profile. Vertisols can be problematic both in agriculture (where they are exceedingly difficult to till because they tend to puddle and then to harden upon drying) and in engineering (where their successive heaving and subsidence tends to warp road pavements and to damage buildings).

BOX 4.2 Clay Soils That Till Themselves

Vertisols behave strangely: They perform a kind of tillage on their own. The part of the profile where we would normally expect to find more or less distinct differentiation between an A horizon and a B horizon in fact exhibits very little in the way of profile development. The entire soil, sometimes to a depth of 1 m or more, actually churns and mixes itself repeatedly.

Here is how it happens: During the wet season, these clayey soils swell markedly. The differential swelling of zones within the profile causes parts of the soil to move relative to adjacent parts, thus creating typically oblique shear planes (called *slickensides*) in the subsoil. During the dry season, the soil shrinks from the top downward and forms an extensive network of deep, wide cracks. The masses of soil between cracks form separated, column-like blocks. The tops of these blocks desiccate and break into numerous small clods, which in turn dribble into the cracks, often to a depth of several decimeters. When the soil is next wetted and in turn desiccated again, new material is exposed at the surface, just as some of the material formerly at the surface is now mixed deep in the profile.

The same process also causes stones (where they are present, as in volcanic areas) to rise in the soil. The small clods that drop into the cracks sometimes dribble past the stones that are embedded in the soil. When the soil is next wetted, these clods swell and push the stones upwards. Farmers in such areas are likely to find a fresh crop of stones on the soil surface every year.

FLOCCULATION AND DISPERSION

As clay particles interact with one another, forces both of repulsion and of attraction can come into play. The one or the other type of force may predominate, depending on physicochemical conditions. When the repulsive forces are dominant, the particles separate and remain apart from each other — and the clay is said to be *dispersed*. On the other hand, when the attractive forces prevail, the clay becomes *flocculated*, a phenomenon analogous to the coagulation of organic colloids, as the particles associate in packets or flocs.

These phenomena can be observed quite readily in dilute aqueous suspensions of clay. For the clay particles to enter into and remain in a state of suspension, they must be dispersed. This can be induced by the addition of a sodic dispersing agent and by mechanical agitation, in the manner described in connection with mechanical analysis. The dispersed suspension is typically turbid and remains so as long as the suspension is stable. This state can be changed rather dramatically by the addition of polyvalent cations or by the addition of salt to increase the overall electrolyte concentration. The turbid suspension suddenly clarifies as the clay particles flocculate and the flocs (behaving like large composite particles) settle to the bottom.

The main repulsive force between clay micelles derives from the like charges of the ionic swarms surrounding the particles, manifested by the swelling tendency of clay–water systems. An attractive force may result, however, if two clay platelets are brought close enough together (within about 1.5 nm) so that their counterions intermingle to form a unified layer of positive charges, which then attracts the negatively charged particles on both sides. This process,

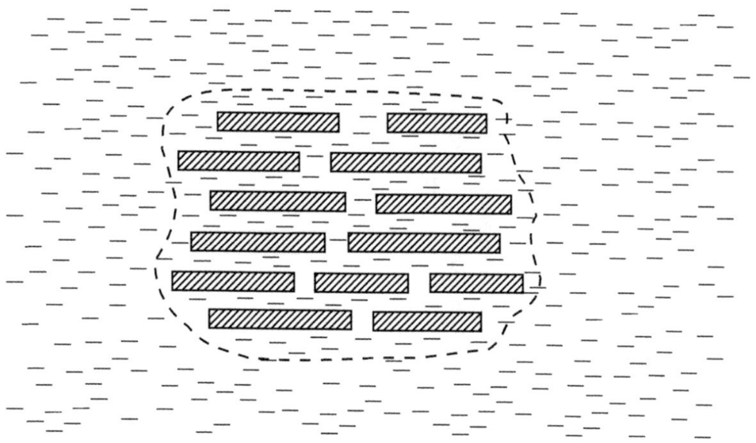

Fig. 4.12. Schematic representation of montmorillonite tactoid.

called *plate condensation*, brings about the formation of *tactoids*, or packets of parallel-oriented platelets (like packs of cards) that form a more or less stable association (Fig. 4.12).

Another type of electrostatic attraction between particles can occur when the plate edges develop positive charges, as they tend to do at low pH values. If the repulsion normally due to the diffuse double layer is not so great as to prevent clay particles from coming together, the positive charges on the edge of one particle may form bonds with the negative charges on the face of another. A "card-house" structure of flocs can then develop, as shown in Fig. 4.13.

The important fact about these various attractive and repulsive forces is that they operate with different intensities and over different ranges. The combined force field, shown schematically in Fig. 4.14, consists of regions within which net attraction prevails and regions over which repulsion predominates. For example, coulombic (electrostatic) forces are inversely proportional to distance squared, whereas London–van der Waals forces are inversely proportional to the seventh power of the distance, so the latter are effective within a narrow space of only a nanometer or so, whereas the former extend to distances 10 times as great.

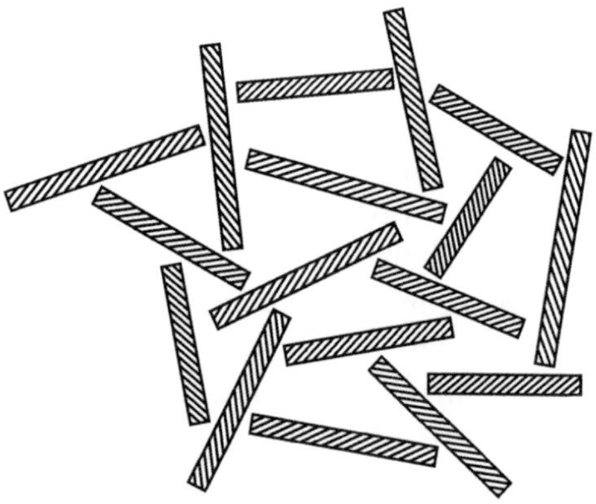

Fig. 4.13. Edge-to-face electrostatic bonding resulting in card-house structure.

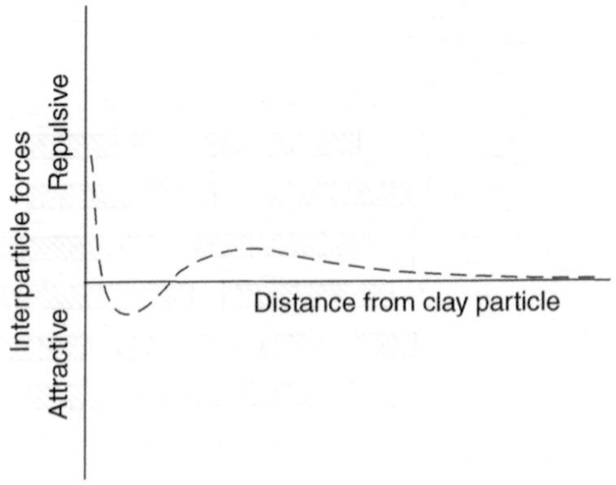

Fig. 4.14. Schematic representation of the combined force field surrounding a hydrated clay particle.

Starting from a dilute stable suspension in which the micelles are far apart so that the repulsive forces prevail, flocculation can only be achieved if these repulsive forces are suppressed sufficiently to allow colliding particles to clump together, rather than bounce apart, as they approach within the range of the attractive forces. Repulsion is maximized, favoring dispersion, when the double layer is fully extended, as when the ambient solution is very dilute and of high pH (thus preventing positive charges from forming at the particle edges), when the dominant cation is monovalent, and when the soil is fully hydrated. Mechanical agitation also inhibits flocculation. In contrast, repulsion is minimized, allowing flocculation, whenever the solution concentration is high and the monovalent cations are replaced by polyvalent ones.

The balance of repulsive versus attractive forces is reversible and, under certain circumstances, can easily shift either way. A soil can thus disperse, flocculate, redisperse, reflocculate, and so forth, several times over, as, for example, when it is irrigated with water of varying salinity and ionic composition.

Sample Problem

If the cation exchange capacity of a soil is 40 meq/100 g (400 mEq/kg), and if sodium ions constitute 25% of the exchange complex, what is the minimum amount of gypsum required per hectare to replace sodium with calcium in the upper 0.20-m layer of the soil? Assume a bulk density of 1.2×10^3 kg/m^3.

The chemical formula for gypsum is $CaSO_4 \cdot 2H_2O$. Hence the molecular weight is 172 and the equivalent weight is 86.

Mass of soil in upper 0.2-m layer = $10^4 m^2$/hectare $\times$ (2×10^{-1} m depth) $\times 1.2 \times 10^3$ kg/m^3 = 2.4×10^6 kg.

Equivalents of sodium to be removed from upper 0.2 m layer = 2.4×10^6 kg $\times$ (400×0.25 mEq/kg) = 2.4×10^5 Eq.

Mass of gypsum needed = 2.4×10^5 Eq $\times$ 86 g/Eq $\approx 2 \times 10^7$ gm $\approx$ 20 metric tons.

HUMUS: THE ORGANIC CONSTITUENT OF SOIL COLLOIDS

Although the main topic of this chapter is the mineral colloidal matter known as clay, we must emphasize the important fact that an entirely different kind of colloidal matter exists in soils, called *humus*. This generally dark-colored material, found mostly in the surface zone (the A horizon) of soils, is defined in the *Glossary of Soil Science Terms* (Soil Science Society of America, 1996) as "the more or less stable fraction of the soil organic matter remaining after the major portion of added plant and animal residues have decomposed." So defined, humus does not include undecomposed or partially decomposed organic residues, such as recent stubble or dead roots.

Like clay, humus particles are negatively charged. During hydration, each particle of humus forms a micelle and acts like a giant, composite anion, capable of adsorbing various organic and inorganic constituents, including cations. The cation exchange capacity of humus is much greater, per unit mass, than that of clay. Unlike most clay, moreover, humus is generally not crystalline but amorphous. Because it is composed mostly of carbon, oxygen, and hydrogen, its charges are due not to the isomorphous substitutions of cations but to the dissociation of carboxylic (–COOH) and phenolic (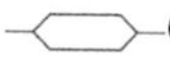OH) groups. Since the cation exchange process depends on replacement of the hydrogen in these groups, it is pH dependent, with the cation exchange capacity generally increasing at higher pH values.

Humus is not a single compound, nor does it have the same composition in different locations. Rather, it is a complex mixture of numerous compounds, including lignoproteins, polysaccharides, polyuronides, and other compounds too varied to list. Furthermore, the organic colloids of humus, although "more or less" stable, are in fact amenable to bacterial action, particularly if the soil's temperature, moisture, or aeration regimes are modified.

The content of humus in mineral soils varies from as high as 10% or even more in the top layer of *chernozem* (the typically black soil that occurs in the American prairie and in the plains of Ukraine) down to nil in desert soils and is of the order of 1–3% in many intermediate soils. The humus content generally diminishes in depth through the B horizon and becomes negligible at the bottom of the normal root zone, unless the soil is a deposit of alluvial material with a high original content of humus. Organic soils such as peat and muck may contain well over 50% organic matter, though not all of that would fit the accepted definition of humus.

The importance of humus goes beyond its effect on cation adsorption or even plant nutrition. Humus often coagulates in association with clay and serves as a cementing agent, binding and stabilizing soil aggregates and thus improving soil structure. This aspect is discussed in Chapter 5.

5. SOIL STRUCTURE AND AGGREGATION

SOIL STRUCTURE DEFINED

Bricks thrown haphazardly atop one another become an unsightly heap. The same bricks, only differently arranged and mutually bonded, can give rise to a home or a school. Similarly, a soil can be merely a loose and unstable assemblage of random particles, or it can consist of a distinctly structured body of interbonded particles associated into aggregates having regular sizes and shapes. Hence it is not enough to study the properties of individual soil particles. To understand how the soil behaves as a composite body, we must consider the manner in which the various particles are packed and held together to form a continuous spatial network that constitutes the *soil matrix* or the *soil fabric* (Gregorich et al., 2002).

The arrangement or organization of the particles in the soil (i.e., the internal configuration of the soil matrix) is called *soil structure*. Since soil particles differ in shape, size, and orientation and can be variously associated and interlinked, the mass of them can form complex and irregular patterns that are difficult to characterize in exact geometric terms. A further complication is the inherently unstable nature of soil structure and its nonuniformity in space. Soil structure is affected by changes in climate, biological activity, and soil management practices, and it is vulnerable to destructive forces of a mechanical and physicochemical nature.

For these various reasons, we have no truly objective or universally applicable way to measure soil structure per se, and the term *soil structure* therefore expresses a qualitative concept rather than a directly quantifiable property. The numerous methods proposed for characterizing soil structure are designed not to measure soil structure itself but soil attributes that are supposed to depend on structure. Many of these methods are specific to the purpose for which they were devised and some are completely arbitrary.

The difficulty of defining soil structure notwithstanding, we can readily perceive its importance, inasmuch as it determines the total porosity as well as the shapes and sizes of the pores in the soil. Soil structure affects the retention and transmission of fluids in the soil, including infiltration and aeration. Moreover, as soil structure influences the mechanical properties of the soil, it may also affect such disparate phenomena as germination, root growth, tillage, overland traffic, and erosion. Agriculturists are usually interested in having the soil, at least in its surface zone, in a loose and highly porous and permeable condition. Engineers, on the other hand, often desire a dense and rigid structure so as to provide maximal stability and resistance to subsequent deformation as well as minimal permeability. In either case, knowledge of basic soil structure relationships is essential for efficient management of the soil.

TYPES OF SOIL STRUCTURE

In general, we can recognize three broad categories of soil structure — *single grained*, *massive*, and *aggregated*. When particles are entirely unattached to one another, the structure is completely loose, as it is in the case of a coarse granular soil or an unconsolidated deposit of desert dust. Such soils were labeled *structureless* in the older literature of soil physics, but, since even a haphazard arrangement is a structure of sorts, we prefer the designation *single-grained* structure. On the other hand, when the soil is tightly packed in large cohesive blocks, as is sometimes the case with dried clay, the structure can be called *massive*. Between these two extremes, we can recognize an intermediate condition in which the soil particles are associated in quasi-stable small clods known as *aggregates* or *peds*.

This last type of structure, called *aggregated*, is usually the most desirable condition for plant growth, especially in the critical early stages of germination and seedling establishment. The presence and maintenance of stable aggregates is the essential feature of soil *tilth*, a qualitative term used by agronomists to describe the highly desirable, yet elusive, physical condition in which the soil is an optimally loose, friable, and porous assemblage of stable aggregates, permitting free entry and movement of water and air, easy cultivation and planting, and unobstructed germination and emergence of seedlings as well as the growth of roots.

STRUCTURE OF GRANULAR SOILS

The structure of most coarse-textured soils is single grained, because there is little tendency for the grains to adhere and form aggregates. The arrangement and internal mode of packing of the grains depends on their distribution of sizes and shapes as well as the manner in which the material had been deposited or formed in place. The two extreme cases of possible packing arrangements are, on the one hand, a system of uniform spherical grains in a state of open packing (and hence minimal density) and, on the other hand, a gradual distribution of grain sizes in which progressively smaller grains fill the voids between larger ones in an "ideal" succession that provides maximal density. An assemblage of grains of uniform size is called *monodisperse*, whereas an assemblage of widely varying grain sizes is called *polydisperse*.

Let us consider a hypothetical system consisting entirely of uniform spheres. Although not realistic, this is a useful exercise, inasmuch as it can help us to establish theoretical limits of porosity by which we may later evaluate real systems. With monodisperse spheres, the minimal density and hence maximal porosity is obtained in the case of cubic (open) packing. In this mode of packing, each grain touches six neighbors on opposite sides of three orthogonal axes and hence is said to have a *coordination number* of 6. The bulk density is then $\pi\rho_s/6$ (where ρ_s is the particle density), and the porosity is 47.6%, regardless of the diameter of the spheres. If $\rho_s = 2.65 \times 10^3$ kg/m^3, we obtain a bulk density of about 1.39×10^3 kg/m^3.

In contrast with the cubic arrangement, we may have the tetrahedral or octahedral arrrangement, both of which provide the densest possible packing of uniform spheres. Both of these conditions have a coordination number of 12, a bulk density of $\pi\rho_s/3(2^{1/2})$ (about 1.97×10^3 kg/m^3 = 1.97 g/cm^3), and a porosity of 25.9%. A schematic representation of modes of packing is given in Fig. 5.1. A quantitative summary is given in Table 5.1.

The analysis of polydisperse systems is obviously more complex than that of monodisperse systems, even if we continue to assume spherical shape. A mixture of two grain sizes depends on the relative concentration of the components. For any particular ratio of grain sizes, there exists an optimal composition that will result in minimal porosity. In principle, the density of a polydisperse system consisting of many particle sizes can be greater than that of a monodisperse system, because smaller particles can fit into the spaces between larger ones (Fig. 5.2). Depending on the array of relative particle radii, coordination numbers as high as 30 or more and porosities lower than 20% are possible. The actual porosities of natural sediments generally lie between the theoretically derivable limits for the ideal packings of monodisperse and polydisperse spheres; that is, they range between 25% and 50%.

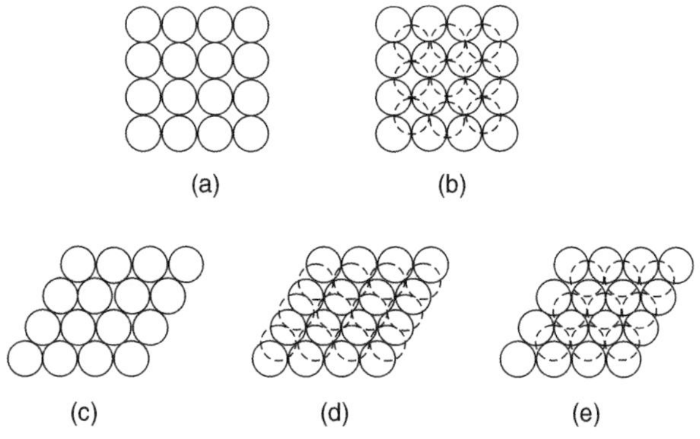

Fig. 5.1. Models of packing of equal spheres. (a) simple cubic, (b) cubical tetrahedral, (c) tetragonal sphenoidal, (d) pyramidal, and (e) tetrahedral. (Deresiewicz, 1958).

TABLE 5.1 Packing of Spheres[a]

Type of packing	Coordination number	Spacing of layers	Volume of unit prism	Density	Porosity (%)
Simple cubic	6	$2R$	$8R^3$	$\pi/6$ (0.5236)	47.64
Cubic tetrahedral	8	$2R$	$4\sqrt{3}\,R^3$	$\pi/3\sqrt{3}$ (0.6046)	39.54
Tetragonal sphenoidal	10	$R\sqrt{3}$	$6R^3$	$2\pi/9$ (0.6981)	30.19
Pyramidal	12	$R\sqrt{2}$	$4\sqrt{2}\,R^3$	$\pi/3\sqrt{2}$ (0.7405)	25.95
Tetrahedral	12	$2R\sqrt{2/3}$	$4\sqrt{2}\,R^3$	$\pi/3\sqrt{2}$	25.95

[a] From Deresiewiez (1958, Table 1).

Loose deposits of granular material cannot be compacted very effectively by application of static pressure (which, however, can be very effective in compacting unsaturated clayey soils). Static pressure is merely transmitted along the soil skeleton and borne by the high frictional resistance to the mutual sliding of the grains. Granular material can, however, be compacted by the application of vibratory action. The vibration pulsates the grains and allows smaller ones to enter between larger ones, thus increasing the packing density. The vibration of granular materials is effective in dry and in saturated states, whereas at unsaturated moisture contents surface tension forces of the water menisci that are wedged between the particles lend cohesiveness and hence greater rigidity to the matrix. The structure of nonideal granular materials (in which the grains are not spherical but oblong or angular) is much more difficult to formulate theoretically, because the orientations of particles must be taken into account.

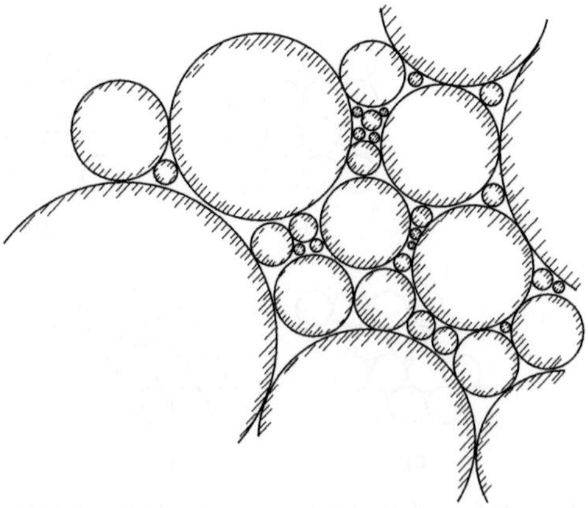

Fig. 5.2. Packing of polydisperse particles (hypothetical).

Sample Problem

Calculate the overall porosity and bulk density values for an assemblage of uniform spherical aggregates in open (cubic) and dense (tetrahedral) packing. Assume that each aggregate has a bulk density ρ_b of 1.8×10^3 kg/m³ (1.8 g/cm³) and that particle density ρ_s is 2.65×10^3 kg/m³. How much of the total porosity is due to macro-(between-aggregate) pores and how much to micro-(within-aggregate) pores in each packing mode?

To determine the microporosity of the aggregates themselves:

Recall that porosity $f = 1 - (\rho_b/\rho_s) = 1 - (1.8/2.65) = 0.32$ and that the volume of a sphere $= (4/3)\pi r^3 = (\pi/6)d^3$, where r is radius and d is diameter.

In cubic packing: Assuming the diameter to be of unit length, each such sphere occupies a cube of unit volume ($d^3 = 1 \times 1 \times 1 = 1$). Therefore the fractional volume of each sphere in its cube $= \pi/6 = 0.5236$.

Hence the macro-(interaggregate) porosity = 1 − 0.5236 = 0.4764.
As a fraction of a unit cube, the microporosity = 0.5236 × 0.32 = 0.1676.
Therefore, total porosity = 0.4764 + 0.1676 = 0.644 = 64.4%.

For dense (tetrahedral) packing:

The fractional volume of the spheres = 0.7405 (see Table 5.1).
The macroporosity = 1 − 0.7405 = 0.2595.
The porosity of each aggregate (as before) = 0.32.
As a fraction of the total volume, microporosity = 0.7405 × 0.32 = 0.237.
Therefore, total porosity = 0.2595 + 0.237 = 0.496 = 49.6%.

Of the total porosity, macropores constituted 0.2595/0.496 = 0.523 and micropores = 0.478.

To compute bulk densities (ρ_b):

From $f = 1 - (\rho_b/\rho_s)$, we obtain: $\rho_b = \rho_s(1 - f)$.
For open packing, $\rho_b = 2.65(1 - 0.644) = 0.934$ g/cm³ $= 0.934 \times 10^3$ kg/m³.
For dense packing, $\rho_b = 2.65(1 - 0.496) = 1.3396$ g/cm³ $= 1.34 \times 10^3$ kg/m³.

Note: Even though the porosity of the aggregates themselves did not change, the overall macroporosity and microporosity (and hence the total porosity and bulk density) changed greatly with the change in packing arrangement.

BOX 5.1 Soil Physics at the Beach

Those who go to the beach merely to swim miss the fun of observing soil physics in action on the sandy shore. The labor of trudging through the soft dry sand is lightened by the realization of why it is so. Sand has such low specific surface that its particles do not bond to one another, so their assemblage has hardly any firmness.

Approaching the shore, however, we notice a zone of firm sand, where walking is much easier. It is the same sand, so what makes it rigid? Aha, it must be the water in the sand. But we know that liquid water is itself a rather "soft" substance, so how can it impart rigidity to the sand? The answer is that water in a moist but unsaturated sand tends to cluster in pendular rings at points where grains touch one another. These rings, also called *menisci*, exhibit surface-tension forces that resist any deformation tending to increase the water-to-air interfacial area.

Now we come to the water's edge and find ourselves walking on saturated sand. Again, it seems soft: In a saturated medium, the water-to-air interfaces disappear and no longer resist deformation. Here, yet another interesting phenomenon comes into play. As each wave of the sea recedes, it leaves behind a strip of glistening, saturated sand. Now if we step on the sand, the area around our foot seems to dry up instantly. That's puzzling: If this were like a saturated sponge, it would spurt water rather than turn dry. So why does saturated sand appear to turn dry when trodden?

The answer lies in a phenomenon called *dilatancy* (pronounced "die late and see" — a good idea in any case). It is the tendency of some materials to expand when subjected to shearing. In the case of sand, the sliding of one plane over another (when a differential force is applied to adjacent areas) causes the rolling of grains over one another. The porosity in the zone of shearing thereby increases. Where this "puffing up" occurs, the volume of water in the sand is no longer sufficient for saturation. Therefore, the sand appears to "dry up."

Now we notice the ripples in the sand under the shallows near the water's edge and the appearance of dark streaks in the troughs between ridgelets. That banding is caused by the segregation of "heavier" minerals (e.g., magnetite, hornblende, zircon) from the "lighter" minerals (e.g., quartz), due to differences in density. The dark minerals are often deposited at the water's edge as the waves — and then the winds — sweep the lighter particles higher up the beach.

Is that all? Certainly not. Notice the frictional sound of the dry sand as it is tread on and the gurgling sound of the sand as it releases entrapped air after each wave's inundation. There is more, much more, to observe at the beach (and everywhere else!).

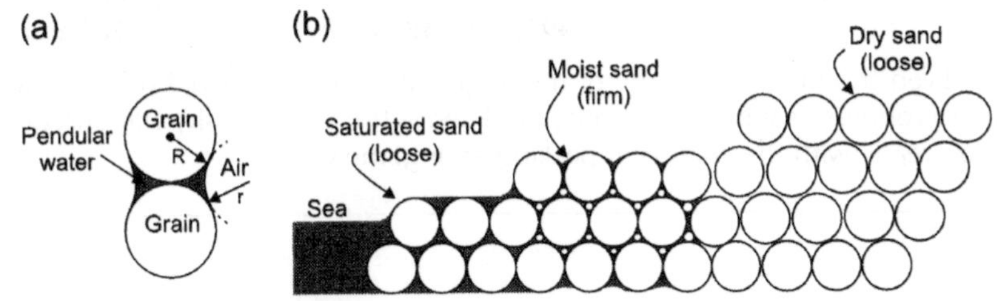

Fig. B5.1. Water in sand: (a) Pendular ring between adjacent grains. (b) The rings disappear in both dry sand and saturated sand.

STRUCTURE OF AGGREGATED SOILS

In soils with an appreciable content of clay, the primary particles tend, under favorable circumstances, to group themselves into composite structural units known as *aggregates*. Such aggregates are not characterized by any universally fixed size, nor are they necessarily stable. The visible aggregates, which are generally of the order of several millimeters to several centimeters in diameter, are often called *peds* or *macroaggregates*. These are usually assemblages of smaller groupings, or *microaggregates*, which themselves are associations of the ultimate structural units, that is, the flocs, clusters, or packets of clay particles. Bundles of these last units attach themselves to, and sometimes engulf, the much larger primary particles of sand and silt. The internal organization of these various groupings can be studied by means of electron microscopy, using both scanning and transmission methods (the latter with thin sections of soil samples embedded in congealed, transparent plastic materials).

A prerequisite for aggregation is that the clay be flocculated. However, flocculation is a necessary but not sufficient condition for aggregation. As stated by Richard Bradfield of Cornell University a half century ago, "Aggregation is flocculation — plus!" That "plus" is *cementation*.

A complex interrelationship of physical, biological, and chemical reactions is involved in the formation and degradation of soil aggregates (Kay and Angers, 2002). An important role is played by the extensive networks of roots that permeate the soil and tend to enmesh soil aggregates. Roots exert pressures that compress aggregates and separate between adjacent ones. Water uptake by roots causes differential dehydration, shrinkage, and the opening of numerous small cracks. Moreover, root exudations and the continual death of roots and particularly of root hairs promote microbial activity, which results in the production of humic cements. Since these binding substances are transitory, being susceptible to further microbial decomposition, organic matter must be replenished and supplied continually if aggregate stability is to be maintained in the long run.

ROLE OF ORGANIC MATTER AND MICROBIAL ACTIVITY

Active humus is accumulated and soil aggregates are stabilized most effectively under perennial sod-forming herbage. Annual cropping systems, on the other hand, hasten the decomposition of humus and the destruction of soil aggregates. The soil surface is especially vulnerable if exposed and desiccated in the absence of a protective cover. The foliage of close-growing vegetation, and its residues, protects surface soil aggregates against slaking by water, particularly under raindrop impact.

Soil structure is affected by the time-variable activity of countless microorganisms, including species of protozoa, bacteria, fungi, actinomycetes, etc. Especially important are rhizospheric bacteria, which flourish in direct association with roots of specific plants, as well as fungi, which often form extensive adhesive networks of fine filaments known as *mycelia* or *hyphae*. The composition of the soil microfauna and microflora depends on the thermal and moisture regimes, on soil pH and oxidation–reduction potential, the nutrient status of the soil substrate, and the type and quantity of organic matter present.

Soil microorganisms bind aggregates by a complex of mechanisms, such as adsorption, physical entanglement and envelopment, and cementation by excreted mucilaginous products. Prominent among the many microbial products capable of binding soil aggregates are polysaccharides, hemicelluloses or uronides, levans, as well as numerous other natural polymers. Such materials are attached to clay surfaces by means of cation bridges, hydrogen bonding, van der Waals forces, and anion adsorption mechanisms. Polysaccharides, in particular, consist of large, linear, and flexible molecules capable of forming multiple bonds with several particles at once. In some cases, organic polymers hardly penetrate between the individual clay particles but form a protective capsule around soil aggregates. In other cases, solutions of active organic agents penetrate into soil aggregates and then precipitate more or less irreversibly as insoluble (though still biologically decomposable) cements.

In addition to increasing the strength and stability of intra-aggregate bonding, organic products may further promote aggregate stability by reducing wettability and swelling. Some of the organic materials are inherently hydrophobic or become so as they dehydrate, so the organo-clay complex may have a reduced affinity for water.

Some inorganic materials can also serve as cementing agents. The importance of clay and of its state of flocculation should be obvious by now. Adhesion between clay particles is the ultimate internal binding force within microaggregates. Calcium carbonate as well as iron and aluminum oxides can impart considerable stability to soil aggregates. The latter are the reason for the remarkable stability of aggregates in tropical soils that may contain little organic matter.

A model of the internal bonding forms that can constitute a soil aggregate is shown in Fig. 5.3. The model is composed of sand or silt-size quartz particles and of "domains" of oriented clay bonded by electrostatic forces. Stability of the aggregate is enhanced by linkage of organic polymers between the quartz particles and the faces or edges of clay crystals.

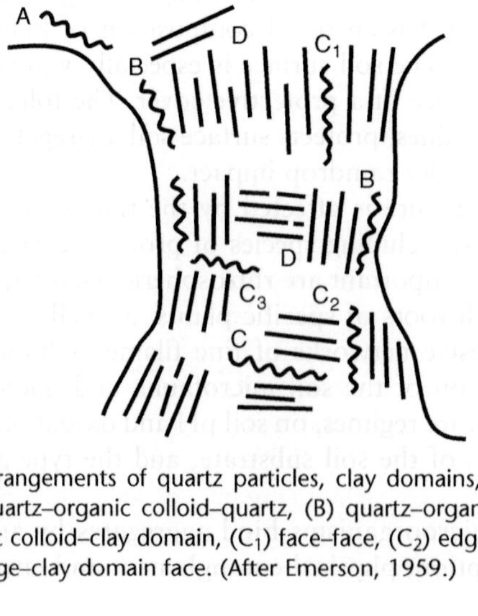

Fig. 5.3. Possible arrangements of quartz particles, clay domains, and organic matter in a soil aggregate: (A) quartz–organic colloid–quartz, (B) quartz–organic colloid–clay domain, (C) clay domain–organic colloid–clay domain, (C_1) face–face, (C_2) edge–face, (C_3) edge–edge, and (D) clay domain edge–clay domain face. (After Emerson, 1959.)

BOX 5.2 The Role of Earthworms

Prominent among the macrobiological fauna affecting soil structure are earthworms, which have aroused the interest of such observers as Aristotle and Darwin. Their role in the soil was described vividly and aptly two centuries ago by George White (quoted by Russel, 1973):

> Worms seem to be the great promoters of vegetation, which would proceed but lamely without them, by boring, perforating, and loosening the soil, and rendering it pervious to rains and fibers of plants, by drawing straws and stalks of leaves and twigs into it, and, most of all, by throwing up such infinite numbers of lumps of earth called wormcasts, which being their excrement, is a fine manure of grain and grass . . . . The earth without worms would soon become cold, hardbound, and void of fermentation, and consequently sterile.

A population of several million earthworms (of various species) per acre is not uncommon in regions where the supply of moisture and fresh organic matter is adequate. Such a population of earthworms can digest and expel as "casts" many tons of soil per acre per year. The earthworm activity may penetrate to a depth of half a meter or more (Blanchart, 1992).

BREAKDOWN OF AGGREGATES

The state of soil aggregation at any time results from the balance between the forces or processes promoting aggregation and those tending to cause its breakdown. A particularly destructive condition may result when thoroughly desiccated aggregates are suddenly submerged in water. The water drawn into each aggregate over its entire periphery may trap and compress the air originally present in the dry aggregate. Because the cohesive strength of the outer part of the clod is reduced by swelling and the pressure of the entrapped air builds up in proportion to its compression, sooner or later the latter exceeds the former and the clod may actually explode. More typically, however, a series of small explosions, each marked by the escape of a bubble of air, shatters the clod into fragments. This destructive process is known as *air slaking* (Fig. 5.4).

Another process by which water can destroy aggregates is the hammering impact of falling raindrops along with the scouring action of flowing water in surface runoff. On the other hand, drying processes, causing shrinkage, can (in themselves) increase aggregate stability by making the aggregates more dense and cohesive as well as by causing gluelike organic gums and gels to "set" irreversibly so as to serve as stable cementing agents. Freezing, incidentally, is itself analogous to drying, in that it causes the extraction of liquid water from regions of the soil toward the freezing sites (*ice lenses*).

With so many factors active simultaneously, aggregate stability is obviously not a simple property but a complex attribute, which depends on the relative strength of the intra-aggregate bonds versus the stresses induced by swelling, scouring, and air entrapment. Accordingly, the task of maintaining aggregation involves strengthening the internal bonding mechanisms while at the same time

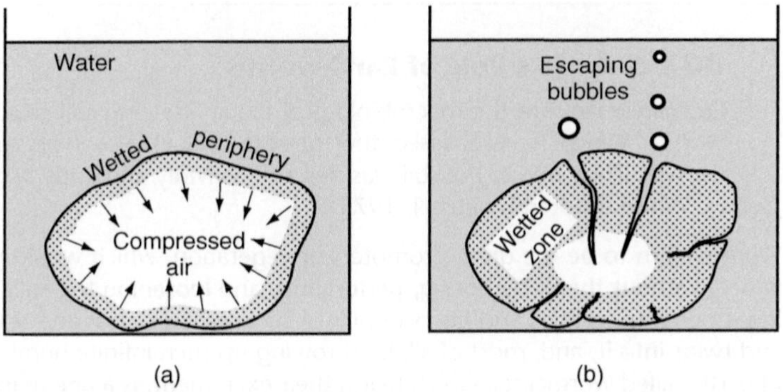

Fig. 5.4. Air slaking of an initially dry aggregate suddenly submerged in water. (a) *Early stage*: The periphery of the aggregate is wetted and water moves into the aggregate, compressing the air inside. (b) *Bursting stage*: Because the wetted zone is weakened by swelling and the pressure of entrapped air increases in proportion to its compression, eventually the aggregate is shattered and air bubbies out. This point may be quite abrupt and result in the collapse of the shattered aggregate.

reducing the destructive forces, for example, preventing sudden submergence and direct exposure of surface aggregates to desiccation, raindrop impact, and running water, as well as avoiding compressive traffic and excessively pulverizing tillage.

CHARACTERIZATION OF SOIL STRUCTURE

The structure of the soil can be studied directly by microscopic observation of thin slices under polarized light. The arrangements of minute clay particles can be examined by means of electron microscopy, using either transmission or scanning techniques. The structure of single-grained soils, as well as of aggregated soils, can be expressed quantitatively in terms of the total porosity and of the pore size distribution. The structure of aggregated soils can, in addition, be characterized qualitatively by specifying the typical shapes of aggregates found in various horizons within the soil profile or quantitatively by measuring their sizes. Additional methods of characterizing soil structure are based on measuring mechanical properties (e.g., resistance to penetration, compression, or shearing) and permeability to fluids. None of these methods has been accepted universally. In each case, the choice of the method to be used depends on the problem, the soil, the equipment available, and, not the least, the soil physicist.

Total porosity f of a soil sample is usually computed from the measured bulk density ρ_b, using the following equation (see Chapter 1):

$$f = 1 - \rho_b/\rho_s \tag{5.1}$$

where ρ_s is the average particle density.

Bulk density is generally measured by means of a *core sampler* designed to extract "undisturbed" samples of known volume from various depths in the profile. An alternative is to measure the volumes and masses of individual clods (*not* including interclod cavities) by *immersion in mercury* or by coating with paraffin wax prior to *immersion in water*. Still other methods are the *sand-funnel* and *balloon technique*, used in engineering, and *gamma-ray attenuation densitometry* (Grossman and Reinsch, 2002).

Pore size distribution measurements can be made in coarse-grained soils by means of the *pressure-intrusion method* (Flint and Flint, 2002), in which a nonwetting liquid, generally mercury, is forced into the pores of a predried sample. The pressure is applied incrementally, and the volume penetrated by the liquid is measured for each pressure step, equivalent (by the capillary theory) to a range of pore diameters. In the case of fine-grained soils, capillary condensation methods or, more commonly, *desorption methods* are used. In the latter method, a presaturated sample is subjected to a stepwise series of incremental suctions, and the capillary theory is used to obtain the equivalent pore size distribution. Water is commonly used as the permeating liquid, though nonpolar liquids have also been tried in an attempt to assess, by comparison, the possible effect of water saturation and desorption in modifying soil structure.

Where the aggregates are fairly distinct, it is sometimes possible to divide pore size distribution into two distinguishable ranges, *macropores* and *micropores*. The macropores are mostly the interaggregate spaces, which serve as the principal avenues for the infiltration and drainage of water and for aeration. The micropores are the intraaggregate capillaries responsible for the retention of water and solutes. However, the demarcation is seldom truly distinct, so differentiating between macropores and micropores is often arbitrary.

SHAPES OF AGGREGATES

The shapes of aggregates observable in the field (Fig. 5.5) can be classified as follows:

1. *Platy*: Horizontally layered, thin and flat aggregates resembling wafers. Such structures occur, for example, in recently deposited clay soils.

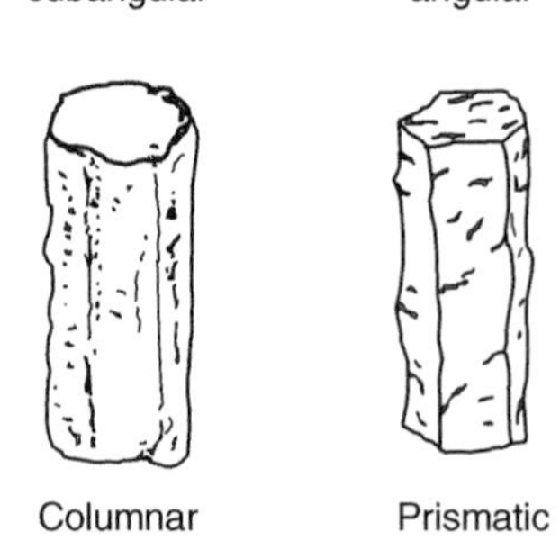

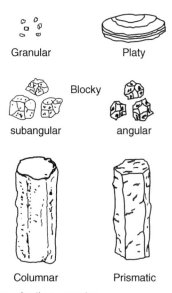

Fig. 5.5. Observable forms of soil aggregation.

TABLE 5.2 Classification of Soil Structure According to Soil Survey Staff (1951; 1993)

	A Type: Shape and arrangement of peds						
				***Blocklike–polyhedral–spheroidal*. Three approximately equal dimensions arranged around a point**			
	***Platelike*. Horizontal axes longer than vertical; arranged around a horizontal plane**	***Prismlike*. Horizontal axes shorter than vertical; arranged around vertical line. Vertices angular**		***Blocklike–polyhedral*. Plane or curved surfaces accommodated to faces of surrounding peds**		***Spheroidal–polyhedral*. Plane or curved surfaces not accommodated to faces of surrounding peds**	
		Without rounded caps	**With rounded caps**	**Faces flattened; vertices sharply angular**	**Mixed rounded, flattened faces; many rounded vertices**	**Relatively nonporous peds**	**Porous peds**
B Class: Size of peds	**Platy**	**Prismatic**	**Columnar**	**Blocky**	**Subangular blocky**	**Granular**	**Crumb**
1. Very fine or very thin	<1 mm	<10 mm	<10 mm	<5 mm	<5 mm	<1 mm	1 mm
2. Fine or thin	1–2 mm	10–20 mm	10–20 mm	5–10 mm	5–10 mm	1–2 mm	1–2 mm
3. Medium	2–5 mm	20–50 mm	10–20 mm	10–20 mm	10–20 mm	2–5 mm	2–5 mm
4. Coarse or thick	5–10 mm	50–100 mm	50–100 mm	20–50 mm	20–50 mm	5–10 mm	
5. Very coarse (very thick)	>10 mm	>100 mm	>100 mm	>50 mm	>50 mm	>10 mm	

C Grade: Durability of peds		
	0. Structureless	No aggregation or orderly arrangement.
	1. Weak	Poorly formed, nondurable, indistinct peds that break into a mixture of a few entire and many broken peds and much unaggregated material.
	2. Moderate	Well-formed, moderately durable peds, indistinct in undisturbed soil, that break into many entire and some broken peds but little unaggregated material.
	3. Strong	Well-formed, durable, distinct peds, weakly attached to each other, that break almost completely into entire peds.

2. *Prismatic* or *columnar*: Vertically oriented pillars, up to 15 cm in diameter. Such structures are common in the B horizon of clayey soils, particularly in semiarid regions. Where the tops are flat, these vertical aggregates are called *prismatic*, and where rounded, *columnar*.
3. *Blocky*: Cubelike blocks of soil, up to 10 cm in size, sometimes angular with well-defined planar faces. These structures occur most commonly in the upper part of the B horizon.
4. *Spherical*: Rounded aggregates, generally not much larger than 2 cm in diameter, often found in a loose condition in the A horizon. Such units are called *granules* and, where particularly porous, *crumbs*.

The shapes of aggregates, as well as their sizes and densities, generally vary within the profile. Because the overburden pressure increases with depth, and inasmuch as the deeper layers do not experience such extreme fluctuations in moisture content as does the alternately saturated and desiccated surface layer, the decrease of swelling and shrinkage activity causes the deeper aggregates to be larger. A typical structural profile in semiarid regions consists of a granulated A horizon underlain by a prismatic B horizon, whereas in humid temperate regions a granulated A horizon may occur with a platy or blocky B horizon. The number of variations found in nature are, however, legion. A detailed classification of aggregate shapes is given in Table 5.2.

AGGREGATE SIZE DISTRIBUTION

Aggregate size distribution is an important determinant of the soil's pore size distribution and has a bearing on the erodibility of the soil surface, particularly by wind. In the field, adjacent aggregates often adhere to one another, though of course not as tenaciously as do the particles within each aggregate. Separating and classifying soil aggregates necessarily involves a disruption of the original, in situ structural arrangement. The application of too great a force may break up the aggregates themselves. Hence the determination of aggregate size distribution depends on the mechanical means employed to separate the aggregates.

Screening through flat sieves is difficult to standardize and entails frequent clogging of the sieve openings. A more practical approach is to use a rotary sieve machine with concentrically nested sieves of different aperture sizes. The operation of such a device can be standardized, thus minimizing the arbitrary personal factor, and clogging can be practically eliminated.

Various indexes have been proposed for the distribution of aggregate sizes (Nimmo and Perkins, 2002). If a single characteristic parameter is desired (so as to allow correlation with such factors as erosion, infiltration, evaporation, and aeration), a method must be adopted for assigning an appropriate weighting factor to each size range of aggregates. One of the most widely used indexes is the *mean weight diameter*, based on weighting the masses of aggregates of the various size classes according to their respective sizes. The mean weight diameter X is thus defined by the following equation:

$$X = \sum_{i=1}^{n} x_i w_i \tag{5.2}$$

Here x_i is the mean diameter of any particular size range of aggregates separated by sieving and w_i is the weight of the aggregates in that size range as a fraction of the total dry weight of the sample. The summation accounts for all size ranges, including the group of aggregates smaller than the openings of the finest sieve.

An alternative index of aggregate size distribution is the *geometric mean diameter* Y, calculated according to the following equation:

$$X = \exp\left[\left(\sum_{i=1}^{n} w_i \log x_i\right) \Big/ \left(\sum_{i=1}^{n} w_i\right)\right] \quad (5.3)$$

where w_i is the weight of aggregates in a size class of average diameter x_i and the denominator $\Sigma\, w_i$ (for i values from 1 to n) is the total weight of the sample.

Sample Problem

Calculate the mean weight diameters of the assemblages of aggregates given in Table 5.3. The percentages refer to the mass fractions of dry soil in each diameter range.

First we determine the mean diameters of the seven aggregate diameter ranges:

Range: 0–0.5, 0.5–1, 1–2, 2–5, 5–10, 10–20, 20–50 mm
Mean: 0.25, 0.75, 1.5, 3.5, 7.5, 15, 35 mm

Recall that the mean weight diameter X is defined by Eq. (5.2):

$$X = \sum_{i=1}^{n} x_i w_i \quad (5.2)$$

Hence, for the dry-sieved virgin soil,

$$Y = (0.25 \times 0.1) + (0.75 \times 0.1) + (1.5 \times 0.15) + (3.5 \times 0.15) + (7.5 \times 0.2) + (15 \times 0.2) + (35 \times 0.1) = 8.85 \text{ mm}$$

For the dry-sieved cultivated soil,

$$X = (0.25 \times 0.25) + (0.75 \times 0.25) + (1.5 \times 0.15) + (3.5 \times 0.15) + (7.5 \times 0.1) + (15 \times 0.07) + (35 \times 0.03) = 4.30 \text{ mm}$$

For the wet-sieved virgin soil,

$$X = (0.25 \times 0.3) + (0.75 \times 0.15) + (1.5 \times 0.15) + (3.5 \times 0.15) + (7.5 \times 0.15) + (15 \times 0.05) + (35 \times 0.05) = 4.56 \text{ mm}$$

For the wet-sieved cultivated soil,

$$X = (0.25 \times 0.5) + (0.75 \times 0.25) + (1.5 \times 0.15) + (3.5 \times 0.05) + (7.5 \times 0.04) + (15 \times 0.01) + (35 \times 0.0) = 1.16 \text{ mm}$$

Note: Wet sieving reduced the mean weight diameter from 8.85 to 4.56 mm in the virgin soil and from 4.30 to 1.16 mm in the cultivated soil. This indicates the degree of instability of the various aggregates under the slaking effect of immersion in water. The influence of cultivation is generally to reduce the water stability of soil aggregates and hence to render the soil more vulnerable to crusting and erosion processes.

TABLE 5.3 Aggregate Data for Sample Problem

Aggregate diameter range (mm)	Dry sieving: Virgin soil (%)	Dry sieving: Cultivated soil (%)	Wet sieving: Virgin soil (%)	Wet sieving: Cultivated soil (%)
0.0–0.5	10	25	30	50
0.5–1.0	10	25	15	25
1–2	15	15	15	15
2–5	15	15	15	5
5–10	20	10	15	4
10–20	20	7	5	1
20–50	10	3	5	0

AGGREGATE STABILITY

Determining the state of aggregation of a soil at any particular moment might not suffice to portray the dynamic nature of soil structure. By whatever measure, the state of aggregation varies over time and space as aggregates form, disintegrate, and reform periodically. For instance, a newly cultivated field may for a time exhibit a nearly optimal array of aggregate sizes, with large interaggregate pores favoring high infiltration rates and unrestricted aeration. However, soil structure may begin to deteriorate quite visibly and rapidly, because the soil is subject to destructive forces resulting from intermittent rainfall (causing swelling and shrinking, slaking and erosion) followed by dry spells (exposing the soil to deflation by wind). Repeated traffic, especially by heavy machinery, further tends to crush the aggregates remaining at the surface and to compact the soil to some depth.

Soils vary, of course, in the degree of their vulnerability to externally imposed destructive forces. *Aggregate stability* is a measure of this vulnerability. More specifically, it expresses the resistance of aggregates to breakdown when subjected to potentially disruptive processes. The reaction of a soil to forces acting on it depends not only on the soil itself but also on the nature of the forces and the manner they are applied.

To test aggregate stability, soil physicists generally subject samples of aggregates to artificially induced forces designed to simulate processes that are likely to occur in the field. The nature of the forces applied during such testing depends on the investigator's perception of the phenomenon to be simulated. The degree of stability is then assessed by determining the fraction of the original sample's mass that has withstood destruction (Nimmo and Perkins, 2002).

If an indication of mechanical stability is sought, measurements can be made of the resistance of aggregates to prolonged dry sieving or to crushing forces. Most frequently, however, the concept of aggregate stability is applied in relation to the destructive action of water. Although mentioned before, it bears repeating that the very wetting of aggregates may cause their collapse, because the bonding substances dissolve or weaken and as the clay swells and possibly disperses. Aggregates are more vulnerable to sudden than to gradual wetting, owing to the air occlusion effect. Raindrops and flowing water provide the energy to detach particles and transport them away. Abrasion by particles carried as suspended matter in runoff water may scour the surface and contributes to the overall breakdown of the aggregated structure at the soil surface.

The classical and still most prevalent procedure for testing the *water stability* of soil aggregates is the *wet sieving* method. A representative sample of air-dry aggregates is placed on the uppermost of a set of graduated sieves and immersed in water to simulate flooding. The sieves are then oscillated vertically so that water is made to flow up and down through the screens and the assemblage of aggregates. At the end of a specified period of sieving (e.g., 20 min) the nest of sieves is removed from the water and the oven-dry weight of material left on each sieve is determined. The results should be corrected for the coarse primary particles retained on each sieve to avoid designating them falsely as aggregates. This is done by dispersing the material collected from each sieve, using a mechanical stirrer and a sodic dispersing agent, and then washing the material back through the same sieve. The weight of sand retained after the second sieving is then subtracted from the total weight of undispersed material retained after the first sieving, and the percentage of stable aggregates %SA is given by

$$\%\mathrm{SA} = 100 \times \frac{(\text{weight retained}) - (\text{weight of sand})}{(\text{total sample weight}) - (\text{weight of sand})} \tag{5.4}$$

An alternative approach is to subject soil aggregates to simulated rain. In the *drop method*, aggregates are bombarded with drops of water in a standardized manner. The number of drops needed for total dissipation of the aggregates or the fractional mass of the aggregates remaining after a given time, can be determined. Another way is to subject a soil in the field to simulated rainfall of controllable raindrop sizes and velocities. The condition of the soil surface can then be compared to the initial condition.

To determine the stability of microaggregates, comparative *sedimentation analysis* can be carried out with dispersed versus undispersed samples. An index indicating the fractional amount of clay associated in microaggregates can then be calculated. Still another index of soil structural stability is obtained by comparing the *permeability* of the soil to an inert fluid with its permeability to water. The permeability of an appropriately packed soil sample is first measured by using air, a fluid that is presumed to have no effect on structure, and then by using water. A ratio of unity, if such were to occur, would indicate perfect stability. Values greater than unity indicate a scale of increasing instability.

SOIL CRUSTING

The aggregates at the soil surface are the most vulnerable to destructive forces. The aggregates that collapse during wetting may form a layer of dispersed mud, typically several millimeters thick, which clogs the macropores of the top layer and thus tends to inhibit the infiltration of water and the exchange of gases between the soil and the atmosphere. Such a layer is often called a *surface seal*. As it dries, this dispersed layer shrinks to become a dense, hard *crust*, which impedes seedling emergence by its hardness and tears seedling roots as it cracks, forming a characteristic polygonal pattern. The effect of soil crusting on seedlings depends on crust thickness and strength as well as on the size and vigor of the seedlings. Most sensitive are small-seeded vegetables and grasses. In soils prone to crusting, seedling emergence may occur only through the crust's cracks.

Attempts have been made to characterize soil crusting, particularly with respect to its effect on seedling emergence, in terms of the resistance of the dry crust to the penetration of a probe (Parker and Taylor, 1965) as well as in terms of its mechanical strength as measured by the *modulus of rupture* test or the *tensile strength* test (see Chapter 13). These tests were designed to imitate the process by which a seedling forces its way upward by penetrating and rupturing the crust. However, the critical crust strength that prevents emergence obviously depends on crust thickness and soil wetness as well as on plant species and depth of seed placement (Hillel, 1972b).

Crust strength increases as the degree of colloidal dispersion increases. The evaporation process charges the soil surface zone with relatively high concentrations of sodic salts and, consequently, with a high exchangeable sodium percentage. With subsequent infiltration of rain or irrigation water, the salts are leached but the exchangeable sodium percentage (ESP) remains high. The resulting combination of high ESP and low salt concentration induces colloidal dispersion, which further contributes to the formation of a hard crust.

SOIL CONDITIONERS

We have already described the beneficial effect on soil aggregation of various natural polymers, products of the microbial decay of organic matter in the soil. Such substances as polysaccharides and polyuronides promote aggregate stability by gluing particles together within aggregates as well as by coating aggregate surfaces. As the soil dries, the gel-like glues undergo practically irreversible dehydration, becoming more or less stable cementing agents that bind flocs of clay to one another as well as to silt and sand grains.

Considerable work has been done to develop and apply synthetic compounds capable of duplicating the effect of natural polymers. Where natural aggregation or aggregate stability is lacking, such synthetic polymers, called *soil conditioners*, can help in the stabilization of aggregates (Wallace and Terry, 1997). Materials have indeed been produced that are effective in relatively small quantities (e.g., 0.1% of the treated soil mass) and can produce a dramatic improvement of soil structure, especially in the soil's upper layer, with consequent beneficial effects upon infiltration, aeration, and the prevention of crusting and erosion. The practical application of such products in the field, however, is still constrained by their high cost.

Part III

THE LIQUID PHASE

6. WATER CONTENT AND POTENTIAL

SOIL WETNESS

The variable amount of water contained in a unit mass or volume of soil and the energy state of water in the soil are important factors affecting the growth of plants. Numerous other soil properties depend on water content. Among these are mechanical properties, such as consistency, plasticity, strength, compactibility, penetrability, stickiness, and trafficability. In clayey soils, swelling and shrinking associated with addition or extraction of water change the bulk density, porosity, and pore size distribution. Soil water content also governs the air content and gas exchange of the soil, thus affecting the respiration of roots, the activity of microorganisms, and the chemical state of the soil (e.g., oxidation–reduction potential).

The per-mass or per-volume fraction of water in the soil is expressed in terms of *soil wetness*. The physicochemical condition or state of soil water is characterized in terms of its *free energy* per unit mass, called the *potential* (a term that alludes to the "potential energy" of water in the soil). Of the various components of this potential, the *matric potential* manifests the tenacity with which soil water is held by the soil matrix.

Wetness and matric potential are functionally related, and the graphical representation of this relationship is termed the *soil-moisture characteristic curve*. The relationship is not unique, however, because it is affected by the direction and rate of change of soil moisture and is sensitive to changes in soil structure. Wetness and matric potential vary in space and time as the soil is repeatedly wetted by rain, drained by gravity, and dried by evaporation and root extraction.

The lowest wetness we are likely to encounter in nature is a variable state called *air dryness*; in the laboratory it is an arbitrary state known as the *oven-dry* condition. On the other hand, the wettest possible condition of a soil is that of *saturation*, defined as a condition in which all soil pores are filled with water. Saturation is relatively easy to define in the case of nonswelling (e.g., sandy) soils. It can be difficult to define in the case of swelling soils, because such soils may continue to imbibe water and swell even after all pores have been filled with water.

In the field, however, the soil seldom attains complete saturation, because bubbles of air may form and remain occluded within the matrix even when the soil is flooded with excess water. Moreover, air bubbles may effervesce within the soil due to microbial action or whenever the temperature rises and the solubility of gases is exceeded. For this reason, some investigators prefer the term *satiation* to describe the condition in the field in which the soil, though not completely saturated, is as wet as it can get under the circumstances.

Here, we prefer the term *soil wetness* to the older (and still prevalent) term *soil water content*, not only for reasons of verbal economy but also because "wetness" implies intensity, whereas "content" implies extensity. One can speak of the water content when referring to the total amount of water in a bucket. In contrast, the wetness of the soil pertains to the relative amount rather than the absolute amount of water in a soil body, independent of its size.

MASS AND VOLUME RATIOS

The fractional content of water in the soil can be expressed in terms of either mass or volume ratios. As given in Chapter 1,

$$w = M_w/M_s \tag{6.1}$$

$$\theta = V_w/V_t = V_w/(V_s + V_w + V_a) \tag{6.2}$$

where w, the mass wetness, is the dimensionless ratio of water mass m_w to dry soil mass M_s and θ, the volume wetness, is the ratio of water volume V_w to total (bulk) soil volume V_t. The latter is equal to the sum of the volumes of solids (V_s), water (V_w), and air (V_a).

Both θ and w are usually multiplied by 100 and reported as percentages by volume or mass, respectively. The two expressions can be related to each other from knowledge of the bulk density ρ_b of the soil and the density of water:

$$\theta = w(\rho_b/\rho_w) = w\Gamma_b \tag{6.3}$$

where Γ_b is the bulk specific gravity of the soil (a dimensionless ratio between the soil bulk density and the water density, Γ_b generally varies between 1.1 and 1.7). The conversion is simple for nonswelling soils in which bulk density, hence bulk specific gravity, are constant regardless of wetness. However, the conversion can be difficult in the case of swelling soils, for which the bulk density is a function of mass wetness. The relation of bulk density (or of its reciprocal, the *bulk specific volume*) to mass wetness is illustrated schematically in Fig. 6.1.

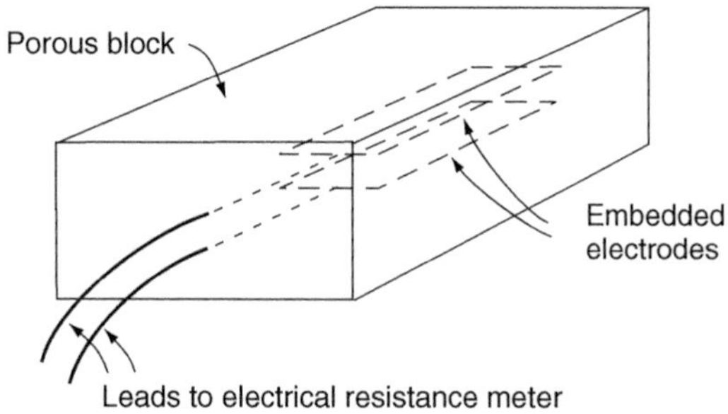

Fig. 6.1. An electrical resistance block. The embedded electrodes may be plates, screens, or wires in a parallel or concentric arrangement.

In many cases, it is useful to express the water content of a soil profile in units of *depth*, that is, as the volume of water contained in a specified total depth of soil d_t per unit area of land. This indicates the equivalent depth d_w that soil water would have if it were extracted and then ponded over the soil surface. Thus

$$d_w = \theta d_t = w\Gamma_b d_t \tag{6.4}$$

Usually d_w is given in millimeters, as are rainfall and evaporation.

Another expression of soil wetness is the liquid ratio θ_r, defined as the volume of water per unit volume of the solid phase:

$$\theta_r = w(\rho_s/\rho_w) = \theta(\rho_w/\rho_b)(\rho_s/\rho_w) = \theta(\rho_s/\rho_b) \tag{6.5}$$

This expression, along with *void ratio* (defined in Chapter 1), is useful mainly with soils having a nonrigid (swelling and shrinking) matrix.

Sample Problem

The data in Table 6.1 were obtained before and after an irrigation. From these data, calculate the mass and volume wetness values of each layer before and after the irrigation, and determine the amount of water (in millimeters) added to each layer and to the profile as a whole.

Using Eq. (6.6), we obtained the following mass wetness values:

$$w_1 = (0.160 - 0.150)/(0.150 - 0.050) = 0.1$$
$$w_2 = (0.146 - 0.130)/(0.130 - 0.050) = 0.2$$
$$w_3 = (0.230 - 0.200)/(0.200 - 0.050) = 0.2$$
$$w_4 = (0.206 - 0.170)/(0.170 - 0.050) = 0.3$$

Using Eq. (6.3), we obtain the following volume wetness values:

$$\theta_1 = 1.2 \times 0.1 = 0.12$$
$$\theta_2 = 1.5 \times 0.2 = 0.30$$
$$\theta_3 = 1.2 \times 0.2 = 0.24$$
$$\theta_4 = 1.5 \times 0.3 = 0.45$$

TABLE 6.1 Data for Sample Problem

Sampling time	Sample number	Depth (m)	Bulk density (kg/m³)	Wet sample + container (kg)	Dry sample + container (kg)	Container (kg)
Before irrigation	1	0.0–0.4	1.2×10^3	0.160	0.150	0.05
	2	0.6–1.0	1.5×10^3	0.146	0.130	0.05
After irrigation	3	0.0–0.4	1.2×10^3	0.230	0.200	0.05
	4	0.6–1.0	1.5×10^3	0.206	0.170	0.05

Using Eq. (6.4), we obtain the following water depths per layer:

$$d_{w_1} = 0.12 \times 400 \text{ mm} = 48 \text{ mm}$$
$$d_{w_2} = 0.30 \times 600 \text{ mm} = 180 \text{ mm}$$
$$d_{w_3} = 0.24 \times 400 \text{ mm} = 96 \text{ mm}$$
$$d_{w_4} = 0.45 \times 600 \text{ mm} = 270 \text{ mm}$$

Depth of water in profile before irrigation = 48 + 180 = 228 mm.
Depth of water in profile after irrigation = 96 + 270 = 366 mm.
Depth of water added to top layer = 96 − 48 = 48 mm.
Depth of water added to bottom layer = 270 − 180 = 90 mm.
Depth of water added to entire profile = 48 + 90 = 138 mm.

MEASUREMENT OF SOIL WETNESS

The need to determine the amount of water contained in the soil arises frequently in many agronomic, ecological, and hydrological investigations aimed at understanding the soil's chemical, mechanical, hydrological, and biological relationships. There are direct and indirect methods to measure soil moisture (Gardner, 1986; Topp and Ferré, 2002), and, as we have already pointed out, there are several alternative ways to express it quantitatively. As yet there is no universally recognized standard method of measurement and no uniform way to compute and present the results of soil-moisture measurements. We next describe, briefly, some of the most prevalent methods for this determination.

Sampling and Drying

The traditional (gravimetric) method of measuring mass wetness consists of removing a sample by augering into the soil and then determining its moist and dry weights. The *moist weight* is determined by weighing the sample as it is at the time of sampling; the *dry weight* is obtained after drying the sample to a constant weight in an oven. The more or less standard method of drying is to place the sample in an oven at 105°C for 24 hours. An alternative method of drying, suitable for field use, is to impregnate the sample in a heat-resistant container with alcohol, which is then burned off, thus vaporizing the water. The *mass wetness*, also called *gravimetric wetness*, is the ratio of the weight loss in drying to the dry weight of the sample (mass and weight being proportional):

$$w = \frac{(\text{wet weight}) - (\text{dry weight})}{\text{dry weight}} = \frac{\text{weight loss in drying}}{\text{weight of dried sample}} \tag{6.6}$$

The gravimetric method, depending as it does on sampling, transporting, and repeated weighings, entails practically inevitable errors. It is also laborious and time consuming, since the samples must be transported from the field to the laboratory and a period of at least 24 hr is usually considered necessary for complete oven drying. The standard method of oven drying is itself arbitrary. Some clays may still contain appreciable amounts of adsorbed water even at 105°C. On the other hand, some organic matter may oxidize and decompose at this temperature, so the weight loss may not be due entirely to the evaporation of water.

The errors of the gravimetric method can be reduced by increasing the sizes and number of samples. However, the extraction of samples from the field is an invasive and destructive exercise, which may disturb an observation or experimental plot sufficiently to distort the results. Hence many workers prefer indirect methods, which, once installed and calibrated, permit repeated or continuous measurements at the same points with much less time, labor, and soil disturbance.

Electrical Resistance

The electrical resistance of a soil body depends not only on its water content, but also on its composition, texture, and soluble-salt concentration. On the other hand, the electrical resistance of porous bodies placed in the soil and left to equilibrate with soil moisture can sometimes be calibrated against soil wetness. Such units (called *electrical resistance blocks*) generally contain a pair of electrodes embedded in a porous material such as gypsum, nylon, or fiberglass. (See Fig. 6.1.)

Porous blocks placed in the soil tend to equilibrate with the matric suction (tension) of soil water (see later in this chapter) rather than with the water content per se. Different soils can have greatly differing wetness versus suction relationships (e.g., a sandy soil may retain less than 5% moisture at, say, 15-bar suctions, whereas a clayey soil may retain three times as much). Calibration of porous blocks against suction is therefore preferable to calibration against soil wetness, particularly when the soil used for the purpose is a disturbed sample differing in structure from the soil in situ.

The equilibrium of porous blocks with soil moisture may be affected by hysteresis, that is, by the direction of change of soil moisture (whether increasing or decreasing) prior to the equilibration. Moreover, the hydraulic properties of the blocks themselves or inadequate contact with the soil may prevent the rapid attainment of equilibrium and cause a time lag between the state of water in the soil and the state of water being measured in the block. This effect, as well as the sensitivity of the block, may not be constant over the entire range of variation in soil wetness. Gypsum blocks are generally more responsive in the dry range, whereas porous nylon blocks, because of their larger pore sizes, are more sensitive in the wet range of soil-moisture variation.

The electrical conductivity of a porous block made of inert material is due primarily to the permeating fluid rather than to the block's solid matrix. Thus it depends on the electrolytic solutes present in the fluid as well as on the volume content of the fluid. Blocks made of such materials as fiberglass, for instance, are highly sensitive to even small variations in salinity of the soil solution. On the other hand, blocks made of plaster of Paris (gypsum) maintain a nearly constant electrolyte concentration corresponding primarily to that of a saturated solution of calcium sulfate. This tends to mask, or buffer, the effect of small or even moderate variations in the soil solution (such as those due to fertilization or low levels of salinity). However, an undesirable consequence of the solubility of gypsum is that these blocks eventually deteriorate in the soil. Hence the relationship between electrical resistance and moisture suction varies not only from block to block but also for each block over time, because the gradual dissolution of the gypsum changes the internal porosity.

For these and other reasons (e.g., temperature sensitivity), the evaluation of soil wetness by means of electrical resistance blocks is likely to be of limited accuracy. On the other hand, an advantage of such blocks is that they can be connected to a recorder to obtain a continuous indication of soil moisture changes in situ.

Neutron Scattering

This method has gained widespread acceptance as an efficient and reliable technique for monitoring soil moisture in the field. Its principal advantages over the gravimetric method are that it allows less laborious, more rapid, nondestructive (after initial installation), and periodically repeatable measurements, in the same locations and depths, of the volumetric wetness of a representative volume of soil. The method is practically independent of temperature and pressure. Its main disadvantages, however, are the high initial cost of the instrument, low degree of spatial resolution, difficulty of measuring moisture in the soil surface zone, and the health hazard associated with exposure to neutron and gamma radiation.

The instrument, known as a *neutron moisture meter* (Fig. 6.2), consists of two main components: (a) a *probe* (containing a *source of fast neutrons* and a *detector of slow neutrons*), which is lowered into an *access tube* inserted vertically into the soil, and (b) a *scaler* or *rate meter* (usually battery powered and portable) to monitor the flux of the slow neutrons that are scattered and attenuated in the soil.

The purpose of the access tube is both to maintain the bore hole into which the probe is lowered and to standardize measuring conditions. Aluminum tubing is usually the preferred material for access tubes since it is nearly transparent to a neutron flux.

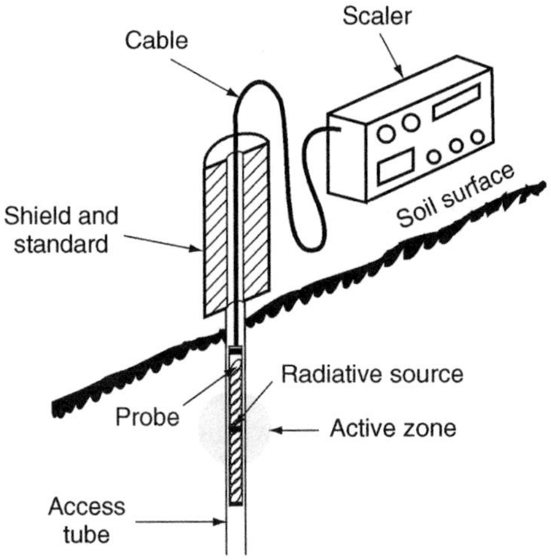

Fig. 6.2. Components of a portable neutron soil-moisture meter, including a probe (with a source of fast neutrons and a detector of slow neutrons) lowered from a shield containing hydrogenous material (e.g., paraffin, polyethylene) into the soil via an access tube. A scaler-rate meter is shown alongside the probe. Recent models incorporate the scaler into the shield body, and the integrated unit is lightweight for easy portability.

A source of fast neutrons is generally obtained by mixing a radioactive emitter of *alpha particles* (helium nuclei) with beryllium. Frequently used is a 2- to 5-millicurie pelletized mixture of radium and beryllium. An Ra-Be source emits about 16,000 neutrons per second per milligram (or millicurie) of radium. The energies of the neutrons emitted by this source vary from 1 to 15 MeV (million electron volts), with a preponderant energy range of 2–4 MeV and an average speed of about 1600 km/sec. Hence, they are called *fast neutrons*. An alternative source of fast neutrons is a mixture of americium and beryllium. Both radium and americium incidentally also emit gamma radiation, but that of the americium is lower in energy and hence less hazardous than that of the radium. The source materials are chosen for their longevity (e.g., radium-beryllium has a half-life of about 1620 yr) so that they can be used for a number of years without appreciable change in radiation flux.

The fast neutrons are emitted radially into the soil, where they encounter and collide elastically (as do billiard balls) with various atomic nuclei. Through repeated collisions, the neutrons are deflected and "scattered," and they gradually lose some of their kinetic energy. As the speed of the initially fast neutrons diminishes, it approaches a speed that is characteristic for particles at the ambient temperature. For neutrons this is about 2.7 km/sec, equivalent to an energy of about 0.03 eV. Neutrons slowed to such a speed are said to be *thermalized* and are called *slow neutrons*. Such neutrons continue to interact with the soil and are eventually absorbed by the nuclei present.

The effectiveness of various nuclei present in the soil in moderating or thermalizing fast neutrons varies widely. The average loss of energy is maximal for collisions between particles of approximately the same mass. Of all nuclei encountered in the soil, the ones most nearly equal in mass to neutrons are the nuclei of hydrogen (protons), which are therefore the most effective moderators of fast neutrons in the soil. The average number of collisions required to slow a neutron from 2 MeV to thermal energies is 18 for hydrogen, 114 for carbon, 150 for oxygen, and $9N + 6$ for nuclei of larger mass number N. If the soil contains an appreciable concentration of hydrogen, the emitted fast neutrons are thermalized within close proximity of the source.

The slow neutrons thus produced scatter randomly in the soil, quickly forming a swarm or cloud of constant density around the probe. The equilibrium density of the slow neutron cloud is determined by the rate of emission by the source and the rates of thermalization and absorption by the medium (i.e., soil) and is established within a small fraction of a second. Certain elements that might be present in the soil (e.g., boron, cadmium, and chlorine) exhibit a high absorption capacity for slow neutrons, and their presence in nonnegligible concentrations might tend to reduce the density of slow neutrons. By and large, however, the density of slow neutrons formed around the probe is nearly proportional to the concentration of hydrogen in the medium surrounding the probe and therefore is approximately proportional to the volume fraction of water present in the soil. Thus

$$N_w = m\theta + b \qquad \text{and} \qquad N_w/N_s = y\theta \tag{6.7}$$

in which N_w is slow-neutron count rate in wet soil, N_s is the count rate in water or in a standard absorber (i.e., the shield of the probe), θ is volumetric wetness, y is a constant, and m and b are the slope and intercept, respectively, of the line indicating N_w as a function of θ.

As the thermalized neutrons repeatedly collide and bounce about randomly, a number of them (proportional to the density of neutrons thus thermalized and scattered and, therefore, approximately linearly related to the concentration of soil moisture) return to the probe. Here they are counted by the detector of slow neutrons. The detector cell is usually filled with BF_3 gas. When a thermalized neutron encounters a ^{10}B nucleus and is absorbed, an alpha particle is emitted, creating an electrical pulse on a charged wire. The number of pulses over a measured time interval is counted by a scaler or indicated by a rate meter (Hignett and Evett, 2002).

The effective volume of soil in which the water content is measured depends on the energy of emitted neutrons as well as on the concentration of hydrogen nuclei; that is, for a given source and soil, it tracks the volume concentration of soil moisture. If the soil is relatively dry, the cloud of slow neutrons surrounding the probe will be less dense and extend farther from the source, and vice versa for wet soil. With commonly used radium-beryllium and americium-beryllium sources, the so-called *sphere of influence*, or effective volume of measurement, varies with a radius of less than 0.10 m in a wet soil to 0.25 m or more in a dry soil. The low and variable degree of spatial resolution makes the neutron moisture meter unsuitable for the detection of moisture profile discontinuities (e.g., wetting fronts or boundaries between layers). Measurements made within 0.20 m of the surface are unreliable because of the possible escape of fast neutrons through the surface. However, a special surface probe is available commercially to allow measurement of the average moisture in the soil's top layer.

For the sake of safety, and also to provide a convenient means of making standard readings, the probe containing the radiation source is normally carried inside a protective shield designed to prevent the escape of gamma-rays as well as of fast neutrons. The shield is usually a cylindrical container lined with lead and filled with some hydrogenous material, such as paraffin or polyethylene. Improper or excessive use of the equipment can be hazardous. The danger from exposure to radiation depends on the strength of the source, the quality of the shield, the distance from source to operator, and the duration of contact. With strict observance of safety rules, however, the equipment can be used without undue risk.

Sample Problem

Calibration of a neutron probe shows that when a soil's volumetric wetness is 15%, we get a reading of 24,000 cpm (counts per minute), and at a wetness of 40% we get 44,000 cpm. Find the equation of the straight line defining the calibration ($Y = mX + b$, where Y is counts per minute, X is volumetric wetness, m is the slope of the line, and b is the intercept on the Y axis). Using the equation derived, find the wetness value corresponding to a count rate of 30,000 cpm.

We first obtain the slope m:

$$m = (Y_2 - Y_1)/(X_2 - X_1) = (44{,}000 - 24{,}000)/(40 - 15) = 800 \text{ cpm per 1\% wetness}$$

We next obtain the Y intercept b:

$$Y = 800X + b, \qquad b = Y - 800X = 44{,}000 - 800 \times 40 = 12{,}000 \text{ cpm}$$
$$(\text{or: } 24{,}000 - 800 \times 15 = 12{,}000 \text{ cpm})$$

The complete equation is therefore

$$Y = 800X + 12{,}000$$

Now, to find the wetness value corresponding to 30,000 cpm, we set

$$30{,}000 = 800X + 12{,}000, \qquad X = (30{,}000 - 12{,}000)/800 = 22.5\%$$

Gamma-Ray Absorption

The gamma-ray scanner for measuring soil moisture generally consists of two spatially separated units, or probes: (1) a source, usually containing a pellet of radioactive cesium (^{137}Cs emitting gamma radiation with an energy of 0.661 MeV), and (2) a detector, normally consisting of a scintillation counter (e.g., a sodium iodide crystal or a synthetic scintillator) connected to a photomultiplier and preamplifier. If the emission of radiation is monoenergetic and radial and if the space between the source and the detector is empty and the two units are a constant distance apart, then the fraction of the emitted radiation received by the detector will depend only on the angular section intercepted, that is, on the distance of separation and the size of the scintillation unit. On the other hand, should the space between the units be filled with some material, a fraction of the original radiation that would otherwise be detected will be absorbed, depending on the interposing mass, that is, on the thickness and density of the intervening material. In the event the material placed between source and detector is a body of soil of constant bulk density, the intensity of the transmitted radiation will vary only with changes in water content. In fact, it will be an exponential function of soil wetness, as follows (Gurr, 1962; Ferguson and Gardner, 1962):

$$N_w/N_d = \exp(-\theta_m \mu_w x) \tag{6.8}$$

where N_w/N_d is the ratio of the monoenergetic radiation flux transmitted through wet soil (N_w) to that transmitted through dry soil (N_d), μ_w is the mass attenuation coefficient for water, x is the thickness of the transmitting soil, and θ_m, is the water mass per unit bulk volume of soil. Equation (6.8) can be transformed to give wetness as a function of the relative transmission rate:

$$\theta_m = \frac{\ln(N_w/N_d)}{\mu_w x} = -\frac{\log_{10}(N_w/N_d)}{0.4343\ \mu_w x} \tag{6.9}$$

where $\ln(N_w/N_d)$ is the natural logarithm of the count ratio for wet to dry soil and $\log_{10}$ is the common logarithm for the same ratio.

The gamma-ray absorption method is used mostly in the laboratory, where the dimensions and density of the soil sample as well as the ambient temperature can be precisely controlled. A high degree of spatial resolution (e.g., 2 mm or so) can be obtained by collimation of the radiation. This is done by drilling a narrow hole or slot into the lead (Pb) wall shielding the source (as well, perhaps, as into a second shield placed in front of the detector), thus allowing passage of only a very narrow beam.

Since the absorption of radiation depends on the entire mass between source and detector, the readings can be related uniquely to soil moisture only if soil bulk density is constant or if its change is monitored simultaneously. To permit concurrent measurement of soil bulk density and moisture changes in swelling or shrinking soils, dual-source scanners have been developed, in which both cesium 137 and americium 241 are used. Analysis of the concurrent transmission of the two beams can allow separation of the change in attenuating mass between that due to bulk density and that due to soil wetness (Gardner, 1986).

The *double-probe gamma-ray method* has also been adapted to field use (Fig. 6.3). In principle, this technique offers several advantages over the neutron moisture meter, in that it allows better depth resolution in the measurement of soil-moisture profiles (i.e., about 1 cm in effective measurement width), sufficient to detect discontinuities between profile layers as well as movement of wetting fronts and conditions near the soil surface. However, the field device is still too cumbersome for general usage. Not the least of the problems is the accurate installation and alignment of two access tubes that must be strictly parallel along with the accurate determination of soil bulk density, which might vary in depth and time.

The health hazard associated with use of gamma-ray equipment is similar in principle to that discussed in connection with the neutron moisture meter. The equipment is considered safe only if strict attention is paid to all the safety rules.

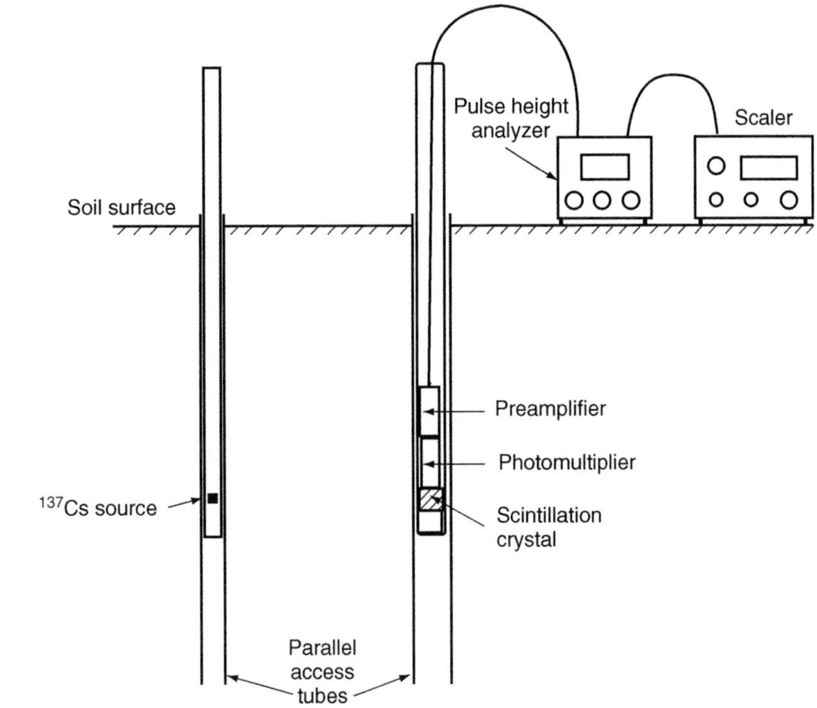

Fig. 6.3. Double-probe gamma-ray apparatus for monitoring soil moisture or density.

Time-Domain Reflectometry (TDR)

This is a relatively new method of measuring soil wetness, based on the unusually high *dielectric constant* of water. A *dielectric*, in general, is a non conductor of electricity, that is, a substance that, when placed between two charged surfaces (a capacitor), allows no net flow of electric charge but only a displacement of charge. (The simplest form of a capacitor consists of two parallel metal plates separated by a layer of air or some other insulating material, such as mica.) The dielectric constant, also called *relative permittivity* (or specific inductive capacity), is defined as the ratio of the *capacitance* of a capacitor with the given substance as dielectric to the capacitance of the same capacitor with air (or a vacuum) as the dielectric.

The value of the dielectric constant (ε_r) depends on the nature of the material. The ε_r value for dry air is usually taken to be 1, for paraffin wax 2.0–2.5, for rubber 2.8–3.0, for porcelain and mica 6.0–8.0. The value for dry soil is roughly 4 (Jackson and Schmugge, 1989). In contrast, the value for water is about 80. Hence the water content of a soil determines its dielectric constant.

In the TDR technique (Topp and Davis, 1985), a pair of parallel metal rods connected to a signal receiver is inserted into the soil. The rods serve as conductors, while the soil between and around the rods is the dielectric medium. When a step voltage pulse or signal is propagated along the parallel transmission lines, the signal is reflected from the ends of those rods and returns to the

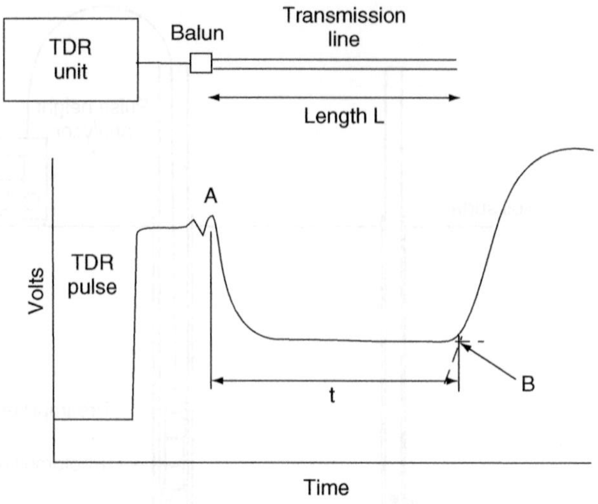

Fig. 6.4. The essential components of a TDR system (above) and an idealized TDR output trace (obtainable with an oscilloscope) showing how the propagation time is determined. (After Topp and Davis, 1985.)

receiver. A device then measures the time between sending and receiving the reflected signal, as shown in Fig. 6.4. For a fixed line length, the time interval relates inversely to the propagation velocity v of the signal in the soil (i.e., the propagation velocity diminishes as the amount of water present increases, so the time interval increases with soil wetness).

The approximate equation given by Topp and Davis (1985) is:

$$v = c/(\varepsilon_r)^{0.5} \tag{6.10}$$

where c is the propagation velocity of an electromagnetic wave in free space (3×10^8 m/sec). The velocity can be determined from the length L of the transmission lines (with the travel distance, forth and back, being $2L$) and the measured signal time t in the soil, using the time-domain reflectometer. The dielectric constant results from inserting $v = 2L/t$ into Eq. (6.10):

$$\varepsilon_r = (ct/2L)^2 \tag{6.11}$$

Topp and Davis (1985) found that the dielectric constant as a function of the soil's volume wetness is only weakly dependent on soil type, bulk density, temperature, or the electrical conductivity of pore water. They offered the following empirical equation:

$$\theta = -5.3 \times 10^{-2} + 2.9 \times 10^{-2}\varepsilon_r - 5.5 \times 10^{-4}\varepsilon_r^2 + 4.3 \times 10^{-6}\varepsilon_r^3 \tag{6.12}$$

The same investigators claimed that the volume wetness of soils can be thus determined with an accuracy of ±2% and a precision (or repeatability) of ±1%. They deemed this accuracy to be sufficient for using the TDR technique for irrigation applications without having to carry out a calibration for each soil or field. They recommended that the transmission rods be spaced 50 mm apart.

The soil "sampled" or measured by TDR is essentially a cylinder whose axis lies midway between the rods and whose diameter is 1.4 times the spacing between the rods. This gives a cross section of 3800 mm^2 for rods spaced 50 mm apart. The volume of the measurement varies directly with the length of the rods. Lengths of 0.1–1.0 m are evidently practical. Davis et al. (1983) even used transmission lines that were 20 m long, placed horizontally in a sandy soil to detect pillars or columns of wet soil that resulted from leaks out of a "sealed" lagoon.

Although the velocity of propagation of the TDR pulse as it travels in the soil is evidently unaffected by the soil solution's electrical conductivity, the intensity of the transmitted signal is affected. The attenuation of the signal amplitude can therefore serve to indicate the soil's salinity (Dalton et al., 1984). However, conductivity measurements using TDR are difficult to carry out with a high level of accuracy in the field.

Other Methods

Additional approaches to the measurement of soil wetness include techniques based on the dependence of soil thermal properties on water content and the use of ultrasonic waves, radar waves, and microwaves (Topp and Ferré, 2002). Some of these and other methods have been tried for the remote sensing of land areas from aircraft or satellites (Jackson and Schmugge, 1989; Engman, 1995; Engman and Chauhan, 1995).

ENERGY STATE OF SOIL WATER

Classical physics recognizes two principal forms of energy, kinetic and potential. Since the movement of water in the soil is quite slow, its kinetic energy, which is proportional to the velocity squared, is generally considered to be negligible. On the other hand, the potential energy, which is due to position or internal condition, is of primary importance in determining the state and movement of water in the soil.

The potential energy of soil water varies over a wide range. Differences in potential energy between one point and another induce water to flow within the soil. The spontaneous and universal tendency of all matter in nature is to move from where the potential energy is higher to where it is lower and for each parcel of matter to equilibrate with its surroundings. Soil water obeys the same universal tendency toward that elusive state known as *equilibrium*, definable as a condition of uniform potential energy throughout. In the soil, water moves constantly in the direction of decreasing potential energy. The rate of decrease of potential energy with distance is in fact the moving force causing flow.

Knowledge of the relative potential energy state of soil water at each point within the soil can allow us to evaluate the forces acting on soil water in all directions and to determine how far the water in a soil system is from equilibrium. This is analogous to the well-known fact that an object will tend to fall spontaneously from a higher to a lower elevation and that in so doing it performs work but that lifting it requires an investment of work. Since potential energy is a measure of the amount of work a body can perform by virtue of the energy stored in it, knowing the potential energy state of water in the soil and in the plant growing in that soil can help us to estimate how much work the plant must expend to extract a unit amount of water.

Clearly, it is not the absolute amount of potential energy "contained" in the water that is important in itself, but rather the relative level of that energy in different regions within the soil. The concept of soil-water potential is a criterion, or yardstick, for this energy. It expresses the specific potential energy of soil water relative to that of water in a standard reference state. The standard state generally assumed is a hypothetical reservoir of pure water, at atmospheric pressure, at the same temperature as that of soil water (or at any other specified temperature), and at a given and constant elevation. Since the elevation of this hypothetical reservoir can be set at will, it follows that the potential determined by comparison with this standard is not absolute, but by employing even so arbitrary a criterion we can determine the relative magnitude of the specific potential energy of water at different locations or times within the soil.

Just as an energy increment can be viewed as the product of a force and a distance increment, so the ratio of a potential energy increment to a distance increment can be viewed as constituting a force. Accordingly, the force acting on soil water, directed from a zone of higher to a zone of lower potential, is equal to the negative *potential gradient* ($-d\phi/dx$), which is the change of energy potential ϕ with distance x. The negative sign indicates that the force acts in the direction of decreasing potential.

The concept of soil-water potential is of great fundamental importance. This concept replaces the arbitrary categorizations that prevailed in the early stages of soil physics and that purported to recognize and classify different forms of soil water, for example, "gravitational water," "capillary water," "hygroscopic water". The possible values of soil-water potential are continuous and do not exhibit any abrupt transitions from one energy level to another.

When the soil is saturated and its water is at a hydrostatic pressure greater than the atmospheric pressure (as, for instance, under a water table), the potential energy of that water may be greater than that of the reference-state reservoir described, and is therefore considered positive. Water then tends to move spontaneously from the soil into such a reservoir. If, on the other hand, the soil is unsaturated, its water is no longer free to flow out toward a reservoir at atmospheric pressure. On the contrary, the spontaneous tendency is for the soil to draw water from such a reservoir if placed in contact with it, much as a blotter draws ink (or as a napkin absorbs spilled coffee). The energy potential of soil moisture is taken to be negative.

The magnitude of the potential at any point depends not only on hydrostatic pressure, however, but also on such additional physical factors as elevation (relative to that of the reference elevation), temperature, and concentration of solutes. When solutes are present that are not uniformly distributed, osmotic effects occur. The potential energy of soil water tends to be lower where the concentration of solutes is higher.

TOTAL SOIL-WATER POTENTIAL

Thus far, we have described the energy potential of soil water in a qualitative way. More precisely, this energy potential can be regarded in terms of the difference in *partial specific free energy* between soil water and standard water. According to a soil physics terminology committee of the International Soil Science Society (Aslyng, 1963), the *total potential of soil water* is "the amount of work that must be done per unit quantity of pure water to transport reversibly and isothermally an infinitesimal quantity of water from a pool of pure water at a specified elevation at atmospheric pressure to the soil water (at the point under consideration)."

This is merely a formal definition, since in actual practice the potential is not measured by transporting water per the definition, but by measuring some other property or properties related to the potential in some known way (e.g., hydrostatic pressure, vapor pressure, elevation). The definition specifies transporting an infinitesimal quantity, in any case, to ensure that the determination procedure does not change either the reference state (i.e., the pool of pure, free water) or the soil-water potential being measured. In any case this definition provides a conceptual rather than an actual working tool.

Soil water is subject to a number of possible forces, each of which may cause its potential to differ from that of pure, free water at a reference elevation. Such force fields result from the attraction of the solid matrix for water as well as from the presence of solutes and the action of external gas pressure and gravitation. Accordingly, the total potential of soil water can be thought of as the sum of the separate contributions of these various factors, as follows:

$$\phi_t = \phi_g + \phi_p + \phi_o + \ldots \tag{6.13}$$

where ϕ_t is the total potential, ϕ_g is the gravitational potential, ϕ_p is the pressure (or matric) potential, ϕ_o is the osmotic potential, and the ellipsis signifies that additional terms are theoretically possible.

Not all of the separate potentials given in Eq. (6.13) act in the same way, and their separate gradients may not always be equally effective in causing flow (e.g., the osmotic potential gradient requires a semipermeable membrane to induce liquid flow). The gravity and pressure potentials generally pertain to the soil solution, whereas the osmotic potential pertains, strictly speaking, to the pure water substance. (The latter difference can be significant whenever the salt content of the soil solution is appreciable.) Despite these differences among the component potentials, the advantage of the total-potential concept is that it provides a unified measure by which the state of water can be evaluated in most cases everywhere within the soil–plant–atmosphere continuum.

GRAVITATIONAL POTENTIAL

Every body on the earth's surface is attracted toward the earth's center by a gravitational force equal to the weight of the body, that weight being the product of the mass of the body and the gravitational acceleration. To raise a body against this attraction, work must be expended, and this work is stored by the raised body in the form of *gravitational potential energy*. The amount of this energy depends on the body's position in the gravitational force field.

The gravitational potential of soil water at each point is determined by the elevation of the point relative to some arbitrary reference level. For the sake of convenience, it is customary to set the reference level at the elevation of a pertinent point within the soil or below the soil profile being considered (e.g., at the water table, where such exists at some definable depth) so that the gravitational potential can always be taken as positive or zero. On the other hand, if the soil surface is chosen as the reference level, as is sometimes done, then the gravitational potential for all points below the surface is negative with respect to that reference level.

At a height z above a reference, the gravitational potential energy E_g of a mass M of water occupying a volume V is

$$E_g = Mgz = \rho_w Vgz \tag{6.14}$$

where ρ_w is the density of water and g is the acceleration of gravity. Accordingly, the gravitational potential in terms of the potential energy per unit mass is

$$\phi_{g,m} = gz \tag{6.15}$$

and in terms of potential energy per unit volume is

$$\phi_{g,v} = \rho_w gz \tag{6.16}$$

The gravitational potential is independent of the chemical and pressure conditions of soil water and dependent only on relative elevation.

PRESSURE POTENTIAL

When soil water is at hydrostatic pressure greater than atmospheric, its pressure potential is considered positive. When at a pressure lower than atmospheric (a subpressure commonly known as *tension* or *suction*), the pressure potential is considered negative. Thus, water at a free-water surface is at a zero pressure potential, water below that surface is at a positive pressure potential, while water that has risen in a capillary tube (or in the pores of the soil) above that surface is at a negative pressure potential. This principle is illustrated in Fig. 6.5.

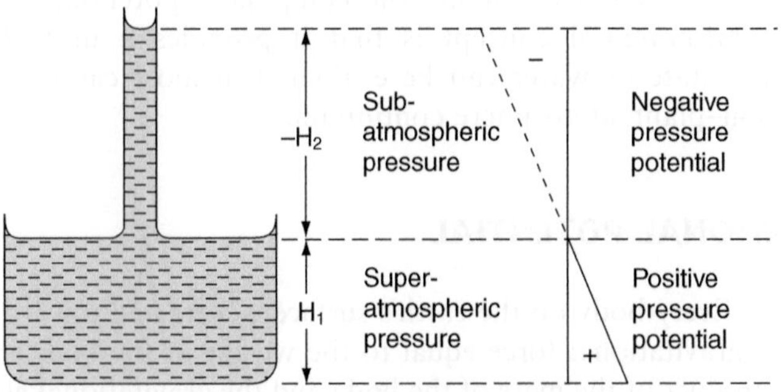

Fig. 6.5. Superatmospheric and subatmospheric pressures below and above a free-water level.

The pressure potential below the groundwater level has been termed the *submergence potential*. The hydrostatic pressure P of water in reference to atmospheric pressure is

$$P = \rho g h \qquad (6.17)$$

Here h is the piezometric head (submergence depth below the free-water surface). The potential energy of this water is then

$$E = PdV \qquad (6.18)$$

and thus the submergence potential, taken as the potential energy per unit volume, is equal to the hydrostatic pressure, P:

$$\phi_{ps} = P \qquad (6.19)$$

The negative pressure potential of soil moisture was termed *capillary potential* in the older literature of soil physics and is now called *matric potential*. It results from the interactive capillary and adsorptive forces between water and the soil matrix. These forces bind water in the soil and lower its potential energy below that of bulk water.

The terms *matric potential, matric suction*, and *soil-water suction* have been used interchangeably. According to the ISSS. committee cited earlier, the *matric suction* is defined as the negative gauge pressure, relative to the external gas pressure on soil water, to which a solution identical in composition with the soil solution must be subjected in order to be in equilibrium through a porous membrane wall with the water in the soil. This definition implies the use of either a *tensiometer* relative to the prevailing atmospheric pressure (conventionally taken to be zero) of the gas phase or a pressure plate extraction apparatus in which the gas phase is pressurized sufficiently to bring the water phase to atmospheric pressure.

As shown in Chapter 2, capillarity results from the surface tension of water and its contact angle with the solid particles. In an unsaturated (three-phase) soil system, curved menisci form, which obey the equation of capillarity:

$$P_0 - P_c = \Delta P = \gamma(1/R_1 + 1/R_2) \qquad (6.20)$$

where P_0 is the reference atmospheric pressure (taken as zero), P_c is the pressure of soil water (which can be smaller than atmospheric), ΔP is the pressure deficit (subpressure) of soil water, γ is the surface tension of water, and R_1, R_2 are the principal radii of curvature of the meniscus.

If the soil were like a simple bundle of capillary tubes, the equations of capillarity might suffice to describe the relation of the negative pressure potential (tension, or suction) to the radii of the soil pores in which the menisci are contained. However, in addition to the capillary phenomenon, the soil also exhibits adsorption, which forms hydration envelopes over the particle surfaces. These two mechanisms of soil-water affinity are illustrated in Fig. 6.6.

The presence of water in films as well as under concave menisci is most important in clayey soil and at high suctions, and it is influenced by the electric double layer and the exchangeable cations present. In sandy soils, adsorption is relatively unimportant and the capillary effect predominates. In general, however, the negative pressure potential results from the combined

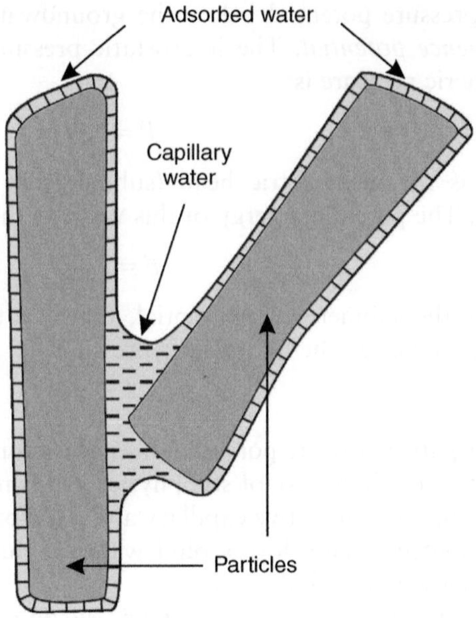

Fig. 6.6. Water in an unsaturated soil is subject to capillarity and adsorption, which combine to produce a "negative" matric potential, or a matric suction.

effect of the two mechanisms, which cannot easily be separated. The capillary "wedges" are at a state of internal equilibrium with the adsorption "films," and the ones cannot be changed without affecting the others. Hence, the older term, *capillary potential*, is inadequate and the better term is *matric potential*, because it denotes the total effect resulting from the affinity of the water to the whole matrix of the soil, including its pores and particle surfaces together.

In the absence of solute effects, the liquid and vapor phases in an unsaturated porous medium are related at equilibrium by

$$\phi_m = RT \ln(p/p_0) \tag{6.21}$$

where R is the gas constant for water vapor, T is the absolute temperature, and p/p_0 is the relative humidity (i.e., the ratio of the atmosphere's vapor pressure at equilibrium with the unsaturated medium to the "saturated" vapor pressure at equilibrium with a body of pure free water).

OSMOTIC POTENTIAL

The presence of solutes affects the thermodynamic properties of water and lowers its potential energy. In particular, solutes lower the vapor pressure of soil water. Thus,

$$\phi_m + \phi_o = RT \ln(p/p_0) \tag{6.22}$$

where ϕ_o is the osmotic potential. While this phenomenon may not affect liquid flow in the soil significantly, it does come into play whenever a membrane or diffusion barrier is present that transmits water more readily than salts. The osmotic effect is important in the interaction between plant roots and soil as well as in processes involving vapor diffusion.

There is a difference in principle between the osmotic potential and the other potential terms defined. Whereas the pressure and gravitational potentials refer to the *soil solution* (i.e., soil water along with its dissolved constituents), the osmotic potential applies to the water substance alone. Strictly speaking, therefore, ϕ_o should not simply be added to ϕ_m and ϕ_g as if those terms were similarly applicable and mutually independent. This fundamental difference can, however, be ignored in practice as long as the soil solution is dilute enough and the solutes it contains do not affect the matrix (and hence the matric potential) significantly.

QUANTITATIVE EXPRESSION OF SOIL-WATER POTENTIAL

The soil-water potential (Table 6.2) is expressible physically in at least three ways:

1. *Energy per unit mass*: This is the fundamental expression of potential, using units of ergs per gram or joules per kilogram. The dimensions of energy per unit mass are length squared per time squared: L^2T^{-2}.

2. *Energy per unit volume*: Since liquid water is only very slightly compressible in the pressure range ordinarily encountered at or near the earth's

TABLE 6.2 Energy Levels of Soil Water

Soil-water potential				Soil-water suction[a]		Vapor pressure (torr) 20°C (%)	Relative humidity[b] at 20°C (%)
Per unit mass		Per unit volume		Pressure	Head		
(erg/g)	(joule/kg)	(bar)	(mm H_2O)	(bar)	(mm H_2O)		
0	0	0	0	0	0	17.5350	100.00
-1×10^4	−1	−0.01	−102	0.01	102	17.5349	100.00
-5×10^4	−5	−0.05	−510	0.05	510	17.5344	99.997
-1×10^5	−10	−0.1	−1,020	0.1	1,020	17.5337	99.993
-2×10^5	−20	−0.02	−2,040	0.2	2,040	17.5324	99.985
-3×10^5	−30	−0.3	−3,060	0.3	3,060	17.5312	99.978
-4×10^5	−40	−0.4	−4,080	0.4	4,080	17.5299	99.971
-5×10^5	−50	−0.5	−5,100	0.5	5,100	17.5286	99.964
-6×10^5	−60	−0.6	−6,120	0.6	6,120	17.5273	99.965
-7×10^5	−70	−0.7	−7,140	0.7	7,140	17.5260	99.949
-8×10^5	−80	−0.8	−8,160	0.8	8,160	17.5247	99.941
-9×10^5	−90	−0.9	−9,180	0.9	9,180	17.5234	99.934
-1×10^6	−100	−1.0	−10,200	1.0	10,200	17.5222	99.927
-2×10^6	−200	−2	−20,400	2	20,400	17.5089	99.851
-3×10^6	−300	−3	−30,600	3	30,600	17.4961	99.778
-4×10^6	−400	−4	−40,800	4	40,800	17.4833	99.705
-5×10^6	−500	−5	−51,000	5	51,000	17.4704	99.637
-6×10^6	−600	−6	−61,200	6	61,200	17.4572	99.556

[a] In the absence of osmotic effects (soluble salts), soil-water suction equals matric suction; otherwise, it is the sum of matric and osmotic suctions.

[b] Relative humidity of air in equilibrium with the soil at different suction values.

surface, its density is practically independent of potential. Hence, there is a direct proportion between the expression of the potential as energy per unit mass and its expression as energy per unit volume. The latter expression yields the dimensions of pressure (for just as energy can be expressed as the product of pressure and volume, so the ratio of energy to volume is equivalent dimensionally to pressure). This equivalent pressure can be expressed as dynes per square centimeter, newtons per square meter, bars, atmospheres, or pascals. The basic dimensions are those of force per unit area (or mass per length per time squared): $ML^{-1}T^{-2}$. This mode of expression is convenient for the osmotic and pressure potentials, but it is seldom used for the gravitational potential.

3. *Energy per unit weight* (hydraulic head): Whatever can be expressed in units of hydrostatic pressure can also be expressed in terms of an equivalent head of water, which is the height of a liquid column corresponding to the given pressure. For example, a pressure of 1 atm is equivalent to a vertical water column, or hydraulic head, of 10.33 m and to a mercury head of 0.76 m. A pressure of 1 MPa is equivalent to a hydraulic head of 102.27 m, and 1 kPa to about 0.1 m. This mode of expression is certainly simpler, and often more convenient, than the preceding mode. Therefore, it is common to characterize the state of soil water in terms of the *total potential head*, the *gravitational potential head*, and the *pressure potential head*, which are usually expressible in meters. Accordingly, instead of

$$\phi = \phi_g + \phi_p \tag{6.23}$$

we could write

$$H = H_g + H_p \tag{6.24}$$

which reads: The total potential head of soil water (H) is the sum of the gravitational (H_g) and pressure (H_p) potential heads. H is commonly called the *hydraulic head.*

The use of various expressions for the soil-water potential can be perplexing to the uninitiated. However, these alternative expressions are in fact equivalent, and each of them can readily be translated into any of the others. If we use ϕ to designate the potential in terms of energy per mass, P for the potential in terms of pressure, and H for the potential head, then

$$\phi = P/\rho_w \tag{6.25}$$

$$H = P/\rho_w g = \phi/g \tag{6.26}$$

where ρ_w is the density of water and g is the acceleration of gravity.

A remark is in order concerning the use of the synonymous terms *tension* and *suction* in lieu of "negative" (or "subatmospheric") pressure. Tension and suction are merely semantic devices to avoid the use of the unesthetic negative sign, which generally characterizes the pressure of water in an unsaturated soil. These terms allow us to speak of the osmotic and matric potentials in positive terms. The two potentials are illustrated schematically in Fig. 6.7.

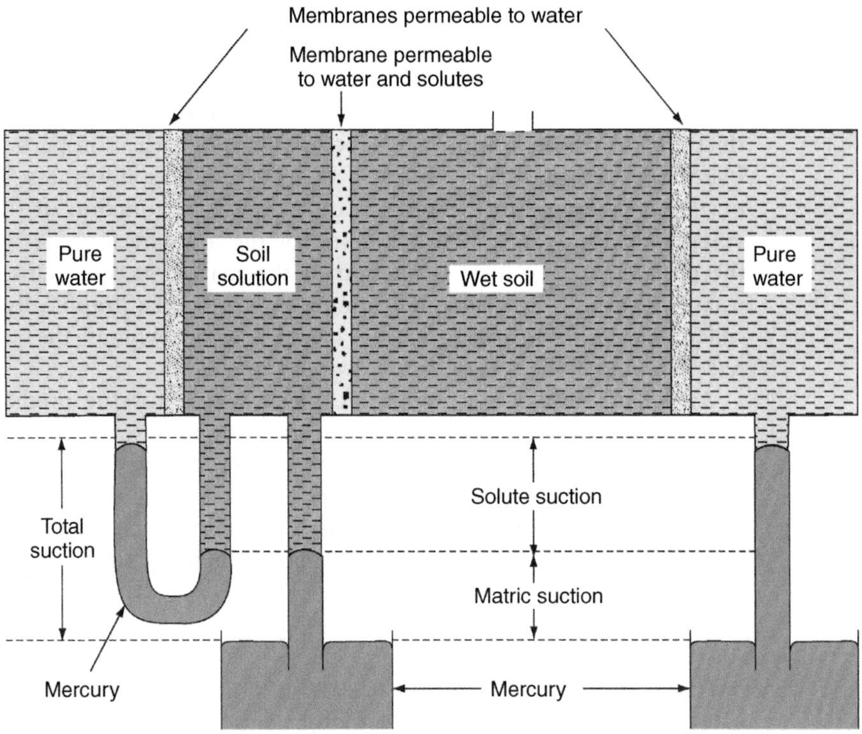

Fig. 6.7. In an isothermal system at equilibrium, matric suction is the pressure difference across a membrane separating the soil solution in situ from the same solution in bulk, with the membrane being permeable to the solution but not to solid particles or air. Osmotic (solute) suction is the pressure difference across a semipermeable membrane separating the bulk phases of pure water and the soil solution. Total suction is the sum of the matric and osmotic suction values and is thus the pressure difference across a semipermeable membrane separating pure water from a soil that contains a solution (After Richards, S. J., 1965.)

SOIL-MOISTURE CHARACTERISTIC CURVE

Water in a saturated soil is at atmospheric pressure if it is at equilibrium with free water at the same elevation. That implies zero hydrostatic pressure and hence zero suction. If a slight suction (i.e., a water pressure slightly subatmospheric) is applied to water in a saturated soil, no outflow may occur until, as suction is increased, a critical value is exceeded at which the largest surface pore begins to empty and its water content is displaced by air. This critical suction is called the *air-entry suction.* Being the threshold of desaturation, air-entry suction is generally small in coarse-textured and in well-aggregated soils having large pores; but it tends to be larger in dense, poorly aggregated, medium-textured or fine-textured soils. Since in coarse-textured soils the pores tend to be fairly uniform in size, these soils typically exhibit critical air-entry phenomena more distinctly than do soils with a wider array of pore sizes.

As suction is applied incrementally, the first pores to be emptied are the relatively large ones that cannot retain water against the suction applied.

Recalling the capillary equation ($-P = \psi = 2\gamma/r$), we can readily predict that a gradual increase in suction will result in the emptying of progressively smaller pores until, at high suction values, only the very narrow pores retain water. Similarly, an increase in soil-water suction is associated with decreasing thickness of the hydration envelopes adsorbed to the soil-particle surfaces. Increasing suction is thus associated with decreasing soil wetness. The amount of water remaining in the soil at equilibrium is a function of the sizes and volumes of the water-filled pores and of the amount of water adsorbed to the particles; hence it is a function of the matric suction. This function is usually measured experimentally, and it is represented graphically by a curve called the *soil-moisture retention curve*, also known as the *soil-moisture release curve* or the *soil-moisture characteristic.*

As yet, no universally applicable theory exists for the prediction of the matric suction versus wetness relationship from basic soil properties (i.e., texture and structure). The adsorption and pore geometry effects are generally too complex to be described by a simple model. Several empirical equations have been proposed that describe the soil-moisture characteristic for some soils and within limited suction ranges. One such equation was advanced by Visser (1966):

$$\psi = a(f - \theta)^b/\theta^c \tag{6.27}$$

Here ψ is matric suction, f is porosity, θ is volumetric wetness, and *a, b, c* are constants. Use of this equation is hampered by the difficulty of evaluating its constants. Visser found that *b* varied from 0 to 10, *a* from 0 to 3, and *f* from 0.4 to 0.6.

Alternative equations to describe the relationship between wetness and matric suction have been proposed by Laliberte (1969), White et al. (1970), Su and Brooks (1975), and van Genuchten (1978). An equation presented by Brooks and Corey (1966a) is

$$(\theta - \theta_r)/(\theta_m - \theta_r) = (\psi_e/\psi)^\lambda \tag{6.28}$$

for suction values greater than the air-entry suction ψ_e. The exponent λ has been termed the *pore size distribution index*. In this equation, θ is the volume wetness (a function of the suction ψ), θ_m is the maximum wetness (saturation), and θ_r is the “residual” wetness remaining at high suction in the small pores that do not form a continuous network (the intra-aggregate pores).

Physically based water retention prediction models have recently been described by Haverkamp and Reggiami (2002) and by Nimmo (2002).

The amount of water retained at low values of matric suction (say, between 0 and 100 kPa) depends on capillarity and the pore size distribution. Hence it is strongly affected by the soil’s structure. At higher suctions, water retention is due increasingly to adsorption, so it is influenced less by the structure and more by the soil’s texture and specific surface. The greater the clay content, in general, the greater the water retention at any particular suction and the more gradual the slope of the curve. In a sandy soil, most of the pores are relatively large, and once these large pores are emptied at a given suction, only a small amount of water remains. In a clayey soil, more of the water is adsorbed, so increasing the matric suction causes a more gradual decrease in wetness (Fig. 6.8).

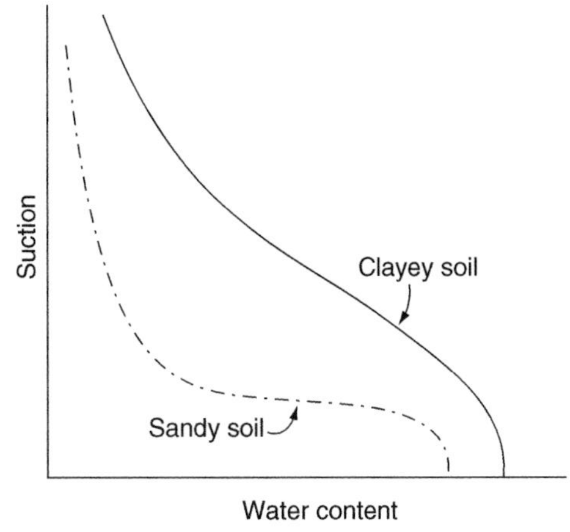

Fig. 6.8. The effect of texture on soil-water retention.

Because soil structure influences the shape of the soil-moisture characteristic curve primarily in the low-suction range, we may expect that the effect of compaction (which destroys the aggregated structure) is to reduce the total porosity and especially the volume of the large interaggregate pores. As a result of compaction, the saturation water content as well as the initial decrease of water content with the application of low suction are diminished. On the other hand, the volume of intermediate-size pores is likely to be greater in a compact soil (because some of the originally large pores have been squeezed into intermediate size by compaction), while the micropores remain unaffected, and thus the curves for the compacted and uncompacted soil tend to converge in the high-suction range (Fig. 6.9).

If two soil bodies differing in texture, structure, and initial wetness are brought into direct physical contact, water from one will normally move into

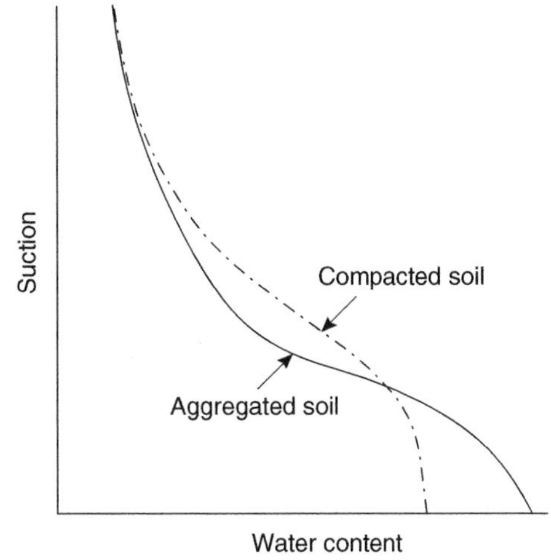

Fig. 6.9. The effect of soil structure on soil-water retention.

the other in a tendency toward potential energy equilibrium. Even at equilibrium, however, the two different soils will not generally be equal in wetness. Rather, each soil will end up with a content of water determined by its own soil-moisture characteristic. Two soil bodies or layers can thus attain equilibrium and yet exhibit a marked difference in wetness. We can easily envision a situation in which a drier soil layer will contribute water to a wetter one if the water potential of the former is higher than that of the latter, owing to textural, structural, elevational, osmotic, or thermal differences.

The slope of the soil-moisture characteristic curve, which is the change of water content per unit change of matric potential, is generally termed the *differential* (or *specific*) *water capacity* c_θ:

$$c_\theta = d\theta/d\phi_p \qquad \text{or} \qquad c_\theta = -d\theta/d\psi \tag{6.29}$$

The c_θ term is analogous to the well-known *differential heat capacity*, which is the change in the heat content of a body per unit change in the thermal potential (temperature). However, while the differential heat capacity is fairly constant with temperature for many materials, the differential water capacity in soils is strongly dependent on the matric potential.

HYSTERESIS

The relation between matric potential and soil wetness can be obtained in two ways: (1) in desorption, by taking an initially saturated sample and applying increasing suction, in stepwise manner, to gradually dry the soil while taking successive measurements of wetness versus suction; and (2) in sorption, by gradually wetting up an initially dry soil sample while reducing the suction incrementally. Each of these two methods yields a continuous curve, but the two curves will in general not be identical. The equilibrium soil wetness at a given suction is greater in desorption (drying) than in sorption (wetting). This dependence of the equilibrium content and state of soil water upon the direction of the process leading up to it is called *hysteresis* (Haines, 1930; Miller and Miller, 1956; Mualem, 1984).

Figure 6.10 shows a typical soil-moisture characteristic curve and illustrates the hysteresis effect in the relationship between soil wetness and matric suction at equilibrium.

The hysteresis effect may be attributed to several causes:

1. The geometric nonuniformity of the individual pores (which are generally irregularly shaped voids interconnected by smaller passages), resulting in the "ink bottle" effect, illustrated in Fig. 6.11.
2. The contact-angle effect, mentioned in Chapter 2, by which the contact angle is greater and hence the radius of curvature greater in the case of an advancing meniscus than in the case of a receding one. A given water content will tend therefore to exhibit greater suction in desorption than in sorption. (Contact-angle hysteresis can arise because of surface roughness, the presence of adsorbed impurities on the solid surface, and the mechanism by which liquid molecules adsorb or desorb when the interface is displaced.)

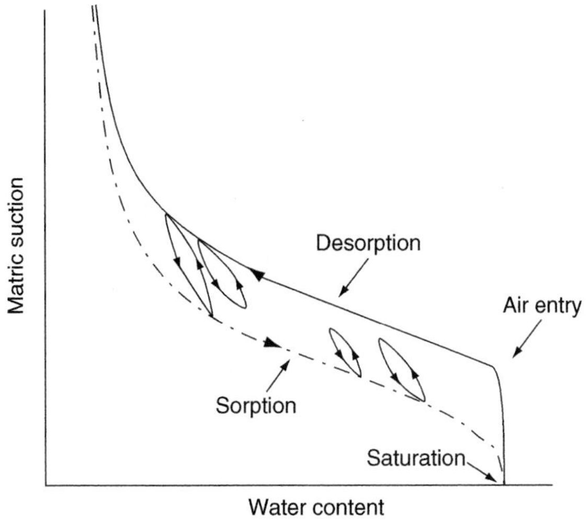

Fig. 6.10. Suction vs. water content curves in sorption and desorption. The intermediate loops are scanning curves, representing complete or partial transitions between the main branches.

3. Entrapped air, which further decreases the water content of newly wetted soil. Failure to attain true equilibrium (although not, strictly speaking, true hysteresis) can accentuate the hysteresis effect.
4. Swelling, shrinking, or aging phenomena, which result in differential changes of soil structure, depending on the wetting and drying history of the sample (Hillel and Mottes, 1966). The gradual solution of air or the release of dissolved air from soil water can also have a differential effect upon the suction–wetness relationship in wetting and drying systems.

Of particular interest is the ink bottle effect. Consider the hypothetical pore shown in Fig. 6.11. This pore consists of a relatively wide void of radius R, bounded by narrow channels of radius r. If initially saturated, this pore drains abruptly the moment the suction exceeds ψ_r, where $\psi_r = 2\gamma/r$. For this pore to rewet, however, the suction must decrease to below ψ_R, where $\psi_R = 2\gamma/R$, whereupon the pore abruptly fills. Since $R > r$, it follows that $\psi_r > \psi_R$.

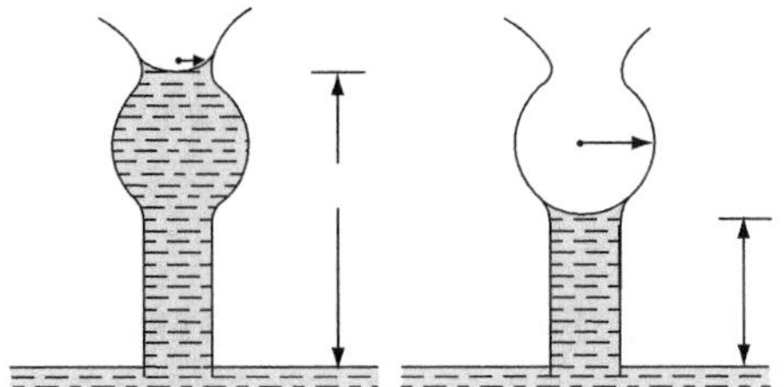

Fig. 6.11. The "ink bottle" effect determines the equilibrium height of water in a variable-width pore: (a) in capillary drainage (desorption) and (b) in capillary rise (sorption).

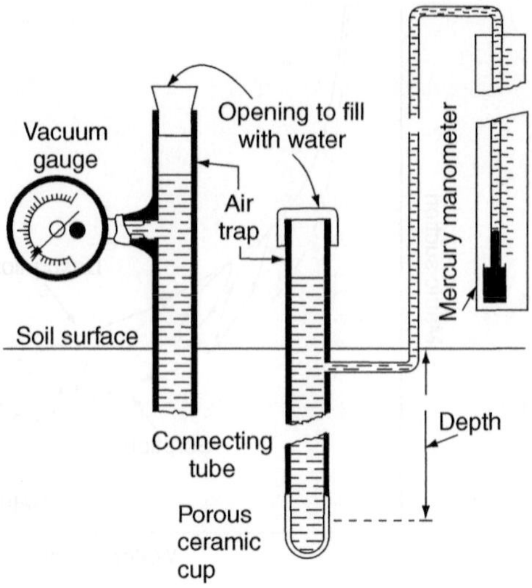

Fig. 6.12. The essential parts of a tensiometer. (After Richards, S. J., 1965.)

Desorption depends on the narrow radii of the connecting channels, whereas sorption depends on the diameters of the large pores. These discontinuous spurts of water, called *Haines jumps* (after W. B. Haines, who first noted the phenomenon in 1930), can be observed readily in coarse sands. The hysteresis effect is generally more pronounced in coarse-textured soils in the low-suction range, where pores empty at an appreciably larger suction than that at which they fill.

The two complete characteristic curves, the one describing the gradual transition from saturation to dryness and the other portraying the transition from dryness to saturation, are called the *main branches* of the hysteretic soil-moisture characteristic. When a partially wetted soil commences to drain or when a partially desorbed soil is rewetted, the relation of suction to moisture content follows some intermediate curve as it "scans" from one main branch to the other. Such intermediate spurs are called *scanning curves*. Cyclic changes often entail wetting and drying scanning curves, which may form loops between the main branches (Fig. 6.12). The ψ–θ relationship can thus become very complicated. Indeed, because of its complexity, the hysteresis phenomenon is too often ignored. The soil-moisture characteristic commonly reported in the literature is the *desorption curve*, also known as the *soil-moisture release curve*. The *sorption curve* is equally important but more difficult to determine, so it is seldom reported.

MEASUREMENT OF SOIL-MOISTURE POTENTIAL

The measurement of soil wetness, described earlier in this chapter, though essential in many soil physical and engineering investigations, is obviously not sufficient to provide a description of the state of soil water. To obtain such a description, evaluation of the energy status of soil water (soil-moisture potential or suction) is necessary. In general, the twin variables of wetness and potential should each be measured directly, because the translation of one to the other on the basis of calibration curves of soil samples is too often unreliable.

Total soil-moisture potential is often thought of as the sum of matric and osmotic (solute) potentials and is a useful index for characterizing the energy status of soil water with respect to plant water uptake. The sum of the matric and gravitational (elevation) heads is generally called the *hydraulic head* (or hydraulic potential) and is useful in evaluating the directions and magnitudes of the water-moving forces throughout the soil profile. Methods are available for measuring matric potential as well as total soil-moisture potential, separately or together (Dane and Hopmans, 2002). A schematic representation of the components of the soil-moisture potential is shown in Fig. 6.7. To measure matric potential in the field, an instrument known as a *tensiometer* is used; in the laboratory, use is often made of tension plates and of air-pressure extraction cells. Total soil-moisture potential can be obtained by measuring the equilibrium vapor pressure of soil water by means of thermocouple psychrometers.

We next describe the tensiometer, which has won acceptance as a practical device for in situ measurement of matric suction, hydraulic head, and hydraulic gradients.

The Tensiometer

The tensiometer is an instrument designed to provide a continuous indication of the soil's matric suction (also called *soil-moisture tension*) in situ. The essential parts of a tensiometer are shown in Fig. 6.12. The instrument consists of a porous cup, generally of ceramic material, connected through a tube to a manometer, with all parts filled with water. When the cup is placed in the soil where the suction measurement is to be made, the bulk water inside the cup comes into hydraulic contact and tends to equilibrate with soil water through the pores in the ceramic walls. When the tensiometer is initially placed in the soil, the water contained in it is generally at atmospheric pressure. Soil water, being generally at subatmospheric pressure, exercises a suction, which draws out a certain amount of water from the rigid and airtight tensiometer, thereby causing the pressure inside the tensiometer to fall below atmospheric pressure. This subpressure is indicated by a manometer, which may be a simple water- or mercury-filled U tube, a vacuum gauge, or an electrical transducer (Young and Sisson, 2002).

A tensiometer kept in place tends to track the changes in the soil's matric suction. As soil moisture is depleted by drainage or plant uptake or as it is replenished by rain or irrigation, corresponding readings on the tensiometer gauge occur. Owing to the hydraulic resistance of the cup and the surrounding soil and of the contact zone between the cup and the soil, the tensiometer response may lag behind suction changes in the soil. The lag time can be minimized by the use of a null-type device or a transducer-type manometer with rigid tubing, so practically no flow of water need take place as the instrument responds to changes in the soil.

Since the walls of the tensiometer's porous cup are permeable to both water and solutes, the solutes in the soil solution diffuse freely into the porous cup, so water inside the tensiometer tends to assume the same solute composition and concentration as soil water. Therefore the instrument does not indicate the osmotic suction of soil water (unless it is equipped with some type of an auxiliary salt sensor).

Suction measurements by tensiometry are limited to matric suction values below 1 atm (about 1 bar, or 100 kPa). This is due first of all to the fact that the vacuum gauge or manometer measures a partial vacuum relative to the external atmospheric pressure. Even if the gauging of pressure is independent of atmospheric pressure (e.g., if it is done by means of a strain-gauge manometer), we are still faced with the general failure of water columns in macroscopic systems to withstand tensions exceeding 1 atm. (Water in narrow capillaries, however, can maintain continuity at much higher tensions; witness, for example, the continuity of liquid water in the xylem vessels of tall trees.) Furthermore, because the ceramic cup is generally made of rather permeable and porous material, in the interest of promoting rapid equilibrium with soil moisture, higher suction may cause the entry of air from the soil into the cup. Such air entry equalizes the internal tensiometric pressure to the atmospheric pressure. Consequently, soil suction may continue to increase even though the tensiometer fails to show it.

In practice, the useful limit of most tensiometers is a maximal tension of about 0.8 atm (80 kPa). However, the limited range of suction measurable by the tensiometer is not as serious a problem as it may seem. Though the suction range of 0–0.8 atm is a small part of the total range of suction variation encountered in the field, it actually encompasses the greater part of the soil wetness range. In many agricultural soils the tensiometer range accounts for more than 50% (and in coarse-textured soils 75% or more) of the amount of soil water available to plants. Thus, where soil management (as in irrigation) is aimed at maintaining the low-suction conditions that are most favorable for plant growth, tensiometers are definitely useful.

The Thermocouple Psychrometer

At equilibrium, the potential of soil moisture is equal to the potential of the water vapor in the ambient air. If thermal equilibrium is ensured and the gravitational effect is neglected, the vapor potential can be taken to be equal to the sum of the matric and osmotic potentials, since air acts as an ideal semipermeable membrane in allowing only water molecules to pass (provided that the solutes are nonvolatile).

A *psychrometer* is an instrument designed to indicate the *relative humidity*, that is, the ratio of the partial pressure of water vapor in the air to the equilibrium partial pressure of vapor in vapor-saturated air at the same temperature. The instrument generally measures the difference between the temperatures registered by a wet bulb and a dry bulb thermometer. The *dry bulb thermometer* indicates the temperature of a nonevaporating surface in thermal equilibrium with the ambient air. The *wet bulb thermometer* indicates the generally lower temperature of an evaporating surface, where latent heat is absorbed in proportion to the rate of evaporation.

The relative humidity of a body of air in equilibrium with a moist soil will depend on the temperature as well as on the state of water in the soil — that is, on the constraining effects of adsorption, capillarity, and solutes, all of which act to reduce the evaporability of soil water relative to that of pure, free water at the same temperature. Hence the relative humidity of an unsaturated soil's atmosphere will generally be under 100%, and the deficit to "saturation" will depend on the soil moisture potential, or suction, due to the combined effects of the osmotic and matric potentials.

Recent decades have witnessed the development of highly precise, miniaturized thermocouple psychrometers that make possible the in situ measurement of soil moisture potential (Rawlins and Campbell, 1986; Andraski and Scanlon, 2002). A *thermocouple* is a double junction of two dissimilar metals. If the two junctions are subjected to different temperatures, they will generate a voltage difference. If, on the other hand, an electromotive force (emf) is applied between the junctions, a difference in temperature will result. Depending on which way a direct current is applied, one junction can be heated while the other is cooled, and vice versa. The soil psychrometer (Fig. 6.13) consists of a fine wire thermocouple, one junction of which is equilibrated with the soil atmosphere by placing it inside a hollow porous cup embedded in the soil while the other junction is kept in an insulated medium to provide a temperature lag.

During operation, an electromotive force is applied so the junction exposed to the soil atmosphere is cooled to a temperature below the dew point of that atmosphere. At this point a droplet of water condenses on the junction, allowing it to serve as a wet bulb thermometer. This is a consequence of the so-called *Peltier effect* (Yavorsky and Detlaf, 1972). The cooling is then stopped; as the water from the droplet reevaporates, the junction attains a wet bulb temperature. That temperature remains nearly constant until the junction dries out, after which it returns to the ambient soil temperature. While evaporation takes place, the difference in temperature between the "wet bulb" and the insulated

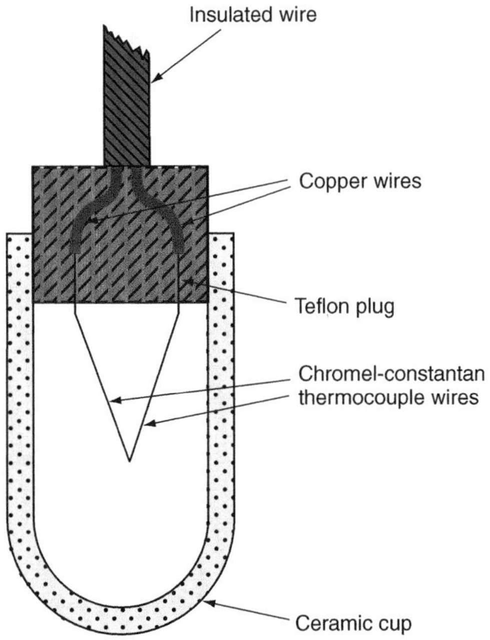

Fig. 6.13. Cross section of a thermocouple psychrometer contained in an air-filled ceramic cup.

junction serving as "dry bulb" generates an emf, which is indicative of the soil moisture potential. The relative humidity (i.e., the vapor pressure relative to that of pure, free water) is related to the soil-water potential according to

$$\phi = RT \ln(p/p_0) \tag{6.30}$$

where p is the vapor pressure of soil water, p_0 is the vapor pressure of pure, free water at the same temperature and air pressure, and R is the specific gas constant for water vapor.

Thermocouple psychrometry can be quite useful considerably beyond the suction range of the tensiometer (i.e., from 0.2 to 5 MPa, or 2 to 50 bar), where the change of wetness per unit change of suction is quite small. This instrument is used mostly in research and is now manufactured commercially. It is also applied to the measurement of plant-water potential.

Measurement of the Soil-Moisture Characteristic Curve

The fundamental relation between soil wetness and matric suction is often determined by means of a tension-plate assembly (Fig. 6.14) in the

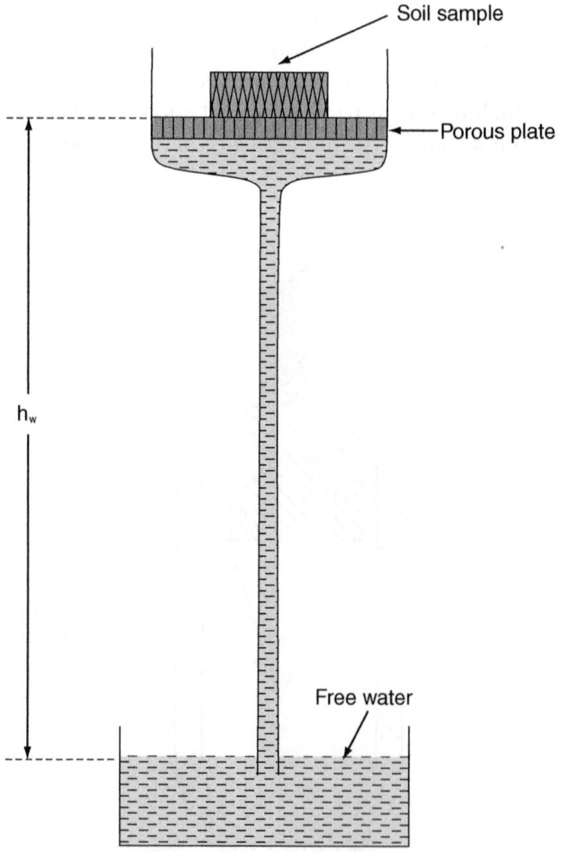

Fig. 6.14. Tension-plate assembly for equilibrating a soil sample with a known matric suction value. This assembly is applicable in the range of 0–1 bar only.

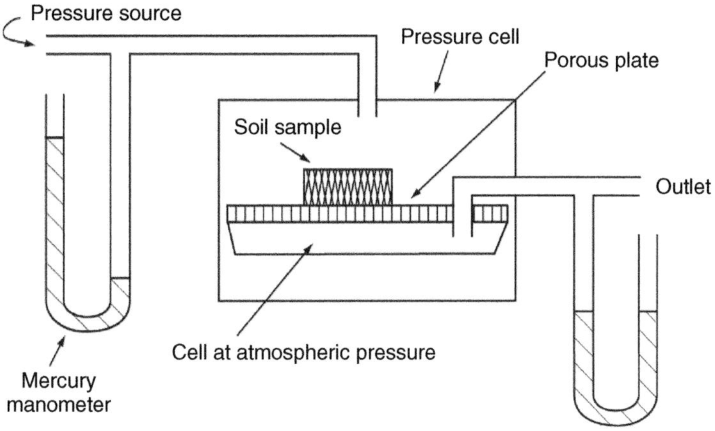

Fig. 6.15. Pressure-plate apparatus for moisture characteristic measurements in the high-suction range. The lower side of the porous plate is in contact with water at atmospheric pressure. Pressurized air is used to extract water from initially saturated soil samples.

low-suction (<1 bar) range and by means of a pressure-plate or pressure-membrane apparatus (Fig. 6.15) in the higher-suction range. These instruments allow the application of successive suction values and the repeated measurement of the equilibrium soil wetness at each suction (Dane and Hopmans, 2002).

The suction value obtainable by tension-plate devices is limited to 1 atm (1 bar, or 100 kPa) if the soil air is kept at atmospheric pressure and the pressure difference across the plate is controlled by vacuum or by a hanging water column. Matric suction values considerably greater than I bar (say, 20 bar or even more) can be obtained by increasing the pressure of the air phase. This requires placing the porous-plate assembly inside a pressure chamber, as shown in Fig. 6.15. The limit of matric suction obtainable with such a device is determined by the design of the chamber (i.e., its safe working pressure) and by the maximal air pressure difference the saturated porous plate can bear without allowing air to bubble through its pores. Ceramic plates generally do not hold pressures greater than about 2 MPa (20 bar), but cellulose acetate membranes can be used with pressures as high as 10 MPa (100 bar).

Soil-moisture retention in the low-suction range (0–100 kPa, equivalent to 0–1 bar) is strongly influenced by soil structure and pore size distribution. Hence, measurements made with disturbed samples (e.g., dried, screened, and artificially packed samples) cannot be expected to represent field conditions. The use of undisturbed soil cores is therefore preferable. Even better, in principle, is the in situ determination of the soil moisture characteristic curve by making simultaneous measurements of wetness (e.g., with the neutron moisture meter or with TDR) and of suction (using tensiometers) in the field (Bruce and Luxmoore, 1986). However, this approach is often frustrated by soil heterogeneity and by uncertainties over equilibrium as well as hysteretic phenomena as they might occur in the field.

REMOTE SENSING OF SURFACE-ZONE SOIL MOISTURE

Remote sensing is the collection of information regarding an object of interest, conducted from some distance without actual contact with that object. It is usually accomplished by detecting and measuring various portions (or bands) of the electromagnetic spectrum, using airborne or satellite-borne electronic scanning devices. Remote sensing of the earth's surface includes aerial photography, multispectral imagery, infrared imagery, radar, and microwave scanning. These techniques may be passive or active. Passive techniques measure signals emitted spontaneously from the ground. Active sensing techniques consist of generating a signal that is sent to the ground and of measuring its response.

Remote sensing is a promising technique in environmental soil physics and hydrology (Engman, 1995). It permits measurements that are not generally possible with traditional techniques, especially in areas where on-site stations are sparse and where data are difficult to obtain otherwise.

Information regarding the varying moisture content of the soil's surface zone is of particular interest. However insignificant this thin layer, with the small amount of water contained in it, may seem in comparison with the total amount of water on earth, it is indicative of the entire interaction between the land surface and the atmosphere as it regulates such processes as the exchange of energy and the allocation of precipitated water among infiltration, runoff, evaporation, and transpiration. Moreover, topsoil moisture affects seed germination, subsequent plant growth, and — ultimately — the success or failure of agriculture.

Of the various techniques suggested for measuring soil moisture, microwave technology appears to be the most promising at present. It can be used from a space platform (as well as from aircraft and truck-mounted devices) and can provide quantitative data of moisture in the soil's top layer (of the order of 5 cm) under a variety of topographic and vegetative conditions. The theoretical basis for measuring soil moisture by microwave techniques is based on the contrast between the dielectric constant of liquid water and of dry soil particles (namely, 80 compared to 3–5). That contrast is due to the tendency of the polarity of water molecules (i.e., their ability to align themselves in an applied electromagnetic field). As soil moisture increases, so does the soil's dielectric constant.

Passive microwave remote sensing of top-layer moisture has been shown to be practical. In the case of a bare surface, a radiometer can be used directly to measure the intensity of emission from the soil. According to Schmugge (1990), this emission is proportional to the product of surface temperature and surface emissivity, commonly referred to as the *microwave brightness temperature* (T_b):

$$T_b = \tau(H)[rT_{sky} + (1 - r)T_{soil}] + T_{atm} \tag{6.31}$$

where $\tau(H)$ is the atmospheric transmissivity for a radiometer at height H above the ground, r is the smooth-surface reflectivity, T_{soil} is the temperature of the soil, T_{atm} is the average temperature of the atmosphere, and T_{sky} is the contribution from the reflected sky brightness. For microwave wavelengths greater than 5 cm, the atmospheric transmissivity approaches 99% and the values of T_{sky} and T_{atm} are both less than 5 K (i.e., small compared to the soil contribution). Assuming the last terms to be negligible, we can simplify the equation to

$$T_b = (1 - r)T_{soil} = \varepsilon T_{soil} \tag{6.32}$$

where $\varepsilon = (1 - r)$ is the emissivity, which depends primarily on the dielectric constant of the soil and secondarily on the surface roughness.

The reflection coefficient of the ground is related nonlinearly to the dielectric constant and hence also to soil moisture. Therefore, most algorithms used for this type of remote sensing of soil moisture are calibrated with reference to empirical ground data (Engman, 1995). There is also some uncertainty over the exact thickness of the soil layer that is effectively monitored. The *penetration depth*, defined as the distance in the soil over which microwave radiated power is attenuated by a factor of e (= 2.718), also depends on the wavelength as well as on the dielectric constant (and hence on soil moisture). For a wavelength of 5 cm, the penetration depth varies from 10 cm for a soil with 1% volumetric wetness to less than 1 cm for a soil with 30% moisture. For a wavelength of 20 cm, the penetration depth varies from about 40 cm with 1% moisture to 5 cm with 30% moisture by volume (Njoku and Kong, 1977).

For the active microwave approach, the measured radar backscatter, σ_t, is made up of the backscatter from vegetation, σ_v, and from the soil, σ_s, and the attenuation caused by the vegetation canopy L:

$$\sigma_t = \sigma_v + \sigma_s/L \tag{6.33}$$

The soil backscatter can be related directly to soil moisture by

$$\sigma_s = R\alpha\theta \tag{6.34}$$

where R is a surface roughness term, α is a soil moisture sensitivity term, and θ is the volumetric moisture content. Because R and α are known to vary with wavelength, polarizataion, and angle of incidence, it is difficult to estimate these terms independently. Hence, as in the case of the passive microwave approach, the relationship between measured backscatter and soil moisture requires an empirical correlation with ground data, even for bare soils.

The effect of vegetation is, on the one hand, to attenuate the microwave emission from the soil and, on the other hand, to contribute its own emission to the total radiative flux. The net effect depends on the structure and density of the vegetative cover as well as on its degree of hydration. The interference of vegetation with the measurement of soil moisture can be minimized by choosing the optimal spectral band, because plant canopies are evidently more transparent to longer wavelengths than to shorter wavelengths in the range of microwave studies (Jackson and Schmugge, 1991).

7. WATER FLOW IN SATURATED SOIL

LAMINAR FLOW IN NARROW TUBES

Before beginning a discussion of flow in so complex a medium as soil, let us first consider the basics of fluid flow in narrow tubes. Early theories of fluid dynamics were based on the hypothetical concept of a "perfect" fluid, both frictionless and incompressible. In such a fluid, contacting layers exhibit no *tangential forces* (shearing stresses), only *normal forces* (pressures). Such fluids do not in fact exist. In the flow of real fluids, adjacent layers do transmit tangential stresses (drag), and the existence of intermolecular attraction causes the molecules in contact with a solid wall to adhere to it rather than slip over it. The flow of a real fluid is associated with the property of *viscosity*, defined in Chapter 2.

We can visualize the nature of viscosity by considering the motion of a fluid between two parallel plates, the lower one at rest, the upper one moving at a constant velocity (Fig. 7.1). Experience shows that the fluid adheres to both walls, so its velocity at the lower plate is zero, and that at the upper plate is equal to the velocity of the plate. Furthermore, the velocity distribution in the fluid between the plates is linear, so the fluid velocity at any plane between the plates is proportional to the distance y from the lower plate.

Maintaining the relative motion of the plates at a constant velocity requires applying a tangential force to overcome the frictional resistance in the fluid. This resistance, per unit area of the plate, is proportional to the velocity of the upper plate U and inversely proportional to the distance h between the plates. The shearing stress τ_s at any point is proportional to the velocity gradient du/dy. The viscosity η is the proportionality factor between τ_s and du/dy:

$$\tau_s = \eta \, du/dy \tag{7.1}$$

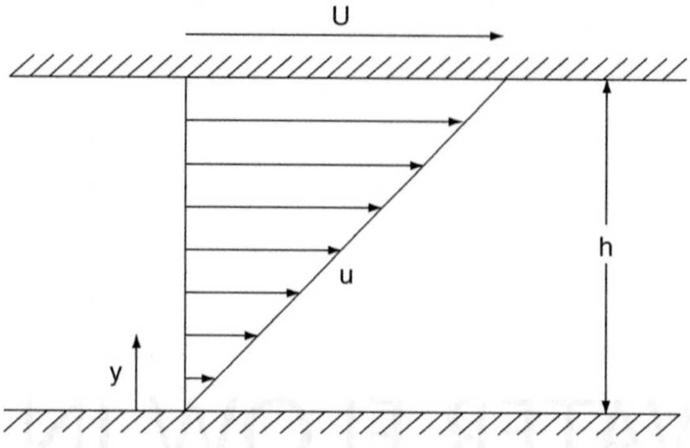

Fig. 7.1. Velocity distribution in a viscous fluid between two parallel plates, the upper one moving at a velocity U relative to the lower one.

Equation (7.1) bears a formal similarity to Hooke's law of elasticity. However, in an elastic solid the shearing stress is proportional to the strain, whereas in a viscous fluid the shearing stress is proportional to the time rate of the strain (i.e., the velocity gradient).

Now let us consider flow through a straight, cylindrical tube of uniform diameter $D = 2R$ (Fig. 7.2). The velocity is zero at the wall (because of adhesion), maximal at the axis, and constant on cylindrical surfaces that are concentric about the axis. Adjacent cylindrical laminae, moving at different velocities, slide over each other. A parallel motion of this kind is called *laminar*. Fluid flow in a horizontal tube is normally caused by a pressure gradient acting in the axial direction. A fluid "particle" is accelerated by the pressure gradient and retarded by the frictional resistance.

Imagine a coaxial fluid cylinder of length L and radius y. For flow velocity to be constant (no net force, hence no acceleration), the pressure force acting on the face of the cylinder $\Delta p\ \pi y^2$ (in which $\Delta p = p_1 - p_2$) must be equal to the frictional resistance due to the shear force $2\pi y L \tau_s$ acting on the circumferential area. Thus,

$$\tau_s = (\Delta p/L)(y/2)$$

Recalling Eq. (7.1), $\tau_s = -\eta\ (du/dy)$ (the negative sign arises because in this case u decreases with increasing y), we obtain

$$du/dy = -(\Delta p/\eta L)(y/2)$$

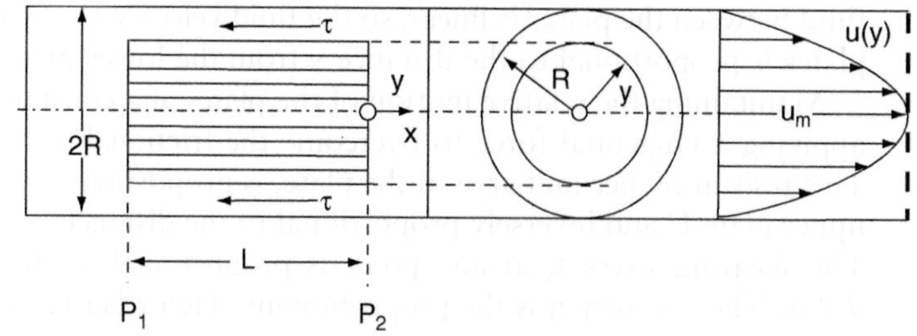

Fig. 7.2. Laminar flow through a cylindrical tube.

which, on integration, gives

$$u(y) = -(\Delta p/\eta L)(c - y^2/4)$$

The constant of integration c is evaluated from the boundary condition of no slip at the wall ($u = 0$ at $y = R$), so $c = R^2/4$. Therefore,

$$u(y) = (\Delta p/4\eta L)(R^2 - y^2) \tag{7.2}$$

Equation (7.2) indicates that the velocity is distributed parabolically over the radius, with maximum velocity u_{max} on the axis ($y = 0$):

$$u_{max} = \Delta p\ R^2/4\eta L$$

The discharge Q, that is, the volume flowing through a section of length L per unit time, can now be evaluated. The volume of a paraboloid of revolution is (1/2)(base × height). Hence

$$Q = (1/2)\pi R^2 u_{max} = \pi R^4 \Delta p/8\eta L \tag{7.3}$$

This equation, known as *Poiseuille's law*, indicates that the volume flow rate is proportional to the pressure drop per unit distance ($\Delta p/L$) and the fourth power of the radius of the tube.

The mean velocity over the cross section is

$$u_{mean} = \Delta p R^2/8\eta L = (R^2/a\eta)\nabla p \tag{7.4}$$

where ∇p is the pressure gradient ($\Delta p/L$). Parameter a, equal to 8 in a circular tube, varies with the shape of the conducting passage.

Laminar flow prevails only at relatively low flow velocities and in narrow tubes. As the radius of the tube and the flow velocity are increased, the point is reached at which the mean flow velocity is no longer proportional to the pressure drop, and the parallel *laminar flow* changes into a *turbulent flow* with fluctuating eddies. Conveniently, however, laminar flow is the rule rather than the exception in most water flow processes taking place in soils, because of the narrowness of soil pores.

Sample Problem

The water in an irrigation hose is at a hydrostatic pressure of 100 kPa (1 bar). Five drip-irrigation emitters are inserted into the wall of the hose. Calculate the drip rate (L/hr) from the emitters if each contained a coiled capillary tube 1 m long and the capillary diameters are 0.2, 0.4, 0.6, 0.8, and 1.0 mm. Assuming laminar flow, what fraction of the total discharge is due to the single largest emitter?

We use Poiseuille's law to calculate the discharge: $Q = \pi R^4\ \Delta P/8\eta L$. Substituting the values for π (3.14), the pressure differential ΔP (10^5 N/m^2 = 10^5 kg/m sec^2), the viscosity η (10^{-3} kg/m sec, at 20°C), the capillary tube length (1 m), and the appropriate tube radii (0.1, 0.2, 0.3, 0.4, and 0.5 mm), we obtain

$$Q_1 = \frac{3.14 \times (10^{-4}\ \text{m})^4 \times 10^5\ \text{kg/m sec}^2}{8 \times 10^{-3}\ \text{kg/m sec} \times 1\ \text{m}} = 3.9 \times 10^{-9}\ \text{m}^3/\text{sec}$$

$$= 0.014\ \text{L/hr}$$

$$Q_2 = \frac{3.14 \times (2 \times 10^{-4}\ \text{m})^4 \times 10^5\ \text{kg/m sec}^2}{8 \times 10^{-3}\ \text{kg/m sec} \times 1\ \text{m}} = 2^4\ Q_1 = 16 \times 1.4 \times 10^{-2}$$

$$= 0.226\ \text{L/hr}$$

$$Q_3 = 3^4 Q_1 = 81 Q_1 = 1.14\ \text{L/hr}$$

$$Q_4 = 4^4 Q_1 = 256 Q_1 = 3.61\ \text{L/hr}$$

$$Q_5 = 5^4 Q_1 = 625 Q_1 = 8.81\ \text{L/hr}$$

Total discharge from all five emitters:

$$Q_{total} = 0.014 + 0.226 + 1.14 + 3.61 + 8.81 = 13.8\ \text{L/hr}$$

Fractional contribution of the single largest emitter:

$$Q_5 / Q_{total} = 8.81/13.8 = 0.639 = 63.8\%$$

The single largest emitter thus accounts for nearly two-thirds of the total discharge, while the smallest emitter accounts for only 0.1% (though its diameter is only one-fifth that of the largest emitter).

Note: Modern drip-irrigation emitters generally depend on partially turbulent (rather than completely laminar) flow, to reduce sensitivity to pressure fluctuations and vulnerability to clogging by particles.

DARCY'S LAW

If the soil were merely a bundle of straight and smooth tubes, each uniform in radius, we could assume the overall flow rate to equal the sum of the separate flows through the individual tubes. Knowledge of the size distribution of the tube radii would then enable us to calculate the total flow, caused by a known pressure difference, using Poiseuille's equation.

Unfortunately from the standpoint of physical simplicity, however, soil pores do not resemble uniform, smooth, cylindrical tubes but are irregularly shaped, tortuous, and intricately interconnected. Flow through soil pores is limited by numerous constrictions, or "necks," with occasional "dead-end" spaces. The fluid velocity varies drastically from point to point, even along the same passage, and varies even more among different passages. Hence, the actual geometry and flow pattern in a typical soil specimen are too complicated to be described in microscopic detail. For this reason, flow through a complex porous medium such as a soil is generally described in terms of a *macroscopic flow-velocity vector*, which is the overall average of the microscopic velocities over the total volume considered. The detailed flow pattern is thus ignored, and the conducting body is treated as though it were a uniform medium, with the flow spread out over the entire cross section, solid and pore space alike. (An implicit assumption here is that the soil volume taken is sufficiently large relative to the pore sizes and microscopic heterogeneities to permit the averaging of velocity and potential over the cross section.)

We now examine the flow of water in a macroscopically uniform, saturated soil body, and attempt to describe the quantitative relations connecting the rate of flow, the dimensions of the body, and the hydraulic conditions at the inflow

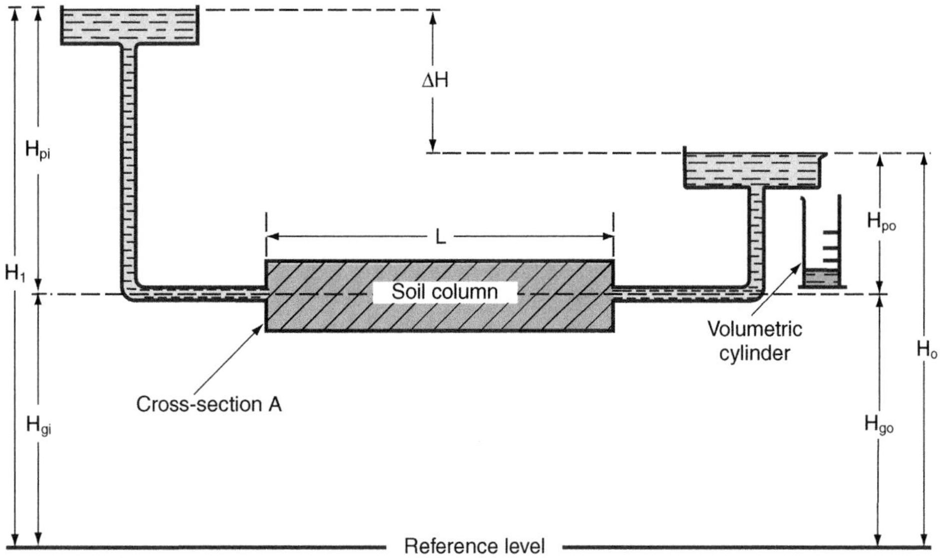

Fig. 7.3. Flow in a horizontal saturated column.

and outflow boundaries. Figure 7.3 shows a horizontal column of soil through which a steady flow of water is occurring from left to right, from an upper reservoir to a lower one, in each of which the water level is maintained constant.

Experience shows that the discharge rate Q, being the volume V flowing through the column per unit time, is directly proportional to the cross-sectional area and to the hydraulic head drop ΔH and inversely proportional to the length of the column L:

$$Q = V/t \propto A\ \Delta H/L$$

The usual way to determine the hydraulic head drop across the system is to measure the head at the inflow boundary H_i and at the outflow boundary H_o relative to some reference level. ΔH is the difference between these two heads:

$$\Delta H = H_i - H_o$$

Obviously, no flow occurs in the absence of a hydraulic head difference, that is, when $\Delta H = 0$.

The head drop per unit distance in the direction of flow ($\Delta H/L$) is the *hydraulic gradient*, which is, in fact, the driving force. The specific discharge rate Q/A (i.e., the volume of water flowing through a unit cross-sectional area per unit time t) is called the *flux density* (or simply the flux) and is indicated by q. Thus, the flux is proportional to the hydraulic gradient:

$$q = Q/A = V/At \propto \Delta H/L$$

The proportionality factor K is termed the *hydraulic conductivity*:

$$q = K\ \Delta H/L \tag{7.5}$$

This equation is known as *Darcy's law*, after Henry Darcy, the French engineer who discovered it over a century ago in the course of his classic investigation of seepage rates through sand filters in the city of Dijon (Darcy, 1856; Hubbert, 1956; Philip, 1995).

Where flow is unsteady (the flux changes with time) or the soil nonuniform, the hydraulic head may not decrease linearly along the direction of flow. Where the hydraulic head gradient or the conductivity is variable over time and space, we must consider the instantaneous localized gradient, flux, and conductivity values rather than average values for the soil system as a whole. A more exact expression of Darcy's law is, therefore, in differential form. Slichter (1899) generalized Darcy's law into a three-dimensional macroscopic differential equation of the form

$$q = -K\,\nabla H \tag{7.6}$$

where ∇H is the three-dimensional gradient of the hydraulic head. Since $K\,\nabla H$ is the product of a scalar K and a vector ∇H, the flux q is a vector, the direction of which is determined by ∇H. In an isotropic medium it is orthogonal to surfaces of equal hydraulic potential H.

Stated verbally, Darcy's law postulates that the flow of a viscous liquid through a porous medium is in the direction of, and proportional to, the *driving force* acting on the liquid (i.e., the *hydraulic gradient*) and is also proportional to a transmitting property of the conducting medium (the *hydraulic conductivity*). In a one-dimensional system, Eq. (7.6) takes the form

$$q = -K\,dH/dx \tag{7.7}$$

Mathematically, Darcy's law is similar to the linear transport equations of classical physics, such as *Ohm's law* (stating that the current, or flow rate of electricity, is proportional to the electrical potential gradient), *Fourier's law* (the rate of heat conduction is proportional to the temperature gradient), and *Fick's law* (the rate of diffusion is proportional to the concentration gradient).

BOX 7.1 Henry Cut the Mustard

The city of Dijon had long been known for its production of mustard, an important ingredient of France's celebrated *haute cuisine.* The pungent wastes of that industry, however, contributed to the fouling of the city's water supply. Even though the denizens of Dijon, being stalwart Frenchmen and Frenchwomen, did not drink much water (preferring, as they still do, the taste of *le vin*), they still needed water for cooking and sanitation. So they invited an engineer named Henry Darcy to design a filtration system. Being an engineer, Darcy must have looked for a soil physics handbook on filtration, but found none, so he had to experiment from scratch. He ended up formulating a new law and achieving immortal, if posthumous, fame. The use of filters packed with inert sand was but a first step in the greatly needed improvement of municipal water supplies. (It led eventually to the use of active materials such as clay, charcoal, diatomaceous earth, and ion exchange resins, along with chlorination, oxidation, flocculation, and — lately — reverse osmosis.) Henry Darcy was a pioneer in the task of purifying drinking water. His ultimate contribution to fundamental soil physics and hydrology was an unintended byproduct of that direct goal.

Sample Problem

Let us suppose we were given the task of purifying Dijon's water, with its 10,000 denizens. Since they drink mostly wine, their daily water requirements are modest, say, no more than 20 liters per day per person, on average. Let us suppose further that we knew (with the benefits of hindsight) that a column thickness of 0.30 m was adequate for filtration and that the hydraulic conductivity of the available sand was 2×10^{-5} m/sec. Could we calculate the area of the filter bed needed under a hydrostatic pressure head of 0.7 m? Consider the flow to be vertically downward to a fixed drainage plane.

We begin by calculating the discharge Q needed:

$$Q = \frac{10^4 \text{persons} \times 20 \text{ L/person day} \times 10^{-3} \text{m}^3/\text{L}}{8.64 \times 10^4 \text{ sec/day}} = 2.31 \times 10^{-3} \text{m}^3/\text{sec}$$

We recall Darcy's law:

$$Q = AK\,\Delta H/L$$

Hence, the area needed is:

$$A = QL/K\,\Delta H$$

The hydraulic head drop ΔH equals the sum of the pressure head and gravitational head drops:

$$\Delta H = 0.7 + 0.3 = 1.0 \text{ m}$$

Substituting these values for L (0.3 m), ΔH (1 m), and K (2×10^{-5} m/sec), we obtain

$$A = \frac{2.31 \times 10^{-3} \text{m}^3/\text{sec} \times 0.3 \text{ m}}{2 \times 10^{-5} \text{m/sec} \times 1 \text{m}} = 34.7 \text{ m}^2$$

Note: Since populations and per capita water use tend to increase and filter beds tend to clog, it might be wise to apply a factor of safety to our calculations and increase the filtration capacity severalfold (especially to accommodate peak demand hours). Incidentally, per capita water use in the U.S. (with running toilets, showers, and dishwashers) ranges from 100 to 400 L/day.

GRAVITATIONAL, PRESSURE, AND TOTAL HYDRAULIC HEADS

The water entering the column of Fig. 7.3 is under a pressure, which is the sum of the hydrostatic pressure and the atmospheric pressure acting on the surface of the water in the reservoir. Since the atmospheric pressure is the same at both ends of the system, we can disregard it and consider only the hydrostatic pressure. Accordingly, the water pressure at the inflow boundary $P_i = \rho_w g H_{pi}$. Since ρ_w and g are both nearly constant, we can express this pressure in terms of the pressure head H_{pi}.

Water flow in a horizontal column occurs in response to a pressure head gradient. Flow in a vertical column may be caused by gravitation as well as pressure. The *gravitational head* H_g at any point is determined by the height of the point relative to some reference plane, while the pressure head is determined by the height of the water column resting on that point. The total hydraulic head H is the sum of these two heads:

$$H = H_p + H_g \tag{7.8}$$

To apply Darcy's law to vertical flow, we must consider the total hydraulic head at the inflow and at the outflow boundaries (H_I and H_o, respectively):

$$H_i = H_{pi} + H_{gi} \quad \text{and} \quad H_o = H_{po} + H_{go}$$

Darcy's law thus becomes

$$q = K[(H_{pi} + H_{gi}) - (H_{po} + H_{go})]/L$$

The gravitational head is often designated as z, which is the vertical distance in the rectangular coordinate system x, y, z. It is convenient to set the reference level as the point $z = 0$ at the bottom of a vertical column or at the center of a horizontal column. However, the exact elevation of this hypothetical level is unimportant, since the absolute values of the hydraulic heads determined in reference to it are immaterial and only their differences from one point in the soil to another affect flow.

The pressure and gravity heads can be represented graphically in a simple way. To illustrate this, we shall immerse and equilibrate a vertical soil column in a water reservoir so that the upper surface of the column will be level with the water surface, as shown in Fig. 7.4. The coordinates of Fig. 7.4 are arranged so that the height above the bottom of the column is indicated by the vertical axis z; and the pressure, gravity, and hydraulic heads are indicated on the horizontal axis. The gravity head is determined with respect to the reference level $z = 0$, and it increases with height at the ratio of 1:1. The pressure head is determined with reference to the free-water surface, at which the hydrostatic pressure is zero. Accordingly, the hydrostatic pressure head at the top of the column is zero and at the bottom of the column is equal to L, the column length. Just as the gravity head diminishes from top to bottom, so the pressure head increases.

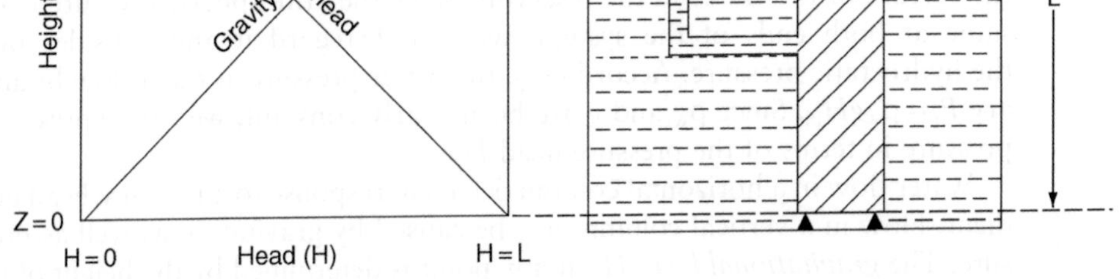

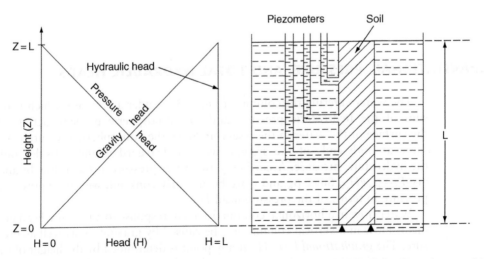

Fig. 7.4. Distribution of pressure, gravity, and total hydraulic heads in a vertical column immersed in water, at equilibrium.

Consequently their sum, which is the hydraulic head, remains constant all along the column. This is a state of equilibrium, at which no flow occurs.

This statement deserves to be further elaborated. The water pressure is not equal along the column, being greater at the bottom than at the top of the column. Why, then, will the water not flow from a zone of higher pressure to one of lower pressure? If the pressure gradient were the only force causing flow (as it is, in fact, in a horizontal column), the water would tend to flow upward. However, opposing the pressure gradient is a gravitational gradient of equal magnitude, resulting from the fact that the water at the top is at a higher gravitational potential than that at the bottom. In the illustration given, the two opposing gradients in effect cancel each other, so the total hydraulic head is constant throughout the column, as indicated by the standpipes (*piezometers*) connected to the column at the left.

As we already pointed out, it is convenient to set the reference level at the bottom of the column so that the gravitational potential can always positive. On the other hand, the pressure head of water can be either positive or negative: It is positive under a free-water surface (i.e., a water table) and negative above it.

FLOW IN A VERTICAL COLUMN

Figure 7.5 shows a uniform, saturated vertical column, the upper surface of which is ponded under a constant head of water H_1, and the bottom surface of which is set in a lower, constant-level reservoir. Flow is thus taking place from the higher to the lower reservoir through a column of length L.

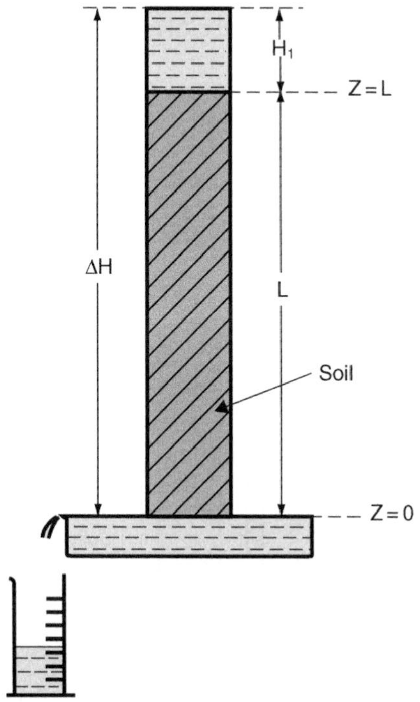

Fig. 7.5. Downward flow of water in a vertical saturated column.

In order to calculate the flux according to Darcy's law, we must know the hydraulic head gradient, which is the ratio of the hydraulic head drop (between the inflow and outflow boundaries) to the column length, as shown here:

		Pressure head		Gravity head
Hydraulic head at inflow boundary H_i	=	H_1	+	L
Hydraulic head at outflow boundary H_o	=	0	+	0
Hydraulic head difference $\Delta H = H_i - H_o$	=	H_1	+	L

The Darcy equation for this case is

$$q = K\,\Delta H/L = K(H_1 + L)/L = KH_1/L + K \tag{7.9}$$

Comparison of this case with the horizontal one shows that the rate of downward flow of water in a vertical column is greater than in a horizontal column by the magnitude of the hydraulic conductivity. It is also apparent that, if the ponding depth H_1 is negligible, the flux is equal to the hydraulic conductivity. This is due to the fact that, in the absence of a pressure gradient, the only driving force is the gravitational head gradient, which, in a vertical column, has the value of unity (since this head varies with height at the ratio of 1:1).

We now examine the case of upward flow in a vertical column, as shown in Fig. 7.6. In this case, the direction of flow is opposite to the direction of the gravitational gradient, and the hydraulic gradient becomes

		Pressure head		Gravity head
Hydraulic head at inflow boundary H_i	=	H_1	+	0
Hydraulic head at outflow boundary H_o	=	0	+	L
Hydraulic head difference $\Delta H = H_i - H_o$	=	H_1	−	L

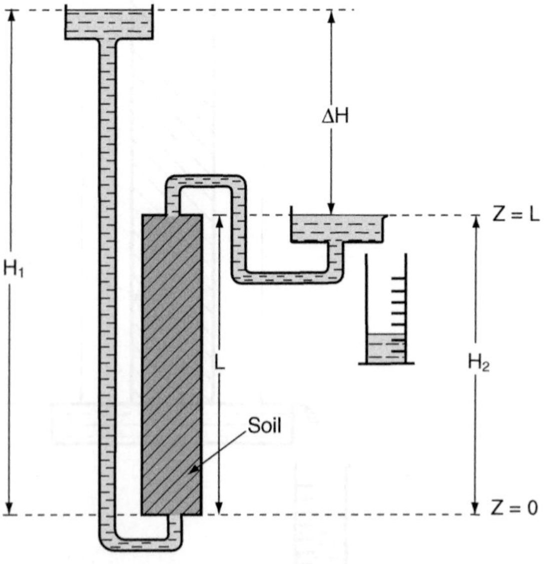

Fig. 7.6. Steady upward flow in a saturated vertical column.

The Darcy equation is thus:

$$q = K(H_1 - L)/L = KH_1/L - K = K\,\Delta H/L$$

FLOW IN A COMPOSITE COLUMN

Figure 7.7 depicts steady flow through a soil column consisting of two distinct layers, each homogeneous within itself and differing from the other layer in thickness and hydraulic conductivity. Layer 1 is at the inlet and layer 2 is at the outlet side of the column. The hydraulic head values at the inlet surface, at the interlayer boundary, and at the outlet end are designated H_1, H_2, and H_3, respectively. At steady flow, the flux through both layers must be equal:

$$q = K_1(H_1 - H_2)/L_1 = K_2(H_2 - H_3)/L_2 \qquad (7.10)$$

where q is the flux, K_1 and L_1 are the conductivity and thickness (respectively) of the first layer, and K_2 and L_2 are the same for the second layer. Here we have disregarded any possible contact resistance between the layers. Thus,

$$H_2 = H_1 - qL_1/K_1 \quad \text{and} \quad qL_2/K_2 = H_2 - H_3$$

Therefore,

$$qL_2/K_2 = H_1 - qL_1/K_1 - H_3 \quad \text{and} \quad q = (H_1 - H_3)/(L_2/K_2 + L_1/K_1) \qquad (7.11)$$

The reciprocal of the conductivity has been called the *hydraulic resistivity*, and the ratio of the thickness to the conductivity ($R_h = L/K$) has been called the hydraulic resistance per unit area. Hence,

$$q = \Delta H/(R_{h1} + R_{h2}) \qquad (7.12)$$

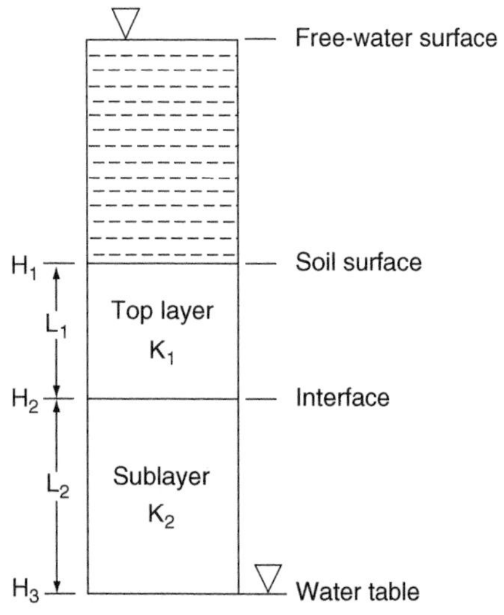

Fig. 7.7. Downward flow through a composite column.

where ΔH is the total hydraulic head drop across the entire system and R_{h1}, R_{h2} are the hydraulic resistances of layers 1 and 2. Equation (7.12) is in complete analogy to Ohm's law for an electric current through two constant resistances in series.

Sample Problem

Let us consider two cases of steady downward percolation through a two-layer soil profile, the top of which is submerged under a 1-m head of water and the bottom of which is defined by a water table. Each of the two layers is 0.50 m thick. In the one case, the conductivity of the top layer is 10^{-6} m/sec and that of the sublayer is 10^{-7} m/sec. In the second case, the same layers are reversed.

To calculate the flux and the hydraulic and pressure heads at the interface between the layers, we use the Ohm's law analogy for steady flow through two resistors in series:

$$q = \Delta H/(R_1 + R_2)$$

where q is the flux, ΔH is the total hydraulic head drop across the profile, and R_1, R_2 are the hydraulic resistances of the top layer and sublayer, respectively.

Each resistance is proportional directly to the layer's thickness and inversely to its conductivity (i.e., $R = L/K$). The pressure head at the soil surface is 1 m, and the gravity head (with reference to the soil's bottom) is also 1 m. Both the pressure and gravity heads at the bottom boundary are zero. Hence,

$$q = (1 \text{ m} + 1 \text{ m})/(0.5 \text{ m}/10^{-6} \text{ m/sec} + 0.5 \text{ m}/10^{-7} \text{ m/sec}) = 3.64 \times 10^{-7} \text{ m/sec}$$

We can now apply Darcy's equation to the top layer alone to obtain the hydraulic head at the interlayer interface:

$$q = K_1\ \Delta H_1/L_1 = K_1(H_{\text{surface}} - H_{\text{interface}})/L_1$$

Hence:

$$H_{\text{interface}} = H_{\text{surface}} - qL_1/K_1$$
$$= 2 \text{ m} - (3.64 \times 10^{-7} \text{ m/sec})(0.5 \text{ m})/10^{-6} \text{ m/sec} = 1.818 \text{ m}$$

Since the gravity head at the interface is 0.5 m (above our reference datum at the bottom of the profile), the pressure head H_p is

$$H_p = H - H_g = 1.818 - 0.50 = 1.318 \text{ m}$$

We now reverse the order of the layers so that the less conductive layer overlies the more conductive. The total head drop remains the same and so does the total resistance. Therefore the flux remains the same (assuming that both layers are still saturated and the conductivity of each does not change). Applying Darcy's equation to the top layer, we have, as previously,

$$H_{\text{interface}} = 2.0 - (3.64 \times 10^{-7})(0.5)/10^{-7} = 0.18 \text{ m}$$

In this case the pressure head at the interface is

$$H_p = H - H_g = 0.18 - 0.50 = -0.32 \text{ m}$$

Note: Comparison between the interface pressures of the two cases illustrates an important principle regarding flow in layered profiles. With the more conductive layer on top, flow is impeded at the interface and there is a pressure buildup, which in our case increased the pressure head from 1 m at the surface to 1.218 m at the interface. The opposite occurs when the upper layer is less conductive. In this case the pressure is dissipated through the top layer, often to the extent that a negative pressure develops at the interface.

FLUX, FLOW VELOCITY, AND TORTUOSITY

As stated earlier, the *flux density* (herein, simply *flux*) is the volume of water V passing through a unit cross-sectional area A (perpendicular to the flow direction) per unit time t, the dimensions of which are equivalent to length per time:

$$q = V/At = L^3/L^2T = LT^{-1}$$

(In SI units, the flux is expressed in cubic meters per square meter per second, equivalent to meters per second). These are the dimensions of velocity, yet we prefer the term *flux* to *flow velocity*, the latter being too ambiguous. Since soil pores vary in shape, width, and direction, the actual flow velocity in the soil is highly variable (e.g., wider pores conduct water more rapidly, and the liquid in the center of each pore moves faster than the liquid in close proximity to the particles). Strictly speaking, therefore, one cannot refer to a single velocity of liquid flow, but at best to an average velocity.

Even the average velocity of the flowing liquid differs from the flux as we have defined it. Flow does not in fact take place through the entire cross-sectional area A, because part of this area is plugged by particles and only the porosity fraction permits flow. Since the real area through which flow takes place is smaller than A, the average velocity of the liquid must be greater than the flux q. Furthermore, the actual length of the path traversed by an average parcel of liquid is greater than the soil column length L, owing to the labyrinthine, or tortuous, nature of the pore passages, as shown in Fig. 7.8.

Tortuosity can be defined as the ratio of the average roundabout path to the apparent, or straight, flow path; that is, the ratio of the average length of the pore passages (as if they were stretched out in the manner one might stretch out a coiled or tangled telephone wire) to the length of the soil specimen. Tortuosity is thus a dimensionless geometric parameter of porous media, which, though difficult to measure precisely, is always greater than 1 and may exceed 2. The *tortuosity factor* is sometimes defined as the inverse of what we defined as the tortuosity.

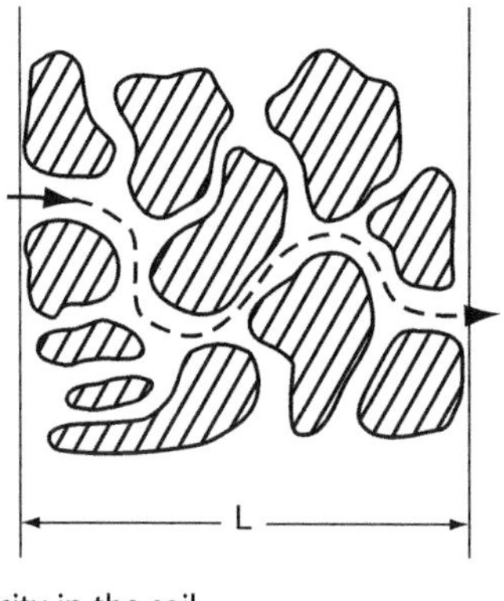

Fig. 7.8. Flow path tortuosity in the soil.

HYDRAULIC CONDUCTIVITY, PERMEABILITY, AND FLUIDITY

Recalling that the hydraulic conductivity K is the ratio of the flux to the potential gradient (i.e., the slope of flux vs. gradient, Fig. 7.9) and that the dimensions of flux are LT^{-1}, we note that the dimensions of hydraulic conductivity depend on those assigned to the potential gradient. In Chapter 6, we showed that the simplest way to express the potential is by use of length, or head, units (though, strictly speaking, H is not a true length but a pressure equivalent in terms of a water column height $H = P/\rho g$). Therefore, the hydraulic head gradient H/L, being the ratio of length to length, is dimensionless. Accordingly, the dimensions of hydraulic conductivity are the same as those of flux, namely, LT^{-1}. However, If the gradient is expressed in terms of the variation of hydraulic pressure with length, then the hydraulic conductivity assumes the dimensions of $M^{-1}L^3T$. Since the latter is cumbersome, the use of head units is generally preferred.

In a saturated soil of stable structure, as well as in a rigid porous medium such as sandstone, the hydraulic conductivity is usually found to be nearly constant. Its order of magnitude is about 10^{-4}–10^{-5} m/sec in a sandy soil and 10^{-6}–10^{-9} m/sec in a clayey soil.

To appreciate the practical significance of these values in more familiar terms, consider the hypothetical case of an unlined (earth-bottom) reservoir or pond in which one wishes to retain water against losses caused by downward seepage. Assuming, for the sake of simplicity, that seepage into and through the underlying soil is by gravity alone (i.e., no pressure or suction gradients in the soil), it will take place at a rate approximately equal to the hydraulic conductivity. A coarse sandy soil might have a K value of, say, 10^{-4} m/sec and would therefore lose water at the enormous rate of nearly 9 m/day (there being 8.64×10^4 sec/day). A loam soil with a K value of 10^{-6} m/sec would lose "only" about 0.1 m/day (100 mm/day). Finally, and in contrast, a bed of clay with a conductivity of 10^{-8} m/sec would allow the seepage of no more than 1 mm/day, less than the generally expectable rate of evaporation.

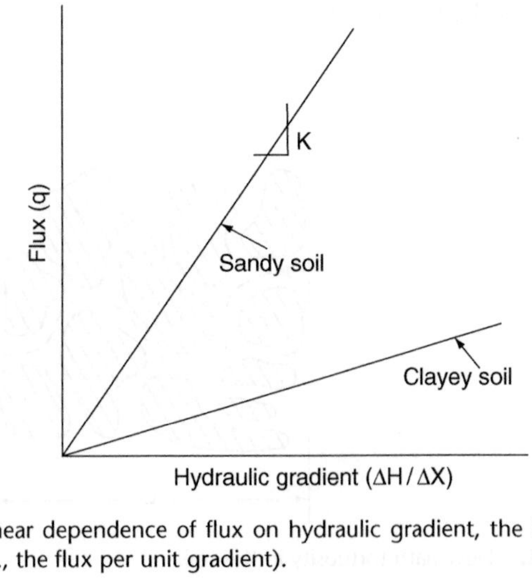

Fig. 7.9. The linear dependence of flux on hydraulic gradient, the hydraulic conductivity being the slope (i.e., the flux per unit gradient).

The hydraulic conductivity is obviously affected by structure as well as by texture, being greater if the soil is highly porous, fractured, or aggregated than if it is tightly compacted and dense. Hydraulic conductivity depends not only on total porosity but also, and primarily, on the sizes of the conducting pores.

Contrary to the convenient assumption (just stated) that the hydraulic conductivity of a saturated soil remains constant, it is often found to vary (generally to diminish) over time. Because of various chemical, physical, and biological processes, the hydraulic conductivity may change as water permeates the soil and flows in it. Changes occurring in the composition of the exchange complex (as when water entering the soil differs in composition or concentration of solutes from the original soil solution) can greatly affect the hydraulic conductivity. In general, the conductivity decreases with decreasing concentration of electrolytic solutes. This is due to swelling and dispersion phenomena, which are also affected by the species of cations present. Detachment and migration of clay particles during flow may result in the clogging of pores.

In practice, it is extremely difficult to saturate a soil with water without trapping some air. Encapsulated air bubbles may block pore passages, as shown in Fig. 7.10. Temperature changes may cause the flowing water to dissolve or to release gas bubbles and will also cause a change in their volume, thus affecting conductivity.

The hydraulic conductivity K is not a property of the soil alone. Rather, it depends jointly on the attributes of the soil and of the fluid. The soil characteristics that affect K are the total porosity, the distribution of pore sizes, and tortuosity — in short, the soil's pore geometry. The fluid attributes that affect conductivity are density and viscosity.

It is possible in theory, and sometimes in practice, to separate K into two factors: *intrinsic permeability* of the soil k and *fluidity* of the permeating liquid (or gas) f:

$$K = kf \qquad (7.13)$$

If K is given in terms of m/sec (LT^{-1}), then k is in units of m^2 (L^2) and f is in units of m^{-1} sec^{-1} ($L^{-1}T^{-1}$).

Fluidity is related directly to density and inversely to viscosity:

$$f = \rho g/\eta \qquad (7.14)$$

Hence,

$$k = K\eta/\rho g \qquad (7.15)$$

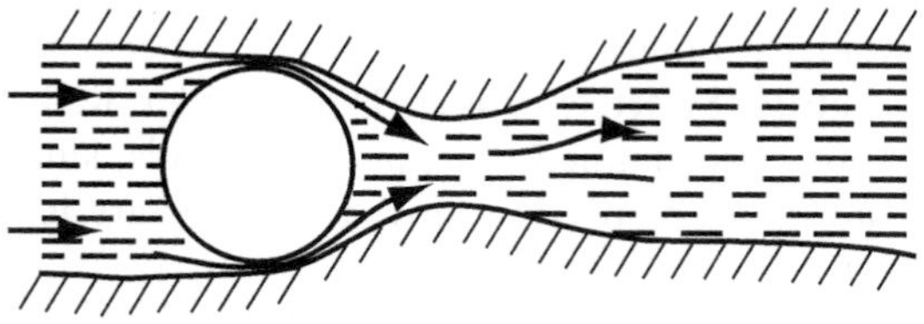

Fig. 7.10. An entrapped air bubble plugging flow.

where η is the dynamic viscosity (N sec/m^2, or Pa sec), ρ is the fluid density (kg/m^3), and g is the gravitational acceleration (m/sec^2).

While fluidity varies with temperature and composition of the fluid, permeability is ideally a property of the porous medium's pore geometry alone, provided the fluid and the solid matrix do not interact in such a way as to change the properties of either. In a stable porous body, the same permeability will be obtained with different fluids (water, air, or oil). However, in many cases water does interact with the solid matrix to modify its permeability, so hydraulic conductivity cannot be resolved into separate and independent properties of water and of soil.

LIMITATIONS OF DARCY'S LAW

Darcy's law is not universally valid for all conditions of liquid flow in porous media. It has long been recognized that the linearity of the flux versus hydraulic gradient relationship fails at high flow velocities, where inertial forces are no longer negligible compared to viscous forces. Darcy's law applies only as long as flow is laminar (i.e., nonturbulent movement of adjacent layers of the fluid relative to one another) and where soil–water interaction does not result in a change of fluidity or of permeability with a change in gradient. Laminar flow prevails in silts and finer materials for most commonly occurring hydraulic gradients found in nature. In coarse sands and gravels, however, hydraulic gradients much in excess of unity may result in nonlaminar flow conditions, so in such cases Darcy's law may not be applicable.

The quantitative criterion for the onset of turbulent flow is the *Reynolds number* N_{Re}:

$$N_{\mathrm{Re}} = du\rho/\eta \tag{7.16}$$

where u is the mean flow velocity, d is the effective pore diameter, ρ is the liquid density, and η is its viscosity. In straight tubes, the critical value of N_{Re} beyond which turbulence sets in has variously been reported to be of the order of 1000–2200 (Childs, 1969). However, the critical Reynolds number at which water flowing in a tube becomes turbulent is apparently reduced greatly when the tube is curved and its diameter varies. For porous media, therefore, it is safe to assume that flux remains linear with hydraulic gradient only as long as N_{Re} is smaller than unity. As flow velocity increases, especially in systems of large pores, the occurrence of turbulent eddies or nonlinear laminar flow results in "waste" of effective energy; that is, some energy is dissipated by the internal churning of the liquid so that the hydraulic potential gradient becomes less effective in inducing flow . This is illustrated in Fig. 7.11.

Deviations from Darcy's law may also occur at the opposite end of the flow-velocity range, namely, at low gradients and in narrow pores. Some investigators (Swartzendruber, 1962; Miller, R. J. and Low, 1963; Nerpin et al., 1966) have claimed that, in clayey soils, small hydraulic gradients may cause no flow or only low flow rates that are less than proportional to the gradient.

A possible reason for this anomaly is that the water in close proximity to the particle surfaces and subject to their adsorptive force fields may be more

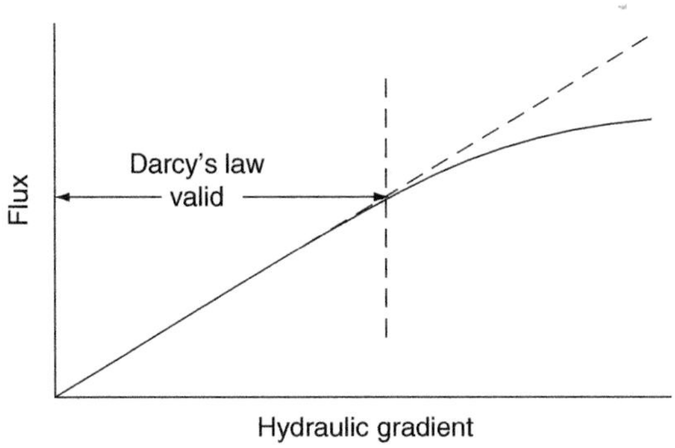

Fig. 7.11. The deviation from Darcy's law at high flux, where flow becomes turbulent.

rigid than ordinary bulk water and may exhibit the properties of a *Bingham liquid* (having a *yield value*) rather than a *Newtonian liquid* (Hillel, 1980a). The adsorbed (or "bound") water may even have a quasi-crystalline structure similar to that of ice. Some soils may exhibit a *threshold gradient* (Miller, R. J. and Low, 1963), below which the flux is either zero (the water remaining apparently immobile) or at least lower than predicted by the Darcy relation, and only at gradients exceeding the threshold value does the flux become proportional to the gradient (Fig. 7.12). These phenomena and their possible explanations, though highly interesting, are generally of little importance in practice, so Darcy's law can be employed in the vast majority of cases pertaining to the flow of water in soil.

Another possible cause for apparent flow anomalies in clay soils is their tendency to swell or compress (Smiles, 1976). As commonly formulated, Darcy's law applies to flow relative to a geometrically fixed porous matrix, and it may seem to fail when the particles composing the matrix are themselves moving relative to a fixed frame of reference.

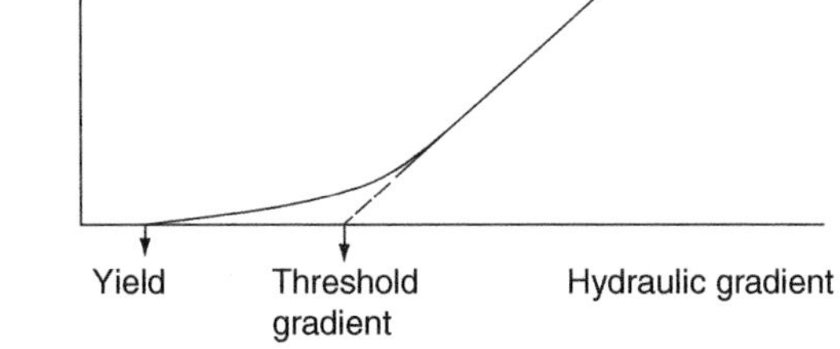

Fig. 7.12. Possible deviations from Darcy's law at low gradients.

HOMOGENEITY AND ISOTROPY

The hydraulic conductivity may be uniform throughout a soil, or it may vary from point to point, in which case the soil is said to be hydraulically *inhomogeneous*. If the conductivity is the same in all directions, the soil is *isotropic*. However, the conductivity at each point may differ for different directions (e.g., the horizontal conductivity may be greater, or smaller, than the vertical), a condition known as *anisotropy*.

A soil may be homogeneous and nonetheless anisotropic, or it may be inhomogeneous (e.g., layered) and yet isotropic at each point. Some soils exhibit both inhomogeneity and anisotropy. In certain cases, K may also be asymmetrical (or directional); that is to say, K may have a different value for opposite directions of flow along a given line. A basic review of anisotropy and layering was given by Bear et al. (1968). Anisotropy is generally due to the structure of the soil, which may be laminar, or platy, or columnar, etc., thus exhibiting a pattern of micropores or macropores with a distinct directional bias.

To illustrate the meaning of the concepts of homogeneity and isotropy, let us consider the hypothetical situation of two continuous bodies of soil or, better yet, two realms within the same body of soil, as shown in Fig. 7.13. The hydraulic conductivities for the three principal axes of the x, y, z coordinate system are designated K_x, K_y, K_z, respectively, for the soil body on the left; and the corresponding conductivities are designated K_x^*, K_y^*, K_z^* for the body on the right. Four possible cases exist:

Case I. The soil is homogeneous and isotropic throughout both realms.

$$\left.\begin{array}{ll}\text{Homogeneous:} & K_x = K_x^*, K_y = K_y^*, K_z = K_z^* \\ \text{Isotropic:} & K_x = K_y = K_z, K_x^* = K_y^* = K_z^*\end{array}\right\} \text{single value of } K$$

Case II. The soil is homogeneous but anisotropic.

$$\left.\begin{array}{ll}\text{Homogeneous:} & K_x = K_x^*, K_y = K_y^*, K_z = K_z^* \\ \text{Anisotropic:} & K_x \neq K_y \neq K_z, K_x^* \neq K_y^* \neq K_z^*\end{array}\right\} \text{3 values of } K$$

Case III. The soil is inhomogeneous but isotropic.

$$\left.\begin{array}{ll}\text{Inhomogeneous:} & K_x \neq K_x^*, K_y \neq K_y^*, K_z \neq K_z^* \\ \text{Isotropic:} & K_x = K_y = K_z, K_x^* = K_y^* = K_z^*\end{array}\right\} \text{2 values of } K$$

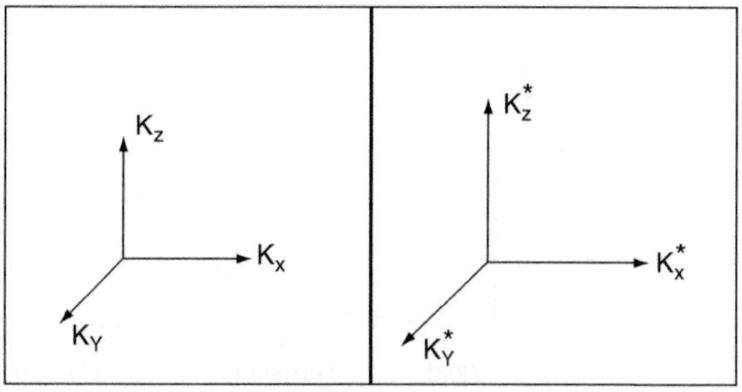

Fig. 7.13. Schematic illustration of homogeneity and isotropy for two adjacent realms in the soil.

Case IV. The soil is both inhomogeneous and anisotropic.

$$\left.\begin{array}{ll}\text{Inhomogeneous:} & K_x \neq K_x^*, K_y \neq K_y^*, K_z \neq K_z^* \\ \text{Anisotropic:} & K_x \neq K_y \neq K_z, K_x^* \neq K_y^* \neq K_z^*\end{array}\right\} \text{6 values of } K$$

Analysis of anisotropic flow systems is also complicated by the fact that the flow direction may not be orthogonal to the equipotential lines or planes; that is to say, the flow does not necessarily take place in the direction of the steepest potential gradient.

MEASURING HYDRAULIC CONDUCTIVITY OF SATURATED SOILS

Methods for measuring hydraulic conductivity in the laboratory were reviewed by Klute and Dirksen (1986) and by Reynolds et al. (2002), and for measurement in the field, by Amoozegar and Warrick (1986). The use of *laboratory permeameters* is illustrated in Figs. 7.14 and 7.15. Measurements are made either with dried and fragmented specimens, packed into the flow cells in a standardized manner, or, preferably, with *undisturbed core samples* taken directly from the field. In either case, provision must be made to avoid boundary flow along the walls of the container.

Field measurements can be made below the water table, as by the *auger-hole method* or by the *piezometer method* (Reynolds et al., 2002). Techniques have also been proposed for measurements above the water table, as by the *double-tube method* (Bouwer, 1961, 1962), the *shallow-well pump-in method*, and the *field-permeameter* method (Winger, 1960). Topp and Sattlecker (1983) described a rapid measurement of horizontal and vertical components of a soil's saturated hydraulic conductivity. Mallants et al.

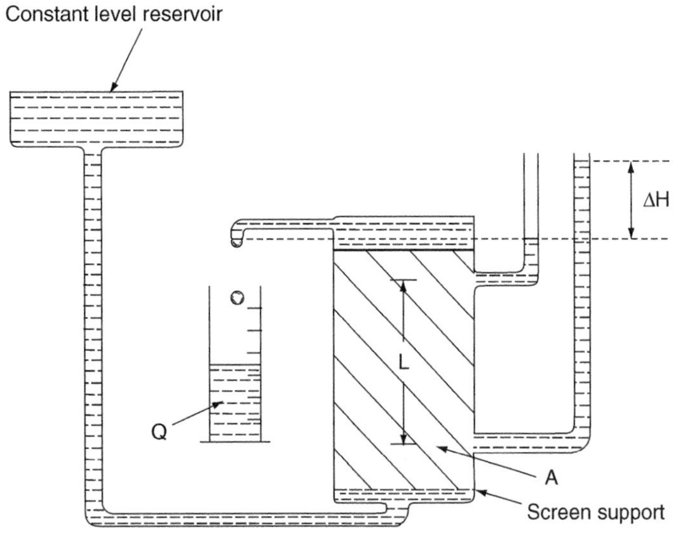

Fig. 7.14. The measurement of saturated hydraulic conductivity with a constant-head

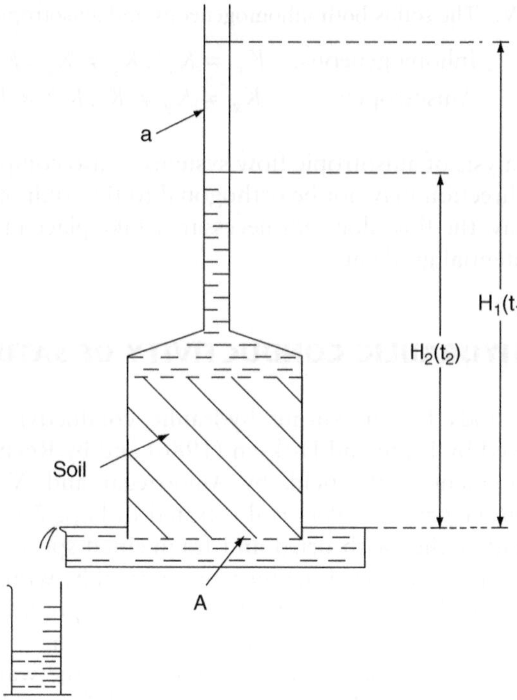

Fig. 7.15. The measurement of saturated hydraulic conductivity with a falling head permeameter: $K = [2.3aL/A(t_2 - t_1)](\log H_1 - \log H_2)$. Note that H_1, H_2 are the values of hydraulic head at times t_1 and t_2, respectively.

(1997) analyzed the spatial dependence of saturated hydraulic conductivity in a soil with macropores.

Typical values of saturated hydraulic conductivity for soils of different texture are listed in Table 7.1. Owing to soil heterogeneity, the apparent hydraulic conductivity measured often depends on the scale of the measurement. Thus, the K value measured on a cubic centimeter or decimeter may differ from the average value measured on a cubic meter. Too often, this is ignored and K values are reported without specifying the scale of the measurement.

TABLE 7.1 Hydraulic Conductivity at Saturation (K_s), Particle Diameters (d), and Height of Capillary Rise (h_c) for some Solids[a]

	K_s (m/sec)	d (mm)	h_c (m)
Clay	10^{-10}–10^{-8}	<0.002	2–4
Silt	10^{-8}–10^{-6}	0.002–0.05	0.7–1.5
Sand	10^{-5}–10^{-3}	0.05–2	0.12–0.35
Gravel	10^{-2}–10^{-1}	>2	Nil

[a] After Boeker and van Grandelle (1995).

EQUATIONS OF SATURATED FLOW

Darcy's law, by itself, is sufficient only to describe *steady*, or *stationary*, flow processes, in which the flux remains constant and equal throughout the conducting medium (and hence the potential and gradient at each point remain constant in time). *Unsteady*, or *transient*, flow processes, in which the magnitude and possibly even the direction of the flux and potential gradient vary in time, require the inclusion of an additional law, namely, the *law of conservation of matter*. To understand how this law applies, consider a small volume element (say, a cube) of soil, into and out of which flow takes place at possibly differing rates. The mass conservation law, expressed in the *equation of continuity*, states that if the rate of inflow into the volume element exceeds the rate of outflow, then the volume element must be storing the excess and increasing its water content. Conversely, if outflow exceeds inflow, storage must be decreasing.

Consider first the simplest case, one-dimensional flow, with q_x as the flux in the direction of x. Any increase of q_x with x must equal the decrease of water content w with time t:

$$\partial\theta/\partial t = -\partial q_x/\partial x \tag{7.17}$$

which in multidimensional systems becomes

$$\partial\theta/\partial \mathrm{t} = -\nabla \cdot q \tag{7.18}$$

We recall Darcy's law,

$$q = -K\, \nabla H \tag{7.19}$$

which in one dimension is

$$q_x = -K\,(dH/dx) \tag{7.20}$$

(where H is the hydraulic head and K is the hydraulic conductivity). Now we combine equation (7.19) with the continuity equation (7.18) to obtain the *general flow equation*:

$$\partial\theta/\partial t = \nabla \cdot K\, \nabla H \tag{7.21}$$

In applying this equation, the assumptions are usually made that inertial forces are negligible in comparison with viscous forces, that water is continuously connected throughout the flow region, that isothermal conditions prevail, and that no chemical or biological phenomena change the fluid or the porous medium during the flow process.

In one dimension, Eq. (7.21) becomes

$$\frac{\partial\theta}{\partial t} = \frac{\partial}{\partial x}\left(K\frac{\partial H}{\partial x}\right) \tag{7.22}$$

Note that Eq. (7.22) is a partial differential equation with two independent variables, t and x.

Since the hydraulic head can be resolved into a pressure head H_p and a gravitational head (an elevation z above some reference datum), we can rewrite Eq. (7.21) as

$$\frac{\partial\theta}{\partial t} = \nabla \cdot [K(\nabla H_p + \nabla z)] \tag{7.23}$$

In horizontal flow, $\nabla z = 0$, so for this case,

$$\frac{\partial\theta}{\partial t} = \frac{\partial}{\partial x}\left[K\left(\frac{\partial H_p}{\partial x}\right)\right] \tag{7.24}$$

while in vertical flow, $\delta z/\delta z = 1$, and therefore, for this case,

$$\frac{\partial\theta}{\partial t} = \frac{\partial}{\partial z}\left[K\left(\frac{\partial H_p}{\partial z} + 1\right)\right] \tag{7.25}$$

In a saturated soil with an incompressible matrix, $\delta\theta/\delta t = 0$, the conductivity is usually assumed to remain constant; hence Eq. (7.22) becomes

$$K_s = \partial^2 H/\partial x^2 = 0 \tag{7.26}$$

where K_s is the hydraulic conductivity of the saturated soil (the "saturated conductivity"). For three-dimensional flow conditions, and allowing for anisotropy, the equation is

$$K_x \frac{\partial^2 H}{\partial x^2} + K_y \frac{\partial^2 H}{\partial y^2} + K_z \frac{\partial^2 H}{\partial z^2} = 0 \tag{7.27}$$

where K_x, K_y, and K_z represent the hydraulic conductivity values in the three principal directions x, y, z.

In an isotropic soil (where $K_x = K_y = K_z$ at each point) that is also homogeneous (the K values of all points are equal), we obtain the well-known *Laplace equation*:

$$\partial^2 H/\partial x^2 + \partial^2 H/\partial y^2 + \partial^2 H/\partial z^2 = 0 \tag{7.28}$$

This second-order partial differential equation of the elliptical type can be solved in some cases to obtain a quantitative description of steady-state flow in various geometric configurations.

8. WATER FLOW IN UNSATURATED SOIL

FLOW IN UNSATURATED VERSUS SATURATED SOIL

Most of the processes involving soil–water interactions in the field, including the supply of moisture and nutrients to plant roots as well as the transport of water and solutes beyond the root zone, occur while the soil is in an unsaturated condition. Unsaturated flow processes are in general complicated and difficult to describe quantitatively, since they often entail changes in the state and content of soil water during flow. Such changes involve complex relations among such variables as soil wetness, suction, and conductivity, whose interrelations are further complicated by hysteresis as well as by spatial variability.

In the preceding chapter, we stated that the flow of water in the soil is driven by a hydraulic potential gradient, that it occurs in the direction of decreasing hydraulic potential, and that its rate (flux) is proportional to the potential gradient and is affected by the geometric properties of the pore channels through which flow takes place. These principles apply in unsaturated as well as saturated soils. However, the nature of the moving force and the effective geometry of the conducting pores may be very different.

Apart from the gravitational force (which is completely independent of soil wetness), the primary moving force in a saturated soil is the gradient of a positive pressure potential. On the other hand, water in an unsaturated soil is subject to a subatmospheric pressure, or matric suction, which is equivalent to a negative pressure potential. The gradient of this potential likewise constitutes a moving force.

Matric suction is due, as we have noted, to the physical affinity between water and the matrix of the soil, including both the soil-particle surfaces and the capillary pores. When suction is uniform all along a horizontal column of soil, the column is at equilibrium and there is no moving force. Not so when a suction gradient exists. In that case, water will be drawn from a zone where the hydration envelopes surrounding the particles are thicker to where they are thinner, and from a zone where the capillary menisci are less curved to where they are more strongly curved. In other words, water will flow spontaneously from where matric suction is lower to where it is higher. It will flow in the pores that are water filled at the existing suction and creep along the hydration films over the particle surfaces, in a tendency to equilibrate the potential. (The ideal state of equilibrium, like human happiness, may never be achieved in practice, but its natural pursuit is a universal rule.)

Perhaps the most important difference between unsaturated and saturated flow is in the hydraulic conductivity. When the soil is saturated, all of the pores are water filled and conducting. The water phase is then continuous and the conductivity is maximal. When the soil desaturates, some of the pores become air filled, so the conductive portion of the soil's cross-sectional area diminishes. Furthermore, as suction develops, the first pores to empty are the largest ones, which are the most conductive (remember Poiseuille's law!), thus relegating flow to the smaller pores. At the same time, the large empty pores must be circumvented, so, with progressive desaturation, tortuosity increases. In coarse-textured soils, water may be confined almost entirely to the capillary wedges at the contact points of the particles, thus forming separate and discontinuous pockets of water (see Fig. 8.1). In aggregated soils, too, the large interaggregate spaces that confer high conductivity at saturation become (when emptied) barriers to liquid flow from one aggregate to another.

For all these reasons, the transition from saturation to unsaturation generally entails a steep drop in hydraulic conductivity, which may decrease by several orders of magnitude (sometimes down to one-millionth of its value at saturation) as suction increases from 0 to 1 MPa. At still higher suctions, or lower wetness values, the conductivity may be so low that very steep suction gradients, or very long times, are required for any appreciable flow to occur at all.

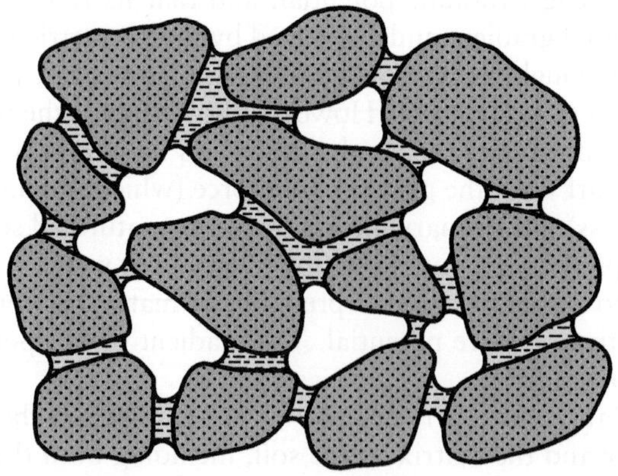

Fig. 8.1. Water in an unsaturated coarse-textured soil.

RELATION OF CONDUCTIVITY TO SUCTION AND WETNESS

Consider an unsaturated, horizontal body of soil in which water is flowing horizontally (i.e., in the absence of gravity gradients) in response to a suction gradient, as illustrated schematically in Fig. 8.2. In this model, the potential difference between the inflow and outflow ends is maintained not by different levels of positive hydrostatic pressure, but by different imposed suctions. In general, as the suction varies along the sample, so will the wetness and conductivity. If the suction head at each end of the sample is maintained constant, the flow rate will be steady but the suction gradient will vary along the sample's axis. Since the product of gradient and conductivity must be constant for steady flow, the hydraulic potential gradient must steepen as the conductivity diminishes with the increase in suction along the length of the sample. This phenomenon is illustrated in Fig. 8.3.

In view of the fact that the gradient along an unsaturated column is generally variable, we should not, strictly speaking, divide the flux by the overall ratio of the head drop to the distance ($\Delta H/\Delta x$) to obtain the conductivity. Rather, we should divide the flux by the local gradient at each point to evaluate the exact conductivity and its variation with suction. In the following treatment, however, we assume that the column of Fig. 8.2 is sufficiently short (and the hydraulic gradient is not too steep) to allow us to evaluate at least an average conductivity for the sample as a whole (i.e., $K = q\ \Delta x/\Delta H$).

The average negative head, or suction, acting in the column is

$$-H_{\text{average}} = \psi_{\text{average}} = -(H_1 + H_2)/2$$

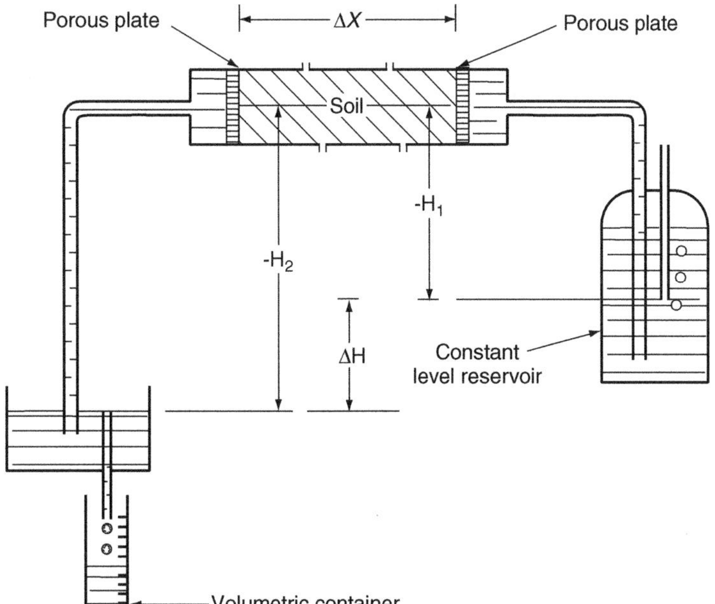

Fig. 8.2. A model illustraning unsaturated flow (under a suction gradient) in a horizontal column.

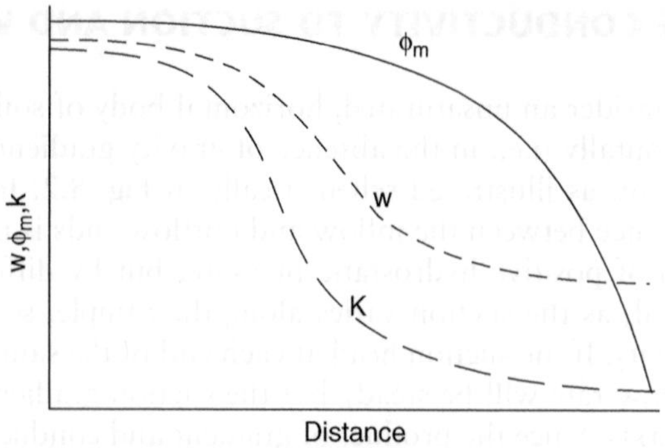

Fig. 8.3. The variation of wetness *w*, matric potential ϕ_m, and conductivity *K* along a hypothetical column of unsaturated soil conducting a steady flow of water.

We further assume that the suction everywhere exceeds the air-entry value so that the soil is unsaturated throughout.

Let us now make successive and systematic measurements of flux versus suction gradient for different values of average suction. The hypothetical results of such a series of measurements are shown schematically in Fig. 8.4. As in the case of saturated flow, we find that the flux is proportional to the gradient. However, the slope of the flux versus gradient line, being the hydraulic conductivity, varies with the average suction. In a saturated soil, by way of contrast, the hydraulic conductivity is usually independent of the water pressure.

Figure 8.5 shows the general trend of the dependence of conductivity on suction in soils of different texture. (Note that curves of *K* versus suction are usually drawn on a log–log scale, because both *K* and ψ vary over several orders of magnitude within the suction range of general interest, say, 0–100 m of suction head). It is seen that, although the saturated conductivity of the

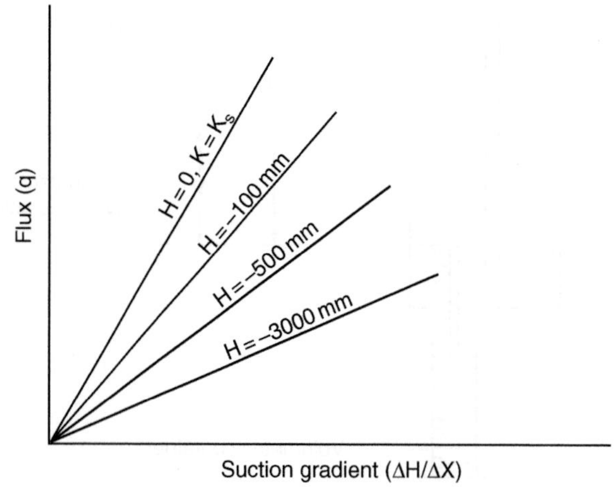

Fig. 8.4. Hydraulic conductivity, the slope of the flux versus gradient relation, depends on the average suction in an unsaturated soil.

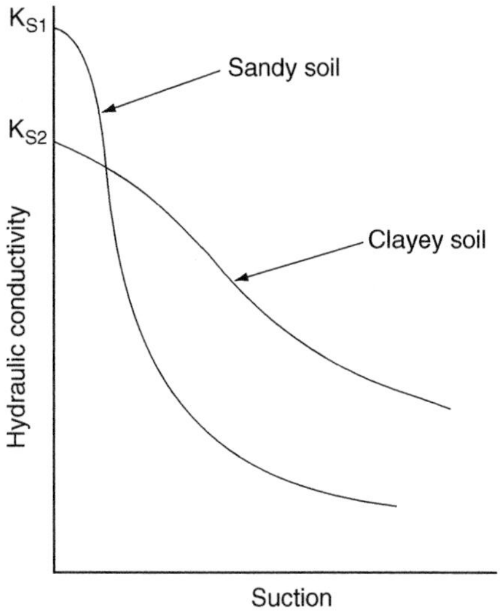

Fig. 8.5. Dependence of conductivity on suction in different soils (log–log scale).

sandy soil K_{s1} is typically greater than that of the clayey soil K_{s2}, the unsaturated conductivity of the former decreases more steeply with increasing suction and eventually becomes lower.

The steep decline of hydraulic conductivity with rising matric suction carries important implications regarding soil-water dynamics. It suggests that processes taking place in wet soil conditions are inherently faster than those occurring in drier soil conditions. Thus, the process of infiltration, during which water moves into the soil profile through a generally saturated surface zone, is much more rapid than evaporation, which typically involves the transfer of water from the interior of the soil to the atmosphere through a drying surface zone.

The fact that a sandy soil is more conductive than a clayey soil at low suction values but is less conductive in the higher suction range is the reason why a sandy soil absorbs water more rapidly during infiltration (while the surface is wetted to saturation or nearly so) but then does not sustain the evaporation process (during which the surface zone becomes desiccated) as can a clay soil.

Various empirical equations have been proposed for the relation of conductivity to suction or to wetness (e.g., Gardner, 1960b), including the following:

$$K(\psi) = a/\psi^n \tag{8.1a}$$

$$K(\psi) = a/(b + \psi^n) \tag{8.1b}$$

$$K(\psi) = K_s/[1 + (\psi/\psi_c)^n] \tag{8.1c}$$

$$K(\theta) = a\theta^n \tag{8.1d}$$

$$K(\theta) = K_s s^n = K_s(\theta/f)^n \tag{8.1e}$$

where K is the hydraulic conductivity at any degree of saturation (or unsaturation), K_s is the saturated conductivity of the same soil, a, b, and n are empirical constants (different in each equation), ψ is matric suction head, θ is volumetric wetness, s is the degree of saturation, and ψ_c is the suction head at which $K = K_s/2$. Note that $s = \theta/f$, where f is porosity.

In all of the equations, the most important parameter is the exponential constant, since it controls the steepness with which conductivity decreases with increasing suction or with decreasing water content. The n value of the first two equations is about 2 or less for clayey soils and may be 4 or more for sandy soils. For each soil, the equation of best fit and the values of the parameters must be determined experimentally.

Sample Problem

The hydraulic conductivity versus matric suction functions of two hypothetical soils (sandy and clayey loams) conform to the empirically based equation

$$K = a/[b + (\psi - \psi_a)^n] \qquad \text{for} \qquad \psi \geq \psi_a$$

where K is hydraulic conductivity (cm/sec), ψ is suction head (cm H_2O) and ψ_a is air-entry suction, and a, b, n are constants, with a/b representing the saturated soil's hydraulic conductivity K_s. The exponential parameter n characterizes the steepness with which K decreases with increasing ψ. Assume that in the sandy soil $K_s = 10^{-3}$ cm/sec, $a = 1$, $b = 10^3$, $\psi_a = 10$ cm, and $n = 3$, whereas in the clayey soil $K_s = 2 \times 10^{-5}$, $a = 0.2$, $b = 10^4$, $\psi_a = 20$, and $n = 2$.

Plot the K versus ψ curves. Note that the curves cross, and the relative conductivities are reversed: The sandy soil with the higher K_s exhibits a steeper decrease of K and falls below the clayey soil beyond a certain suction value ψ_c. Calculate the values of ψ (designated $\psi_{1/2}$) at which each K equals $(1/2)K_s$, and estimate the common value of ψ_c at which the two curves intersect.

For the sandy soil:

$$K = [10^3 + (\psi - 10)^3]^{-1}$$

The suction $\psi_{1/2}$ at which $K = 0.5K_s = a/2b$ can be obtained by substituting $\tilde{\psi}$ for $(\psi - \psi_a)$ and setting $\tilde{\psi}^3 = 10^3$ (thus doubling the denominator). Therefore,

$$\tilde{\psi} = 10 \quad \text{and} \quad \psi_{1/2} = \tilde{\psi} + \psi_a = 10 + 10 = 20 \text{ cm}$$

For the clayey soil:

$$K = 0.2/[10^4 + (\psi - 20)^2]$$

Following the previous procedure, and setting $\tilde{\psi}^2 = 10^4$, we obtain

$$\tilde{\psi} = 10^2 \quad \text{and} \quad \psi_{1/2} = \tilde{\psi} + \psi_a = 100 + 20 = 120 \text{ cm}$$

Crossover suction value (ψ_c): We can attempt to obtain this value algebrically (by setting the two expressions for K equal to each other and solving for the common ψ value) or graphically (Fig. 8.6) by reading the ψ value at which the $K(\psi)$ curves intersect. The latter procedure is easier in this case, and it shows ψ_c to be about 48 cm.

(*Note*: The soils depicted are completely hypothetical and are not to be taken as typical of real sandy and clayey soils.)

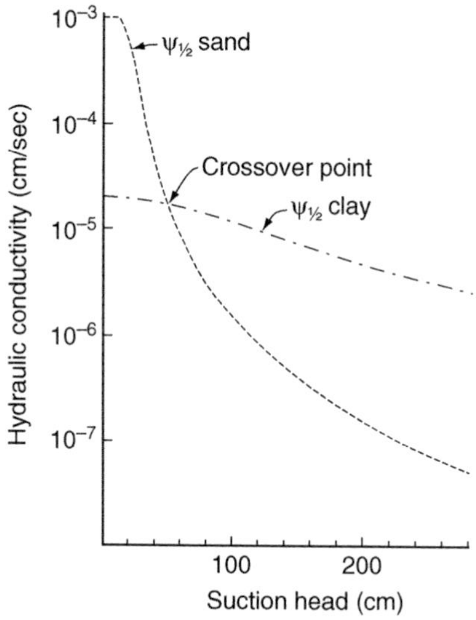

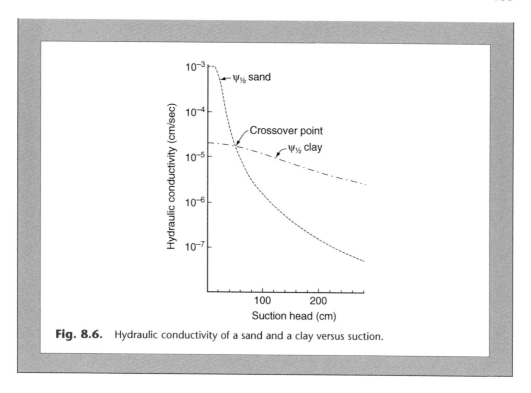

Fig. 8.6. Hydraulic conductivity of a sand and a clay versus suction.

EQUATIONS OF UNSATURATED FLOW

The Continuity Equation

Students may feel that the *equation of continuity*, based on the mass conservation law, is intuitively understandable or logically self-evident so that it requires no formal proof. It is a simple statement, in mathematical form, that for a conserved substance such as water (i.e., a substance that is neither created nor destroyed in the soil), the amount entering minus the amount exiting a specified volume of soil must be equal to the change in water content of the same body. For those who like formalism, however, we offer the following.

Consider a volume element of soil as a rectangular parallelepiped inside a space defined by the rectangular coordinates x, y, z (Fig. 8.7). Assume that the sides of the volume element are Δx, Δy, and Δz and that its volume is $\Delta x\ \Delta y\ \Delta z$. Now consider the net flow in the x direction. If the flux emerging from the right-hand face exceeds the flux entering the left-hand face by the amount $(\delta q/\delta x)\Delta x$, then the difference in discharge (volume per unit time) flowing through the two faces must be (discharge being equal to flux multiplied by area $\Delta y\ \Delta z$)

$$\text{Change in volume discharge} = q\ \Delta y\ \Delta z - [q + (\partial q/\partial x)\Delta x]\Delta y\ \Delta z \quad (8.2)$$
$$\text{(net inflow rate)} = \text{(inflow rate minus outflow rate)}$$

The net inflow rate must equal the rate of gain of water by the volume element of soil per unit time:

$$\text{Net inflow rate} = -(\partial q/\partial x)\Delta x\ \Delta y\ \Delta z \quad (8.3)$$

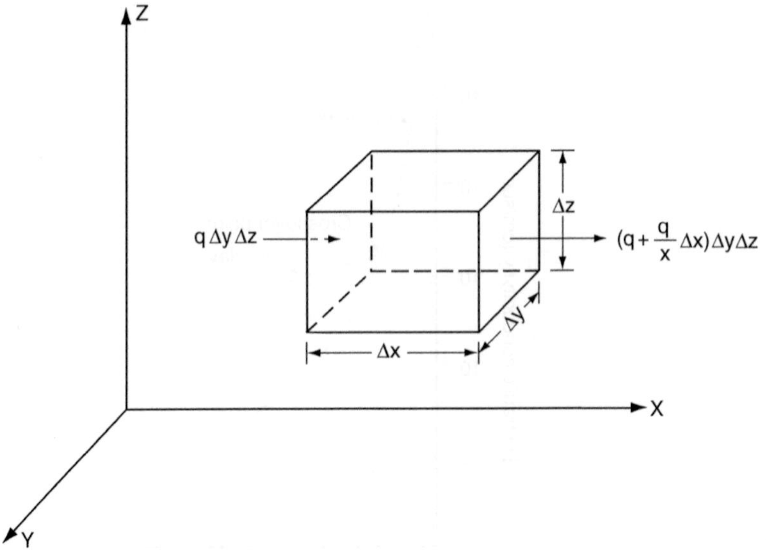

Fig. 8.7. The continuity principle: a volume element of soil gaining or losing water in accordance with the divergence of the flux.

The rate of gain of water by the volume element of soil can also be expressed in terms of the time rate of change of the volume concentration of water, θ, multiplied by the volume of the element:

$$\text{Rate of gain} = (\partial\theta/\partial t)\Delta x\ \Delta y\ \Delta z$$

Setting the two alternative expressions equal to each other, we get

$$(\partial\theta/\partial t)\Delta x\ \Delta y\ \Delta z = -(\partial q/\partial x)\Delta x\ \Delta y\ \Delta z$$

$$\text{or}\quad \partial\theta/\partial t = -\partial q/\partial x \tag{8.4}$$

If we also consider the fluxes in the y and z directions, we obtain the three-dimensional form of the continuity equation:

$$\frac{\partial\theta}{\partial t}\Delta x\Delta y\Delta z = -\left(\frac{\partial q_x}{\partial x} + \frac{\partial q_y}{\partial y} + \frac{\partial q_z}{\partial z}\right)\Delta x\Delta y\Delta z \tag{8.5}$$

$$\frac{\partial\theta}{\partial t} = -\left(\frac{\partial q_x}{\partial x} + \frac{\partial q_y}{\partial y} + \frac{\partial q_z}{\partial z}\right) \tag{8.6}$$

where q_x, q_y, q_z are the fluxes in the x, y, z directions, respectively.

In shorthand mathematical notation, the last equation is written as

$$\partial\theta/\partial t = -\nabla \cdot q \tag{8.7}$$

where the symbol ∇ (del) is the *vector differential operator*, representing the three-dimensional gradient in space (in our case it is the spatial gradient of the flux q). The scalar product of the del operator and a vector function is called the *divergence* and is designated *div*. An alternative expression of Eq. (9.7) is therefore

$$\partial\theta/\partial t = -\text{div } q \tag{8.8}$$

The Combined Flow Equation

Darcy's law, though originally conceived for flow in saturated porous media, has been extended to unsaturated flow, with the provision that the conductivity is now a function of the matric suction head [i.e., $K = K(\psi)$]:

$$q = -K(\psi)\ \nabla H \tag{8.9}$$

where ∇H is the hydraulic head gradient, which may include both suction and gravitational components. This equation, alons with its alternative formulations, is known as the Darcy-Buckingham equation.

Equation (8.9) as written here fails to take into account the hysteresis of soil-water characteristics. In practice, the hysteresis problem can sometimes be evaded by limiting the use of this equation to cases in which the suction (or wetness) change is monotonic — either increasing or decreasing continuously. However, in processes involving successive wetting and drying phases, the $K(\psi)$ function may be highly hysteretic. However, the relation of conductivity to volumetric wetness $K(\theta)$ or to degree of saturation $K(S)$ is affected by hysteresis to a lesser extent than is the $K(\psi)$ function. Darcy's law for unsaturated soil can thus be written

$$q = -K(\theta)\ \nabla H \tag{8.10}$$

which, however, still leaves us the problem of dealing with the hysteresis between ψ and θ.

To account for transient flow processes, we introduce the continuity principle:

$$\partial\theta/\partial t = -\nabla \cdot q$$

Thus, we obtain the Richards equation

$$\partial\theta/\partial t = \nabla \cdot [K(\psi)\nabla H] \tag{8.11}$$

Remembering that the hydraulic head is, in general, the sum of the pressure head (or its negative, the suction head ψ) and the gravitational head (or elevation z), we can write

$$\partial\theta/\partial t = \nabla \cdot [K(\psi)\nabla(\psi - z)] \tag{8.12}$$

Since ∇z is zero for horizontal flow and unity for vertical flow, we can rewrite Eg. (8.6) as follows:

$$\frac{\partial\theta}{\partial t} = -\nabla \cdot (K(\psi)\nabla\psi) + \frac{\partial K}{\partial z} \tag{8.12a}$$

or

$$\frac{\partial\theta}{\partial t} = -\frac{\partial}{\partial x}\left(K\frac{\partial\psi}{\partial x}\right) - \frac{\partial}{\partial y}\left(K\frac{\partial\psi}{\partial y}\right) - \frac{\partial}{\partial z}\left(K\frac{\partial\psi}{\partial z}\right) + \frac{\partial K}{\partial z} \tag{8.13}$$

BOX 8.1 Two-Phase Flow

Most theoretical treatments of water flow ignore the presence of the air phase in an unsaturated soil or at least imply that soil air does not hinder water movement. As mentioned, transient-state flow processes generally involve changes in the amount of water contained in a given volume element of soil. If the bulk density of the soil body does not change, then a net inflow of water implies expulsion of air, and an outflow of water implies entry of air. As long as the soil's air phase is continuous (i.e., not encapsulated in the form of isolated pockets or bubbles) and open to the external atmosphere, it presents no appreciable resistance to water flow. This is so because the fluidity of air (inversely related to the viscosity) is some 50 times greater than that of water.

However, when the free entry or escape of air is obstructed by some barrier (e.g., a saturated or tightly compacted layer with constricted pores), the air phase is confined and necessarily affects water movement. Under such conditions, the simultaneous flows of the two fluids should be considered jointly (Kutilek and Nielsen, 1994).

Processes may occur in which ∇z (the gravity gradient) is negligible compared to the strong matric suction gradient $\nabla\psi$. In such cases,

$$\partial\theta/\partial t = \nabla \cdot [K(\psi)\nabla\psi] \tag{8.14}$$

or, in a one-dimensional horizontal system,

$$\frac{\partial\theta}{\partial t} = \frac{\partial}{\partial x}\left[K(\psi)\frac{\partial\psi}{\partial x}\right] \tag{8.15}$$

HYDRAULIC DIFFUSIVITY

Efforts have been made to simplify the mathematical treatment of unsaturated flow processes by casting the flow equations into a form analogous to the equations of diffusion and heat conduction, for which ready solutions are available (e.g., Carslaw and Jaeger, 1959; Crank, 1975), in some cases involving boundary conditions applicable to soil-water flow processes. It is sometimes possible to relate the flux to the water content (wetness) gradient rather than to the potential (suction) gradient. We illustrate the approach using the one-dimensional form of the flow equation, disregarding gravity.

The matric suction gradient $\delta\psi/\delta x$ can be expanded by the chain rule:

$$\frac{\partial\psi}{\partial x} = \frac{d\psi}{d\theta}\frac{\partial\theta}{\partial x} \tag{8.16}$$

Here $\partial\theta/\partial x$ is the wetness gradient and $\partial\psi/\partial\theta$ is the reciprocal of *specific water capacity*, $C(\theta)$:

$$C(\theta) = d\theta/d\psi \tag{8.17}$$

which is the slope of the soil-moisture characteristic curve at a particular value of wetness θ.

We can now rewrite the Darcy equation as follows:

$$q = K(\theta)\frac{\partial\psi}{\partial x} = -\frac{K(\theta)}{c(\theta)}\frac{\partial\theta}{\partial x} \tag{8.18}$$

To cast this equation into a form analogous to Fick's law of diffusion, a function is introduced called the *diffusivity*, D:

$$D(\theta) = K(\theta)/C(\theta) = K(\theta)(d\psi/d\theta) \tag{8.19}$$

D is thus defined as the ratio of the hydraulic conductivity K to the specific water capacity C, and since both are functions of soil wetness, so D must be also. To avoid confusion between the classical concept of diffusivity pertaining to the diffusive transfer of components in the gaseous and liquid phases (see Chapters 9 and 11 on solute movement and gas exchange in the soil) and this borrowed application of the same term to describe convective flow, we propose to qualify it with the adjective *hydraulic*. Here, therefore, we shall employ the term *hydraulic diffusivity* when referring to D of Eq. (8.19). We can now rewrite Eq. (8.10):

$$q = -D(\theta)\nabla\theta \tag{8.20}$$

or, in one dimension,

$$q = -D(\theta)(\partial\theta/\partial x) \tag{8.21}$$

which is mathematically identical to Fick's first law of diffusion.

Hydraulic diffusivity can thus be viewed as the ratio of the flux (in the absence of gravity and of hysteresis effects) to the soil-water content (wetness) gradient. As such, D has dimensions of length squared per unit time (L^2T^{-1}), since K has the dimensions of volume per unit area per time (LT^{-1}) and the specific water capacity C has dimensions of volume of water per unit volume of soil per unit change in matric suction head (L^{-1}). In the use of Eq. (8.21), the gradient of wetness is taken to represent, implicitly, a gradient of matric potential, which is the true driving force.

Introducing the hydraulic diffusivity into Eq. (8.15), for one-dimensional flow in the absence of gravity, we obtain

$$\frac{\partial\theta}{\partial t} = \frac{\partial}{\partial x}\left[D(\theta)\frac{\partial\theta}{\partial x}\right] \tag{8.22}$$

which has only one dependent variable (θ) rather than the two (θ and ψ) of Eq. (9.15).

In the special case that the hydraulic diffusivity remains constant (though it is not generally safe to assume this except for a very small range of wetness), Eq. (8.22) can be written in the form of Fick's second diffusion equation:

$$\partial\theta/\partial t = D(\partial^2\theta/\partial x^2) \tag{8.23}$$

A word of caution is now in order. In employing the diffusivity concept and all relationships derived from it, we must remember that the process of liquid water movement in the soil is not one of diffusion but of *mass flow*, also termed *convection*. As we have already suggested, the borrowed term *diffusivity*, if taken literally, can be misleading. Furthermore, the diffusivity equations become awkward whenever the hysteresis effect is appreciable or where the soil is layered or in the presence of thermal gradients. Under such conditions, flow may not occur down a water-content gradient or bear any simple or consistent relation to that gradient, and it may actually be in the opposite direction to it.

Whenever the use of hydraulic diffusivity is appropriate, an important advantage to doing so is that its range of variation is generally much smaller than that of hydraulic conductivity. The maximum value of D found in practice is of the order of 1 m^2/day (10^4 cm^2/day). D generally diminishes to about 10^{-3}–10^{-4} m^2/day (1–10 cm^2/day) at the lower limit of wetness normally encountered in the root zone. It thus varies about a thousandfold rather than about a millionfold, as does the hydraulic conductivity in the same wetness range. Hence small changes in θ are likely to affect the value of $D(\theta)$ to a much small degree than that of $K(\theta)$. Conversely, the use of an approximate or "average" value of D is less likely to entail large errors in calculation of flux than the use of an inaccurate value of K.

To take account of gravity, a diffusivity equation can be written in the form

$$\frac{\partial \theta}{\partial t} = \nabla \cdot [D(\theta)\nabla\theta] + \frac{\partial K(\theta)}{\partial z} = \nabla \cdot [D(\theta)\nabla\theta] + \frac{dK}{d\theta}\frac{\partial \theta}{\partial z} \tag{8.24}$$

The relation of hydraulic diffusivity to wetness is shown in Fig. 8.8.
Note that the right-hand section of the curve, shows a rise in diffusivity with wetness. In the very dry range, however, the diffusivity often indicates an opposite trend — a rise with diminishing soil wetness, apparently due to the contribution of vapor movement. In the very wet range, as the soil approaches complete saturation, the diffusivity becomes indeterminate [since $C(\theta)$ tends to zero]. The latter effect is due to the shape of the soil moisture characteristic, θ versus ψ, in the near-saturation range.

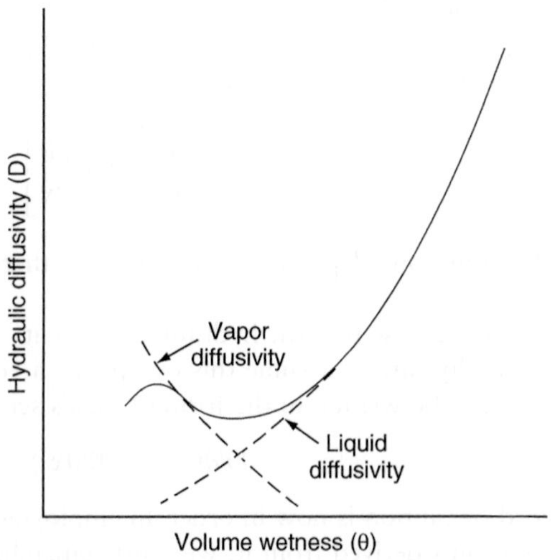

Fig. 8.8. Relation of diffusivity to soil wetness.

LABORATORY MEASUREMENT OF CONDUCTIVITY AND DIFFUSIVITY

Knowledge of the unsaturated hydraulic conductivity and diffusivity at different suction and wetness values is generally required before any of the mathematical theories of water flow can be applied in practice. Since there is as yet no universally accepted way to predict these values from more basic or easily obtainable soil properties, K and D must be measured experimentally. In principle, these parameters can be obtained from either *steady-state* or *transient-state* flow systems. In steady flow systems, the flux, gradient, and water content are constant in time, whereas in transient flow systems they vary. Hence measurements based on steady flow may seem to be more convenient and more accurate. The difficulty, however, lies in setting up a steady flow system, which may take a very long time to stabilize.

Techniques for measurement of conductivity and diffusivity of soil samples or models in the laboratory were described by Klute and Dirksen (1986). The conductivity is usually measured by applying a constant hydraulic head difference across the sample and measuring the resulting steady flux of water. Soil samples can be desaturated either by tension-plate devices or in a pressure chamber. The measurements are made at successive levels of suction and wetness so as to obtain the functions $K(\psi)$, $K(\theta)$, and $D(\theta)$. The $K(\psi)$ relation may be strongly hysteretic, so a complete determination requires measurements in desorption and in sorption (as well, perhaps, as in intermediate scanning). This is difficult, however, and requires specialized apparatus, so all too often only the desorption curve is measured (starting at saturation and proceeding to increase the suction in increments).

Such laboratory techniques can be applied to the measurement of undisturbed soil cores taken from the field. This is certainly preferable to measurements taken on fragmented and artificially packed samples, though it should be understood that no field sampling technique yet available provides truly undisturbed samples. Moreover, any attempt to represent a field soil by means of extracted samples incurs the problem of field soil heterogeneity as well as the associated problem of determining the appropriate scale (i.e., the representative volume) for realistic measurement of parameters.

A widely used transient flow method for measurement of conductivity and diffusivity in the laboratory is the outflow method. It is based on measuring the diminishing rate of outflow from a sample in a pressure cell after the pressure is raised by a certain increment. In the application of this method, the *hydraulic resistance* of the porous plate or membrane on which the sample is placed must be taken into account. Laboratory measurements of conductivity and diffusivity can also be made on long columns of soil, not only on small samples contained in cells. In such columns, steady-state flow can be induced by evaporation or by infiltration. If the column is long enough to allow the measurement of suction gradients (e.g., by a series of tensiometers) and of wetness gradients (by sectioning or, preferably, by a nondestructive technique such as gamma-ray scanning), the $K(\theta)$ and $K(\psi)$ relationships can be obtained for a range of θ values with a single column or with a series of similarly packed columns.

IN SITU MEASUREMENT OF UNSATURATED HYDRAULIC CONDUCTIVITY

Application of the theories of soil physics to the description or prediction of actual processes in the field (processes involved in infiltration, runoff control, irrigation, drainage, water conservation, groundwater recharge and pollution, etc.) depends on knowledge of soil hydraulic characteristics, including the functional relation of hydraulic conductivity and of matric suction to soil wetness, as well as the their spatial and temporal variation.

It seems fundamentally unrealistic to measure the unsaturated hydraulic conductivity of field soil by making laboratory determinations on discrete samples removed from their natural continuum. Such samples are generally dried, fragmented, and repacked into experimental containers, so the original structure is destroyed. Hence it is necessary to devise and test practical methods for measuring soil hydraulic conductivity on a realistic scale in situ. We now proceed to give a brief description of several of the methods available for this purpose. A comprehensive review was given by Green et al. (1986) and more recently by Clothier and Scotter (2002).

Sprinkling Infiltration

The principle of this method is that a continued supply of water to the soil at a constant rate lower than the saturated hydraulic conductivity of the soil eventually establishes a steady moisture distribution in the conducting profile. Once steady-state conditions are established, a constant flux exists. In a uniform soil the suction gradients will tend to zero, and with only a unit gravitational gradient in effect the hydraulic conductivity becomes effectively equal to the flux. This test can be performed on an initially dry soil to which successively increasing sprinkling intensities are applied incrementally. It then becomes possible to obtain different values of hydraulic conductivity corresponding to different values of soil wetness.

The difficulty of the steady sprinkling infiltration test in the field is that it requires rather elaborate equipment, which must be maintained in continuous operation for long periods of time. The requirement of maintaining continuous operation becomes increasingly important, and difficult, as one attempts to extend the test toward the greater suction range by reducing the application rate below 1 mm/hr. Another difficulty is to avoid the raindrop impact effect, which can cause the exposed surface soil to disperse and seal, thus reducing infiltrability. This problem can be avoided by mulching the soil surface with straw. A more serious problem is the possible presence at some depth of soil layers that might impede flow, thus preventing the attainment throughout the profile of a condition of a zero matric suction gradient. Ideally, therefore, this test applies to uniform (or nearly uniform) soil profiles rather than to distinctly layered ones.

Infiltration Through an Impeding Layer

This method was first suggested by Hillel and Gardner (1970a), who showed that an impeding layer at the surface of the soil can be used to achieve the desired boundary conditions for measuring the underlying soil's unsaturated hydraulic conductivity and diffusivity as functions of soil wetness. The effect of an impeding layer present over the top of the profile during infiltration is to decrease the hydraulic potential in the profile under the impeding layer. Thus, the soil wetness and, correspondingly, the conductivity and diffusivity values of the infiltrating profile are reduced.

When the surface is covered by an impeding layer ("crust") with a saturated conductivity smaller than that of the underlying soil, steady flow conditions are eventually established. At that stage, the hydraulic head gradient in the subcrust soil is unity (i.e., gravitational only), while the head gradient through the impeding layer (owing to its low conductivity) is necessarily greater than unity. If the ponding depth of water over the surface is negligible, such an impeding layer induces the development of suction in the sublayer, the magnitude of which increases with increasing hydraulic resistance of the crust. After steady infiltration is achieved, the conductivity of the subcrust soil becomes equal to the flux.

A variant of the crust method is the use of a so-called *tension infiltrometer* (Elrick and Reynolds, 1992; Warrick, 1992; Bootlink and Bouma, 2002). It consists of a reservoir of water designed to maintain a constant subatmospheric pressure by means of a Mariotte tube, at the bottom of which is a fixed a porous plate. That plate introduces hydraulic resistance (analogous to a surface crust of the soil itself), and the hydraulic gradient through the plate when it is placed on the soil surface creates a suction at its contact place with the soil. This instrument has been used in soils with macropores to assess their potential contribution to the process of infiltration under ponding (see Chapter 14).

A serious problem that may invalidate the crust test in some cases pertains to the so-called *unstable flow* phenomenon. It has been observed that in transition from a fine-textured zone to a coarse-textured zone during infiltration, the advance of the water may not be even (or planar, as it tends to be in a uniform profile), but that sudden breakaway flows may occur in specific locations where fingerlike intrusions take place. This phenomenon is discussed in Chapter 14.

Internal Drainage

This method is based on monitoring the transient-state internal drainage of a profile in the field (Fig. 8.9). The method can help to minimize the alteration of soil hydraulics that can result from methods that disrupt soil structure. The method requires frequent and simultaneous measurements of the soil wetness and matric suction profiles under conditions of drainage alone (evapotranspiration prevented). The use of strain gauge pressure transducers in tensiometry facilitate the rapid and automatic acquisition of suction data, while the required soil wetness data are obtainable with a neutron moisture gauge or a time-domain reflectometer. From these measurements it is possible to obtain instantaneous values of the potential gradients and fluxes operating within the profile and hence also of hydraulic conductivity values. Once the hydraulic conductivity at each elevation within the profile is known in relation to wetness, the data can be applied to the analysis of drainage and evaporation in a bare field or of evapotranspiration in a vegetated field.

To apply this method in the field, one must choose a characteristic fallow plot that is large enough (say, at least 5×5 m) so that processes at its center are

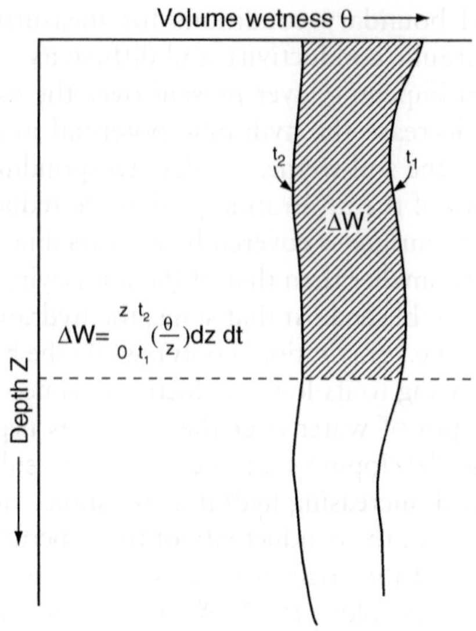

Fig. 8.9. Change in profile water content ΔW in the depth interval $Z = 0$ (soil surface) to depth Z in the time interval t_1 to t_2 during internal drainage, assuming no evaporation or lateral flow.

unaffected by the boundaries. Within this plot, at least one neutron access tube is installed through and below the root zone. The desirable depth will sometimes exceed 2 m. A series of tensiometers is installed at various depths near the access tube so as to represent profile horizons. (The depth interval between succeeding tensiometers should not exceed 30 cm.) Water is then ponded on the surface and the plot is irrigated long enough so that the entire profile becomes as wet as it can be. After this, the soil surface is covered by an opaque plastic sheet so as to minimize soil heating and to prevent evaporation from the soil surface. As internal drainage proceeds, periodic measurements are made of distribution and tension of soil moisture throughout the profile. The handling of the data and a sample calculation of the hydraulic conductivity were described by Hillel et al. (1972) and more recently by Vachaud and Dane (2002).

The method as described is not applicable where lateral movement of soil moisture is appreciable. This movement is not normally significant when the soil profile is unsaturated, but lateral movement can become significant wherever an impeding layer occurs on which saturated conditions might prevail for some time. In practice, the moisture range for which conductivity can be measured by the internal drainage method is generally limited to suctions not exceeding about 0.5 bar, since the drainage process often slows down within a few days or weeks to become practically imperceptible thereafter.

VAPOR MOVEMENT

As already stated, liquid water moves in the soil by mass flow, a process by which the entire fluid flows in response to differences in hydraulic potential. In certain circumstances, the movement of water vapor can also occur as mass flow, for instance, when wind gusts induce bulk movement of air (containing vapor) in the surface zone of the soil. In general, however, vapor movement within the soil profile occurs by diffusion, a process in which different components of a mixed fluid move independently, and at times in opposite directions, in response to differences in concentration (or partial pressure) from one location to another. Water vapor is always present in the gaseous phase of an unsaturated soil, and vapor diffusion occurs whenever differences in vapor pressure develop within the soil.

The diffusion equation for water vapor is

$$q_v = -D_v(\partial \rho_v/\partial x) \tag{8.25}$$

where ρv is the vapor density (or concentration) in the gaseous phase and *D*v is the diffusion coefficient for water vapor. *D*v in the soil is lower than in open air because of the restricted volume and the tortuosity of air-filled pores (Currie, 1961).

By considering that the liquid water in the soil serves as both a source and a sink for water vapor, and assuming that changes in liquid water content with time are much greater than the changes in vapor density with time, Jackson (1964a) described nonsteady vapor transfer in terms of the liquid water content (volumetric wetness) θ:

$$\frac{\partial \theta}{\partial t} = \frac{\partial}{\partial x}\left[D_v \frac{\partial \rho_v}{\partial \theta}\frac{\partial \theta}{\partial x}\right] \tag{8.26}$$

For the simultaneous transfer of both liquid and vapor, the following equation applies:

$$\frac{\partial \theta}{\partial t} = \frac{\partial}{\partial x}\left[\left(D_v \frac{\partial \rho_v}{\partial \theta} + D_\theta\right)\frac{\partial \theta}{\partial x}\right] \tag{8.27}$$

where D_θ is the hydraulic diffusivity for liquid water.

The foregoing equations consider water vapor diffusion as an isothermal process, assuming that both viscous flow in the liquid phase and diffusion of vapor are impelled by the forces of adsorption and of capillarity. No explicit account is thus taken of osmotic or solute effects on vapor pressure, though such effects can sometimes be significant. More importantly, this approach disregards the simultaneous and interactive transport of both water and heat in nonisothermal situations, to be described in a subsequent chapter.

At constant temperature, vapor-pressure differences in a nonsaline soil are likely to be very small. For example, at standard temperature a change in matric suction from 0 to 10 MPa (100 bar) is accompanied by a vapor pressure reduction from 17.54 to 16.34 torr, i.e. a difference of only 1.2 torr (1.6×10^{-4} MPa, or 1.6 mbar). (The torr is a unit of pressure equal to 1 mm of mercury under standard conditions.) Under normal field conditions (except at extreme dryness) soil air is nearly vapor saturated at almost all times.

When temperature differences do occur, they can cause considerable differences in vapor pressure. For example, a change of 1 degree in water temperature, from 19 to 20°C, results in an increase in vapor pressure of 1.1 torr.

Thus, a change in temperature of just 1°C has nearly the same effect on vapor pressure as a change in suction of 10 MPa (100 bar)!

In the range of temperatures prevailing in the field, the variation of saturated vapor pressure (that is, the vapor pressure in equilibrium with pure, free water) is as follows:

Temperature (°C)	0	20	30	40
Vapor pressure (torr)	4.6	17.5	38.0	55.8

Vapor movement tends to take place from warmer to cooler zones in the soil. Since the soil surface is warmer during the daytime and colder during the night than the deeper layers, vapor movement tends to be downward during the day and upward during the night. Temperature gradients can also induce liquid flow, because higher temperatures tend to reduce matric suction and thus can induce hydraulic potential gradients. As commonly observed, liquid flow is the dominant mode in moist, nearly isothermal soils; hence the contribution of vapor diffusion to overall water movement is probably negligible in the main part of the root zone, where diurnal temperature fluctuations are slight.

9. SOLUTE MOVEMENT AND SOIL SALINITY

THE SOIL SOLUTION

In the preceding chapters, we discussed soil water with only occasional reference to solutes. However, water present in the soil and constituting its liquid phase is never chemically pure. In the first place, the water entering the soil as rain or irrigation is itself a solution. Rainwater is of course distilled and essentially pure when it first condenses to form clouds, but as it descends through the atmosphere it generally dissolves such atmospheric gases as carbon dioxide and oxygen, often together with such products of our industrial civilization as oxides of sulfur and nitrogen, as well as — along the coast — appreciable quantities of salt that enter the air as sea spray. Solute concentrations (mostly electrolytic salts) typically found in rainwater are of the order of 5–20 mg/L (units roughly equivalent to mg/kg, also designated parts per million, or ppm).

During its residence in the soil, the infiltrated water tends to dissolve additional solutes, including soluble products of mineral and organic matter decomposition as well as fertilizers and pesticides. The concentration of salts in irrigation water (generally obtained from surface reservoirs or subterranean aquifers) may range between 100 and 1000 mg/L. Finally, the concentration may be as high as 10,000 mg/L in drainage from saline soil. (*Note*: The concentration of ocean water is of the order of 35 grams per liter.)

As it moves through the profile, soil water carries its solute load in its convective stream, leaving some of it behind to the extent that the component salts are adsorbed, taken up by plants, or precipitated whenever their concentration exceeds their solubility (mainly at the soil surface during evaporation). Solutes move not only with soil water, but also within it, in response to concentration gradients. At the same time, solutes react among themselves and interact with the solid matrix of the soil in a continuous cyclic succession of interrelated physical and chemical processes. These interactions involve and are strongly influenced by such variable factors as acidity, temperature, oxidation–reduction potential, composition, and concentration of the soil solution.

In former times, processes involving solutes were considered to belong in the exclusive realm of soil chemistry and outside the scope of a treatise on soil physics. Nowadays, however, we have come to recognize that arbitrary separations among "traditional" disciplines that are in fact complementary and overlapping should not be allowed to hinder our quest for an ever more comprehensive understanding of interactive phenomena in the environment.

A better understanding of the simultaneous movement and interactions of water and solutes in the soil is essential to the improvement of soil fertility through control of nutrients in the root zone as well as to the prevention of soil salinity, alkalinity, and other occurrences of toxicity. Such an understanding has also become crucial in the area of environmental management whenever soil-borne solutes migrate to, and affect the quality of, groundwater or surface water resources (Hendrickx et al., 2002).

CONVECTIVE TRANSPORT OF SOLUTES

The convection (or *mass flow*) of soil water, sometimes called the *Darcian flow*, carries with it a flux of solutes J_c proportional to their concentration c:

$$J_c = qc = -c[K\ (dH/dx)] \tag{9.1}$$

where $q = -K(dH/dx)$ is Darcy's law, discussed in the preceding two chapters. Since q is usually expressed as volume of liquid flowing through a unit area (perpendicular to the flow direction) per unit time and c as mass of solute per unit volume of solution, J_c is given in terms of mass of solute passing through a unit cross-sectional area of soil per unit time.

To estimate the distance of travel of a solute per unit time, we consider the *average apparent velocity* $\bar{v}$ of the flowing solution:

$$\bar{v} = q/\theta \tag{9.2}$$

where θ is volumetric wetness and $\bar{v}$ is taken as the straight-line length of path traversed in the soil in unit time. In this formulation we disregard the roundabout path caused by the geometric tortuosity of the water-filled soil pores. Since actual velocities vary over several orders of magnitude within pores and between pores, the concept of *average velocity* is obviously a gross approximation (much like lumping insects and elephants in computing an "average" size animal). However, assuming $\bar{v}$ to be a working approximation, we get

$$J_c = \bar{v}\theta c \tag{9.3}$$

It is sometimes useful to have an estimate of the average *residence time* t_r of a solute within a layer of soil of thickness L (especially if we are concerned with some time-dependent interactive process involving the solute under consideration). Accordingly,

$$t_r = L/\bar{v} \tag{9.4}$$

If the flow is impelled by gravity alone, with no pressure gradients, then the downward flux of liquid is equal to the hydraulic conductivity K of the medium, a function of wetness θ:

$$q = K(\theta)$$

Using Eq. (9.2), we thus obtain for Eq. (9.4)

$$t_r = L\ \theta/K(\theta) \tag{9.5}$$

The foregoing equations allow us to estimate, for instance, the distance of travel of a soluble pollutant from the bottom of, say, a septic tank or sanitary landfill to the water table, through the so-called *unsaturated zone*, as follows:

$$L_t = tK(\theta)/\theta \tag{9.6}$$

where L_t is the average distance of convective transport in time t. If the values of wetness and hydraulic conductivity vary within the soil profile, as often happens, the foregoing calculations must be carried out layer by layer to determine the solution's space-variable flux, average distance of travel, and per-layer residence time.

The serious shortcoming of this approach is that one cannot be sure that the transport of solutes occurs by convection alone. In fact, we know that solutes do not merely move with the water as sedentary passengers in a train, but also move within the flowing water in response to concentration gradients in the twin processes of *diffusion* and *hydrodynamic dispersion*. Moreover, solutes are not always as inert as we have thus far assumed, because they tend to interact with the biological system within the soil (e.g., to be taken up or released by microbes and the roots of plants) and with the physicochemical system (e.g., to be adsorbed or exchanged within or by the soil). Additionally, solutes may undergo chemical reactions and may also be removed from solution by precipitation or volatilization.

Sample Problem

A soluble pollutant was inadvertently spilled on the ground. Suppose that it is nondegradable, nonvolatile, not taken up by plants, not adsorbed by the soil, and not immobilized by any other mechanism. If the annual rainfall is 1500 mm, the annual evapotranspiration is 1250 mm, the water table is 20 m deep, and the so-called unsaturated zone underlying the soil has a constant volumetric wetness of 25%, estimate the residence time in the unsaturated zone and the time required for the pollutant to reach the groundwater.

As a rough approximation, we assume that the solute is transported only by the convective stream of water draining out of the soil, vertically downward through the unsaturated zone toward the water table. The possible effects of diffusion and dispersion are thus disregarded. We further assume that this drainage occurs under steady-state conditions, that is, that temporal perturbations at the soil surface resulting from intermittent periods of rainfall and evapotranspiration are damped out in the upper layer of the soil.

To estimate the distance of travel of a solute per unit time, we take the average velocity $\bar{v}$ of the flowing solution to be [Eq. (9.2)]

$$\bar{v} = q/\theta$$

where $\bar{v}$ is the straight-line length of path traversed through the soil or subsoil by the solution in unit time, θ is the volumetric wetness, and q is the flux (volume flowing through unit area per unit time). Substituting the given values we obtain

$$\bar{v} = (1500 - 1250)/0.25 = 1000 \text{ mm/yr}$$

To estimate the residence time t_r of the solute within the unsaturated zone, we refer to Eq. (9.4):

$$t_r = L/\bar{v}$$

where L is the thickness of the zone considered. Thus,

$$t_r = 20 \text{ m}/1 \text{ m/yr} = 20 \text{ y}$$

Thus, the bulk of the solute can be expected to reach the water table and enter the groundwater in about 20 years. Since in fact diffusion and dispersion phenomena do operate, we can expect some of the solute to move faster than the bulk of the solution and some to lag behind. However, the calculation based on convective transport alone can serve as a useful rough estimation of the movement rate.

DIFFUSION OF SOLUTES

Diffusion processes commonly occur within multicomponent gaseous or liquid phases, in consequence of the random thermal motion (often called *Brownian motion*) and repeated collisions and deflections of molecules in the fluid. The net effect is a tendency toward equalizing the spatial distribution of the components in a nonhomogeneous fluid.

Diffusion processes are very important in the soil. As we show in Chapter 11 on soil aeration, diffusion in the air phase of such gases as oxygen, carbon dioxide, nitrogen, and water vapor can have a decisive influence on the soil's chemical and biological processes. Equally important are diffusion processes involving solutes in the soil's liquid phase, including nutrients as well as salts and toxic compounds that affect plant growth.

Wherever solutes are not distributed uniformly throughout a solution, concentration gradients exist. Consequently, solutes tend to diffuse from zones where their concentration is higher to where it is lower. In bulk water at rest, the rate of diffusion J_d is related by *Fick's first law* to the gradient of the concentration c:

$$J_d = -D_0(dc/dx) \tag{9.7}$$

in which D_0 is the diffusion coefficient for a particular solute diffusing in bulk water and dc/dx is that solute's effective concentration gradient.

For diffusion in the soil's liquid phase, the effective diffusion coefficient is generally less than the diffusion coefficient in bulk water, D_0, for several reasons. In the first place, the liquid phase occupies only a fraction of the soil volume; at most, in a state of saturation, its volume fraction equals the soil's porosity. Secondly, the soil's pore passages are tortuous, so the actual path length of diffusion is significantly greater than the apparent straight-line distance. In an unsaturated soil, as soil wetness diminishes, the fractional volume available for diffusion in the liquid phase decreases still further and the tortuous length of path increases.

If the sole factors affecting the diffusion coefficient in the soil (D_s) are fractional water volume θ and *tortuosity* ξ, we can write

$$D_s = D_0\theta\xi \tag{9.8}$$

where ξ, the tortuosity factor, is an empirical parameter smaller than unity, expressing the ratio of the straight-line distance to the average roundabout-path length through the water-filled pores for a diffusing substance. This parameter has been found to depend on both the fractional volume and the geometric configuration of the water phase and, hence, to decrease with decreasing θ. Thus, D_s is strongly dependent on θ, both directly and through its dependence on ξ, which itself is a function of θ. To show this dependence, we write $D_s(\theta)$.

Other factors in addition to the geometric ones considered in Eq. (9.8) also tend to decrease the effective diffusion coefficient, particularly in unsaturated soils with an appreciable content of clay. As soil wetness is reduced and the water films coating the particles contract, the increasing density of the exchangeable cations adsorbed to the clay surfaces and the corresponding exclusion of anions, as well as the possible increase in viscosity of the adsorbed liquid phase, might combine to further retard diffusion. Because these and other complicating factors are not mutually independent, it has seemed impossible thus far to formulate them separately. Hence, it is tempting to lump them all together into a single *complexity factor*, which we can designate α. Thus,

$$D_s = D_0\theta\alpha \tag{9.9}$$

We can now rewrite Eq. (9.7) for diffusion in the liquid phase of an unsaturated soil:

$$J_d = -D_s(\theta)(dc/dx) \tag{9.10}$$

By itself, this equation can only describe *steady-state diffusion processes.* In order to proceed to a more generalized diffusion equation capable of describing *transient-state processes* (in which the rate and concentration vary in time), we must (as in the case of water flow alone) invoke the *mass conservation law*, expressed in the *continuity equation.*

Let us assume that there are no sources or sinks for the diffusing solute in the soil body in which diffusion is occurring. Now let us consider a rectangular volume element of soil that contains a liquid phase and is bounded by two parallel square planes, of area A, separated by a distance Δx. If diffusion takes place from left to right, the amount of solute diffusing through the left-hand plane into the volume element per unit time is AJ_d, and the amount diffusing out through the right-hand plane is $A[J_d + (\partial J_d/\partial x)\Delta x]$. The rate of accumulation of the solute in the volume element is $\theta A(\partial c/\partial t)\Delta x$, where $\partial c/\partial t$ is the time rate of change of concentration. Thus,

$$\theta A(\partial c/\partial t)\Delta x = A[J_d + (\partial J_d/\partial x)\Delta x] - AJ_d$$

which reduces to

$$\theta(\partial c/\partial t) = -\partial J_d/\partial x \qquad (9.11)$$

The negative sign indicates that any increase of diffusive flux along the direction of diffusion from the entry face to the outlet face of the volume element necessarily depletes the concentration in the volume element; and vice versa (the concentration increases if the rate of out-diffusion is less than the in-diffusion).

Combining this with Eq. (9.10), we obtain a second-order equation as follows:

$$\theta(\partial c/\partial t) = \partial[D_s(\theta)(\partial c/\partial x)]/\partial x \qquad (9.12)$$

In the special case when the diffusion coefficient D_s is constant, Eq. (9.12) assumes the classical form of *Fick's second law of diffusion*:

$$\theta(\partial c/\partial t) = D_s(\partial^2 c/\partial x^2) \qquad (9.13)$$

However, D_s is generally not constant and in fact might be dependent not only on wetness but also on concentration, i.e., be a function of both θ and c.

Attempts to apply the foregoing equations to describe the diffusion of solutes in the soil solution, particularly in unsaturated soil conditions, have difficulties. The soil varies both in space and time. Solutes interact with, and modify, the solid matrix — and hence may affect the pore space. Different species of solutes in the liquid phase interact with one another as well as with the adsorbed phase. Evaporation and condensation of water in pores further modify the concentration gradients and the pattern of diffusion. Finally, the convective flow of the solution also affects the diffusion process by changing the distribution of solutes and by inducing a process called *hydrodynamic dispersion* (Skaggs and Leij, 2002).

HYDRODYNAMIC DISPERSION

The motion of any inhomogeneous solution in a porous body brings about another process that differs from diffusion in its mechanism but that tends to produce an analogous or synergetic tendency to mix and eventually to even-out the differences in concentration or composition among different portions of the flowing solution. This process, which at times predominates over diffusion, is called *hydrodynamic dispersion.* It results from the microscopic nonuniformity of flow velocity in the soil's conducting pores. Since water moves faster in wide pores than in narrow ones and faster in the center of each pore than along its walls, some portions of the flowing solution move ahead while other portions lag behind.

Consider, for example, the laminar flow of a solution through a single capillary pore hypothetically shaped like a cylindrical tube. From our earlier derivation of Poiseuille's law (Chapter 7) we know that flow velocity v in such a tube is a decreasing function of radial distance r from the center:

$$v = 2\bar{v}(1 - r^2/R^2) \tag{9.14}$$

Here $\bar{v}$ is average velocity and R is the radius of the tube. Thus the velocity of a solute molecule carried by the convective stream depends on its position within the pore passage. At $r = R$, which is at the wall of the tube, the velocity is zero, whereas at $r = 0$, in the center of the tube, the velocity is maximal and equal to twice the average velocity.

Added to the nonuniformity of velocity within each pore is the fact that pores vary widely in radius over several orders of magnitude, say from 1 μm to 1 mm. We recall Poiseuille's law [Eq. (7.3)],

$$Q = 4R^4(\Delta p/8\eta L)$$

which states that the discharge Q (volume flow per unit time) is proportional to the fourth power of the radius R. Hence a pore with an effective radius of 1 mm will conduct a volume of water $(10^3)^4 = 10^{12}$ (a million million) times greater than a pore having an effective radius of 1 μm. Great indeed is the microscopic-scale variation of pore water velocity in the soil!

The fact that some portions of a flowing solution move faster than other portions causes an incoming solution to mix or disperse within the antecedent solution. The degree of mixing depends on such factors as average flow velocity, pore size distribution, degree of saturation, and concentration gradients. When the convective velocity is sufficiently high, the relative effect of hydrodynamic dispersion can greatly exceed that of molecular diffusion and the latter can be neglected in the analysis of solute movement. On the other hand, when the soil solution is at rest, hydrodynamic dispersion does not come into play at all.

Mathematically, hydrodynamic dispersion can be formulated in a manner analogous to diffusion, per Eqs. (9.10) and (9.12), except that a *dispersion coefficient* D_h is used instead of a diffusion coefficient. D_h has been found (Bresler, 1972b) to depend more or less linearly on the average flow velocity $\bar{v}$. Thus

$$D_h = a\bar{v} \tag{9.15}$$

with a an empirical parameter.

Because of the similarity in effect (though not in mechanism) between diffusion and dispersion, it is tempting to assume the two effects to be additive. Accordingly, the diffusion and dispersion coefficients are often combined into a single term, called the *diffusion–dispersion coefficient* (D_{sh}, which is a function of both the fractional water volume θ and the average velocity $\bar{v}$:

$$D_{sh}(\theta, \bar{v}) = D_s(\theta) + D_h(\bar{v}) \tag{9.16}$$

MISCIBLE DISPLACEMENT AND BREAKTHROUGH CURVES

When a liquid different in composition or concentration from the preexisting pore liquid is introduced into a column of soil, and the outflow from the end of the column is collected and analyzed, its composition is seen to change in time as the old liquid is displaced and replaced by the new one. If the two liquids are not mutually soluble (as is the case, for example, with oil and water), then the process is called *immiscible displacement.* If, on the other hand, the two liquids mix readily — as do many aqueous solutions — the process just described is referred to as *miscible displacement.* Plots of the outflowing solution's composition versus time or versus cumulative discharge are called *breakthrough curves.*

If the soil were saturated throughout the experiment and if neither diffusion nor hydrodynamic dispersion were to take place at the boundary between the displacing and the displaced solutions, then that boundary would remain a sharp front moving along the length of the column at a velocity equal to the flux. If we were to monitor the composition at the column's outlet, we would notice an abrupt change in composition at the moment the last portion of the old solution was completely driven out and the new solution arrived. Such a pattern of displacement without mixing is commonly called *piston flow.* It is seldom if ever encountered in practice. What normally happens at the boundary or front between the two solutions is a gradual mixing resulting from both diffusion and hydrodynamic dispersion, so that the boundary becomes increasingly diffuse about the mean position of the advancing front, as shown in Figs. 9.1 and 9.2.

Ideally, breakthrough curves should be sigmoidal in shape and symmetrical about the front of the advancing solution, with the inflection representing 50% displacement at a cumulative flow of one "pore volume" if the soil is saturated. A typical curve of this sort is shown in Fig. 9.2 for a sand. Curves obtained with finer-textured soils differ from the symmetrical ideal, particularly in the case of structured (aggregated) soils, owing to various interactions between solutes and the soil matrix (e.g., positive or negative adsorption) and to the possible existence of pore spaces where water movement is so sluggish that portions of the antecedent solution lag behind and are displaced only very slowly.

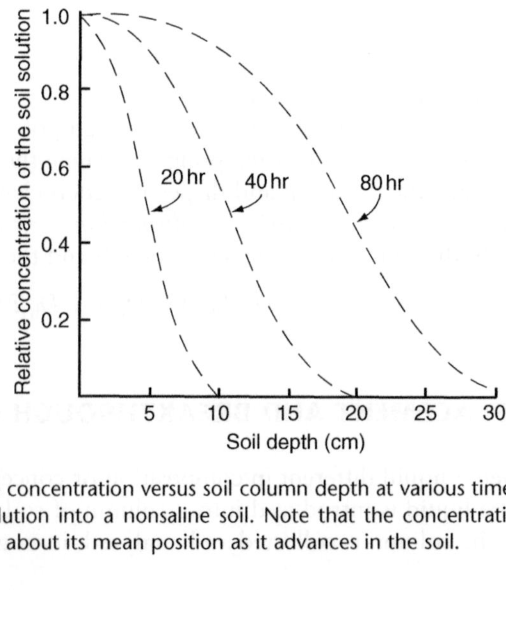

Fig. 9.1. Solute concentration versus soil column depth at various times during the infiltration of a saline solution into a nonsaline soil. Note that the concentration "front" becomes increasingly diffuse about its mean position as it advances in the soil.

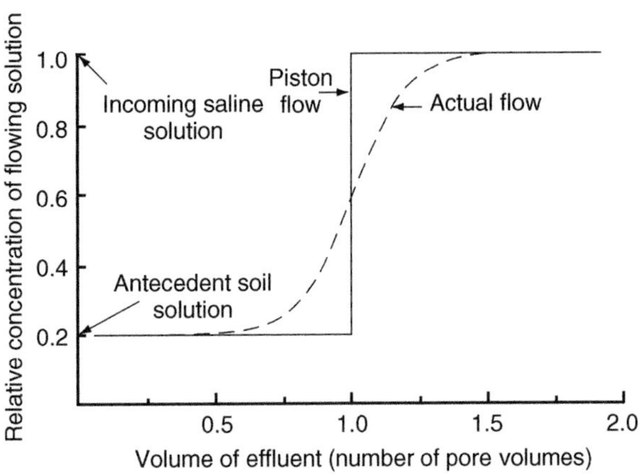

Fig. 9.2. Schematic illustration of a sigmodial breakthrough curve in a saturated sandy soil during the replacement of a dilute antecedent solution by an entering saline solution of fivefold greater concentration.

Miscible displacement phenomena and breakthrough curves are not merely of theoretical interest but are indeed relevant to the solution of many real-world problems, such as the leaching of excess salts from saline soils, the distribution of nutrient solutions, and the pollution of groundwater by the migration of soil-borne solutes of various types, including radioactive wastes, toxic chemicals, and agrochemical (pesticide and fertilizer) residues.

COMBINED TRANSPORT OF SOLUTES

To take into account the three mechanisms of solute movement described thus far [mass flow (convection), molecular diffusion, and hydrodynamic dispersion], we combine Eqs. (9.3), (9.10), and (9.16):

$$J = \bar{v}\theta c - D_s(\theta)(dc/dx) + D_h(\bar{v})(dc/dx)$$

which can be stated in words:

[combined solute flux] = [flux due to convection] + [flux due to diffusion]
+ [flux due to hydrodynamic dispersion]

Since in practice the diffusion and dispersion phenomena cannot be separated, the foregoing equation is usually written in the form

$$J = v\theta c - D_{sh}(\theta, \bar{v})(dc/dx) \tag{9.17}$$

Here J is the total mass of a solute transported across a unit cross-sectional area of soil per unit time, D_{sh} is the lumped diffusion–dispersion coefficient (a function of volumetric wetness θ and average pore-water velocity $\bar{v}$), c is the solute concentration, and dc/dx is the solute gradient.

Equation (9.17) by itself can describe only steady-state (time-invariant) processes. Moreover, it is limited to "noninteracting" (inert) solutes, by which term we designate solutes not subject to adsorption by soil solids or subject to chemical or biological reactions. Strictly speaking, truly noninteracting solutes scarcely exist. Moreover, parameters D_s, D_h, $\bar{v}$, θ, and c can only be defined in macroscopic terms as gross spatial averages. Hence Eq. (9.17) is only an approximation of the process it is meant to depict.

Turning now to transient-state processes, in which fluxes and concentrations can vary both in time and in space, we once again invoke the conservation ("continuity") principle, which for combined convective–diffusive–dispersive transport can be written

$$\partial(c\theta)/\partial t = -\partial J/\partial x \tag{9.18}$$

The rate of change of the solute mass in a volume element of soil should equal the difference between the incoming and outgoing fluxes of the solute, provided there are no gains or losses of the solute by any mechanisms operating within the volume element itself (i.e., there are no sources or sinks).

Combining Eqs. (9.8) and (9.17), we get

$$\frac{\partial(c\theta)}{\partial t} = -\frac{\partial(v\theta c)}{\partial x} + \frac{\partial}{\partial x}\left(D_{sh}\frac{\partial c}{\partial x}\right) \tag{9.19}$$

For steady water flow (which, however, does not necessarily imply steady solute movement) θ, $\bar{v}$, and D_{sh} can be taken as constant, and Eq. (9.19) simplifies to

$$\frac{\partial c}{\partial t} = -\bar{v}\frac{\partial c}{\partial x} + \frac{D_{sh}}{\theta}\frac{\partial^2 c}{\partial x^2} \tag{9.20}$$

Certain solutes are generated within the soil. An example is the nitrate ion, which evolves under appropriate circumstances from nitrification of organic matter. In contrast, some solutes may disappear from the soil volume (as in the case of nitrates being taken up by plants or undergoing denitrification). To account for such "sources" and "sinks," we must modify Eq. (9.18) to include a composite source–sink term, S, expressing the rate of production or disappearance of the particular solute:

$$\partial(c\theta)/\partial t = -\partial J/\partial x + S \tag{9.21}$$

The term S represents the net sum of all n possible sources $\Sigma\, s_i$ and all m possible sinks $\Sigma\, s_j$. Accordingly,

$$S = \sum_{i=1}^{n} s_i - \sum_{j=1}^{m} s_j \tag{9.22}$$

The terms of this equation must be specified as quantities of the solute generated or dissipated per unit volume of soil and per unit length of time.

An additional possibility to consider is the existence of a dynamic storage for the solute inside the soil body but outside the mobile liquid phase, as, for instance, in precipitated form or in the soil's exchange complex. In this case, the left-hand side of Eq. (9.21) can be expanded to include the quantity of solute in storage (σ_s). Accordingly, the left-hand side of Eq. (9.21) becomes $\partial(\theta c + \sigma_s)/\partial t$. The time derivative of σ_s (namely, $\partial\sigma_s/\partial t$)

expresses the rate of increase of storage outside the solution phase for the solute under consideration.

We can now write a comprehensive equation of transient-state solute dynamics to include convective–dispersive–diffusive movement as well as sources, sinks, and storage changes. For a vertical soil profile, we substitute the axis z (representing depth below the soil surface) for x, to obtain

$$\frac{\partial(\theta c + \sigma_s)}{\partial t} = \frac{\partial}{\partial z}\left(D_{sh}\frac{\partial c}{\partial z}\right) - \frac{\partial(qc)}{\partial z} + S \tag{9.23}$$

wherein, as previously defined, the solution's convective flux q is equal to the product of its average velocity and volume fraction in the soil; that is, $q = \bar{v}\theta$.

The adsorption of ions in the soil can be positive, as is the attachment of cations to clay surfaces, or it can be negative, as is the repulsion or partial exclusion of anions from the electrostatic double layer of the same clay. Since, in consequence, the anions are relegated to the centers of the pores, they tend to travel somewhat faster than cations when an electrolytic solution is passed through a soil, even as the cations undergo adsorption and exchange phenomena within the soil's cation exchange complex and their motion is thereby retarded. If the attainment of equilibrium between the ions in the solution phase and the soil's exchange complex is rapid enough to be considered instantaneous, then the amount adsorbed A can be taken as dependent only on the concentration c of the ion species in the soil solution. Assuming that storage outside the solution phase is due entirely to adsorption (i.e., there is no other storage mechanisms, such as precipitation of a component of limited solubility), then the amount in storage σs is equal to A. Using these assumptions, Eq. (9.23) can be rewritten

$$[\theta + A(c)]\frac{\partial c}{\partial t} = \frac{\partial}{\partial z}\left(D_{sh}\frac{\partial c}{\partial z}\right) - \frac{\partial(qc)}{\partial z} + S \tag{9.24}$$

Here $A(c)$ is the slope of the *adsorption isotherm*, a function of concentration; θ, c, s, and q are functions of depth and time; and D_{sh} is a function of θ and average velocity $\bar{v} = q/\theta$.

EFFECTS OF SOLUTES ON WATER MOVEMENT

€ The first phenomenon to consider in this regard is the hydration of clay particles and consequent swelling of the soil. Recall that negatively charged clay particles attract cations, with which they form an *electrostatic double layer* (see Chapter 4). This process of imbibition, causing swelling, is especially pronounced when the ambient solution (i.e., the soil solution away from the particle surfaces) is the more dilute. Imbibition between clay platelets is constrained by interparticle bonds and usually ceases at ambient solution concentrations above 200–400 meq/L (McNeal, 1974). With more dilute solutions, continued swelling — in a tendency to relieve the osmotic pressure differential between the clay domains and the ambient solution — drives the particles apart and weakens the interparticle bonds. A combination of this osmotic swelling with mechanical disturbance of the soil system ("puddling") can lead to a rupturing of interparticle bonds, so adjacent particles separate and the clay fraction undergoes dispersion, which in turn alters the geometry of soil pores and results in a decrease of intrinsic permeability.

Combinations of low salt concentration and high *exchangeable sodium percentage* (ESP) are the conditions most likely to cause swelling, dispersion, and reduction of permeability. The collapse of aggregates resulting from dispersion of clay tends to plug the large interaggregate pores, particularly in the top layer, so an "open" soil surface can become sealed. Moreover, dispersed particles can move with percolating water and migrate into the soil profile. Evidence of such migration can be seen in the occurrence of clay skins over aggregates deeper in the profile and in the natural deposition of clay in the formation of distinct layers called *clay pans*.

Loss of permeability resulting from the adverse effect of salt concentration and exchangeable sodium percentage can become a major problem in irrigated agriculture. It can cause problems, as well, in the operation of municipal and industrial waste disposal systems, such as septic tank drainage fields, where decreases in the hydraulic conductivity by one or more orders of magnitude have been reported.

A widely accepted index for characterizing the soil solution with respect to its likely influence on the exchangeable sodium percentage is the *sodium adsorption ratio* (SAR). It has also been used to assess the quality of irrigation water. SAR is defined as follows:

$$\mathrm{SAR} = [\mathrm{Na}^+]/\{([\mathrm{Ca}^{2+}] + [\mathrm{Mg}^{2+}])/2\}^{0.5} \qquad (9.25)$$

In words, it is the ratio of the sodium ion concentration to the square root of the average concentration of the divalent calcium and magnesium ions. In this context, all concentrations are expressed in milliequivalents per liter. SAR is thus an approximate expression for the relative activity of Na^+ ions in exchange reactions in soils. A high SAR, particularly at low concentrations of the soil solution, causes high ESP and a decrease of soil hydraulic conductivity. The relation of hydraulic conductivity to composition and concentration of the soil solution and to the composition and capacity of the soil's exchange complex is influenced by the quantity and nature of the predominant clay (e.g., whether kaolinitic or montmorillonitic) as well as by the presence and content of cementing agents (e.g., humus and sesquioxides).

SOIL SALINITY AND ALKALINITY

An excessive accumulation of salts in the soil causes a decline in productivity. *Soil salinity* is the term used to designate a condition in which the soluble salt content of the soil reaches a level harmful to crops. Soil salinity affects plants directly through the reduced osmotic potential of the soil solution and the toxicity of specific ions, (such as boron, chloride, and sodium. If the salts are mainly sodic salts, their accumulation increases the concentration of sodium ions in the soil's exchange complex, which in turn affects soil properties and behavior. Thus, salinity can also have indirect effects on plant growth through deleterious modification of such soil properties as swelling, porosity, water retention, and permeability.

Soluble salts may be present in the original soil material or be brought to the soil by invading surface water or rising groundwater. The ocean is a source of salt for low-lying soils along the coast, where it may also contribute salt to rainfall by sea spray. Except along the seacoast, however, saline soils seldom occur in humid regions, thanks to the typically downward net seepage of water through the soil profile brought about by the excess of rainfall over evapotranspiration. In arid regions, however, there may be no net downward percolation of water and no effective leaching, so salts can accumulate in the root zone, especially in its top layer. Here the combined effects of meager rainfall, a high evaporation rate, the occurrence of salt-bearing sediments, and the presence of shallow, brackish groundwater give rise to a group of soils known in classical pedology as *solonchaks*.

Less obvious than the appearance of naturally saline soils, but perhaps more ominous, is the *induced salination* of originally productive soils by the injudicious practice of irrigation in certain arid regions. Typically located in the river valleys of the dry zone, irrigated soils are particularly vulnerable to the insidious, and for a time invisible, rise of the water table caused by failure to provide adequate groundwater drainage. *Proper irrigation* maintains a supply of good-quality moisture needed by plants to answer the climatically imposed evaporational demand while at the same time ensuring a favorable salt balance and nutrient supply together with adequate aeration and temperature regimes throughout the root zone. *Efficient irrigation*, furthermore, avoids wasting water through runoff or excessive drainage, except insofar as some drainage is necessary to flush out potentially harmful salts that would otherwise accumulate in the root zone. Crop plants extract water from the soil while leaving most of the salt behind. Unless leached away (preferably continuously, but at least periodically) such salts will sooner or later begin to hinder crop growth.

The classical and still pertinent publication on soil salinity is the U.S. Department of Agriculture *Handbook 60* (L. A. Richards, editor, 1954). Although newer and much revised publications have since been issued (e.g., Bresler et al., 1982; Tanji, 1990; Rhoades et al., 1999; and Hillel, 2000), some of the definitions and concepts of the original publication are still used widely. The accepted classification of soil salinity (as well as of irrigation water quality) is based on total salt concentration, as measured by its electrical conductivity, and on the relative concentration of sodium. *Soil alkalinity* (or *sodicity*) is characterized by the ESP, the content of exchangeable sodium ions (expressed in milliequivalents per 100 grams of soil) as a fraction (percent) of the total cation exchange capacity.

According to these criteria, a *saline soil* is one whose saturation extract indicates an electrical conductivity exceeding 0.4 S/m (4 mmho/cm) at 25°C but whose ESP is less than 15. (A saturation extract is obtained after adding an amount of water just sufficient to saturate a soil sample, which is then stirred to form a slurry or "paste" in a standard manner. The water is later sampled by filtration for determination of salt concentration.) Such soils usually have pH values less than 8.5 and are ordinarily well flocculated.

Reclamation of saline soils requires removal of excess salts by leaching, with due care to prevent a rise in the exchangeable sodium percentage. *Saline-alkali soils* have been defined as having a saturation-extract conductivity greater than 0.4 S/m and an ESP greater than 15. Such soils, when leached, become highly dispersed and exhibit higher pH values. To be reclaimed, these soils require, in addition to leaching, treatment with soluble calcium salts (e.g., gypsum) to replace the excess of exchangeable sodium. Finally, *nonsaline alkali soils* exhibit an ESP greater than 15 but a salinity less than the 0.4-S/m level. Such soils, which often exhibit very high pH values (8.5–10) have traditionally been called *solonetz* by pedologists and sometimes *black alkali* (in reference to the black color resulting from the highly dispersed state of organic matter at the surface).

SALT BALANCE OF THE SOIL PROFILE

The *salt balance* is a quantified summary of all salt inputs and outputs for a defined volume or depth of soil during a specified period of time. If salts are conserved (that is to say, if they are neither generated nor decomposed chemically in the soil), then the difference between the total input and output must equal the change in salt content of the soil zone monitored. Accordingly, if input exceeds output, then salt must be accumulating, and vice versa. The salt balance has been used as an indicator of salinity trends and the need for salinity control measures in large-scale irrigation projects as well as in single irrigated fields.

The following equation applies to the amount of salt in the liquid phase of the root zone per unit area of land:

$$[\rho_w(V_r c_r + V_i c_i + V_g c_g) + M_s + M_a] - (M_p + M_c + \rho_w V_d c_d) = \Delta M_{sw} \qquad (9.26)$$

Here V_r is the volume of rainwater entering the soil with a salt concentration c_r; V_i is the volume of irrigation with a concentration c_i; V_g is the volume of groundwater with a concentration c_g entering the root zone by capillary rise; V_d is the volume of water drained from the soil with a concentration c_d; M_s and M_a are masses of salt dissolved from the soil and from agricultural inputs (fertilizers and soil amendments), respectively; M_p and M_c are the mass of salt precipitated (or adsorbed) in situ and the mass removed by the crop, respectively; ρ_w is the density of water; and, ΔM_{sw} is the change in mass of salt in the soil's liquid phase during the period considered.

A simplified form of the salt balance equation was given by Buras (1974) for a complete annual period. Assuming that the net change in total soil water content from one year to the next (as monitored either at the end of the wet season or at the end of the dry season) is close to zero; disregarding surface runoff, the *water balance* per unit area is:

$$V_i + V_r + V_g = V_{et} + V_d \qquad (9.27)$$

Here V_i, V_r, V_g, V_{et}, and V_d are total annual volumes of irrigation, atmospheric precipitation, capillary rise of groundwater, evapotranspiration, and drainage, respectively. As evapotranspiration removes no salt and crops generally remove only a negligible amount, and if we disregard agricultural inputs and in situ precipitation and dissolution of salts, the salt balance corresponding to Eq. (9.29), assuming no accumulation, is

$$V_i c_i + V_r c_r = (V_d - V_g) c_d \qquad (9.28)$$

where c_i, c_r, and c_d are the average concentrations of salt in the irrigation, rainfall, and drainage waters, respectively. If the water table is kept deep enough so that no substantial capillary rise into the root zone occurs, and since the salt content of atmospheric precipitation is usually negligible, Eq. (9.30) further simplifies to

$$V_i c_i = V_d c_d \tag{9.29}$$

Such an overall "black box" approach disregards the mechanisms and rates of salt and water interactions in the root zone as well as the changing pattern of salt distribution throughout the soil profile. A more detailed treatment of the dynamic balance of salts in the root zone under irrigation was given by Bresler (1972b), based on the following equation:

$$\frac{\partial}{\partial t}\int_0^z [c(z,t)\,\theta(z,t)]\,dz = q(o,t)c_0(t) - q(z,t) \tag{9.30}$$

where the flux q, soil wetness θ, and salt concentration c are considered as functions of depth z and time t, and c_0 is the salt concentration of the applied water. This equation was solved explicitly for $c(z, t)$, via numerical methods, for different initial and boundary conditions.

Figure 9.3 demonstrates the results obtained from these computations for a fallow (unvegetated) loam soil during cycles of infiltration, redistribution of soil moisture, and evaporation. The computation of salt and water dynamics in the presence of an active root system, though more complicated, is amenable to the same type of quantitative approach (e.g., Hillel et al., 1975).

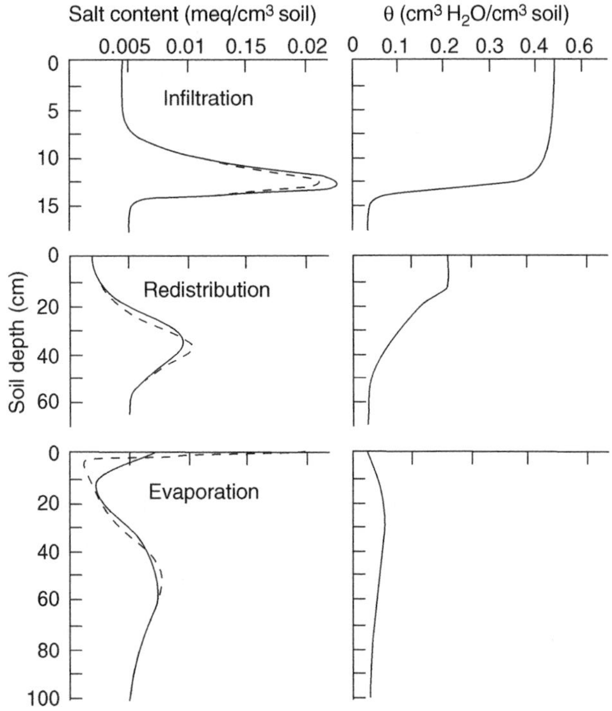

Fig. 9.3. Computed salt and water content distributions following successive stages of the field water cycle. The dispersion term was considered (solid lines) and omitted (dashed lines). (From Bresler, 1972b.)

Sample Problem

The following data were obtained in a field in an arid zone:

(a) Rainfall occurred only in winter and amounted to 300 mm, with a total salt concentration of 40 ppm.
(b) Capillary rise from shallow, saline groundwater during spring and autumn totaled 100 mm at a concentration of 1000 ppm.
(c) Irrigation was applied during the summer season and amounted to 900 mm (400 ppm salts).
(d) Drainage occurred only during the irrigation season and amounted to 200 mm, with a soluble salt concentration of 800 ppm.
(e) An additional increment 0.12 kg/m^2 of soluble salts was added in the form of fertilizers and soil amendments, whereas an amount of 0.1 kg/m^2 was removed by the harvested crops.

Disregarding dissolution and/or precipitation of salts within the soil, compute the annual salt balance. Is there a net accumulation or release of salts by the soil?

We begin with a slightly modified version of Eq. (9.28) to calculate the salt balance of the root zone per unit of land area:

$$\Delta M_s = \rho_w(V_r c_r + V_i c_i + V_g c_g - V_d c_d) + M_a - M_c$$

where ΔM_s is the change in mass of salt in the root zone (in the liquid phase as well as in the solid phase, i.e., the sum of all dissolved, adsorbed, and precipitated salt); ρ_w is the density of water; V_r, V_i, V_g, and V_d are the volumes of rain, irrigation, groundwater rise, and drainage, respectively, with corresponding salt concentrations of c_r, c_i, c_g, c_d; and M_a, M_c are the masses of salts added agriculturally and removed by the crops, respectively.

Using 1 m^2 as the unit "field" area, calculating water volumes in terms of cubic meters and masses in terms of kilograms (water density being 1000 kg/m^3), we can substitute the given quantities into the previous equation to obtain the change in mass content of salt in the soil:

$$\begin{aligned}\Delta M_s &= [10^3\ \text{kg/m}^3(0.3\ \text{m} \times 40 \times 10^{-6} + 0.1\ \text{m} \times 1000 \times 10^{-6} \\ &\quad + 0.9\ \text{m} \times 400 \times 10^{-6} - 0.2\ \text{m} \times 800 \times 10^{-6}) \\ &\quad + 0.12\ \text{kg/m}^2 - 0.1\ \text{kg/m}^2] \\ &= 0.322\ \text{kg/m}^2\ \text{yr}\end{aligned}$$

Thus, the soil's root zone is accumulating salt at a rate equivalent to 3220 kg (3.22 metric tons) per hectare per year.

BOX 9.1 Silt and Salt in Ancient Mesopotamia

Ancient Mesopotamia owed its prominence to its agricultural productivity, based on its soil and water resources and its favorable climate. The soils of this extensive alluvial valley are deep and fertile. The topography is level and the climate is warm and dry, with abundant sunshine year-round. The scant rainfall poses no appreciable problems of water erosion or of nutrient leaching, such as occur on sloping lands in more humid regions. The main constraint to crop production in this arid valley is, of course, water. Fortunately, water is abundant in the twin rivers, the Euphrates and the Tigris.

However, the diversion of river water onto valley land led to a series of interrelated problems. The first problem was sedimentation. Early in history, the upland watersheds of the twin rivers were deforested and overgrazed. Erosion resulting from seasonal rains proceeded to strip off the soil of sloping uplands and pour it into the streams. As the floodwaters wound their way toward the lower reaches of the valley, the sediment settled along the bottoms and sides of the rivers, raising their beds and banks above the adjacent plain. During periodic floods the rivers overtopped their banks and inundated large tracts of land.

The second insidious problem was salt. It resulted from the inexorable rise in the water table, which, in the absence of adequate drainage, naturally follows the flood irrigation of low-lying lands. Elevated rivers continually seep into the groundwater. So do diversion canals and distribution ditches. Finally, irrigation itself causes seepage from the entire surface of the land. Since all irrigation waters contain dissolved salts, and since crop roots normally exclude salts while extracting soil moisture, the salts tend to accumulate in the soil. Unless leached out, the salts eventually poison the root zone.

In arid regions, natural rainfall is generally insufficient for annual leaching; irrigation must hence be applied in excess of crop water requirements so as to remove harmful salts by downward percolation beyond the root zone. For some years or even generations, the processes of groundwater salination and water table rise are invisible and go unnoticed. Then, when the water table comes close to the ground surface, a secondary process of capillary rise comes into play. The rising groundwater evaporates at the surface, thus infusing the topsoil with salt. This was the process that eventually doomed the civilizations of Mesopotamia. Those civilizations — Sumer, Akkad, Babylonia, and Assyria — each in turn rose and then declined, as the center of population and culture shifted over the centuries from the lower to the central to the upper parts of the Tigris–Euphrates valley (Adams, 1981; Jacobsen, 1982; Artzy and Hillel, 1988).

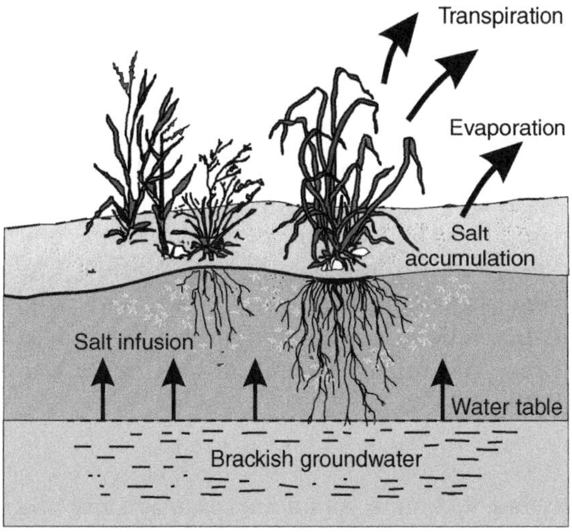

Fig. B9.1. The process of water table rise and salination.

LEACHING OF EXCESS SALTS

It is a startling fact that 1-m depth of even reasonably good-quality irrigation water (an amount normally applied in a single irrigation season) contains sufficient salt to salinize an initially salt-free soil (i.e., about 3000–5000 kg, or 3–5 tons, per hectare).

To prevent salts from accumulating during repeated irrigation–evapotranspiration cycles, the obvious remedy is to apply water in an amount greater than evapotranspiration so as to cause a portion of the applied water to flow through the root zone and wash away the excess salts. However, unless the water table is very deep or lateral groundwater drainage is sufficiently rapid, the excess irrigation often causes a progressive rise of the water table. Once the water table comes within a meter or so of the soil surface, groundwater tends to seep upward into the root zone by capillary action and to reinfuse the soil with salt.

From the foregoing it should be obvious that any attempt to leach without provision of adequate drainage is not merely doomed to fail but can indeed exacerbate the problem. In many areas where natural drainage is slow and artificial drainage is not provided, it becomes impossible to sustain irrigation and the land must sooner or later be abandoned owing to progressive salination.

Much attention has been devoted to assessing the optimal quantity of water that must be applied to cause leaching. The application of too much water can be as harmful as the application of too little. Exaggerated leaching not only wastes water but also tends to remove essential nutrients and to impede aeration by waterlogging the soil. The *leaching requirement* concept was developed by the U.S. Salinity Laboratory. It has been defined as the fraction of the irrigation water that must be leached out of the bottom of the root zone so as to prevent average soil salinity from rising above some specifiable limit.

According to the standards of the Handbook of the U.S. Salinity Laboratory, the maximum concentration of the soil solution, expressed in terms of electrical conductivity, should be kept below 0.4 S/m (4 mmhos/cm) for sensitive crops. Tolerant crops such as beets, alfalfa, and cotton may give good yields at values up to 0.8 S/m, while very tolerant crops such as barley may give good yields at salinity values as high as 1.2 S/m.

The leaching requirement depends on the salt concentration of the irrigation water, on the amount of water extracted from the soil by the crop (the evapotranspiration), and on the salt tolerance of the crop, which determines the maximum allowable concentration of the soil solution in the root zone. Assuming steady-state conditions of through-flow (thus disregarding short-term changes in soil moisture content, flux, and salinity) and assuming no appreciable dissolution or precipitation of salts in the soil and no removal of salts by the crop or capillary rise of salt-bearing water from below, we obtain from Eq. (9.29),

$$V_d/V_i = c_i/c_d \tag{9.31}$$

in which V_d and V_i are the volumes of drainage and of irrigation, respectively, and c_d and c_i are the corresponding concentrations of salt. Water volumes are normally expressed per unit area of land as equivalent depths of water, and salt concentrations are generally measured and often reported in terms of electrical conductivity. Since the volume of water drained is the difference between the volumes of irrigation and evapotranspiration (i.e., $V_d = V_i - V_{et}$), we can transform the last equation as follows:

$$V_i = [c_d/(c_d - c_i)]V_{et} \tag{9.32}$$

which is equivalent to the formulation by Richards (1954):

$$d_i = [E_d/(E_d - E_i)]d_{et} \tag{9.33}$$

Here d_i is depth of irrigation, d_{et} is the equivalent depth of the crop's "consumptive use," and E_d, E_i are the electrical conductivities of the drainage and irrigation waters, respectively.

The leaching requirement equation implies that by varying the fraction of applied water percolated through the root zone, it is possible to control the concentration of salts in the drainage water and hence to maintain the concentration of the soil solution in the main part of the root zone at some intermediate (optimal) level between c_i and c_d.

Leaching soils at a water content below saturation (e.g., under low-intensity sprinkling irrigation or under intermittent irrigation) can produce more efficient leaching and thereby reduce the amount of water required as well as mitigate drainage problems in areas of high water tables. If a soil with large vertical cracks is ponded with water, much of the water moves through the cracks and is ineffective in leaching. Under rainfall or slow sprinkling, in contrast, a greater portion of the applied water moves through the soil blocks and micropores, thus producing more efficient leaching of the soil matrix per unit volume of water infiltrated.

Sample Problem

Estimate the "leaching requirements" of a field subject to a seasonal evapotranspiration of 1000 mm, if the electric conductivity of the irrigation water is 0.1 S/m (equivalent to a salt concentration of about 650 ppm) and that of the drainage water is allowed to reach 0.4 S/m (about 2600 ppm salts). What would be the leaching requirement if the irrigation water were half as concentrated? And what if the drainage water were allowed to become twice as concentrated? Finally, what would be the electrical conductivity of the drainage water if the amount of irrigation were 1500 mm?

Using Eq. (9.33):

$$d_i = [E_d/(E_d - E_i)]d_{et}$$

where d_i and d_{et} are the volumes of water per unit land area (in depth units) of irrigation and of evapotranspiration, respectively, and E_d and E_i are the electrical conductivities of the drainage and irrigation waters, respectively. Substituting the appropriate values, we get

$$d_i = [0.4/(0.4 - 0.1)]\ 1000\ \text{mm} = 1333\ \text{mm}$$

The "leaching depth" is

$$d_e = d_i - d_{et} = 1333 - 1000 = 333\ \text{mm}$$

If the irrigation water were half as concentrated ($E_i = 0.05$ S/m),

$$d_i = [0.4/(0.4 - 0.05)]\ 1000\ \text{mm} = 1143\ \text{mm}$$

Thus, the leaching depth would be only 143 mm, which is less than half the previously required leaching volume. If the drainage water were permitted to be twice as concentrated, that is, if E_d were 0.8 S/m instead of 0.4, then

$$d_i = [0.8/(0.8 - 0.1)]\ 1000\ \text{mm} = 1143\ \text{mm}$$

In words: Doubling the allowable concentration of the drainage water is equivalent to halving the concentration of the applied irrigation water, in terms of its effect on reducing leaching requirements. If the depth of irrigation water applied were 1500 mm and its electric conductivity were 0.1 S/m, the electric conductivity of the drainage water would be

$$E_d = E_i/[1 - (d_{et}/d_i)]$$

which is obtained by simple transformation of Eq. (9.33). Thus,

$$E_d = 0.1\ \text{S/m}/[1 - (1000/1500)] = 0.3\ \text{S/m}$$

BOX 9.2 How Ancient Egypt Escaped the Scourge of Salinity

In sharp contrast to Mesopotamia, the civilization of Egypt survived and continued in the same location for several millennia. What explains the persistence of irrigated farming in Egypt in the face of its demise in Mesopotamia?

The answer lies in the different soil and water regimes of the two river valleys. Neither clogging by silt nor poisoning by salt was as severe along the Nile as in the Tigris–Euphrates valley. The silt of Egypt is delivered by the Blue Nile from the volcanic highlands of Ethiopia, and it is mixed with the organic matter delivered by the White Nile from its swampy sources. It was not so excessive as to choke the irrigation canals, but it was fertile enough to add nutrients to the fields. Whereas in Mesopotamia the inundation usually comes in the spring and summer evaporation tends to make the soil saline, the Nile crests at a much more favorable time: after the summer heat has killed the weeds and aerated the soil, and in time to wet the soil for the prewinter planting.

The narrow floodplain of the Nile (except in the Delta) precluded the widespread rise of the water table. The water table was controlled by the stage of the river, which, over most of its length, normally lies below the level of the adjacent land. When the river inundated the land, the seepage naturally raised the water table. As the river's level dropped, it pulled the water table down after it. This annual pulsation of the river and the associated fluctuation of the water table under a free-draining floodplain created an automatically self-flushing cycle by which the salts were leached from the soil and carried away by the Nile itself (Hillel, 1994).

Unfortunately, the soil of Egypt is now threatened with degradation. The Aswan High Dam has blocked the fertile silt that had formerly been delivered by the Nile. Hence Egyptian farmers must rely increasingly on chemical fertilizers. The river, now running clear of silt, has increased its erosivity and has been scouring its own banks. Along the estuaries of the Delta there is no more deposition, so coastal erosion has begun, along with seawater intrusion. The artificial maintenance of a nearly constant water level in the river, needed to allow easy pumping of irrigation water throughout the year, has resulted in raising the water table and making drainage more difficult. So Egypt is now experiencing the maladies of waterlogging and salination from which it had for so long seemed immune.

Part IV

THE GASEOUS PHASE

10. GAS CONTENT AND COMPOSITION

THE VITAL ROLE OF SOIL AERATION

The process of *soil aeration* is an important determinant of soil productivity. Plant roots absorb oxygen and release carbon dioxide in the process of *respiration*. In most terrestrial plants (excepting such specialized plants as rice), the internal transfer of oxygen from the above-ground parts (leaves and stems) to the roots cannot take place at a rate sufficient to supply the oxygen requirements of the latter. Adequate root respiration therefore requires that the soil itself be aerated, that is to say, that gaseous exchange take place between soil air and the atmosphere at a rate sufficient to prevent a deficiency of oxygen and an excess of carbon dioxide from developing in the root zone (Fig. 10.1). Microorganisms in the soil also respire and, under restricted aeration, might compete with the roots of higher plants for scarce oxygen.

Gases can move either in the air phase (i.e., in the pores that are drained of water, provided they are interconnected and open to the atmosphere) or in dissolved form through the water phase. The rate of transfer of gases in the air phase is generally much greater than in the water phase; hence soil aeration depends largely on the volume fraction of air-filled pores.

Impeded aeration resulting from poor drainage and waterlogging or from mechanical compaction of the soil surface zone can strongly inhibit crop growth. The problem of soil compaction seems to have worsened in modern times, along with the growing trend to use larger and heavier machinery and the tendency to tread over the field repeatedly for such purposes as seeding, fertilization, pest control, and harvesting. With greater use of fertilizers and irrigation, shortages of nutrients and water have been obviated in many places so that, by default, soil aeration has gained in relative importance as a major

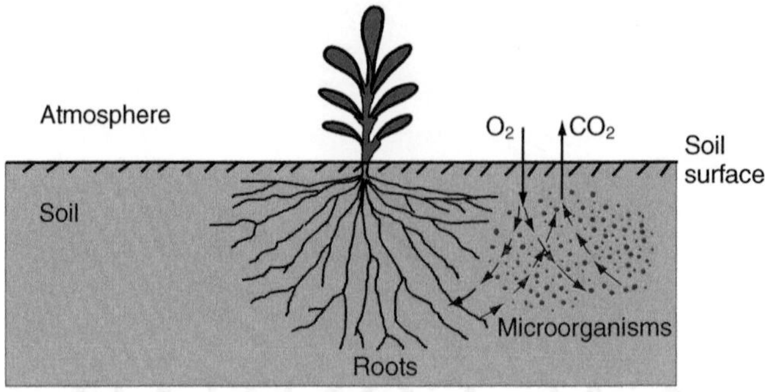

Fig. 10.1. Soil aeration as a process of O_2 and CO_2 exchange with the atmosphere. Among other gases involved in the soil atmosphere exchange are various volatile forms of nitrogen, sulfur, and hydrocarbons.

constraint to the attainment of maximal productivity. Root systems are commonly restricted in extent by the progressive decrease of aeration in the deeper regions of the soil profile.

Anaerobic conditions in the soil induce a series of chemical and biochemical *reduction reactions*. Among these reactions are *denitrification* (the processes by which nitrate is reduced to nitrite, thence to nitrous oxide, and finally to elemental nitrogen: $NO_3 > NO_2 > N_2O > N_2$); *manganese reduction* from the manganic to the manganous form; *iron reduction* from the ferric to the ferrous form; and *sulfate reduction* to form hydrogen sulfide. Many organic compounds are formed in the anaerobic decomposition of organic matter, some of which are toxic to plants (e.g., ferrous sulfide, ethylene, and acetic, butyric, and phenolic acids).

The subject of soil aeration has been reviewed repeatedly over the years. The evolving state of our knowledge is reflected in the successive publications by Currie (1975), Cannell (1977), Phene (1986), Rolston (1986), and Scanlon et al. (2002).

VOLUME FRACTION OF SOIL AIR

In most natural soils, the volume ratios of the three constituent phases — solid particles, water, and air — are continually changing as the soil undergoes wetting and drying, swelling and shrinkage, tillage and compaction, aggregation and dispersion, etc. Specifically, since the twin fluids of water and air together occupy the pore space, their volume fractions are so related that an increase of the one generally entails a decrease of the other. Thus,

$$f_a = f - \theta \tag{10.1}$$

where f_a is the volume fraction of air, f is the total porosity (the fractional volume of soil not occupied by solids), and θ is the volume fraction of water (volume wetness).

Since the volume fraction of air is largely dependent on the content of water in the soil, it can only be used as an index of soil aeration in conjunction with some specifiable and reproducible wetness value. The wetness value generally chosen for this purpose is the so-called "field capacity" (see Chapter 15). Although the term defies exact physical definition, it is usually defined in an approximate sense as the volume fraction of water retained by a freely draining soil profile after the initially rapid stage of internal drainage. Objections to this concept notwithstanding for the moment, we can use it as the basis for an analogous soil aeration index, which we might call *field-air capacity*, definable as the fractional volume of air in a soil at the field-capacity water content. Though under different names (e.g., "noncapillary porosity," "air-filled porosity," and "air content at field capacity"), this index or something very similar to it has been used for many years to characterize the state of soil aeration. It has been found to depend on numerous factors, not all of which are amenable to human control.

In the first place, air capacity depends on soil texture. In sandy soils it is of the order of 25% or more; in loamy soils it is typically in the range of 15–20%; and in clayey soils, which tend to retain the most water, it is likely to fall below 10% of total soil volume. In fine-textured soils, however, soil structure (as well as soil texture) has much to do with determining the air capacity. Strongly aggregated soils generally have a considerable volume of macroscopic (interaggregate) pores that drain quickly and remain air filled most of the time. Hence such soils exhibit an air capacity of 20–30%. When the clay fraction is dispersed and when the aggregates are broken down by physicochemical processes or mechanical forces, the macroscopic pores tend to disappear, so a strongly compacted soil may contain less than 5% air by volume at its characteristic field-capacity value of soil moisture. The effect of compaction (bulk density) on a soil's capacity to retain air and water is shown hypothetically in Fig. 10.2.

There are fundamental limitations to the air capacity index as a means of characterizing soil aeration. First, the value is difficult to determine accurately, since it generally depends on prior determination of two highly variable

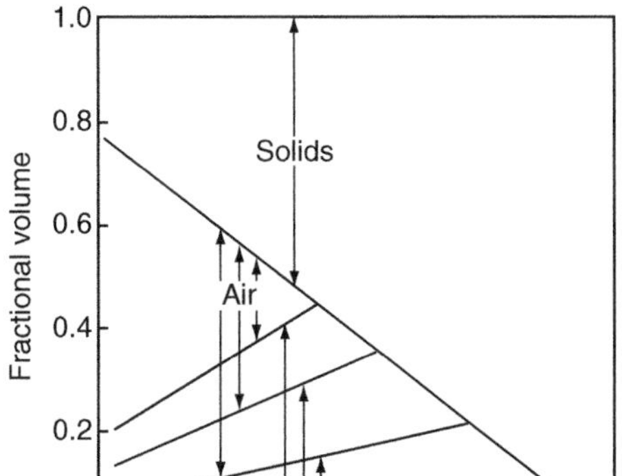

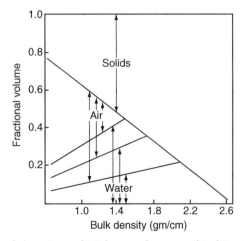

Fig. 10.2. Fractional air, water, and total pore volumes as related to soil bulk density for different mass wetness values (0.1, 0.2, and 0.3).

parameters that are themselves rather cumbersome to measure, namely, the field capacity and the total porosity. Both determinations depend on the method of sampling, and their results can be grossly inaccurate. An even more serious objection to the use of air capacity as an index of soil aeration is that in principle the rate of exchange of soil air rather than simply the content of soil air constitutes the decisive factor. At high wetness values, soils often contain isolated pockets of occluded air that, though forming a part of the air-filled volume, do not contribute to active gas exchange (Fayer and Hillel, 1986a,b). At times, even a thin surface crust, if highly compact or saturated, can act as a bottleneck limiting aeration to the entire soil profile, regardless of the air-filled porosity beneath. We are thus led to the necessity of characterizing aeration in more dynamic terms.

COMPOSITION OF SOIL AIR

In a well-aerated soil, the composition of soil air is close to that of the external ("open") atmosphere, because the oxygen consumed in the soil is readily replaced from the atmosphere and the carbon dioxide generated is readily vented to the atmosphere. Not so in a poorly aerated soil. Analyses of the actual composition of soil air in the field reveal it to be much more variable than the external atmosphere. Depending on such factors as time of year, temperature, soil moisture, depth below the soil surface, root growth, microbial activity, pH, and — above all — the rate of exchange of gases through the soil surface, soil air can differ to a greater or lesser degree from the composition of the external atmosphere.

The greatest difference is in concentration of carbon dioxide (CO_2), the principal product of aerobic respiration by the roots of higher plants as well as by numerous macroorganisms and microorganisms in the soil. The CO_2 concentration of the atmosphere is about 0.036% (some 360 parts per million, in terms of partial volume). In the soil, however, it frequently reaches levels that are 10 or more times as great.

Because CO_2 is produced in the soil by oxidation of carbonaceous (organic) matter, an increase in CO_2 concentration is generally associated with a decrease in elemental oxygen (O_2) concentration (though not necessarily to an exactly commensurate degree, since additional sources of oxygen may exist in dissolved form and in easily reducible compounds). Since the O_2 concentration of air is normally above 20%, it would seem that even a hundredfold increase of CO_2 concentration, say, from 0.036% to 3.6%, can diminish the O_2 concentration to only about 17%. However, even before they begin to suffer from lack of oxygen per se, some plants may suffer from excessive concentrations of CO_2 and other gases in both the gaseous and aqueous phases. In more extreme cases of aeration restriction, the O_2 concentration can fall to near zero. Prolonged anaerobic conditions can result in the development of a chemical environment characterized by reduction reactions such as denitrification, the evolution of such gases as hydrogen sulfide (H_2S), methane (CH_4), and ethylene (C_2H_4), and the reduction of mineral oxides such as those of iron and manganese.

Both carbon dioxide (released from the soil by the decomposition of organic matter under aerobic conditions) and methane (released under anaerobic conditions) contribute to the atmospheric greenhouse effect. In contrast, practices that build up the organic matter content of the soil sequester carbon from the atmosphere and thus may help to mitigate the potentially disruptive effects of the same greenhouse effect (i.e., global warming).

Sample Problem

A tropical forest is cleared and cultivated. The topsoil, depth 0.4 m, contains 2% readily decomposable organic residues having a carbon content of 40%. If the O_2 consumption rate (due to soil respiration) is constant at 0.08 kg/m^2 per day, how much carbon is released to the atmosphere in 1 month (30 days) and in 4 months, and what fraction of the organic matter decomposes in each period? Assume soil bulk density = 1250 kg/m^3.

Calculations:

Mass of soil in toplayer: 0.4 m × 1250 kg/m^3 = 500 kg/m^2.
Mass decomposable organic matter: 2% × 500 kg/m^2 = 10 kg/m^2.
Mass of carbon in toplayer: 40% × 10 kg/m^2 = 4 kg/m^2.

Note: As one C atom (atomic weight 12) combines with two O atoms (atomic weight 16) to form CO_2, the C release rate is 12/32 = 0.375 times the O_2 consumption rate. Hence:

Daily carbon release = 0.375 × 0.08 kg/m^2 = 0.03 kg/m^2
Daily organic matter decomposed = 0.03/0.4 = 0.075 kg/m^2
Mass of C released in 1 month (30 days) = 0.03 × 30 = 0.9 kg/m^2
Organic matter decomposed in 1 month = 0.9/0.4 = 2.25 kg/m^2
% organic matter decomposed in 1 month = 2.25/10 = 22.5%
Organic matter decomposed in 4 months = 4 × 2.25 = 9 kg/m^2
% organic matter decomposed in 4 months = 9/10 = 90%
Organic matter remaining after 4 months = 1 kg/m^2
Mass of C released in 4 months 4 × 0.9 = 3.6 kg/m^2
Mass of CO_2 released in 4 months = 3.6 × 44/12 = 13.2 kg/m^2

Note: 44/12 is the ratio of the molecular weight of CO_2 to the atomic weight of C.

SOIL RESPIRATION AND AERATION REQUIREMENTS

The overall rate of respiration due to all biological activity in the soil — that is, the amount of oxygen consumed and the amount of carbon dioxide produced by the entire profile — determines the aeration requirement of the soil. Quantitative knowledge of the aeration requirements of different crops and soils in varying circumstances is essential if we are to devise means to ensure that these requirements are indeed met. However, the information is difficult to obtain, because the rate and spatial distribution of soil respiration, as well as its temporal variation, depend on numerous factors. Prominent among these factors are soil temperature, soil wetness, pH, and organic matter content and composition (whether fresh or well decayed) — all of which influence the time-variable respiratory activity of the numerous species of macroorganisms and microorganisms living in the soil.

Anaerobiosis, or oxygen stress, will occur in the soil whenever the rate of supply falls below the demand. This condition can develop quite quickly, since the storage of oxygen in the soil is generally rather low in relation to the quantity required for soil respiration. To illustrate, let us consider a soil with an effective root zone depth of 60 cm and 15% air-filled porosity, containing 90 L of air under each square meter of soil surface. With an initial oxygen concentration of 20% in the gaseous phase, the storage of oxygen can be calculated to be 18 L, equivalent to about 25 g, or 0.025 kg. [An ideal gas at standard temperature and pressure occupies a volume of 22.4 L/mol. The mass of a mole of O_2 (atomic weight 16) is 0.032 kg. The mass contained in 18 L is therefore $0.032 \times 18/22.4 = 0.0257$ kg.]

If the oxygen requirement for soil respiration is of the order of 0.01 kg/day per square meter of ground, the initial oxygen reserve in the soil would last only 2.5 days. Oxygen stress symptoms would probably begin even earlier. However, these figures are given only to provide an order of magnitude, since in actual conditions the aeration rate varies between wide limits. Plant growth probably depends more on the occurrence and duration of periods of oxygen deficiency than on average conditions.

Measurements in the field demonstrate the effect of temperature on soil respiration. Figure 10.3 illustrates the seasonal variation of soil temperature and soil respiration. Respiration rates in summer can be more than 10 times as

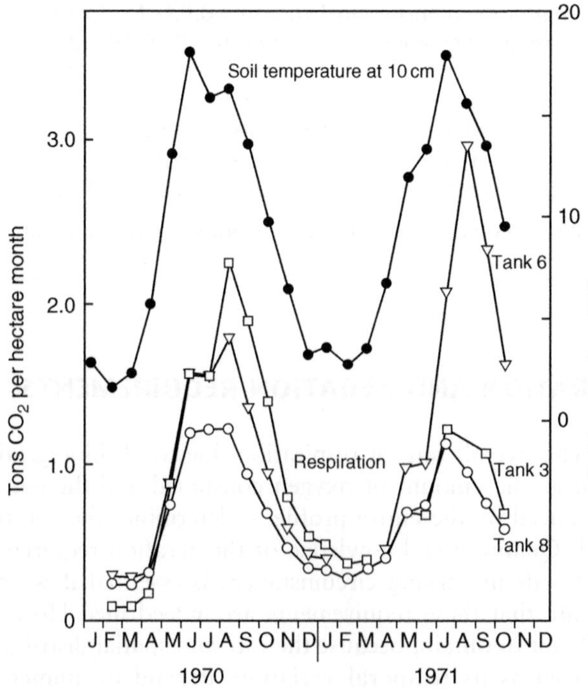

Fig. 10.3. Seasonal variation of soil respiration and soil temperature at Rothamsted. (After Currie, 1975.)

great as in winter. Other factors can also have a strong influence. For example, respiration rates at a given temperature tend to be greater in spring than in autumn, apparently because of the vigorous activity of microorganisms and greater availability of undecomposed organic residues in springtime. Soil respiration is greater in cropped than in fallow land, owing to root respiration and to the enhanced microbial activity resulting from exudations of live roots and the decay of dead roots. The diurnal variation of soil respiration also follows the pattern of soil temperature, with oxygen uptake rates registering a greater than twofold increase from early morning to mid-afternoon. Thus, the respiration rates in soil vary from season to season, from day to day, and from hour to hour and are related to crop growth stage and to microbial activity. The effect of waterlogging the soil is slight in winter, for, although air diffusion is constricted, the oxygen requirements are low. The same waterlogging in summer, however, could be severely damaging to a crop at its most active growth stage.

Sample Problem

Assume the same soil conditions and same first-month organic matter decomposition rate as in the preceding Sample Problem. Thereafter, the decomposition rate does not remain constant but is proportional to the amount remaining in the soil. Again, calculate the mass of C released in 4 months and the amount and percent of the initial organic matter decomposed in that period.

We use the *exponential decay equation*:

$$-dM/dt = kM$$

where M is the mass of organic matter present in the soil and $-dM/dt$ is the time rate of its decomposition. The minus sign indicates that the amount of organic matter is diminishing with time. In this equation the variables can be separated:

$$-dM/M = k\,dt$$

Integrating between the limits M_2 at t_2 and M_1 at t_1, we obtain

$$[-\ln M]_{M_1}^{M_2} = [kt]_{t_1}^{t_2}$$

Here, ln M is the natural logarithm (to the base e) of M:

$$(-\ln M_2) - (-\ln M_1) = k(t_2 - t_1)$$

Solving for k (known as the *specific reaction rate*), we get

$$k = (\ln M_1 - \ln M_2)/(t_2 - t_1) = [2.303/(t_2 - t_1)]\log_{10}(M_1/M_2)$$

We can simplify this equation by taking t_1 as zero time. Then M_1 becomes M_0, the mass of organic matter at zero time. M is the mass at any time t.

Setting $t_1 = 0$, $t_2 = 1$ month, $M_0 = 10$ kg/m^2, and $M = 10 - 2.25 = 7.75$ kg/m^2, we get

$$k = (2.303/t)[\log_{10}(M_0/M)] = (2.303/1)\log(10/7.75) = 0.25$$

We now calculate the organic matter (M) remaining after 4 months:

$$0.25 = (2.303/4)\log(10/M)$$
$$(4 \times 0.25)/2.303 = \log 10 - \log M$$

Hence $\log M = 1 - 0.434 = 0.566$, and $M = \text{antilog } 0.566 = 3.69\ \text{kg/m}^2$.

Organic matter decomposed in 4 months = 10 − 3.69 = 6.31 kg/m^2
Fraction of organic matter decomposed in 4 months = 63.1%
Mass of carbon released in 4 months = 40% of 6.31 = 2.524 kg/m^2
Mass of CO_2 released in 4 months = $2.524 \times 44/12 = 9.25\ \text{kg/m}^2$

Comment: The exponential decay calculation (this problem) may be more realistic than the constant-rate method (preceding Sample Problem). Although the figures given in these problems are hypothetical, the rapid decomposition of organic matter in newly cleared tropical areas is a real problem, resulting not only in the of soil productivity but also in significant contributions to global warming.

MEASUREMENT OF SOIL-AIR CONTENT AND COMPOSITION

An early approach to the problem of measuring aeration was to determine the fractional air space, or air-filled porosity, at a standard value of soil wetness. This was done by taking an "undisturbed" sample from a soil presumed to be at its "field capacity" moisture content. In practice, this has usually meant either sampling the soil in the field two days after a deep wetting or saturating a soil sample in the laboratory and then subjecting it to some specified water tension. By either method, the determination of fractional air space is fraught with uncertainties and gives no real indication of aeration dynamics.

Another traditional approach is to measure the composition of soil air (Bremner and Blackmer, 1982; Farrell et al., 2002). Although still a static measurement, this appears to be a better diagnostic tool than measurement of air volume alone, for it can reveal more directly when a problem might exist — that is, when the oxygen content of soil air falls significantly below that of the atmosphere owing to restricted gas exchange. The difficulty here is how to extract an air sample at once large enough to provide a reliable measurement and yet small enough to represent the sampled point and to avoid disturbance and mixing of soil air or even contamination from the atmosphere. The gas chromatograph technique, using a syringe to extract small samples (only 0.5 mL in volume) may help to make the measurement more reliable.

Gas sampling extracts air mainly from the larger, better aerated pores, and the samples are likely to contain some external air, either from dead space in the sampling equipment or from leakage into the sampling tubes (Payne and Gregory, 1988). Numerous measurements have shown that oxygen concentrations in the soil's air phase tend to be lower than in the open atmosphere, but they are rarely below 10% by volume, while CO_2 concentrations rarely rise above a few percent. However, Blackwell (1983), using very small samples, showed that in waterlogged soils oxygen concentrations can fall to virtually zero.

MEASUREMENT OF SOIL RESPIRATION

Soil respiration values reported by various investigators have varied widely. Papendick and Runkles (1965), working under laboratory conditions, measured respiration rates of 1.7×10^{-5} mol/m^3 sec, equivalent to 0.75 moles of oxygen per cubic meter per 12-hour day, or 2.4×10^{-2} kg/m^3 day. At the other extreme, Grable and Siemer (1968) reported values 10,000 times higher. Under field conditions, respiration values are likely to be smaller than those found in growth chambers. Greenwood (1971), working in England, measured average oxygen consumption rates of 1.3×10^{-7} mL oxygen per second per milliliter of a soil carrying a mature crop and maximum rates three times as high. The average value is equivalent to 8×10^{-3} kg of oxygen per cubic meter of soil per 12-hour day.

Comprehensive data on soil respiration rates were obtained in England from *field respirometers*, first built at Wrest Park and later placed at Rothamsted Experiment Station (Fig. 10.3). These installations were described by Currie (1975). They consist of large containers ($91 \times 91 \times 91$ cm) filled with soil and sealed on top, with provisions to maintain normal atmospheric composition continuously by adding or removing amounts of oxygen and carbon dioxide as required. When the soil is cropped, the plants are grown through holes in the lid, using silicone rubber to seal the gaps around the stems. It is thus possible to measure daily CO_2 output and O_2 uptake and their seasonal variation for soils with and without plants (Fig. 10.4).

The *respiratory quotient*, RQ, that is, the ratio of the volume of carbon dioxide produced to the volume of oxygen consumed (Payne and Gregory,

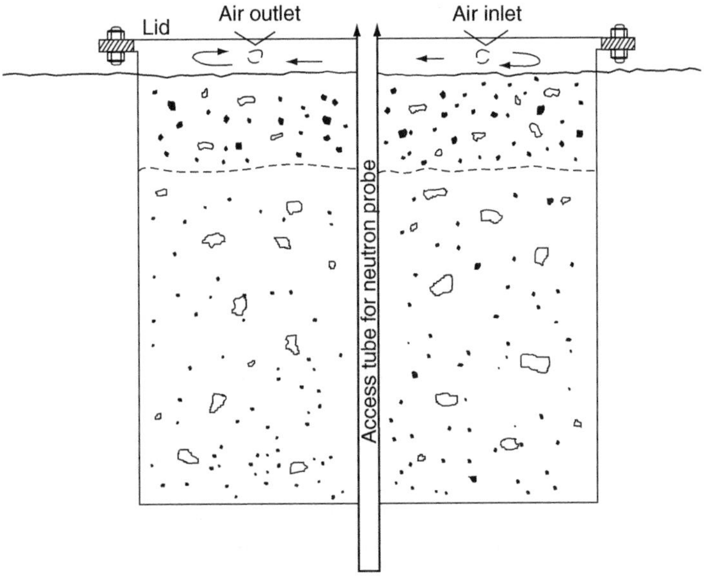

Fig. 10.4. Soil respirometer at Rothamsted. (After Currie, 1975.)

TABLE 10.1 Oxygen Consumption and Carbon Dioxide Production in a Bare Soil and a Soil under a Crop of Kale (g/m^2 per day)

	January		July	
Soil	**Cropped**	**Bare**	**Cropped**	**Bare**
Oxygen consumption	2.0	0.7	24	12
Carbon dioxide release	3.0	1.2	35	16
Soil temperature at 0.1 m	3°C		17°C	

1988), is close to unity in the case of a well-aerated soil. It rises to above unity only when there are anaerobic pockets present in the soil. A determination of this quotient in a soil is therefore a sensitive method for checking if the whole body of the soil is aerobic.

Published data regarding carbon dioxide production and oxygen consumption in different conditions vary widely. Monteith et al. (1964) reported the following relationship:

$$R = R_0 Q^{T/10} \tag{10.2}$$

where R_0 is the flux at 0°C and R that at T°C. He found that Q had a value of about 3. However, Currie (1970), working in the same location (Rothamsted, England) but using a more direct method to measure fluxes, reported much higher figures. Since he measured both the O_2 and CO_2 fluxes, Currie was able to demonstrate that the respiratory quotient was almost exactly unity, except in mid-summer (when the respiratory requirements were especially high), at which time RQ was 1.05. His results are summarized in Table 10.1.

OXIDATION–REDUCTION PROCESSES IN THE SOIL

Soil organisms obtain the energy they need for their vital processes by means of a series of chemical reactions involving the transfer of electrons from substances that serve as sources of energy to substances that may become products of respiration (Rowell, 1981). For example, glucose is a source of electrons when it is oxidized to pyruvic acid in the first stage of its breakdown to carbon dioxide:

$$C_6H_{12}O_6 \rightarrow 2CH_3COCOOH + 4H^+ + 4e^- \tag{10.3}$$

In aerobic respiration, the final electron sink is oxygen, which accepts electrons and combines with hydrogen ions to form water:

$$O_2 + 4H^+ + 4e^- \rightarrow 2H_2O \tag{10.4}$$

Where free oxygen is scarce, several other substances will accept electrons. Some ions containing oxygen (e.g., nitrate and sulphate ions) can accept electrons and lose oxygen:

$$NO_3^- + 2H^+ + 2e^- \rightarrow NO_2^- + H_2O \qquad (10.5)$$

$$2NO_2^- + 8H^+ + 6e^- \rightarrow N_2 + 4H_2O \qquad (10.6)$$

Some variable-valence ions will accept electrons and thus be reduced to a lower-valence state. Examples are the reduction of high-valence iron and manganese ions:

$$Fe^{3+} + e^- \rightarrow Fe^{2+}$$
$$Mn^{4+} + 2e^- \rightarrow Mn^{2+} \qquad (10.7)$$

And hydrogen ion itself can accept an electron to become hydrogen gas:

$$2H^+ + 2e^- \rightarrow H_2 \qquad (10.8)$$

The tendency of a solution to donate electrons to a reducible substance or to accept electrons from an oxidizable substance can be measured in terms of its oxidation–reduction potential, commonly known as *redox potential*, E_h. (It is defined as the potential, in volts, required in an electric cell to produce oxidation at the anode and reduction at the cathode. This potential is measured relative to a standard "hydrogen electrode", which is taken as zero. The hydrogen electrode consists of platinum, over which hydrogen is bubbled, to produce a standard concentration of hydrogen ions.) The more strongly reducing a substance, the lower is its potential. The E_h of a solution also depends on its pH (Rowell, 1981). Values of E_h may vary from +400 mV in a well-aerated soil to −350 mV in poorly aerated, carbon-rich soils (Glinski and Stepniewski, 1985; Paine and Gregory, 1988).

BOX 10.1 Appearance of Poorly Aerated Soils

Poorly aerated soils often occur in humid regions characterized by high year-round rainfall. They may also occur in semihumid or even arid regions, in low-lying river valleys, deltas, and estuaries, as well as along seacoasts. They can generally be recognized by their waterlogged condition, though occasionally a poorly aerated soil may appear to be dry on the surface. A pit dug into such a soil will often emit such gases as hydrogen sulphide, noted for its foul smell, and methane, which is highly flammable. Such gases result from the anaerobic decomposition of organic matter. Nitrogen is often reduced from its nitric state to a nitrous state, and it may even be reduced to the elemental N_2 gas that volatilizes into the atmosphere.

Prolonged anaerobic conditions result in the chemical reduction of mineral elements, particularly iron and manganese. Consequently, an anaerobic soil layer within the profile will typically acquire a bleached, grayish (occasionally even bluish or greenish) hue. A telltale sign of poor aeration is a condition known as *mottling*, characterized by the appearance of spotty discoloration, typically just above the water table. These spots generally indicate that some of the normally reddish ferric oxides have been reduced to ferrous oxides. In extreme cases, manganese is reduced from an oxide to the elemental state and occurs in the form of small blackish concretions. Soils with such symptoms of poor aeration, typically resulting from the blockage of pores by excessive wetness, are called *gley soils* in the classical pedological terminology (the term having originated from the Ukrainian word *glei*, meaning "clay").

Even apparently well-aerated soils may contain zones that are poorly aerated. In some cases, such zones occur as distinct layers, generally composed of tight clay. In other cases, they occur as isolated spots throughout the profile, for example, in the interior of aggregates, where water is retained for long periods. The outer portion of an aggregate may be bright red and seem to be entirely oxidized, while the interior of the same aggregate may be mottled. Plant roots may grow all around the periphery of such aggregates but seldom penetrate into them (Sextone et al., 1985).

11. GAS MOVEMENT AND EXCHANGE

PROCESSES OF GASEOUS TRANSFER

To sustain aerobic life in the soil, the oxygen consumed must be replaced and the carbon dioxide produced must be vented out. Maintenance of soil respiration requires the constant exchange of soil air with the external atmosphere. The main path through which such exchange can take place is the network of air-filled pores that form a continuous system connecting the surface to the deeper layers of the soil. Soil aeration is restricted when this network is partially or wholly blocked, as when the soil is excessively compacted and — especially — when it is excessively wet.

Gaseous transfer in the soil can occur by means of several mechanisms (Payne and Gregory, 1988). Changes in temperature and atmospheric pressure cause soil air to either expand or contract; rainwater entering the soil carries "new" oxygen in dissolved form into the pores, even as it pushes out "old" air. Wind blowing over the soil surface may also pump air into, or suck air out of, the soil surface. The extraction of soil moisture by plants creates a pressure deficit that draws air into the soil.

In all such phenomena, two main processes operate to effect gaseous transport within the soil: *convection* and *diffusion*. Each of these processes can be formulated in terms of a linear rate law, stating that the flux is proportional to the moving force. In the case of convection, also called *mass flow*, the moving force consists of a gradient of *total gas pressure*, and it causes in the entire mass of air to stream from a zone of higher pressure to one of lower pressure. In the case of diffusion, on the other hand, the moving force is a gradient of *partial pressure* (or concentration) of any constituent member of the variable gas mixture that we call air, and it causes the molecules of the unevenly distributed constituent to migrate from a zone of higher concentration to one of lower concentration even while the gas as a whole may remain isobaric and stationary.

CONVECTIVE FLOW OF AIR IN THE SOIL

Pressure differences between soil air and the outer atmosphere, inducing convective flow into or out of the soil, can be caused by barometric pressure changes, temperature gradients, and wind gusts. Additional phenomena affecting the pressure of soil air are the penetration of water during infiltration, causing displacement (and sometimes compression) of antecedent soil air, the fluctuation of a shallow water table pushing air upward or drawing air downward, and the extraction of soil water by plant roots. Short-term changes in soil air pressure can also occur during tillage or compaction by machinery.

The degree to which air pressure fluctuations and the resulting convective flow may contribute to the exchange of gases between the soil and the atmosphere has long been a subject of debate among soil physicists. Most of the time, diffusion rather than convection appears to be the more important mechanism of soil aeration. Under certain conditions, however, convection can contribute significantly to soil aeration, particularly at shallow depths and in soils with large pores. For example, Vomocil and Flocker (1961), assuming that the O_2 consumption rate is 0.2 mL/hr g of root tissue (fresh weight), that rooting density is 0.1 g/cm^3, that the water extraction rate is 7.5 mm/day, and that soil bulk density remains constant, calculated that the process of root water extraction by itself can draw into the soil as much as 70% of the oxygen required for respiration of crop roots.

The convective flow of air in the soil is similar in some ways to the flow of water and different in other ways. The similarity is in the fact that the flow of both fluids is usually impelled by, and is proportional to, a pressure gradient. The dissimilarity results from the relative incompressibility of water in comparison with air, which is highly compressible so that its density and viscosity are strongly dependent on pressure (as well as temperature). The gravitational potential gradient, where it exists (as in vertical systems), is directly important in causing water to flow, but it is hardly involved in air flow. Quite another difference is that water has the greater affinity to the surfaces of mineral particles (i.e., it is the wetting fluid) and is thus drawn into narrow necks and pores, forming capillary films and wedges (menisci).

In a three-phase system, therefore, air tends to occupy the larger pores. The two fluids — water and air — coexist in the soil by occupying different portions of the pore space having different geometric configurations. For this reason, the soil exhibits toward the two fluids different conductivity or permeability functions, as these relate to the different effective diameters, tortuosities, and interconnectedness of the pore sets occupied by each fluid.

Notwithstanding the differences between water flow and air flow, we can formulate the convective flow of air in the soil in an equation analogous to Darcy's law for water flow:

$$q_v = -(k/\eta)\ \nabla P \tag{11.1}$$

where q_v is the *volume convective flux* of air (volume flowing through a unit cross-sectional area per unit time), k is the permeability of the air-filled pore space, η is the viscosity of soil air, and ∇P is the three-dimensional gradient of soil air pressure. In one dimension, this equation takes the form

$$q_v = -(k/\eta)(dP/dx) \tag{11.2}$$

If the flux is expressed in terms of mass (rather than volume) per unit area and per unit time, then the equation is

$$q_m = -(\rho k/\eta)(dP/dx) \tag{11.3}$$

wherein q_m is the mass convective flux and ρ is the density of soil air.

Recalling that the density of a gas depends on its pressure and temperature, we now assume that soil air is an ideal gas in which the relation of mass, volume, and temperature is given by the equation

$$PV = nRT \tag{11.4}$$

where P is pressure, V is volume, n is the number of moles of gas, R is the universal gas constant per mole, and T is absolute temperature. Since the density $\rho = M/V$ and the mass M is equal to the number of moles n times the molecular weight m, we have

$$\rho = (m/RT)P \tag{11.5}$$

We now recall the continuity equation (see Chapters 8 and 9) for a compressible fluid:

$$\partial\rho/\partial t = -\,\partial q_m/\partial x \tag{11.6}$$

Substituting the expression for ρ from Eq. (11.5) and the expression for q_m from Eq. (11.3) into Eq. (11.6), we obtain

$$(m/RT)(\partial P/\partial t) = \partial/\partial x[(\rho k/\eta)\partial P/\partial x] \tag{11.7}$$

If the composite term $\rho k/\eta$ is nearly constant (if pressure differences are small), we can write

$$\partial P/\partial t = \alpha \partial^2 P/\partial x^2 \tag{11.8}$$

where $\alpha = RTk/m$, a composite constant. This is an approximate equation for the transient-state convective flow of air in soil. An assumption underlying this equation is that flow is laminar.

Quite a different mechanism of convective movement of gases in the soil is the transfer of dissolved gases by rain or irrigation water infiltrating into and percolating through soils. Since oxygen-saturated air at atmospheric pressure contains only 6 liters of O_2 per cubic meter, which is only enough to provide for the respiration of 1 kg (dry weight) of active roots, it seems unlikely that the oxygen supplied by infiltrating water can have anything but a very temporary effect in most situations.

Sample Problem

Consider a cropped field with an effective root zone depth of 0.8 m, a daily transpiration rate of 6 mm, and a daily soil respiration rate of 0.1 kg/m^2. Calculate what fraction of the oxygen requirement is supplied by convection if air is drawn from the atmosphere in immediate response to the pressure deficit created in the soil by the extraction of soil moisture.

Calculations:

Volume of water extracted per square meter per day: 1 m^2 × 0.006 m = 0.006 m^3 = 6 L.
Volume of air drawn from atmosphere = volume of water withdrawn from soil = 6 L.
Volume of oxygen drawn from atmosphere (21% O_2) = 6 × 21/100 = 1.26 L.
Mass of oxygen drawn from the atmosphere = (1.26/22.4) × 0.032 = 0.0018 kg.
Percentage of daily oxygen requirement supplied by convection = 0.0018/0.1 = 1.8%.

Note: 0.032 kg = mass of 1 mol O_2, occupying 22.4 L at standard pressure and temperature.

DIFFUSION OF GASES IN THE SOIL

The diffusive transport of gases such as O_2 and CO_2 in the soil occurs partly in the gaseous phase and partly in the liquid phase. Diffusion through the air-filled pores maintains the exchange of gases between the atmosphere and the soil, whereas diffusion through water films of various thicknesses maintains the supply of oxygen to, and disposal of CO_2 from, live tissues, which are typically hydrated. For both portions of the pathway, the diffusion process can be described by Fick's law (see Chapter 9):

$$J_g = -D(dc/dx) \qquad (11.9)$$

where J_g is the diffusive flux of a gas (mass diffusing across a unit area per unit time), D is the diffusion coefficient (generally having the dimensions of area per time), c is concentration (mass of diffusing substance per volume), x is distance, and dc/dx is the concentration gradient. If partial pressure p is used instead of concentration of the diffusing component, we get

$$J_g = -(D/\beta)(dp/dx) \qquad (11.10)$$

where β is the ratio of the partial pressure to the concentration.

Considering first the diffusive path in the air phase, we note that the diffusion coefficient in the soil D_s must be smaller than that in bulk air D_0 because of the limited fraction of the total volume occupied by continuous air-filled pores and also because of the tortuous nature of these pores. Hence we can expect D_s to be some function of the air-filled porosity, f_a.

Different workers have over the decades have reported different relations between D_s and f_a for various soils. For instance,
Penman (1940) found a linear relation:

$$D_s/D_0 = 0.66 f_a \qquad (11.11)$$

where 0.66 is a tortuosity coefficient, suggesting that the apparent (straight-line) path is about two-thirds the length of the mean path of diffusion in the soil. Tortuosity depends on the fractional volume of air-filled pores (i.e., it stands to reason that the tortuous path length should increase as the air-filled pore volume decreases); hence we can expect Penman's constant coefficient to hold for only a limited range of variation of air-filled porosity or of volume wetness.

An analysis by de Vries (1950) showed on the basis of theory that the relation sought between the effective diffusion coefficient and air-filled porosity should be curvilinear and dependent on pore geometry; hence it is not expected to be the same for different soils and water versus air contents. Consequently, Grable and Siemer (1968) found that as air-filled porosity fell to around 10%, the ratio D_s/D_0 fell to about 0.02. At even lower f_a values of 4–5%, Lemon and Erickson (1952) found the D_s/D_0 ratio to be as low as 0.005.

Currie (1961) studied diffusion in aggregated soils. He compared the diffusion coefficient of a gas in a dry soil in its natural structure with the diffusion coefficient within the individual soil crumbs and found that the within-crumb coefficient was only about one-fifth that of the between-crumb coefficient for a given air space. He offered the following equation:

$$D_c/D_0 = f_c/[1 + (k_c - 1)(1 - f_c)] \qquad (11.12)$$

where D_c and f_c are, respectively, the diffusion coefficient and volumetric air space within the crumb and k_c is a constant. Currie showed that k_c is a useful measure of soil structure, being high for poorly structured soils, and that it is in fact a measure of the tortuosity of the larger pores within the crumb. He reported values of k_c that varied from 4.2 to 11.0. More recent measurements of oxygen diffusion in soil aggregates have been reported by Sextone et al. (1985).

Methods of measuring gas diffusivity in soils are described by Rolston and Moldrup (2002).

Sample Problem

Consider a soil profile in which the air-phase O_2 concentration diminishes linearly from 21% at the surface to half that at 1-m depth. If the total porosity is a uniform 45% and the volume wetness 35%, calculate the diffusion rate using Penman's formula for the effective diffusion coefficient of oxygen in the soil (D_s). Assume steady-state diffusion, bulk-air diffusion coefficient (D_0) = 1.89×10^{-5} m²/sec.

Our first step is to estimate the effective diffusion coefficient D_s using Penman's linear relation between D_s and air-filled porosity f_a:

$$D_s = 0.66 f_a D_0 = 0.66 \times (0.45 - 0.35) \times 1.89 \times 10^{-5} = 1.26 \times 10^{-6}\ \text{m}^2/\text{sec}.$$

We use Fick's first law to calculate the steady-state one-dimensional diffusive flux J_g of oxygen through the soil profile from the external atmosphere to a plane at 1-m depth, where O_2 concentration is 10.5%:

$$J_g = D_s(\Delta c/\Delta x)$$
$$= (1.26 \times 10^{-6}\ \text{m}^2/\text{sec})(0.3 - 0.15\ \text{kg/m}^3)/1\ \text{m} = 1.89 \times 10^{-7}\ \text{kg/m}^2\ \text{sec}.$$

Note: The atmospheric concentration of $O_2 = 21\% \times 0.032$ kg/22.4 L = 0.3 kg/m³.

FORMULATION OF DIFFUSION PROCESSES

Having established the variable nature of D_s, we now return to the mathematical formulation of diffusion processes in soils. For transient conditions, we once again introduce the continuity principle:

$$f_a(\partial c/\partial t) = -\partial J_g/\partial x \tag{11.13}$$

which states that the time rate of change of concentration of a diffusing gas equals the rate of change of diffusive flux with distance. The foregoing assumes that the diffusing substance is conserved throughout (i.e., is neither created nor destroyed within the soil body considered).

As O_2 and CO_2 diffuse through the soil, however, O_2 is taken up and CO_2 is generated by aerobic biological activity along the diffusive path. To take account of the amount of a diffusing substance added to or subtracted from the system per unit time, we add a plus-or-minus S term to the right-hand side of Eq. (11.14). Note that a positive sign represents an increment rate (source) and a negative sign represents a decrement rate (sink) for the substance considered. Accordingly,

$$f_a(\partial c/\partial t) = -\partial J_g \partial x \pm S(x,t) \tag{11.14}$$

The designation $S(x,t)$ implies that the source–sink term is a function of both space and time.

We next substitute Eq. (11.9) into Eq. (11.15) and consider only the vertical direction z:

$$f_a(\partial c/\partial t) = \partial(D_s \partial c/\partial z)/\partial z \pm S(z,t) \tag{11.15}$$

Note that we use D_s, for which one may wish to substitute an expression such as the one by Penman ($D_s = 0.66 D_0 f_a$) or any other empirical or theoretically based function.

In the event that D_s is constant, Eq. (11.16) simplifies to

$$f_a(\partial c/\partial t) = D_s(\partial^2 c/\partial x^2) \pm S(z,t) \tag{11.16}$$

In aggregated soils, gaseous diffusion occurs readily in the between-aggregate macropores, which drain rapidly after a rain or irrigation and form a continuous air-filled phase. However, the within-aggregate micropores can remain nearly saturated for extended periods and restrict the internal aeration of aggregates. Therefore plant roots are often confined to the larger pores between aggregates and scarcely penetrate the aggregates themselves, whether because the small internal pores of the aggregates and their mechanical rigidity do not permit penetration or because of aeration restriction. However, microorganisms do penetrate aggregates and, by their demand for oxygen, affect soil aeration.

Equations (11.16) and (11.17) consider diffusion in the soil profile as a whole. If we wish to take a closer look at diffusion into or out of individual soil aggregates (clods), we must recast our equations into three-dimensional form. Assuming an aggregate to be isotropic and approximately spherical, we can use polar coordinates to obtain

$$f_a(\partial c/\partial t) = (1/r^2)\partial[D_s r^2(\partial c/\partial r)]/\partial r \pm S \tag{11.17}$$

where r is radial distance from the center of the sphere. If we further assume steady-state exchange (equal and constant-rate absorption of O_2 and release of CO_2) between the clod and its surroundings, we can simply set c/t equal to zero and solve the equation for appropriate boundary conditions. For a uniform aggregate of radius R, respiring steadily and uniformly, the concentration difference Δc between the surface and the center is given by (Crank, 1975)

$$\Delta c = \pm SR^2/6D_s \tag{11.18}$$

Note that the concentration difference (i.e., expressing the O_2 deficit or CO_2 excess inside the clod) is proportional directly to the respiration rate S and inversely to the diffusion coefficient D_s. As pointed out by Currie (1975), knowledge of both S and D_s is required to quantify aeration.

Thus far we have confined our attention almost entirely to diffusion in the air phase. In the case of root respiration, the final segment of the diffusion path takes place through the hydration envelope surrounding the root. To describe this stage of the aeration process, we can employ the cylindrical form of the diffusion equation (Fig. 11.1):

$$f_a(\partial c/\partial t) = (1/r)\partial(rD_w\partial c/\partial r)/\partial r \tag{11.19}$$

in which r is the radial distance from the root axis. Here we specify the diffusion coefficient for the soluble diffusing substance in water, D_w. Although this stage of the process takes place through a very short distance (the hydration film being, at most, only a few millimeters thick), it may be the rate-limiting stage, because the diffusivity of O_2 in water is only about 1/10,000 its value in air (i.e., about 2.6×10^{-3} mm^2/sec in water as against 22.6 in air). The solubility of oxygen in water (only 4% of the solubility of carbon dioxide) may be limiting also, especially at high temperatures.

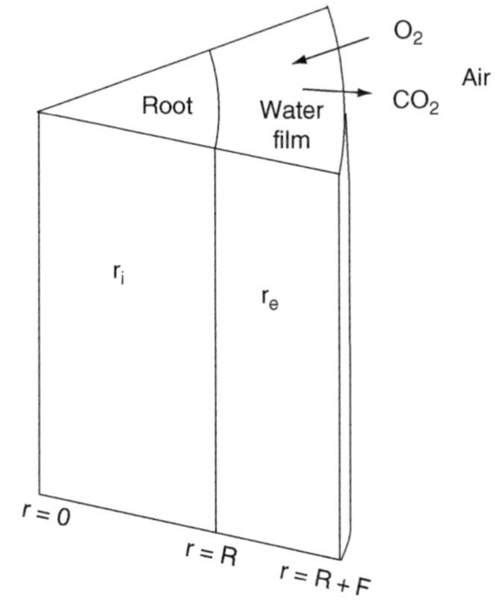

Fig. 11.1. Diffusion of O_2 from soil to root, representating root and hydration film as concentric cylinders. Here r is the radial distance from the root axis, R is the radius of the root, and F is the thickness of the water film.

The diffusion of oxygen to roots was first analyzed by Lemon (1962). His analysis divided the process between two regions: an inner region, the root itself, and an outer region, the water film enveloping the root. For steady-state diffusion, the equations pertaining to the inner region (subscripted *i*) and the outer one (subscripted *e*) are

$$D_i\left(\frac{\partial^2 c_i}{\partial r^2}+\frac{1}{r}+\frac{\partial^2 c_i}{\partial r}\right)-S=0,\qquad D_e\left(\frac{\partial^2 c_e}{\partial r^2}+\frac{1}{r}+\frac{\partial^2 c_e}{\partial r}\right) \tag{11.20}$$

If the diffusion coefficients for the root D_i and for the water film D_e are known and if the oxygen consumption rate per length of root S is also known, it is possible to solve the pair of equations simultaneously for different conditions and to define the conditions likely to limit root respiration (Glinski and Stepniewski, 1985).

There is a fundamental difference between the relation of a soil's permeability to pore geometry and the corresponding relation of the diffusion coefficient. Because the permeability pertains to pressure-induced viscous flow, it obeys Poiseuille's law, which states that flow rate varies as the fourth power of the pore radius. Hence permeability is strongly dependent upon *pore size distribution.* Diffusion, on the other hand, depends mainly on the total volume and tortuosity of continuous pores available for diffusion. The reason that diffusion does not depend on pore size distribution is that for air at atmospheric pressure the *mean free path* of molecules in thermal motion (i.e., the distance an "average" molecule in random motion travels before it collides with another) is of the order of 0.0001–0.0005 mm and thus much smaller than the radii of the pores that generally account for most of a soil's air-filled porosity. A possible exception to this generalization is the diffusion of gases in a porous medium under extremely low gas pressures, at which a phenomenon called *Knudsen diffusion* may come into play (Clifford and Hillel, 1986). Knudsen diffusion occurs when the mean free path of gas molecules is much greater than the pore radius, hence molecule-wall collisions dominate over molecule-molecule collisions (Scanlon et al., 2002). The Knudsen diffusive flux depends on the molecular weight, density, and temperature of the diffusing gas, as well as on pore radii.

MEASUREMENT OF GASEOUS CONVECTION AND DIFFUSION IN SOILS

The set of processes involved in the transport of gases within the soil and in the exchange of gases between the soil and the atmosphere is altogether too complex to be definable by any single measurement. Soil physicists have sought, therefore, to measure soil attributes and component processes considered to be relevant to the problem at hand and to be indicative of soil aeration as whole. The early attempts to characterize soil aeration in terms of the content and composition of soil air were essentially inadequate in that their results were, at best, static snapshots of the situation at a given moment in time, with no indication of process dynamics, directions, and rates of change.

A quite different approach to characterizing soil aeration is to measure the air permeability, that is, the coefficient governing convective transmission of air through the soil in response to a gradient of total gas pressure (Ball and Schjønning, 2002). This measurement can provide useful information on the effective sizes and the continuity of air-filled pores. The techniques that have been proposed include constant-pressure and falling-pressure devices (Fig. 11.2). The method has been applied in the field and found useful for assessing the "openness" of the surface layer to the entry of air, as affected by such cultural practices as tractor traffic and tillage (Fig. 11.3).

The measurement of diffusion rates through the air-filled pores alone gives no indication the possible impedance of the liquid envelope surrounding a root. A soil may have a high O_2 concentration in the network of large pores that are open to the atmosphere and yet not provide an adequate supply to the roots if they are thickly hydrated. An interesting method for measuring oxygen diffusion to a rootlike probe is based on the use of a thin platinum electrode maintained at a constant potential. The reaction that takes place at the negatively charged electrode is that dissolved oxygen molecules reaching the electrode by diffusion take up four electrons and react with hydrogen ions to form water in an acid solution ($O_2 + 4H^+ + 4e^- \rightarrow 2H_2O$) or react with water to form hydroxyl in an alkaline solution ($O_2 + 2H_2O + 4e^- \rightarrow 4OH^-$).

The resulting current measures the flux of oxygen to the moisture-covered electrode, which acts as a sink, thus simulating the action of a respiring root. In practice, one inserts the probe into a moist soil and waits until the reading becomes steady, at which time the flux of oxygen to the probe is taken to represent the oxygen-supplying power of the soil, commonly called the ODR (for oxygen diffusion rate). The technique fails in relatively dry soils, which, however, are unlikely to present aeration problems. Descriptions of the technique and the results obtainable by it have been published by Blackwell (1983), and Payne and Gregory (1988), and Farrell et al. (2002).

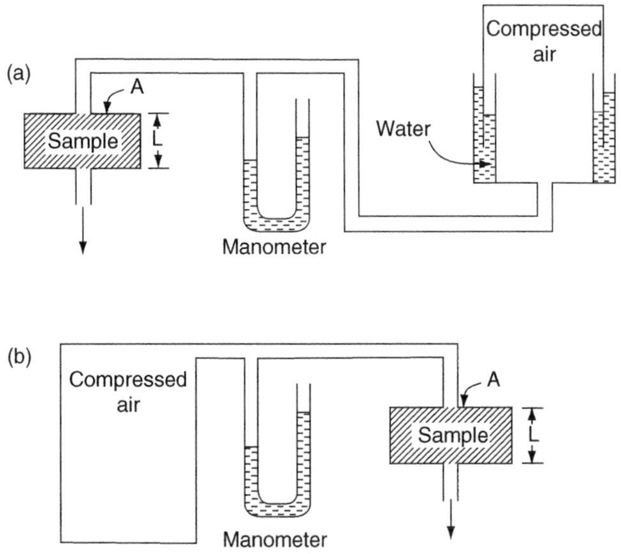

Fig. 11.2. Measurement of air permeability with (a) a constant-pressure, variable-volume permeameter [$k = (L\eta/Af)(\Delta V/t)$] and (b) a falling-pressure, constant-volume permeameter [$k = (2.3L\eta V/AP_a)(\log[p_1/p_2]/\Delta t)$]. (Note: k is air permeability, L is sample length, A is sample cross-sectional area, V is volume of the air cell, P_a is barometric pressure, p is cell air pressure, t is time, and p_1 and p_2 are cell pressure at start and end of time step ΔT, respectively.)

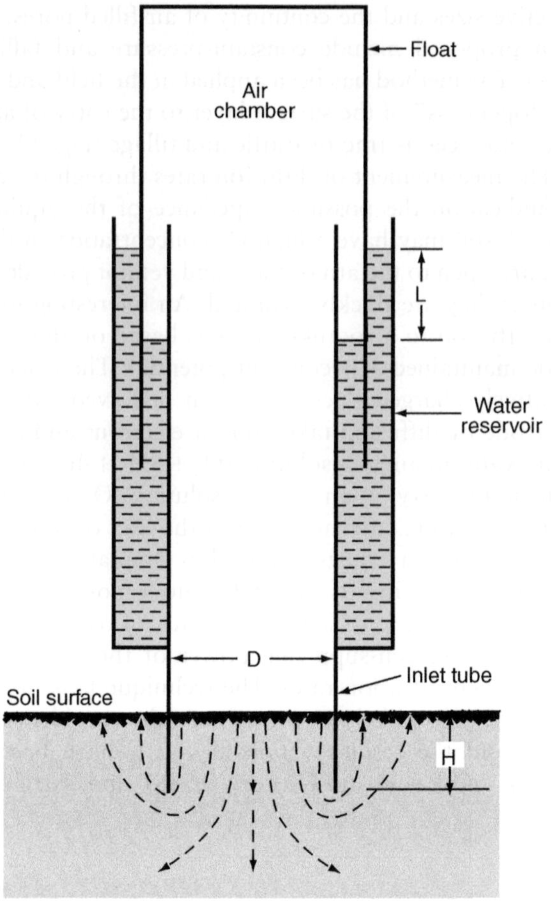

Fig. 11.3. Air permeameter for field use. (After Grover, 1956.)

BOX 11.1 Soil Emissions of Radon Gas

In recent decades, concern has risen over the generation of radon gas by some soils, especially in areas underlain by granite, as a consequence of the radioactive decay of radium. Persistent radon gas may diffuse into the basements of homes that are built over radon-emitting soils. Being colorless, odorless, practically inert chemically, and present in only trace concentrations, radon can nonetheless constitute a serious health hazard because it is radioactive and known to be a powerful carcinogen (Eisenbud, 1987; Boyle, 1988). Since it is difficult to vent houses with appreciable radon concentrations after the fact, the best approach is to test soil emissions prior to construction and to insulate the ground surface at the time of construction.

EMISSIONS OF GREENHOUSE GASES FROM SOILS

Growing attention is now being devoted to the "greenhouse effect" of the atmosphere and its enhancement by the anthropogenic emissions of radiatively active gases. Often cited as the major culprit is the burning of fossil fuels (petroleum, coal, natural gas), which releases carbon dioxide into the atmosphere in excess of the amount that can be absorbed by green plants and the oceans. Lately, agriculture's role in climate change has come to light (Lal et al., 1995b; Rosenzweig and Hillel, 1998). Clearing of forests for fields and pastures, transforming virgin soil into cultivated land, flooding areas for rice and sugarcane production, burning crop residues, raising ruminant animals, and applying nitrogen fertilizers are activities that have been implicated in the release of greenhouse gases to the atmosphere.

The gases emitted are primarily carbon dioxide, methane (Harriss et al., 1988; Cicerone and Orenland, 1988; Knowles, 1993), and nitrous oxide (Denmead et al., 1979; Andreae and Schimel, 1989). Altogether the agricultural sources are now estimated to constitute about 20% of the total anthropogenic emissions of greenhouse gses. Land-use changes for agricultural development may account for an additional 14% (IPCC, 1995).

On a global basis, soil is second only to the ocean in the amount of carbon it contains, holding perhaps 1.5–3 times as much carbon as living terrestrial vegetation. Estimates of total soil carbon (including litter and soil organic matter) are about $1.4–1.6 \times 10^{12}$ metric tons (Schlesinger, 1991; IPCC, 1995). When areas originally covered with natural vegetation are converted to cultivated fields, much of the vegetation initially present is converted to carbon dioxide. The material above ground either is burned or decomposes rapidly, whereas the organic matter in the soil decays more slowly but over a period of time releases significant quantities of CO_2.

However, the release of CO_2 from different agricultural practices varies. Curtailed tillage practices, such as minimum or no tilling, efficient crop rotation, and periodic fallowing, tend to reduce CO_2 effluxes. Controlling soil erosion also helps to conserve the carbon content of soils. When lands that have degraded from overuse or are salinized are retired from cultivation, they tend to sequester rather than release carbon.

Methane (CH_4) is the second most important greenhouse gas after carbon dioxide (IPCC, 1995; Rosenzweig and Hillel, 1998). Although it is very much less prevalent than CO_2, methane is about 20 times more powerful per molecule in terms of radiative forcing (IPCC, 1990). Hence, its rising concentration in the atmosphere is a matter of concern. Agriculture is the largest anthropogenic source of methane, accounting for about 40% of total emissions. The major agricultural sources of methane are fields of wetland rice (known as rice paddies), herds of ruminant animals, and biomass burning. The conversion of natural wetlands to shallow impoundments for irrigation water management tends to increase methane emissions as well (Harriss et al., 1988). Methane is produced in the process of carbohydrate (including cellulose) decomposition carried out by anaerobic bacteria (Cicerone and Oremland, 1988; Bouwman, 1990).

Nitrous oxide (N_2O) is present in the atmosphere in even smaller quantities than methane. The current concentration is about 0.31 parts per million by volume (ppmv), compared with ~360 ppmv for carbon dioxide and ~1.72

ppmv for methane (IPCC, 1995). The radiative forcing of the N_2O molecule is about 200 times greater than that of CO_2, and its mean lifetime in the atmosphere is about 120 years. The annual rate of rise of nitrous oxide is 0.2–0.3%. Besides being a greenhouse gas, nitrous oxide plays another important role in the atmosphere: It is further oxidized into NO_x in the stratosphere, where it acts to deplete ozone. Nitrous oxide emissions are part of the complex global nitrogen cycle by which various forms of nitrogen are transported and transmuted in soils, plants, animals, and the atmosphere (Spreat, 1987; Brady, 1990; Schlesinger, 1991).

Soils are the major medium that generates N_2O, through microbial processes. Agricultural activities that contribute to nitrous oxide emissions include the liberal application of nitrogenous fertilizers, burning of biomass, and conversion of forests to pastures in tropical regions. The production of nitrous oxide in the soil is an intermediate stage in both nitrification (oxidation of ammonium to nitrate) and denitrification (the reverse of nitrification), but it is the latter that constitutes the main process through which nitrous oxide is released from soils.

A detailed description of soil-borne processes and of agriculture in relation to the greenhouse effect is presented in Rosenzweig and Hillel (1998).

Part V

COMPOSITE PHENOMENA

12. SOIL TEMPERATURE AND HEAT FLOW

IMPORTANCE OF SOIL TEMPERATURE

Soil temperature, its value at any moment and the manner of its variation in time and space, is a factor of primary importance in determining the rates and directions of soil physical processes and of energy and mass exchange with the atmosphere. Temperature governs evaporation and aeration as well as the types and rates of chemical reactions that take place in the soil. Finally, soil temperature strongly influences biological processes, such as seed germination, seedling emergence and growth, root development, and microbial activity.

Soil temperature varies in response to changes in the radiant, thermal, and latent energy exchange processes that take place primarily through the soil surface. The effects of these phenomena are propagated into the soil profile by a complex series of transport processes, the rates of which are affected by time-variable and space-variable soil properties. Hence the quantitative formulation and prediction of the soil thermal regime can be a formidable task. Even beyond passive prediction, the possibility of actively controlling or modifying the thermal regime requires a thorough knowledge of the processes at play and of the environmental and soil parameters determining their rates. The pertinent soil parameters include the specific heat capacity, thermal conductivity and thermal diffusivity (all of which are strongly affected by bulk density and wetness) as well as the internal sources and sinks of heat operating at any time.

Many reviews of soil temperature and heat flow have been published over the years by (among others): de Vries (1975b), Campbell (1977), Taylor and Jackson (1986), Fuchs (1986), Hanks (1992), Evett (2002), and McInnes (2002).

MODES OF ENERGY TRANSFER

We begin with basic physics. There are three principal modes of energy transfer: radiation, convection, and conduction. By *radiation*, we refer to the emission of energy in the form of electromagnetic waves from all bodies above 0 K. According to the *Stefan–Boltzmann law*, the total energy emitted by a body, J_t, integrated over all wavelengths, is proportional to the fourth power of the absolute temperature T of the body's surface. This law is usually formulated

$$J_t = \varepsilon\sigma T^4 \tag{12.1}$$

where σ is a constant and ε is the *emissivity coefficient*, which equals unity for a perfect emitter (generally called a *black body*). The absolute temperature also determines the wavelength distribution of the emitted energy. *Wien's law* states that the wavelength of maximal radiation intensity λ_m is inversely proportional to the absolute temperature:

$$\lambda_m = 2900/T \tag{12.2}$$

where λ_m is in micrometers. The radiative intensity as a function of wavelength and temperature is given by *Planck's law*:

$$E_\lambda = C_1/\lambda^5\ [\exp(C_2/\lambda T) - 1] \tag{12.3}$$

where E_λ is energy flux emitted in a given wavelength range and C_1, C_2 are constants.

Since the temperature of the soil surface averages about 300 K (though it can range from below 273 K, the freezing point, to 330 K or even higher), the radiation emitted by the soil surface has its peak intensity at a wavelength of about 10 μm and its wavelength distribution over the range of 3–50 μm. This is in the realm of infrared, or heat, radiation.

A very different spectrum is emitted by the sun, which acts as a black body at an effective surface temperature of about 6000 K. The sun's radiation includes the visible light range of 0.3–0.7 μm as well as some infrared radiation of greater wavelength (up to about 3 μm) and some ultraviolet radiation ($\lambda < 0.3$ μm). Since there is very little overlap between the two spectra, it is customary to distinguish between them by calling the incoming solar spectrum *short-wave radiation* and the spectrum emitted by the earth *long-wave radiation*.

The second mode of energy transfer, called *convection*, involves the movement of a heat-carrying mass, as in the cases of ocean currents and atmospheric winds. An example more pertinent to soil physics would be the infiltration of warm wastewater into an initially cold soil.

Conduction, the third mode of energy transfer, is the propagation of heat within a body by internal molecular motion. Because temperature is an expression of the kinetic energy of a body's molecules, the existence of a temperature difference within a body will normally cause the transfer of kinetic energy by the collisions of rapidly moving molecules from the warmer region of the body to their neighbors in the colder region. The process of heat conduction is thus analogous to diffusion; and in the same way that diffusion tends in time to equilibrate a mixture's composition throughout, heat conduction tends to equilibrate a body's internal temperature.

In addition to the three modes of energy transfer described, there is a composite phenomenon that one may recognize as a fourth mode, namely, *latent heat transfer*. A prime example is the process of distillation, which includes the heat-absorbing stage of evaporation, followed by the convective or diffusive movement of the vapor, and ending with the heat-releasing stage of condensation. A similar catenary process can also occur in transition back and forth from ice to liquid water in soils subject to freezing and thawing.

ENERGY BALANCE FOR A BARE SOIL

A detailed elucidation of the energy and water balances of vegetated fields is given in Chapter 20. At this stage we outline the energy regime of a bare (unvegetated) soil. We begin with the radiation balance of a bare surface, which can be written thus (van Bavel and Hillel, 1976):

$$J_n = (J_s + J_a)(1 - \alpha) + J_{li} - J_{lo} \tag{12.4}$$

Here J_n is the *net radiation*, that is, the sum of all incoming minus outgoing radiant energy fluxes, J_s is the incoming flux of short-wave radiation directly from the sun, J_a is the short-wave diffuse radiation from the atmosphere (sky), J_{li} is the incoming long-wave radiation flux from the sky, J_{lo} is the outgoing long-wave radiation emitted by the soil, and α is the *albedo*, or *reflectivity coefficient*, which is the fraction of incoming short-wave radiation reflected by the soil surface rather than absorbed by it. We shall disregard in the present context all terms that do not pertain to the soil, namely, J_s, J_a, and J_{li}.

The albedo α is an important characteristic of soil surfaces, and it can vary widely in the range of 0.1–0.4, depending on the soil's basic color (whether dark or light colored), the surface's roughness, and the inclination of the incident radiation relative to the surface (Sellers, 1965). In the short run, the albedo also depends on the changing wetness of the exposed soil (Jackson et al., 1974). The drier the soil, the smoother its surface; the brighter its color, the higher its albedo. To a certain extent, the albedo can be modified by various surface treatments, such as tillage and mulching.

Apart from the reflected short-wave radiation, governed by the albedo, we have another soil-dependent process, namely, the emission of long-wave radiation. In accordance with Eq. (12.1), the emitted flux J_{le} depends on surface temperature but is also affected by emissivity ε. This parameter, in turn, depends on soil wetness and generally varies between 0.9 and 1.0.

The net radiation received by the soil surface is transformed into heat, which warms the soil and air and vaporizes water. We can thus write the surface energy balance as follows:

$$J_n = S + A + LE \tag{12.5}$$

where S is the soil heat flux (the rate at which heat is transferred from the surface downward into the soil profile), A is the "sensible" heat flux transmitted from the surface to the air above, and LE is the evaporative heat flux, a product of the evaporative rate E and the latent heat per unit quantity of water evaporated, L. The total surface energy balance [combining Eqs. (12.4) and (12.5)] is:

$$(J_s + J_a)(1 - \alpha) + J_{li} - J_{lo} - S - A - LE = 0 \tag{12.6}$$

Conventionally, all components of the energy balance are taken as positive if directed toward the surface and negative otherwise.

Sample Problem

Consider the energy balance of a bare-surface soil, assuming the following conditions: The daytime (12-hr) average global (sun and sky) radiation is 3.35×10^4 J/m^2 min (0.8 cal/cm^2 min). The albedo is 0.15. The average soil surface temperature during the diurnal period is 27°C. Advection in daytime balances the outflow of sensible heat during the night, so diurnal net sensible heat exchange with the atmosphere is negligible. Evaporation is 2 mm/day. The emissivity is 0.9, and the atmosphere returns 60% of the long-wave radiation emitted by the ground. Estimate the daytime soil heat transfer term. Is it positive or negative?

Following Eq. (12.6), the diurnal energy balance can be written

$$J_s(1 - \alpha) - J_l - S - A - LE = 0$$

where J_s is incoming global short-wave radiation, α is albedo, J_l is net long-wave emitted radiation, S is heat flow into the soil, A is sensible heat transfer to the air, and LE is latent heat loss. To obtain S, we rearrange this equation to read

$$S = J_s(1 - \alpha) - J_l - A - LE$$

For the net short-wave radiation, we have

$$\begin{aligned} J_s(1 - \alpha) &= (3.35 \times 10^4 \, J/m^2 \text{ min})(720 \text{ min/day})(1 - 0.15) \\ &= 2.05 \times 10^7 \text{ J/m}^2 \text{ daytime} \end{aligned}$$

Net outgoing long-wave radiation (using the *Stefan–Boltzmann law*) is:

$$J_l = 0.4\ (\varepsilon\sigma T)$$

Note: The total energy radiated per unit surface of a "black body" in unit time is proportional to the fourth power of the thermodynamic temperature. The constant of proportionality has the value of 5.67×10^{-8} J sec^{-1} m^{-2} K^{-4}.

$$\begin{aligned} J_l &= 0.4 \times 0.9 \times (5.67 \times 10^{-8} \text{ J sec}^{-1} \text{ m}^{-2} \text{ K}^{-4})(8.64 \times 10^4 \text{ sec/diurnus})(273 + 27)^4 \text{ K} \\ &= 1.43 \times 10^7 \text{ J/m}^2 \end{aligned}$$

Note: The factor 0.4 is used because, as stated, the atmosphere returns 60% of the outgoing J_l. Also, the emission of longwave radiation takes place throughout the 24-hr day.

The net sensible heat transfer A is negligible. For the latent heat loss term, we have

$$\begin{aligned} LE &= 2.43 \times 10^6 \text{ J/kg} \times 2 \text{ kg/m}^2 \text{ day} = 4.86 \times 10^6 \text{ J/m}^2 \text{ per daytime} \\ &(= 580 \text{ cal/g} \times 0.2 \text{ g/cm}^2 \text{ day} = 116 \text{ cal/cm}^2 \text{ day}) \end{aligned}$$

Note: Evaporation = 2 mm/day = 2 L/m^2 day = 2 kg/m^3 (water density = 1000 kg/m^3).

Finally, we can sum up all of these quantities to obtain the soil heat flow:

$$\begin{aligned} S &= 2.05 \times 10^7 - 1.43 \times 10^7 - 0 - 0.49 \times 10^7 \\ &= 1.3 \times 10^6 \text{ J/m}^2 \text{ per diurnus } (\sim 32.1 \text{ cal/cm}^2 \text{ day}) \end{aligned}$$

Ergo, the soil is gaining heat. If this amount of heat is absorbed in the top 0.2 m of the soil with a specific heat capacity of 2000 J/kg and a bulk density of 1600 kg/m^3, it will raise the temperature by 2 degrees.

CONDUCTION OF HEAT IN SOIL

The conduction of heat in solids was analyzed as long ago as 1822 by Fourier, whose name is given to the linear equation that has been used ever since to describe heat conduction. This equation is mathematically analogous to the diffusion equation (Fick's law) as well as to Ohm's law for the conduction of electricity and Darcy's law for water flow in soil.

The first law of heat conduction, known as *Fourier's law*, states that the flux of heat in a homogeneous body is in the direction of, and proportional to, the temperature gradient:

$$q_h = -\kappa \nabla T \tag{12.7}$$

Here q_h is the thermal flux (i.e., the amount of heat conducted across a unit cross-sectional area in unit time), κ (Greek letter kappa) is thermal conductivity, and ∇T is the spatial gradient of temperature T. In one-dimensional form, this law is written

$$q_h = -\kappa_x\ (dT/dx) \qquad \text{or} \qquad q_h = -\kappa_z\ (dT/dz) \tag{12.8}$$

Here dT/dx is the temperature gradient in any direction, designated x, and dT/dz is, specifically, the gradient in the vertical direction representing soil depth ($z = 0$ being the soil surface). The subscripts attached to the thermal conductivity term are meant to account for the possibility that this parameter may have different values in different directions (i.e., that it may be nonisotropic). The negative sign in these equations is due to the fact that heat flows from a higher to a lower temperature (i.e., in the direction of, and in proportion to, a *negative* temperature gradient).

Equation (12.7) is sufficient to describe heat conduction under steady-state conditions, that is to say, where the temperature at each point in the conducting medium is invariant and the flux is constant in time and space. To account for nonsteady (transient) conditions, we need a second law analogous to Fick's second law of diffusion as embodied in Eq. (9.13). To obtain the second law of heat conduction, we invoke the *principle of energy conservation* in the form of the *continuity equation*, stating that, in the absence of internal sources or sinks of heat, the time rate of change in heat content of a volume element must equal the change of flux with distance:

$$\rho c_m\ (\partial T/\partial t) = -\nabla \cdot q_h \tag{12.9}$$

where ρ is mass density and c_m *specific heat capacity per unit mass* (defined as the change in heat content of a unit mass of the body per unit change in temperature). The product ρc_m (often designated C) is the *specific heat capacity per unit volume*, and $\partial T/\partial t$ is the time rate of temperature change. Note that the symbol ρ represents the total mass per unit volume, including the water in the case of a moist soil. The symbol ∇ (del) is the shorthand representation of the three-dimensional gradient. Equation (12.9) can thus be restated as

$$\rho c_m\ (\partial T/\partial t) = -(\partial q_x/\partial x + \partial q_y/\partial y + \partial q_z/\partial z)$$

where x, y, z are the orthogonal direction coordinates.

Combining Eqs. (12.9) and (12.7), we obtain the desired *second law of heat conduction*:

$$\rho c_m\ (\partial T/\partial t) = -\nabla \cdot (\kappa\ \nabla T) \tag{12.10}$$

which, in one-dimensional form, is

$$\rho c_m\ (\partial T/\partial t) = (\partial/\partial \mathrm{x})\ [\kappa\ (\partial T/\partial \mathrm{x})] \tag{12.11}$$

Sometimes we may need to account for the possible occurrence of heat sources or sinks in the realm where heat flow takes place. Heat sources include such phenomena as organic matter decomposition, wetting of initially dry soil material, and condensation of water vapor. Heat sinks are generally associated with evaporation. Lumping all these sources and sinks into a single term S, we can rewrite the last equation as

$$\rho c_m (\partial T/\partial t) = (\partial/\partial x)\ [\kappa\ (\partial T/\partial x)] \pm S(x, t) \qquad (12.12)$$

in which the source–sink term is shown as a function of both space and time.

The ratio of the thermal conductivity κ to the volumetric heat capacity C_v ($= \rho c_m$) is called the *thermal diffusivity*, designated D_T. Thus,

$$D_T = \kappa/C_v \qquad (12.13)$$

Substituting D_T for κ, we can rewrite Eq. (12.8) and (12.11):

$$q_h = -D_T C_v\ (\partial T/\partial x) \qquad (12.14)$$

In the special case where D_T can be taken as constant (not a function of distance), we can write

$$\partial T/\partial t = D_T\ (d^2T/dx^2) \qquad (12.15)$$

To solve the foregoing equations so as to obtain a description of how temperature varies in space and time, we need to know, by measurement or calculation, the pertinent values of the three parameters just defined, namely, the volumetric heat capacity C_v, thermal conductivity κ, and thermal diffusivity D_T. Together, they are called the *thermal properties of soils*.

VOLUMETRIC HEAT CAPACITY OF SOILS

A soil's volumetric heat capacity C is defined as the change of a unit volume's heat content per unit change in temperature. It is expressed as calories per cubic centimeter per degree or joules per cubic meter per degree. Thus, C depends on the composition of the soil's solid phase (mineral and organic components), on bulk density, and on soil wetness (Table 12.1).

TABLE 12.1 Densities and Volumetric Heat Capacities of Soil Constituents (at 10°C) and of Ice (at 0°C)

	Density ρ		Heat capacity C	
Constituent	**(g/cm³)**	**(kg/m³)**	**(cal/cm³K)**	**(J/m³ K)**
Quartz	2.66	2.66×10^3	0.48	2.0×10^6
Other minerals (average)	2.65	2.65×10^3	0.48	2.0×10^6
Organic matter	1.3	1.3×10^3	0.6	2.5×10^6
Water (liquid)	1.0	1.0×10^3	1.0	4.2×10^6
Ice	0.92	0.92×10^3	0.45	1.9×10^6
Air	0.00125	1.25	0.003	1.25×10^3

The value of C can be estimated by summing the heat capacities of the various constituents, weighted according to their volume fractions. As given by de Vries (1975a), it is

$$C = \Sigma(f_{si}C_{si} + f_w C_w + f_a C_a) \tag{12.16}$$

Here, f denotes the volume fraction of each phase: solid (subscripted s), water (w), and air (a).

The solid phase includes a number of components subscripted i, such as various minerals and organic matter, and the symbol Σ indicates the summation of the products of their respective volume fractions and heat capacities. The C value for water, air, and each component of the solid phase is the product of the particular density and the specific heat per unit mass (i.e., $C_w = \rho_w c_{mw}$, $C_a = \rho_a c_{ma}$, $C_{si} = \rho_{si} c_{mi}$).

Most of the minerals composing soils have nearly the same values of density (about 2.65 g/cm^3, or 2.65×10^3 kg/m^3) and of heat capacity (0.48 cal/cm^3 K, or 2.0×10^6 J/m^3 K). Since it is difficult to separate the different kinds of organic matter present in soils, it is tempting to lump them all into a single constituent (with an average density of about 1.3 g/cm^3, or 1.3×10^3 kg/m^3, and an average heat capacity of about 0.6 cal/cm^3 K, or 2.5×10^6 J/m^3 K).

Although the density of water is less than half that of mineral matter (about 1 g/cm^3, or 1.0×10^3 kg/m^3), its specific heat is more than twice as large (1 cal/cm^3 K, or 4.2×10^6 J/m^3 K). Finally, since the density of air is only about 1/1000 that of water, its contribution to the specific heat of the composite soil can generally be neglected.

Thus, Eq. (12.16) can be simplified as follows:

$$C_v = f_m C_m + f_o C_o + f_w C_w \tag{12.17}$$

where subscripts m, o, w refer to mineral matter, organic matter, and water, respectively. Note that $f_m + f_o + f_w = 1 - f_a$ and the total porosity $f = f_a + f_w$. The reader will recall that in preceding chapters we designated the volume fraction of water f_w as θ. Knowing the approximate average values of C_m, C_o and C_w, we can further simplify Eq. (12.17) to give

$$C_v = 0.48 f_m + 0.60 f_o + f_w \tag{12.18}$$

The use of Eq. (12.18) must be qualified in the case of frozen or partially frozen soils, since the properties of ice differ from those of liquid water ($\rho = 0.92$ g/cm^3, or 0.92×10^3 kg/m^3, and $C = 0.45$ cal/cm^3 K, or 1.9×10^6 J/m^3 K). In typical mineral soils, the volume fraction of solids is in the range of 0.45–0.65, and C_v values range from about 1 MJ/m^3 K (less than 0.25 cal/cm^3 K) in the dry state to about 3 MJ/m^3 K, or 0.75 cal/cm^3 K, in the water-saturated state.

The measurement of heat capacity and specific heat is described by Kluitenberg (2002).

THERMAL CONDUCTIVITY AND DIFFUSIVITY

Thermal conductivity κ is defined as the quantity of heat transferred through a unit area of the conducting body in unit time under a unit temperature gradient. As shown in Table 12.2, the thermal conductivities of specific

TABLE 12.2 Thermal Conductivities of Soil Constituents (at 10°C) and of Ice (at 0°C)

Constituent	mcal/cm sec K	(W/m K)
Quartz	21	8.8
Other minerals (average)	7	2.9
Organic matter	0.6	0.25
Water (liquid)	1.37	0.57
Ice	5.2	2.2
Air	0.06	0.025

soil constituents differ widely (see also Table 12.3). Hence the space-averaged (macroscopic) thermal conductivity of a soil depends on its mineral composition and organic matter content as well as on the volume fractions of water and air.

Since the thermal conductivity of air is very much smaller than that of water or solid matter, a high air content (or low water content) corresponds to a low thermal conductivity. Moreover, since the proportions of water and air vary continuously, κ is also time variable. Soil composition is seldom uniform in depth; hence κ is generally a function of depth as well as of time. It also varies with temperature, but under normal conditions this variation is ignored. Unlike heat capacity, thermal conductivity is sensitive not only to the mineral composition of a soil but also to the sizes, shapes, and arrangements of soil particles. In the normal range of soil wetness experienced in the field, *C* may undergo a threefold or fourfold change, whereas the corresponding change in κ may be a hundredfold or more.

TABLE 12.3 Average Thermal Properties of Soils and Snow[a]

Soil type	Porosity f	Volumetric wetness θ	Thermal conductivity (10^{-3} cal/cm sec °C)	Volumetric heat capacity C_v (cal/cm³ sec °C)	Damping depth (diurnal) d (cm)
Sand	0.4	0.0	0.7	0.3	8.0
	0.4	0.2	4.2	0.5	15.2
	0.4	0.4	5.2	0.7	14.3
Clay	0.4	0.0	0.6	0.3	7.4
	0.4	0.2	2.8	0.5	12.4
	0.4	0.4	3.8	0.7	12.2
Peat	0.8	0.0	0.14	0.35	3.3
	0.8	0.4	0.7	0.75	5.1
	0.8	0.8	1.2	1.15	5.4
Snow	0.95	0.05	0.15	0.05	9.1
	0.8	0.2	0.32	0.2	6.6
	0.5	0.5	1.7	0.5	9.7

[a] After van Wijk and de Vries (1963).

The relationship between the overall thermal conductivity of a soil and the specific conductivities and volume fractions of the soil's constituents is very intricate. Two relatively simple alternative cases can be envisaged: a dry soil and a water-saturated soil with the same internal structure. In both cases we have a two-phase system in which the particles are dispersed in a continuous fluid (air or water) with a volume fraction f_0 and thermal conductivity κ_0. The particles then occupy a volume fraction $f_1 = 1 - f_0$ and have a thermal conductivity κ_1. A composite thermal conductivity for the medium can be defined as follows: Consider a representative cube of soil with side L, large in comparison with the diameters of the particles and pores. Assume that the upper face is at a temperature T_1 and the bottom face is at a lower temperature, T_2. A constant heat flux q_h will then pass through the cube, proportional to the temperature gradient, with κ_c as the proportionality factor for the composite medium:

$$q_h = -\kappa_c\,(dT/dx) = \kappa_c(T_1 - T_2)/l$$

Since the cube is a mixture of two phases, the composite thermal conductivity κ_c will be intermediate between κ_0 and κ_1. According to de Vries (1975a),

$$\kappa_c = (f_0\kappa_0 + kf_i\kappa_i)/(f_0 + k_{fi}) \tag{12.19}$$

where the factor k is the ratio of the average temperature gradient in the particles to the corresponding gradient in the continuous fluid:

$$k = (dT/dz)_2/(dT/dz)_1$$

The value of k depends not only on the ratio κ_1/κ_0, but also on the particle sizes, shapes, and mode of packing.

If there are several types of particles with different shapes or conductivities, Eq. (12.20) can be generalized:

$$\kappa_c = \frac{\sum_{i=1}^{n} k_i f_i \kappa_i}{\sum_{i=1}^{n} k_i f_i} \tag{12.20}$$

Here n is the number of particle classes within which all particles have about the same shape and conductivity. The thermal conductivity of soils of widely differing compositions can be estimated by Eq. (12.20).

The following form of Eq. (12.20), for an unsaturated soil, was used by van Bavel and Hillel (1975, 1976):

$$\kappa_c = (f_w\kappa_w + k_s f_s\kappa_s + k_a f_a\kappa_a)/(f_w + k_s f_s + k_a f_a)$$

where κ_w, κ_a, and κ_s are the specific thermal conductivities of the soil constituents (water, air, and an average value for the solids, respectively). The factor k_s represents the ratio between the space average of the temperature gradient in the solidphase and that in the water phase. It depends on grain shapes as well as on mineral composition and organic matter content. The k_a factor represents the corresponding ratio for the thermal gradient in the air and water phases.

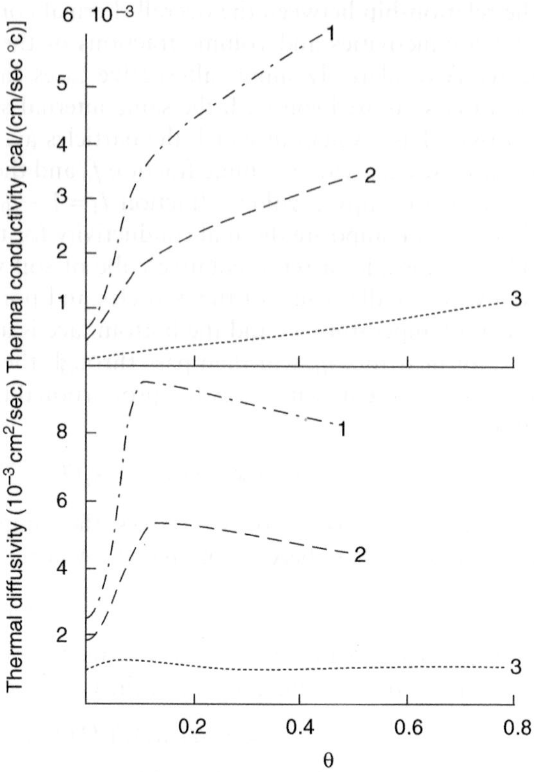

Fig. 12.1. Thermal conductivity and thermal diffusivity as functions of volume wetness (volume fraction of water) for (1) sand (bulk density 1460 kg/m^3, volume fraction of solids 0.55); (2) loam (bulk density 1330 kg/m^3, volume fraction of solids 0.5); and (3) peat (volume fraction of solids (0.2). (After de Vries, 1975.)

The dependence of thermal conductivity and diffusivity on soil wetness is illustrated in Fig. 12.1. The influence of latent heat transfer by the diffusion of water vapor in the air-filled pores is proportional to the temperature gradient in these pores. It can be taken into account (van Bavel and Hillel, 1976; Hillel, 1977) by adding to the thermal conductivity of air an apparent conductivity due to evaporation, transport, and condensation of water vapor (the so-called *vapor enhancement factor*). This value is strongly temperature dependent and rises rapidly with increasing temperature.

Methods of measuring thermal conductivity were summarized by Jackson and Taylor (1986) and more recently by Bristow (2002).

The thermal diffusivity D_T, instead of the conductivity κ, is sometimes desired (Horton, 2002). It can be defined as the change in temperature produced in a unit volume by the quantity of heat flowing through the volume in unit time under a unit temperature gradient. An alternative definition, easier to perceive, is that the thermal diffusivity is the ratio of the conductivity to the product of the specific heat and density:

$$D_T = \kappa / c_s \rho = \kappa C_v \quad (12.21)$$

where C_v is the volumetric heat capacity. As shown in the preceding section, the specific heat and density of both solids and water must be considered when calculating C_v:

$$C_v = \rho_s(c_s + c_w w) \tag{12.22}$$

where ρ_s is the density of dry soil, c_s is the specific heat of dry soil, c_w is the specific heat of water, and w is the ratio of the mass of water to the mass of dry soil. The thermal diffusivity can be measured directly, as described by Jackson and Taylor (1986).

SIMULTANEOUS TRANSPORT OF HEAT AND MOISTURE

The flows of water and of thermal energy under nonisothermal conditions in the soil are interactive phenomena: The one entails the other. Temperature gradients affect the moisture-potential field and induce both liquid and vapor movement. Reciprocally, moisture gradients move water, which carries heat. The simultaneous occurrence of temperature gradients and moisture gradients in the soil therefore brings about the combined transport of heat and moisture. This combined transport can generally be ignored in the extreme cases of a saturated or nearly saturated soil and of a nearly dry soil. In the former, the influence of temperature gradients on liquid water flow is generally small in comparison with the influence of gravity or pressure gradients; in the latter, the movement of heat can entail no significant movement of either liquid water or vapor. Thus, we are left with the problem of how to deal with the wide range of intermediate situations in which transport of liquid water and of vapor can be both significant and mutually influenced.

Two separate approaches to the combined transfer of heat and moisture have been attempted: (1) a *mechanistic approach*, based on a physical model of the soil system, and (2) a *thermodynamic approach*, based on the phenomenology of irreversible processes in terms of coupled forces and fluxes. Though starting from different points of view, the two approaches have been shown to be related and, properly formulated, can be cast into an equivalent mold (Groenevelt and Bolt, 1969; Jury, 1973).

The mechanistic approach was originally formulated by Philip and de Vries (1957). Their model was based on the concept of viscous flow of liquid water under the influence of gravity and of capillary and adsorptive forces and on the concept of vapor movement by diffusion. Local "microscopic-scale" thermodynamic equilibrium between liquid and vapor was assumed to exist at all times and at each point within the soil. The general differential equation describing moisture movement in a porous system under combined temperature and moisture gradients for unidimensional vertical flow is, accordingly,

$$\partial\theta/\partial t = \nabla \cdot (D_T \nabla T) + \nabla \cdot (D_w \nabla\theta) - \partial K/\partial z \tag{12.23}$$

where θ is volumetric wetness, t is time, T is absolute temperature, D_T is the water diffusivity under a temperature gradient (the sum of the liquid and vapor diffusivities), D_w is the water diffusivity under a moisture gradient, K is the hydraulic conductivity, and z is the vertical space coordinate. The last term on the right-hand side is due to the gravity gradient and becomes positive if z is taken to be increasing downward.

The heat transfer equation is, similarly,

$$C_v\ \partial T/\partial t = \nabla \cdot (\kappa\ \nabla T) - L\nabla \cdot (D_{w,vap}\ \nabla\theta) \qquad (12.24)$$

Here C_v is volumetric heat capacity, κ is apparent thermal conductivity of the soil, L is latent heat of vaporization of water, and $D_{w,vap}$ is diffusivity for heat conveyed by water movement (mostly vapor). The preceding equations are both of the diffusion type, involving θ- and T-dependent diffusivities as well as gradients of both θ and T.

Taken together, Eqs. (12.23) and (12.24) describe the coupled transport of moisture and heat in soils. The mechanistic nature of the theory and of the coefficients involved was explained by de Vries (1975b). The assumption of local thermodynamic equilibrium links the vapor pressure p_v to the matric potential ψ by the following relation: $p_v = p_{vs}h = p_{vs} \exp(Mg\psi/RT)$, where p_{vs} is the saturated vapor pressure at the particular temperature T, h is relative humidity, M is molar mass, g is the acceleration of gravity, and R is the universal gas constant. The diffusivities for water and heat by vapor transport are obtained by use of this relationship. However, the difficulty encountered in making the theory operational is in measuring the diffusivities. A more fundamental problem is that, since the two mechanisms of flow represented in each equation do interact, they are not, strictly speaking, simply additive.

To consider the approach based on *irreversible thermodynamics*, we must first understand in principle the difference between this relatively new branch of science and the older, "classical" thermodynamics, which deals with reversible processes and equilibrium states. Classical thermodynamics can predict whether, and in what direction (but not at what rate), a spontaneous process will occur in a system not at equilibrium. However, in a natural system any number of different forces might be operating simultaneously to produce mutually interacting fluxes in a combination of irreversible processes. For instance, a concentration gradient causes diffusion, a pressure gradient causes convection, and a temperature gradient results in the transfer of heat, with each of these fluxes affecting the others. If the system is not too far from equilibrium, the fluxes are taken to be related linearly to the forces causing them.

In application to simultaneous water and heat flow, as an example, the approach based on the thermodynamics of irreversible processes formulates a pair of *phenomenological equations* in which the fluxes of moisture q_w and heat q_h are expressed as linear functions of the moisture potential (e.g., pressure) gradient dP/dz and the temperature gradient dT/dz:

$$\begin{aligned} q_w &= -L_{ww}(1/T)\ (dP/dz) - L_{wh}(1/T^2)(dT/dZ) \\ q_h &= -L_{ww}(1/T)\ (dP/dz) - L_{hh}(1/T^2)(dT/dZ) \end{aligned} \qquad (12.25)$$

The four phenomenological coefficients occurring in these equations (L_{ww}, L_{wh}, L_{hw}, L_{hh}, relating water flow to the water potential gradient, water flow to the thermal potential gradient, heat flow to the water potential gradient, and heat flow to the thermal potential gradient, respectively) are unknown functions of P (or θ) and T. According to *Onsager's theorem* (Katchalsky and Curran, 1965), the cross-coupling coefficients L_{wh} and L_{hw} are equal when the fluxes and forces are properly formulated. Thus, the number of coefficients that must be measured is reduced.

An apparent advantage of the irreversible thermodynamics approach is that it makes no *a priori* assumptions regarding the mechanisms of the transport phenomena formulated. Hence it would seem to be less restrictive than a physical theory, whose validity is constrained at the outset by its mechanistic assumptions. The disadvantage of the approach, however, is precisely in its failure to provide insight into the nature and internal workings of the processes considered.

THERMAL REGIME OF SOIL PROFILES

In nature, soil temperature varies continuously in response to the ever-changing meteorological regime acting on the soil–atmosphere interface. That regime is governed by a regular periodic succession of days and nights and of summers and winters. Yet the regular diurnal and annual cycles are perturbed by such irregular episodic phenomena as cloudiness, cold waves, warm waves, rainstorms or snowstorms, and periods of drought. Added to these external influences are the soil's own changing properties (i.e., temporal changes in reflectivity, heat capacity, and thermal conductivity as the soil alternately wets and dries, and the variation of all these properties with depth), as well as the influences of geographic location, vegetative cover, and — finally — human management. All these labile factors complicate the effort to define the thermal regime of soil profiles.

The simplest mathematical representation of nature's fluctuating thermal regime is to assume that at all depths in the soil the temperature oscillates as a pure harmonic (sinusoidal) function of time around an average value. Since nature's actual variations are not so orderly, this may be a rather crude approximation. Nonetheless, it is an instructive exercise in itself, and when used in conjunction with field data it may lead to a better understanding, and perhaps even provide a basis for the prediction, of a soil's thermal regime.

To begin, let us assume that although soil temperature varies differently at different depths in the soil, the average temperature is the same for all depths. We next choose a starting time ($t = 0$) such that the surface is at the average temperature. The temperature at the surface can then be expressed as a function of time (Fig. 12.2):

$$T(0,t) = T_{\mathrm{ave}} + \mathrm{A}_0 \sin \omega t \qquad (12.26)$$

where $T(0,t)$ is the temperature at $z = 0$ (the soil surface) as a function of time t, T_{ave} is the average temperature of the surface (as well as of the profile), and A_0 is the amplitude of the surface-temperature fluctuation (the range from maximum, or from minimum, to the average temperature). Finally, ω is the radial frequency, which is 2π times the actual frequency. In the case of diurnal variation, the period is 86,400 sec (24 hr), so $\omega = 2\pi/86{,}400 = 7.27 \times 10^{-5}$/sec. Note that the argument of the sine function is expressed in radians rather than in degrees.

The last equation is the boundary condition for $z = 0$. For the sake of convenience, let us assume that at infinite depth ($z = \infty$) the temperature is constant

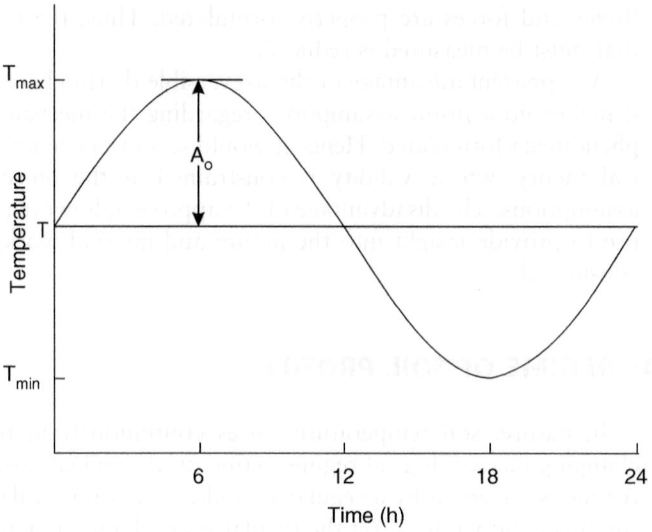

Fig. 12.2. Idealized daily fluctuation of surface soil temperature, according to the equation $T = T_{ave} + A_o \sin(\omega t/p)$, where T is temperature, T_{ave} is average temperature, A_o is amplitude, t is time, and p is the period of the oscillation (in this case, p refers to the diurnal 24 hr).

and equal to T_{ave}. Under these circumstances, the temperature at any depth z can also be represented as a sine function of time, as shown in Fig 12.3:

$$T(z,t) = T_{ave} + A_z \sin[\omega t + \phi(z)] \quad (12.27)$$

in which A_z is the amplitude at depth z. Both A_z and $\phi(z)$ are functions of z but not of t. They can be determined by substituting the solution of Eq. (12.26) in the differential equation $C_v(\partial T/\partial t) = \kappa(\partial^2 T/\partial z^2)$. This leads to the solution

$$T(z,t) = T_{ave} + A_0[\sin(\omega t - z/d)]/e^{-z/d} \quad (12.28)$$

The constant d is a characteristic depth, called the *damping depth*, at which the temperature amplitude equals $1/e$ ($1/2.718 = 0.37$) of the amplitude at the soil surface A_0. The damping depth is related to the thermal properties of the soil and the frequency of the temperature fluctuation:

$$d = (2\kappa/C_v\omega)^{1/2} = (2D_T/\omega)^{1/2} \quad (12.29)$$

It is seen that at any depth the amplitude of the temperature fluctuation A_z is smaller than A_0 by a factor $e^{z/d}$ and that there is a phase shift (a time delay of the temperature peak) equal to $-z/d$. The decrease of amplitude and increase of phase lag with depth are typical phenomena in the propagation of a periodic temperature wave in the soil.

The physical reason for the damping and retarding of the temperature waves with depth is that a certain amount of heat is absorbed or released along the path of heat propagation when the temperature of the conducting soil increases or decreases, respectively. The damping depth is related inversely to the frequency, as can be seen from Eq. (12.29). Hence it depends directly on the period of the temperature fluctuation considered. The damping depth is

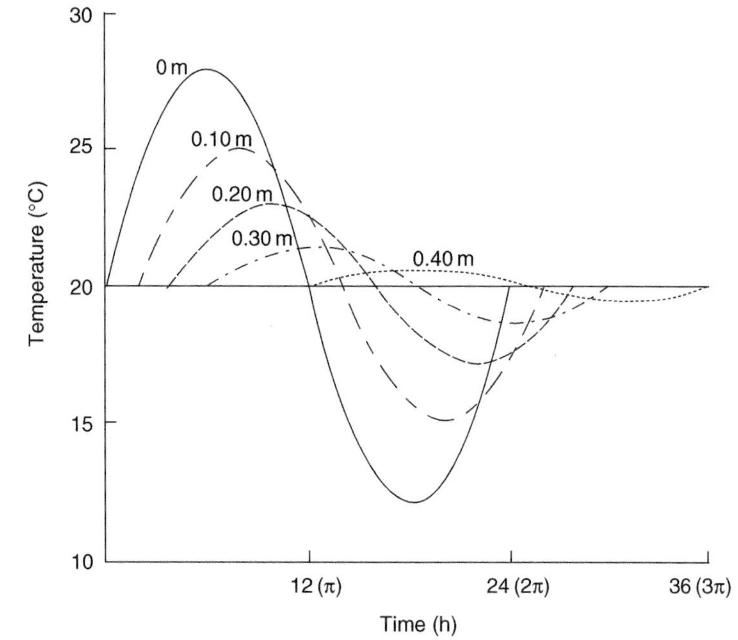

Fig. 12.3. Idealized variation of soil temperature with time for various depths. Note that at each succeeding depth the peak temperature is damped and shifted progressively in time. Thus, the peak at a depth of 0.4 m lags about 12 hr behind the temperature peak at the surface and is only about 1/16 of the latter. In this hypothetical case, a uniform soil was assumed, with a thermal conductivity of 1.68 J/m sec deg (or 4×10^{-3} cal/cm sec deg) and a volumetric heat capacity of 2.1×10^6 J/m^3 deg (0.5 cal/cm^3 deg).

$(365)^{1/2} = 19$ times larger for the annual variation than for the diurnal variation in the same soil. For example, van Wijk and de Vries (1963) calculated the damping depth for a soil with $\kappa = 0.96$ J/m sec deg (equal to 2.3×10^{-3} cal/cm sec deg) and obtained $d = 0.12$ m for the diurnal temperature fluctuation and $d = 2.29$ m for the annual fluctuation. Whereas the amplitude at depth $z = d$ is 0.37 as great as the amplitude at the surface, it is only about 0.05 of the surface amplitude at $z = 3d$ (0.36 m for the diurnal variation in the case of the soil used by these authors). When an arbitrary zero point t_0 is introduced into the time scale, Eq. (12.28) becomes

$$T(z,t) = T_{\text{ave}} + A_0[\sin(\omega t + \phi_0 - z/d)]/e^{z/d} \qquad (12.30)$$

The constant ϕ_0 is called the *phase constant.*

The annual variation of soil temperature down to considerable depth causes deviations from the simplistic assumption that the daily average temperature is the same for all depths in the profile. The combined effect of the annual and diurnal variations of soil temperature can be expressed by

$$\begin{aligned} T(z,t) = T_{\text{ave},y} &+ A_y[\sin(\omega_y t + \phi_y - z/d_y)]/e^{z/d_y} \\ &+ A_d[\sin(\omega_d t + \phi_d - z/d_d)]/e^{z/d_d} \end{aligned} \qquad (12.31)$$

wherein the subscripted indices y and d refer to the yearly and daily temperature waves, respectively. Thus $T_{ave,y}$ is the annual mean temperature. The daily cycles are now seen to be short-term perturbations superimposed on the annual cycle. Vagaries of weather (e.g., spells of cloudiness or rain) can cause considerable deviations from simple harmonic fluctuations, particularly for the daily cycles. Longer-term climatic irregularities can also affect the annual cycle, of course. Also, since the annual temperature wave penetrates much more deeply than the daily wave, the assumptions of soil homogeneity in depth and of the time constancy of soil thermal properties are clearly unrealistic (Gao et al., 2003).

An alternative theoretical approach is possible, one with fewer constraining assumptions. It is based on numerical, rather than analytical, methods for solving the differential equations of heat conduction. Computer-based mathematical simulation models now allow soil thermal properties to vary in time and space (e.g., in response to periodic changes in soil wetness) so as to account for alternating surface saturation and desiccation and for profile layering. They also allow various climatic inputs to follow more realistic and irregular patterns. The surface amplitude of temperature need no longer be taken to be an independent variable, but one that depends on the surface energy balance and thus is affected by both soil properties and above-soil conditions. Examples of the numerical approach can be found in the published works of van Bavel and Hillel (1975, 1976), Hillel (1977), and Evett et al. (1994).

Other developments of practical importance include techniques for monitoring the soil thermal regime more precisely than was possible previously. One such technique is the *infrared radiation thermometer* for scanning or remote sensing of surface temperature for both fallow and vegetated soils without disturbance of the measured surface. Knowledge of the surface temperature and its variation in time is important in assessing energy exchange between soil and atmosphere as well as in determining boundary conditions for within-soil heat transfer.

An additional technique is the use of *heat flux plates*. These are flat and thin plates or disks of constant thermal conductivity, which allow precise measurement of the temperature difference between their two sides so as to yield the heat flux through them. When embedded horizontally in the soil at regular depth intervals, a series of such heat flux plates can provide a continuous record of heat transfer throughout the profile. There are problems, however. The presence of heat flux plates can distort the flow of heat in the surrounding medium if their thermal conductivity is very different from that of the soil. The experimental error can be minimized by constructing plates of maximal thermal conductivity and minimal thickness and by calibrating them in a medium with a thermal conductivity near to that of the soil in which they are to be placed. Another problem is that such plates preclude vapor flow, which can sometimes be an important component of heat transfer. The use of heat flux plates is described by Fuchs (1986) and by Sauer (2002).

The soil-temperature profile as it might vary from season to season in a frost-free region is illustrated in Fig. 12.4. The diurnal variation of temperature and the directions of heat flow within a soil profile are illustrated in Fig. 12.5.

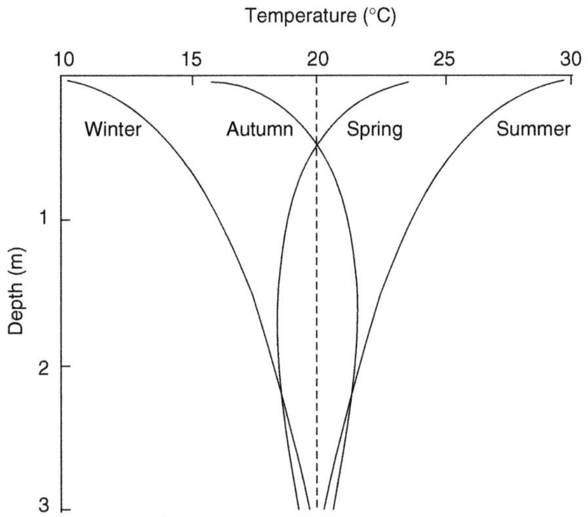

Fig. 12.4. Soil-temperature profile as it varies from season to season in a frost-free region.

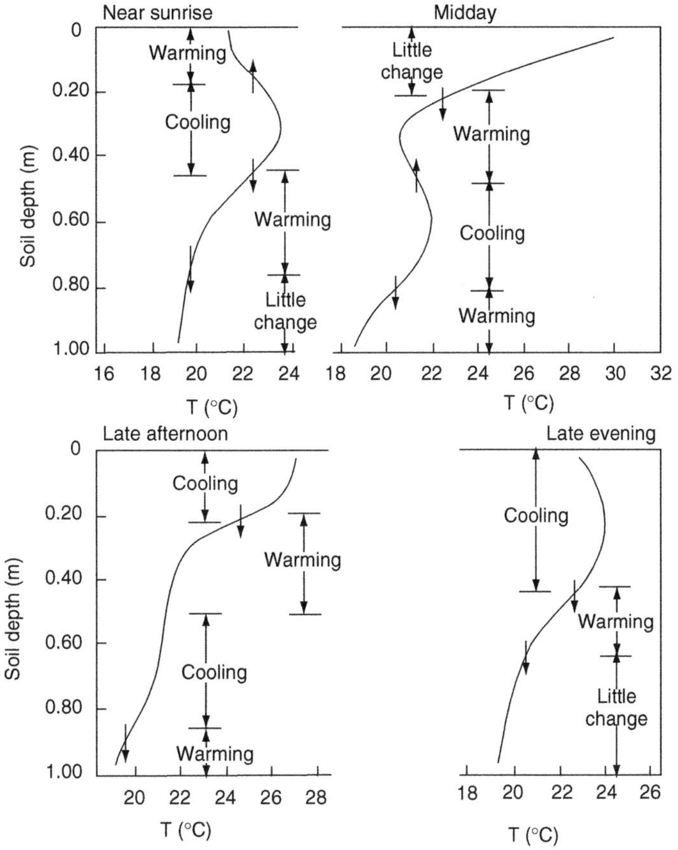

Fig. 12.5. Typical variation of temperature with depth at different times of day in summer. (From Sellers, 1965; data by Carson, 1961.)

Sample Problem

The daily maximum soil-surface temperature is 40°C and the minimum is 10°C. Assume that the diurnal temperature wave is symmetrical, that the mean temperature is equal throughout the profile (with surface temperature equal to the mean value at 6 A.M. and 6 P.M.), and that the "damping depth" is 0.10 m. Calculate the temperatures at noon and midnight for depths 0.05, 0.10, and 0.20 m.

Since the temperature range is 30°C and the mean (T_{ave}) 25°C, and the amplitude at the surface A_0, the maximum value above the mean is 15°.

Use Eq. (12.28) to calculate the temperature T at any depth z and time t:

$$T(z,t) = T_{ave} + A_0[\sin(\omega t - z/d)]/e^{z/d}$$

where ω is the radial frequency ($2\pi/24$ hr) and d is the "damping depth" at which the temperature amplitude is $1/e$ (= 0.37) of A_0. *Note*: Radial angle is expressed in radians, not in degrees (i.e., $\sin \pi/2 = 1$, $\sin 3\pi/2 = -1$).

At the soil surface (depth zero):

Noontime temperature (6 hr after $T = T_{ave}$)

$$T(0.6) = 25 + 15 \times [\sin(\pi/2 - 0)]/e^0 = 25 + 15 = 40°\text{C}$$

Midnight temperature (18 hr after $T = T_{ave}$)

$$T(0,18) = 25 + 15 \times [\sin(3\pi/2 - 0)]/e^0 = 25 - 15 = 10°\text{C}$$

At depth 0.05 m:

Noontime temperature:

$$\begin{aligned} T(0.05,6) &= 25 + 15 \times [\sin(\pi/2 - 0.05/0.1)]/e^{0.05/0.1} \\ &= 25 + 15 \times \sin(1.07)/1.65 = 33°\text{C} \end{aligned}$$

Midnight temperature:

$$\begin{aligned} T(0.05,18) &= 25 + 15 \times [\sin(3\pi/2 - 0.05/0.1)]/e^{0.05/0.1} \\ &= 25 + 15 \times \sin(4.21)/1.65 = 17°\text{C} \end{aligned}$$

At depth 0.1 m (the damping depth):

Noontime temperature:

$$\begin{aligned} T(0.1,6) &= 25 + 15 \times [\sin(\pi/2 - 0.1/0.1)]/e^{0.1/0.1} \\ &= 25 + 15 \times \sin(0.57)/2.72 = 28°\text{C} \end{aligned}$$

Midnight temperature:

$$\begin{aligned} T(0.1,18) &= 25 + 15 \times [\sin(3\pi/2 - 0.1/0.1)]/e^{0.1/0.1} \\ &= 25 + 15 \times \sin(3.71)/2.72 = 22°\text{C} \end{aligned}$$

At depth 20 cm:

Noontime temperature:

$$\begin{aligned} T(0.2,6) &= 25 + 15 \times [\sin(\pi/2 - 0.2/0.1)]/e^{0.2/0.1} \\ &= 25 + 15 \times \sin(-0.43)/7.4 = 24.15°\text{C} \end{aligned}$$

Midnight temperature:

$$T(0.2,18) = 25 + 15 \times [\sin(3\pi/2 - 0.2/0.1)]/e^{0.2/0.1}$$
$$= 25 + 15 \times \sin(2.71)/7.4 = 25.85°\mathrm{C}$$

Note: At a depth of 0.2 m the phase shift is so pronounced that at midnight the temperature is higher than at noon. A useful exercise for students is to plot the sinusoidal course of temperature at each depth, to observe how the phase shift (time lag of maximum and minimum values) increases and the amplitude decreases with depth.

13. STRESS, STRAIN, AND STRENGTH OF SOIL BODIES

SOIL RHEOLOGY

The fundamental concepts of *soil rheology* and the principles of *soil mechanics* based on them pertain to the behavior of soil bodies under applied forces. Engineers have long used these principles in solving problems related to the stability of slopes and foundations and to the use of soil as a construction material. Recently and increasingly the same fundamental concepts have been applied to the behavior and management of soils in traction and tillage.

The term *rheology* originates from the Greek word *rheos*, meaning "flowing." As a branch of science, however, it is concerned with all types of deformation, of which flow is merely one. Broadly speaking, rheology can be described as the mechanics of deformable bodies. In connection with the study of soil deformation, however, the analogous term *soil dynamics* (rather than soil rheology) seems to have won wider acceptance in the literature.

Extensive expositions of soil dynamics were published by Gill and Vanden Berg (1967), Hillel (1980a), Koolen and Kuipers (1983), McKyes (1985), and Chancellor (1994). Here, we deal only briefly with selected highlights of that complex topic.

STRAIN–STRESS RELATIONSHIPS AND FAILURE OF SOIL BODIES

The reaction of a body to a force or a combination of forces acting on or within it can be characterized in terms of its *relative deformation*, or *strain*,

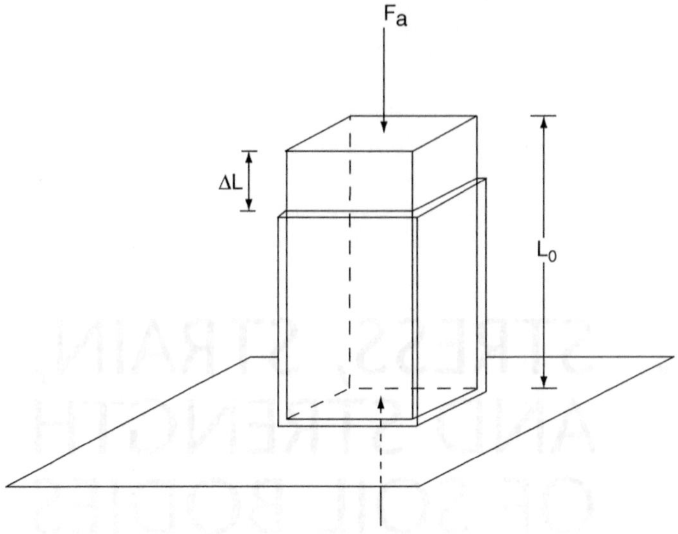

Fig. 13.1. Longitudinal deformation of an elongated body subjected to an axial load.

that is, the ratio of the deformation to the body's initial dimensions. The simplest example is a rectangular or cylindrical body subjected to a force directed along its main axis (Fig. 13.1). If the initial length is L_0 and the change in length ΔL, then the *longitudinal strain* ε is

$$\varepsilon = \Delta L / L_0 \tag{13.1}$$

Another form of deformation occurs when the angles rather than the length dimensions of a body change. The simplest such case, called *simple shear*, is depicted in Fig. 13.2. The measure of *shear strain* γ (or *tangential strain*) is the relative change of the initially right angle:

$$\gamma = u/h \tag{13.2}$$

where u is the lateral (or tangential) displacement and h is the height of the body. The ratio u/h is thus the tangent of the deformation angle. Being a ratio of lengths, strain is dimensionless and is often expressed as a fraction or percentage of the original dimension.

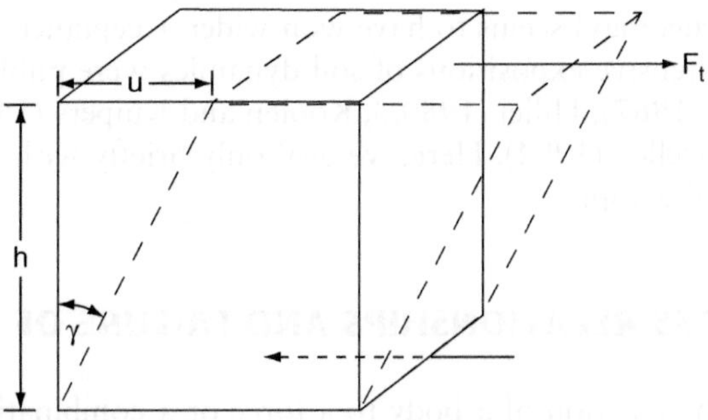

Fig. 13.2. Angular deformation in simple shear of a cube.

The effect of a force in causing deformation is related directly to the magnitude of the force and inversely to the area over which it acts. The ratio of force to area, that is, force per unit area, is called *stress*. A *normal stress* is caused by a force that is perpendicular to the area over which it acts. Such a stress, equivalent to a pressure, is usually designated σ. Thus,

$$\sigma = F_n/A \tag{13.3}$$

where F_n is the force normal to the surface area A. When the direction of a force is parallel to the surface area, it constitutes a *tangential stress*, also called a *shearing stress*, designated τ:

$$\tau = F_t/A \tag{13.4}$$

where F_t is the tangential force acting on area A.

Another important factor in rheology is the time rate of stress application and the time dependence of strain. The stress–strain–time relationships of a material determine its rheological character, that is, whether it is an elastic solid, a plastic solid, a Newtonian or non-Newtonian liquid, or any other of the various types of materials recognized and characterized by rheologists.

The standard physical definition of a solid is a material with an ordered internal structure (e.g., a crystal). The rheological criterion for a solid depends not on internal structure, important as it is, but on mechanical behavior. Solids are bodies of distinct shape that may exhibit various types of response when under stresses. An *elastic solid* deforms instantaneously and retains its new form as long as the stress is maintained; it regains its original dimensions when the stress is released. (Strictly speaking, there are no such materials, since most real bodies exhibit *some* residual deformation after release of stress.)

Three centuries ago, Robert Hooke discovered that loaded springs stretched in proportion to the load applied to them. His observation has been generalized and formulated into an equation known as *Hooke's Law*, which states that, for elastic solids, *strain* ε is *proportional to stress*. It furthermore implies that strain occurs instantaneously when stress is applied and that it disappears immediately and completely when the stress is relieved. Thus,

$$\varepsilon = \sigma/E \tag{13.5}$$

where E is a constant of proportionality known as *Young's modulus*.

The equation of elasticity for shearing stress, which is analogous to Eq. (13.5), is

$$\gamma = \tau/G \tag{13.6}$$

where G is the *modulus of shearing*, a measure of the body's rigidity.

When a body is subjected to isotropic pressure, its volume diminishes in proportion to the pressure applied. The constant of proportionality is termed the *bulk modulus* κ. Thus,

$$\varepsilon_v = P/\kappa \tag{13.7}$$

where ε_v is the volume compression or expansion relative to the original volume. Bulk modulus is the reciprocal of the *coefficient of compressibility* $[C = \rho^{-1}(d\rho/dP)]$, which is the relative change of density ρ with change of pressure P.

In the case of *nonisotropic materials* (e.g., wood, which exhibits different properties in directions parallel to us perpendicular to its fibers), the various coefficients just defined can have different values for the three principal axes as well as for different planes.

In contrast with elasticity, *plasticity* is the property of a body not merely to deform when under stress but also to retain its deformed shape when the stress is relieved. Many materials under stress exhibit an elastic phase until some point, called the *yield point*, at which elasticity gives way to plastic behavior. The transition from elastic to plastic behavior may be affected by the rate of strain and by a phenomenon known as *strain hardening*. Such behavior is typically encountered in the case of ductile metals. An increase in strength of a metal as a consequence of deformation is usually due to internal structural changes and recrystallization. An analogous effect may be observed in soils due to compaction.

The conventional definitions of strain given by Eqs. (13.1) and (13.2) apply to small strains only (normally under 1%). The usefulness of the strain concept can be extended to larger deformations by redefining it in terms of *natural strain*, namely, the incremental change of length or angle divided by the instantaneous value rather than by the original prestrained value.

In soil, strain is often very large. For example, an increase in bulk density from 1.1 to 1.4 metric tons/m^3 represents a volume change of 27%. In small and confined samples of soil, strain may be fairly uniform. Under natural soil conditions, though, deformation can take place in very complicated geometric patterns. Soil deformation is often dependent in ways that may not conform to either elastic or plastic theories. In soils, failure under stress is more complex than in most ductile or brittle materials. Depending on such variables as moisture and clay content, soil properties can vary from those of a viscous liquid to those of a plastic solid and finally to those of a brittle elastic solid. In transition from one state or mode of behavior to another, intermediate states may occur. The modes of failure vary accordingly (Fig. 13.3).

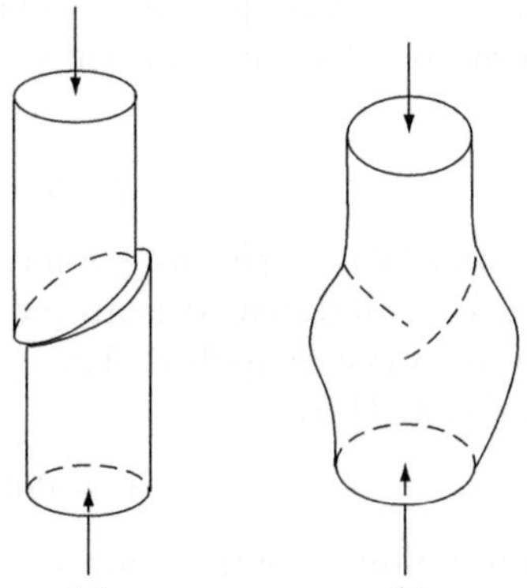

Fig. 13.3. Modes of shear failure that occur when a cylinder of soil is subjected to progressive axial stress: (a) rigid soil, failure by fracture; and (b) soft soil, plastic flow.

SOIL STRENGTH AND ITS MEASUREMENT

Stated qualitatively, *soil strength* is the capacity of a soil body to withstand forces without experiencing failure, whether by rupture, fragmentation, or flow. In quantitative terms, soil strength can be defined as the maximal stress that a given soil body can bear without failing. In practice, it is generally defined as the minimal stress that will cause the body to fail.

Though seemingly easy to define, soil strength is not at all easy to measure, being a highly variable property that often changes during the very process of measurement, because the deformed body of soil might either decrease or increase its resistance to further deformation. In unsaturated soil, for instance, strength may increase as the soil becomes more compact; whereas in saturated soils transient stresses may cause loss of cohesion and even liquefaction (a phenomenon known as *thixotropy*). The manner and rate of stressing can thus influence both the pattern of deformation and the mode of failure (Fredlund and Vanapalli, 2002).

Since in most actual cases soil bodies fail by shear, it is useful at this point to concentrate specifically upon the ability of a soil to withstand shearing stresses, known as *shearing strength*. A very useful approach to the analysis of shearing is the theory of Mohr, based on the functional relationship between σ and τ for any plane within a body of soil that is subject to differential (nonisotropic) stresses. The theory considers the possible distribution of stresses on the surfaces of a small volume element of soil (assumed to be orthogonal to the three coordinate axes x, y, z), as illustrated in Fig. 13.4. It then resolves the stresses (normal and tangential) acting on any arbitrarily inclined plane inside that volume element, as shown, for example, in Fig. 13.5 for a plane inclined to the x and y axes but parallel to the z axis.

The analysis results in a circle, called the *Mohr circle*, plotted on a hypothetical σ-τ plane. This graphic representation indicates the values of the normal and tangential (shearing) stresses on any angle of inclination α. The Mohr

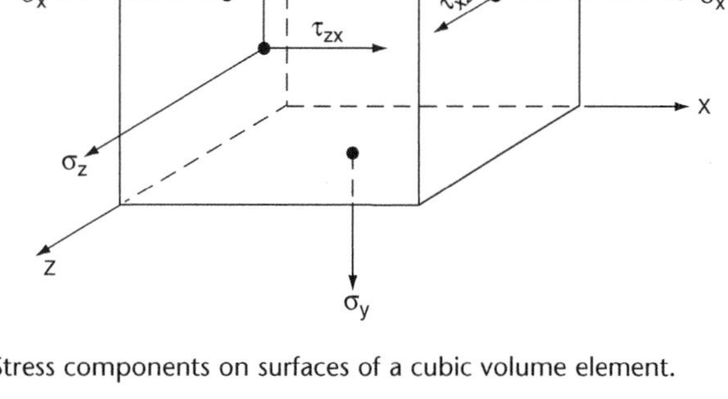

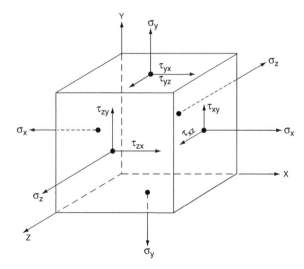

Fig. 13.4. Stress components on surfaces of a cubic volume element.

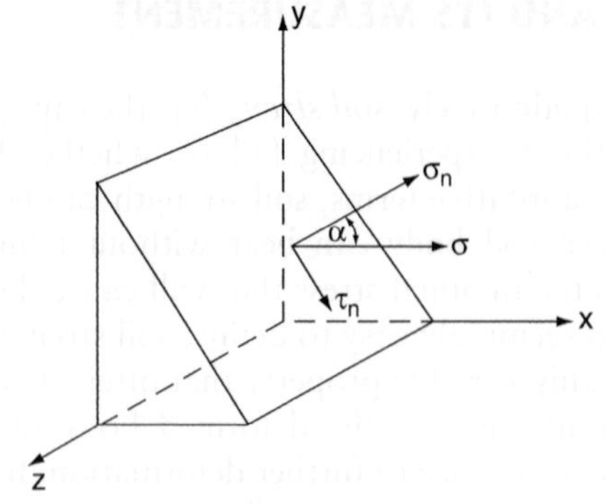

Fig. 13.5. The resolution of σ_x on an inclined plane.

circle is illustrated in Fig. 13.6. A detailed explanation of the principles underlying it is given in Hillel (1980a) as well as in various textbooks on soil mechanics.

If a series of stress states just sufficient to cause failure is imposed on the same material and these states are plotted as a set, or family, of Mohr circles, then the line tangent to these circles, called the *envelope* of the family of Mohr circles, can be used as a criterion of shearing strength. Where this envelope is found to be a rising straight line, or nearly so, it can be described by the equation

$$\tau = a + b\sigma \tag{13.8}$$

where the constant a, indicated in the figure as τ_0, is the intercept of the envelope line on the τ axis and the constant b is the tangent of the angle ϕ that the envelope line makes with the horizontal line (Fig. 13.7).

The linearity of τ versus σ pertaining to the internal shearing strength of bodies is analogous to that of *Coulomb's law* for sliding friction between bodies. Coulomb's law states that the frictional resistance toward a tangential stress tending to slide one planar body over another is proportional to the

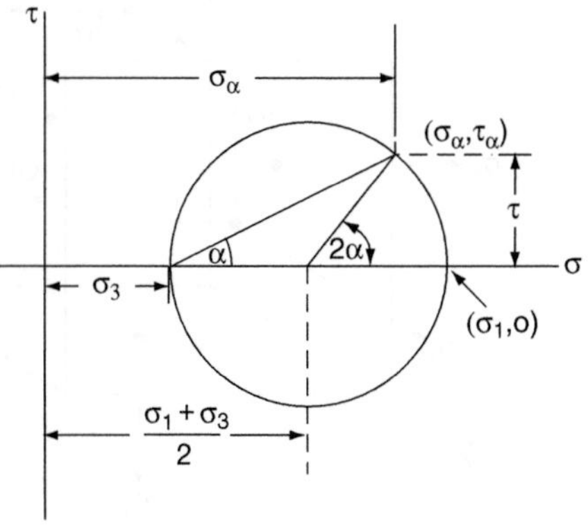

Fig. 13.6. The Mohr circle of stress.

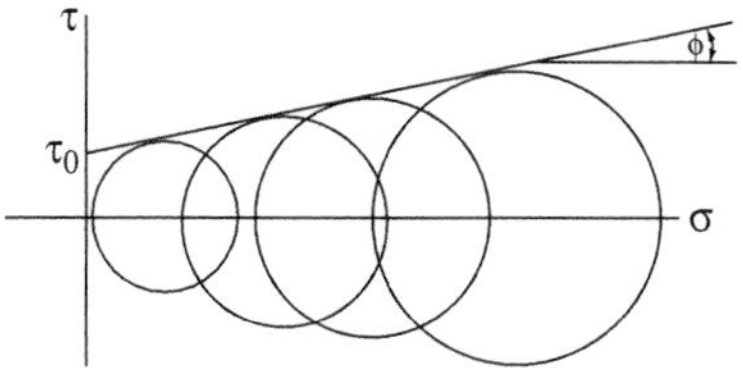

Fig. 13.7. The Mohr envelope of stress circles.

normal force pressing the bodies together (Fig. 13.8). Because of this analogy, the angle ϕ is often termed the *angle of internal friction.* Moreover, the intercept τ_0, visualized as the shear stress needed to cause failure at zero normal load, is generally called *soil cohesion.* The relationship of Eq. (13.8) is generally expressed in these terms as

$$\tau = c + \sigma \tan \phi \tag{13.9}$$

The Mohr theory of the straight-line envelope has been widely accepted in engineering soil mechanics, although it does not necessarily represent shear failure conditions under all circumstances. The Mohr envelope can be determined experimentally by means of the *direct shear test* or, preferably, by means of the so-called *triaxial shear test.* In principle, (1) all points under the envelope represent states of stress that the soil can bear without failure, (2) all points on the envelope are stress states just sufficient to cause failure, and (3) points above the envelope are impossible stress states (i.e., stress states that the soil cannot stably bear) (Barber, 1965).

Measurement of soil strength is generally based on some procedure for determining the stresses acting on or within a body at the moment of its mechanical failure. Soil resistance to applied stresses can be characterized in terms of two parameters: *cohesiveness c*, presumably representing the mutual bonding of soil particles that must be overcome if the soil body is to be fractured or sheared, and the *angle of internal friction* ϕ, representing the frictional resistance encountered when soil is forced to slide over soil along some shear plane. It should be understood that this dual representation of soil strength is merely a conceptual model, because in real cases the two parameters may not be truly independent of each other.

Soils that exhibit appreciable cohesiveness, called *cohesive soils*, generally contain a fair amount of clay, which has the effect of bonding or cementing the soil internally. Dry sand, on the other hand, is generally *noncohesive*,

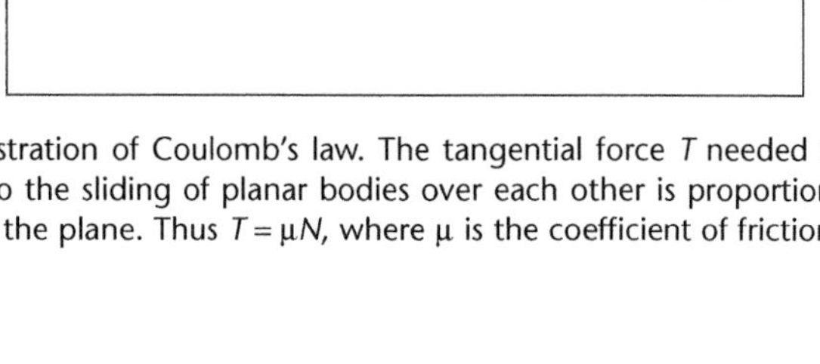

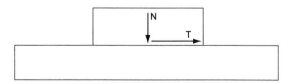

Fig. 13.8. Illustration of Coulomb's law. The tangential force T needed to overcome frictional resistance to the sliding of planar bodies over each other is proportional to the normal force N acting on the plane. Thus $T = \mu N$, where μ is the coefficient of friction.

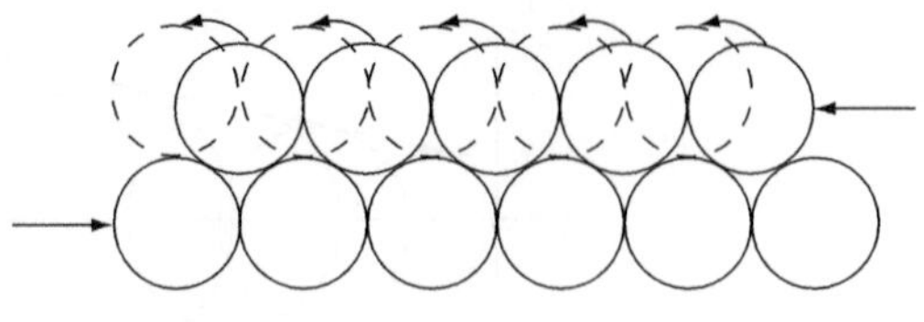

Fig. 13.9. Illustration of dilatancy (volume increase) during shearing of a dense sand, due to the interparticle rolling effect.

hence the only resistance to shear is due to interparticle friction that results from sliding and rolling of particles over one another. Dense sand might actually expand during shear, an attribute known as *dilatancy* and illustrated in Fig. 13.9. In the case of moist (unsaturated) sand, the surface tension effect of the water menisci between grains imparts to the soil an *apparent cohesiveness*, which disappears when the matrix is either dried or saturated with water.

Two principal methods have been used to measure failure by shear in soil and thus to obtain the values of c and ϕ. The direct shear apparatus is illustrated in Fig. 13.10, showing a sectioned container packed with soil to a known bulk density. Normal stress is applied to the shear plane by loading a piston-like porous stone placed over the soil surface. Shearing is induced by gradually increasing the lateral, or tangential, stress until failure occurs. The test can be carried out either at a constant rate of stress increase or at a constant rate of deformation: the so-called *stress-controlled* versus *strain-controlled* methods. The direct shear test has several drawbacks. The shearing plane does not remain constant during the test, so the stresses can vary even if the applied forces remain constant. Moreover, the size and shape of the container influence the test results.

The alternative and more fundamentally based method for determination of shear strength is known as the *cylindrical shearing test*. In this test, the failure surface is not predetermined but allowed to form within a specimen as successive combinations of lateral and axial stresses are applied to a series of samples of the same soil. Thus a series of Mohr circles representing shear failure states are determined, and the envelope of these circles yields the desired c and b values (Bishop and Henkel, 1964; Fredlund and Vanapalli, 2002).

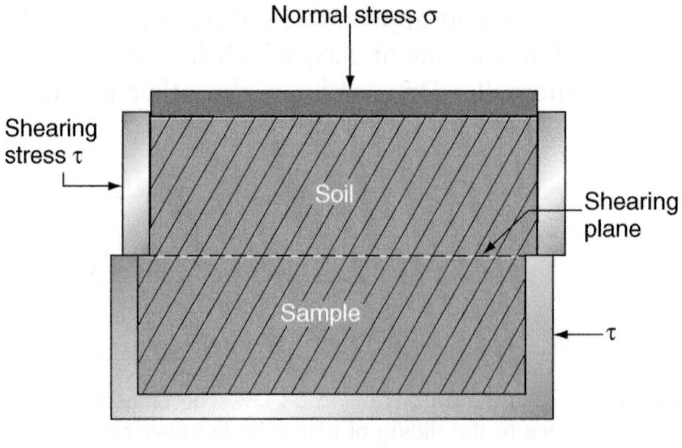

Fig. 13.10. Direct shear apparatus.

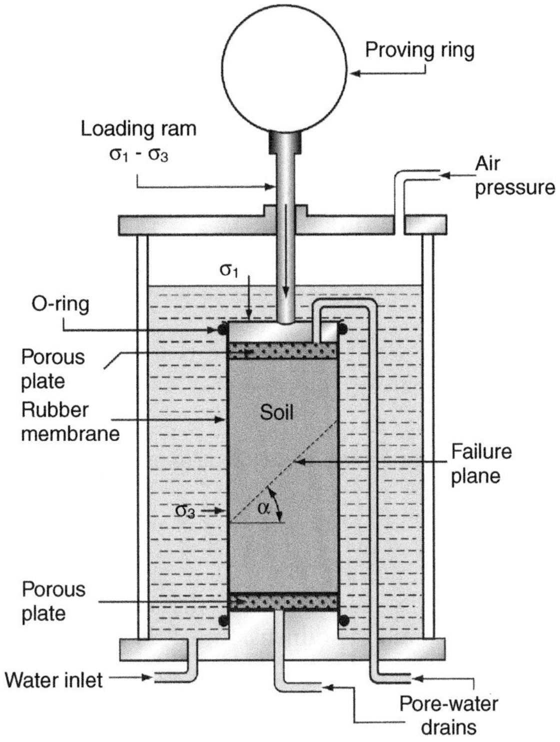

Fig. 13.11. The cylindrical shear apparatus.

The cylindrical shear apparatus is illustrated in Fig. 13.11. A cylindrical sample of soil is wrapped in a thin sleevelike rubber membrane, tightly sealed to parallel metallic plates fitted with porous plugs at the top and bottom of the sample. All this is placed in a water-filled pressurized cell that provides a constant and uniform stress, conventionally designated σ_3. An axial stress σ_1 exceeding the lateral stress is applied. The difference $(\sigma_1 - \sigma_3)$, called *deviatoric stress*, is applied through a loading ram attached to a proving ring. The axial stress is gradually increased until failure occurs, at which moment the axial stress as well as the lateral stress (the cell pressure) are noted.

An alternative laboratory procedure to characterize the cohesive strength of dry soil briquettes is to measure the *modulus of rupture*. This consists of centrally loading a small bean of soil, supported at both ends, until failure occurs (as shown in Fig. 13.12). The equation for determining the modulus of rupture σ_b is

$$\sigma_b = 3FL/2bd^2 \tag{13.10}$$

where F is the force applied to cause failure, L is the length of the briquette between its supports, and d is its thickness.

A very different method of measuring soil cohesiveness, suitable for use in the field, is the *vane shear test* (ASTM, 1956), illustrated in Fig. 13.13. In this test, the vane is driven into the soil to the desired depth and then rotated. The torque required to shear the soil along the surface of a cylinder generated by

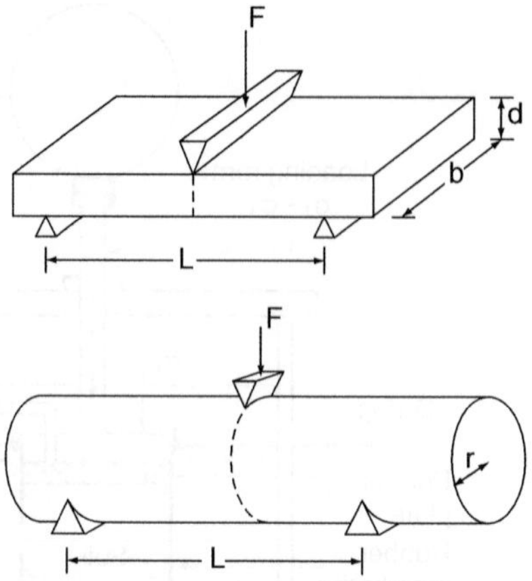

Fig. 13.12. Measurement of the modulus of rupture on a rectangular and on a cylindrical specimen of soil.

the blade edges is measured. The theoretical relation between torque T, soil cohesiveness c, and the dimensions of the blades is

$$T = c\pi(d^2h/2 + d^3/6) \tag{13.11}$$

For the standard height-to-radius ratio of 4:1, cohesiveness can be computed from the equation

$$c = 3T/28\pi r^3 \tag{13.12}$$

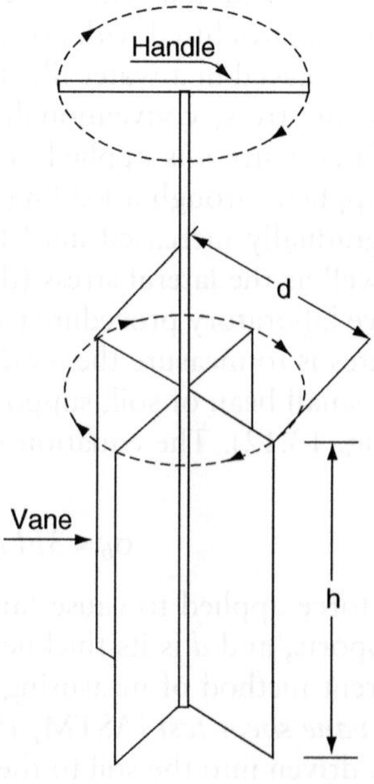

Fig. 13.13. The vane shear apparatus.

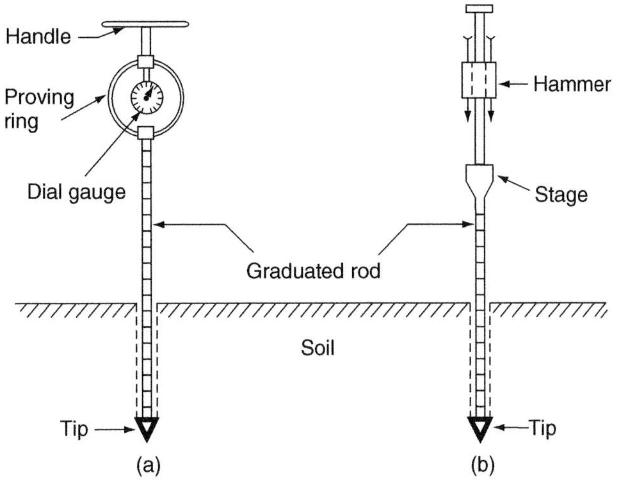

Fig. 13.14. (a) Push-type and (b) hammer-driven penetrometers for field characterization of soil strength.

The advantage of this method is that it can be applied in situ at succeeding depths, without extracting the apparatus, to obtain a strength profile of the natural soil.

An indirect method for assessing soil strength in the field is to use a *penetrometer*. This instrument, illustrated in Fig. 13.14, is designed to evaluate soil resistance to the penetration of a narrow probe. What is measured is not soil strength per se, but a composite parameter considered to be related to soil strength, though the quantitative nature of its relation to soil strength is not exactly defined. Nevertheless, this technique has certain advantages, particularly in the ease and simplicity of in situ measurement. Penetrometers have been used to estimate the force required to plow soils under different conditions as well as to indicate soil *trafficability* (the ability of a soil to provide traction and allow unobstructed overland passage of vehicles such as tractors, automobiles, and tanks). Penetrometers have been used to simulate seedling emergence through a surface crust. Procedures and specifications for using penetrometers for various purposes were provided by the American Society for Testing Materials (1984) by Bradford (1986), and by Lowery and Morrison (2002).

Sample Problem

A triaxial shearing test was performed on a cohesive soil. Three samples of the soil were mounted in an apparatus of the type shown in Fig. 13.11. Each sample was subjected to a different cell pressure and loaded axially at a slow rate so as to allow free drainage and dissipation of pore-water pressure. The axial stress at which failure occurred was recorded in each case. The following data were obtained:

Cell pressure σ_3:	1.0	2.0	3.0 bar
Axial load σ_1:	5.9	8.5	11.2 bar

Plot the set of Mohr circles and their envelope, and determine the soil's cohesiveness and angle of internal friction. Predict the deviator and axial stresses as well as the normal and shear stresses on the failure plane at a lateral stress of 2.5 bar.

The three stress circles are shown in Fig. 13.15. Note that the diameter of each circle is equal to the deviator stress $\sigma_1 - \sigma_3$. A tangent is drawn to the three circles and is seen to be a straight line. When this line is extrapolated leftward, it intercepts the τ axis. According to the Mohr theory of stress, this line can be expressed in terms of Eq. (13.9):

$$s = c + \sigma_n \tan \phi$$

where s is the shearing strength of the soil. From the graph, we determine the intercept c, representing the soil's cohesiveness, to be 1 bar. The slope of the $\tau - \sigma$ function (representing $\tan \phi$) is seen to be about 0.5. Hence the angle of internal friction ϕ is about 26.5°.

We can now use these parameters to determine the normal and shear stresses on any plane in a sample subjected to any pair σ_1, σ_3 under the envelope. For example, if the lateral stress is 2.5 bar, the circle of failure, tangent to the envelope, is as drawn in Fig. 13.15 (the dotted circle). The deviator stress (equal to the circle's diameter) is seen to be 7.4 bar (i.e., an axial load of about 9.9 bar is needed to cause failure). We can determine the normal and shearing stresses on the failure plane graphically by reading the values of σ and τ at the point of the circle's tangency with the envelope line. The values turn out to be $\sigma_n = 4.5$ bar, $\tau = 3.25$ bar.

Note: The Mohr envelope for a confined system becomes horizontal if the soil is saturated. Additional cell and axial pressures cannot add to the effective stress, merely to pore-water pressure.

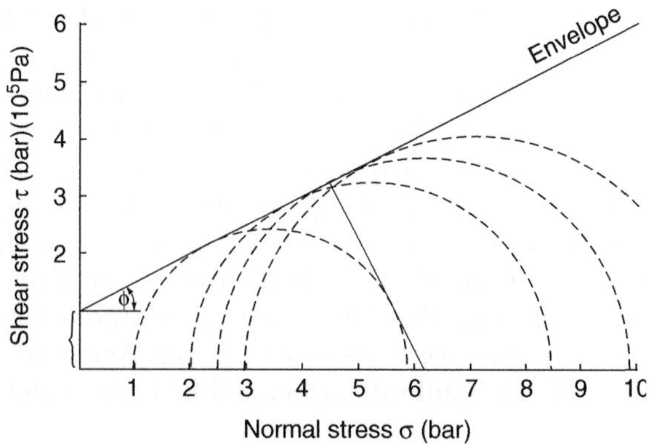

Fig. 13.15. Mohr circles and their envelope for the data in the sample problem.

SOIL CONSOLIDATION

Consolidation is the process by which a body of soil, either initially saturated or compacted to the point of saturation, is compressed in a manner that results in reduction of pore volume by expulsion of water. As such, consolidation is not an immediate response of soil to transient pressures such as those caused by episodic traffic or tillage. Rather, it is manifested in the gradual settlement, or subsidence, of soil under a long-term loading such as that due to a heavy structure (e.g., a single building or even an entire city). Hence the practical importance of consolidation lies in the domain of civil engineering, where the amount and uniformity of foundation settlement are of concern to those interested in the stability and safety of structures.

To arrive at a conceptual understanding of consolidation, let us first consider a sample of water-saturated soil confined in a rigid-walled cylinder and subjected to compressive load applied to a piston in such a manner as to allow no possible outlet for water. Because the confined water is practically incompressible, the applied pressure can cause no compression of the soil matrix and must be borne entirely by the water phase. In other words, while each particle of soil is subject to the hydrostatic pressure of water surrounding it, the soil matrix as a whole is not subject to any stress at all. If, however, we bore a narrow hole through the piston, we can expect the pressured water to begin oozing out, thus gradually relieving the hydrostatic pressure inside the cell and transmitting the load to the soil matrix. The matrix would then tend to compress at a rate commensurate with the extrusion of water from the cell. To flow out of the cell, the water must wend its way through the narrow and tortuous pores of the soil. This accounts for the time-variable nature of the consolidation process.

In general, coarse-grained (sandy) soils are the least compressible soils, and such consolidation as does take place usually occurs relatively rapidly, thanks to these soils' large pores and high permeability. In contrast, fine-grained soils, such as clays, generally have greater total porosity than sands (and hence greater overall compressibility), but the rate of their consolidation is likely to be slow (possibly extending over a period of months or even years), owing to their smaller pores. This difference is illustrated in Fig. 13.16.

A classical mechanical model, shown in Fig. 13.17, has been found useful in illustrating the principles involved in the consolidation process. The model consists of a perforated or porous piston pressed into a water-filled cylinder

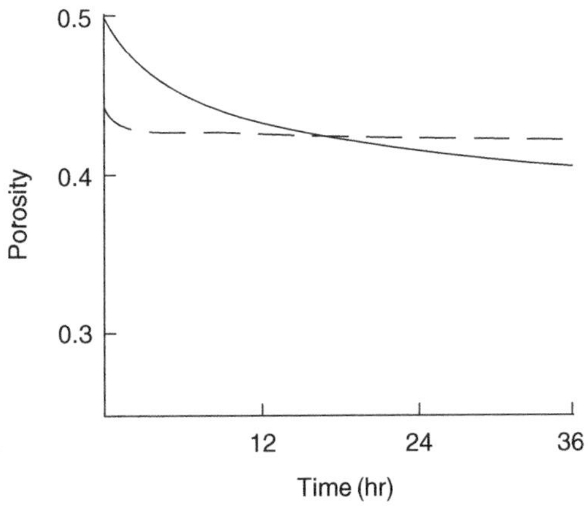

Fig. 13.16. Relative compressibilities and rates of consolidation for a sandy soil (dashed line) and a clayey soil (solid line).

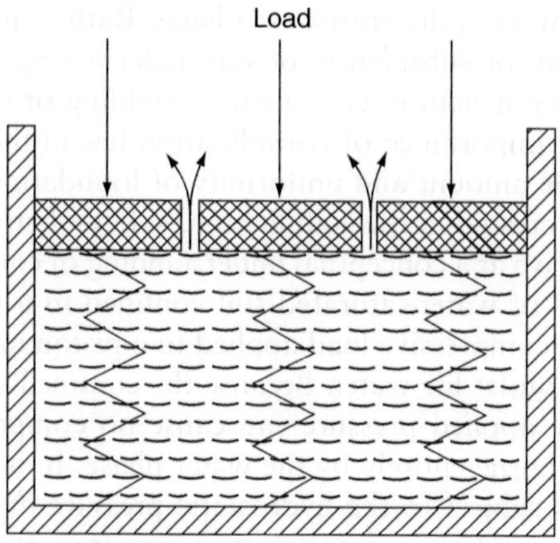

Fig. 13.17. Conceptual model illustrating the process of consolidation.

fitted with springs. When the piston is first loaded and before any downward motion has yet taken place, the pressure applied is transmitted to the water only. However, as water escapes and the piston moves down, the springs begin to contract. As more and more of the water escapes, the springs contract to an increasing degree. If the piston is place under a constant load, the downward movement will finally stop and static equilibrium will prevail when the resistance of the compressed springs just equals the downward force acting on the piston. Thenceforth, the entire load will be borne by the springs. Thus at the beginning of the process the water, not the springs, is stressed and at the end of the process the situation is reversed. At any intermediate point during the process, the load is carried partly by the water and partly by the springs. The springs are quite obviously analogous to the soil matrix, and the water in the cylinder is analogous to the water permeating the pores of a saturated soil that is subject to a compressive stress.

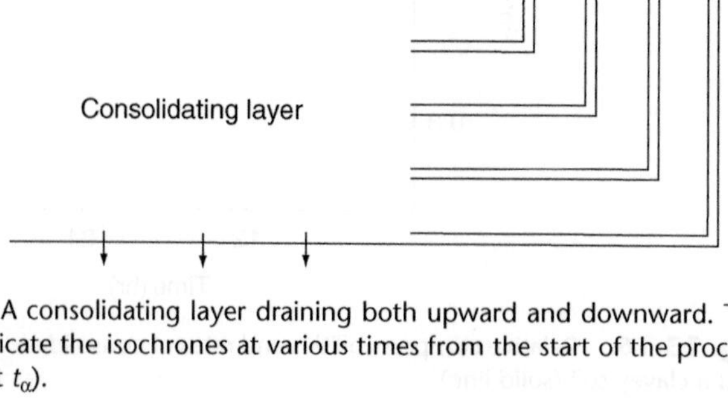

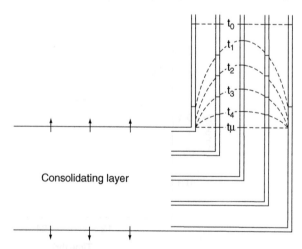

Fig. 13.18. A consolidating layer draining both upward and downward. The piezometers on the right indicate the isochrones at various times from the start of the process (at t_0) to the eventual end (at t_α).

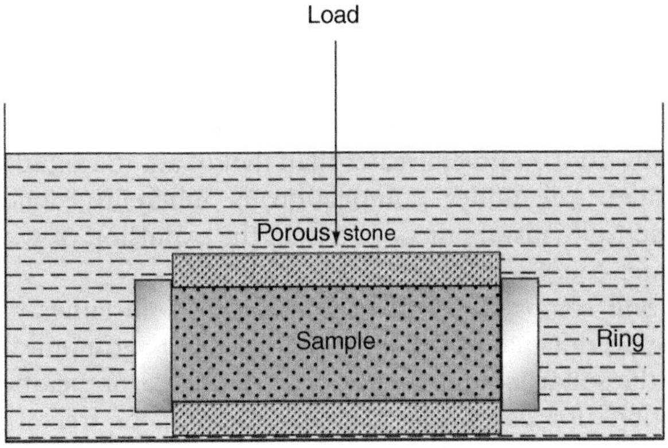

Fig. 13.19. The consolidation test.

The concept described in so many words can be summed up in the form of an equation, known as *Terzaghi's effective stress equation* (Terzaghi, 1953):

$$\sigma_t = p + \sigma_e \quad \text{or} \quad \sigma_e = \sigma_t - p \tag{13.13}$$

Here σ_t is the total stress, p is the hydrostatic pressure known as *pore-water pressure*, and σ_e is the *effective stress* borne by the solid matrix and therefore often termed *intergranular stress*. The term p is a neutral stress, in as much as the liquid pressure in a saturated soil acts equally in all directions and, hence has no effect on compressing the matrix. When a total stress is suddenly applied to a soil and then maintained for an indefinite period, p diminishes gradually and σ_e increases, as shown hypothetically in Fig. 13.18.

To characterize soils with respect to their consolidation behavior, laboratory tests are generally carried out with small samples, preferably "undisturbed" cores, placed in an apparatus such as shown in Fig. 13.19. Extraction of a sample from some depth in the field generally entails release of the original in situ confining pressure (often called the *overburden*). The sample, however, retains what might be called a "memory" of its state of *preconsolidation* pressure. When reconsolidated, it will not compress significantly until its preconsolidation pressure is exceeded. Thereafter, its compression will tend to follow a straight-line relation for void ratio versus the logarithm of pressure. The typical pattern of consolidation is illustrated in Fig. 13.20.

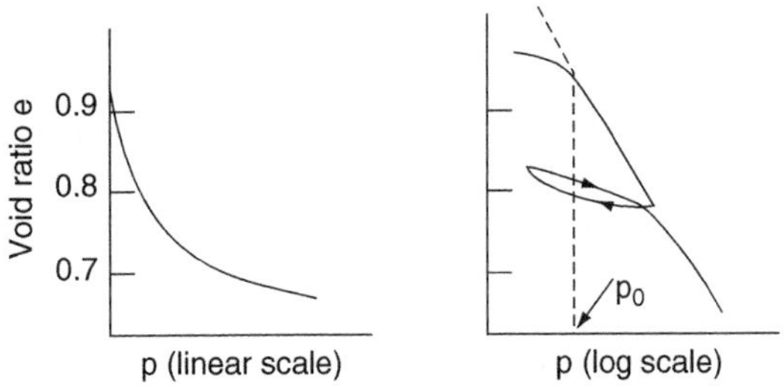

Fig. 13.20. Void ratio versus effective pressure curve for a compressible soil during consolidation. The semilogarithmic plot indicates the apparent preconsolidation pressure p_0 and a pressure release ("rebound") reconsolidation loop.

SOIL COMPACTION IN RELATION TO WETNESS

The term *compaction* applies to the densification of an unsaturated soil by the reduction of the fractional air volume. Since the viscosity of water is generally 50–100 times greater than that of air, the expulsion of air from an unsaturated soil under compaction is inherently much more rapid — being practically instantaneous — than the expulsion of water from a saturated soil under consolidation.

Soil compaction can be considered from either the viewpoint of the civil engineer or that of the plant ecologist and agronomist. The engineer, interested in the soil as a construction material (e.g., for roadbeds, embankments, foundations for dams and various structures, or tractable surfaces for vehicles), wishes to enhance the strength and to reduce the permeability of soil layers. Soil compaction is a means to achieve those ends. In extreme contrast, the ecologist or agronomist, interested in the soil as a medium in which seeds can germinate and roots can proliferate and obtain sustenance, views soil compaction as a scourge, an undesirable consequence of mechanization that reduces the soil's biological productivity and — in extreme cases — makes it unfit for plant growth. So what one professional may wish to achieve, another may wish to avoid. We next describe each of these approaches on its own merits.

The engineer generally seeks ways to achieve the maximum practicable degree of compaction with the least expenditure of energy. An early empirical approach to this problem was developed by R. R. Proctor, whose testing procedure, known as the *Proctor test*, was designed to determine the "optimal" soil wetness at which compaction of a given soil can be achieved most effectively by a given compactive effort. Although modified repeatedly and given different names, the basic procedure has enjoyed universal acceptance and has been adopted as the standard criterion for soil compaction in engineering practice (ASTM, 1984).

Compactive stresses can be administered in several different ways. For testing purposes, perhaps the simplest is to confine a soil sample in a rigid-walled container and apply a *static load* (a weight) by means of a piston resting on the sample's surface. An alternative way is to apply a time-variable dynamic *load* by means of an impacting or hammering device. Still another way, found to be most effective in the case of dry granular materials, is to vibrate the sample. Finally, one could attempt to compact soil by applying a space-variable and time-variable set of compressive and shearing stresses in combination. Such action, which in effect causes a churning (or "puddling") of the soil, is commonly called *kneading compaction*. This last method forms the basis of the Proctor test, which is, in essence, an imitation of the common engineering practice of compacting soil in the field by means of spiked ("sheepsfoot") rollers. The standard test is carried out in a cylindrical mold. The soil material is compacted in three layers, each layer being worked by an impact-driven tamper in a standardized manner.

For any given compactive effort, the resulting bulk density is a function of soil wetness. This functional dependence, illustrated in Fig. 13.21, indicates that, starting from a dry condition, the attainable bulk density at first increases with soil wetness and then reaches a maximal density at a wetness value called *optimum moisture*, beyond which the density decreases.

This behavior is readily explainable, at least qualitatively. A dry soil typically resists compaction because of its stiff matrix and high degree of particle-to-particle bonding, interlocking, and frictional resistance to deformation. As soil

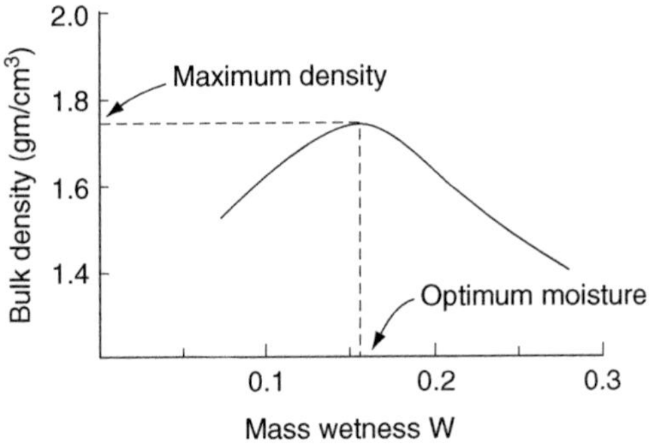

Fig. 13.21. Typical moisture–density curve for a medium-textured soil, indicating the maximum density obtainable with a given compactive effort.

wetness increases, the films of moisture weaken the interparticle bonds and seem to reduce internal friction by "lubricating" the particles. The soil thus becomes more readily compactible. As soil wetness nears saturation, however, the fractional volume of expellable air is reduced and the soil can no longer be compacted by the same compactive effort to the same degree as at the lower wetness. Henceforth, any further increase in soil moisture reduces, rather than enhances, soil compactibility. Finally, at saturation, no amount of kneading can cause any increase in bulk density unless water is expelled. At high wetness values, the water that hydrates the grains pushes them apart and causes swelling, thus preventing the closer packing of the soil matrix.

The function described, relating attainable bulk density to soil moisture, does not constitute a single characteristic curve for a given soil but a family of curves, as shown in Fig. 13.22. For each level of compactive effort, the curve is shifted upward and leftward, indicating higher attainable bulk density at lower values of "optimal moisture." Experience shows that the line connecting the peaks of all the curves of bulk density versus wetness corresponds approximately to the 80% degree-of-saturation line and that the descending portions of these curves tend to converge on a line representing a degree of saturation of about 85–90%.

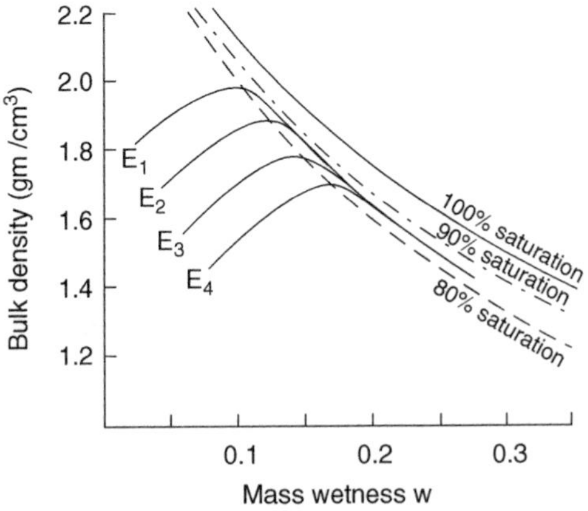

Fig. 13.22. Family of moisture-density curves for different compactive efforts ($E1 > E2 > E3 > E4$). Note that the 100% saturation line, representing zero air-filled porosity, was calculated assuming a particle density value of 2.63 g.cm^3.

To the extent that the Proctor test portrays or simulates compaction in the field, it can serve as a criterion by which to guide earth construction works. A typical job specification might require a contractor to compact an earth fill to, say, 90% of attainable maximal density as defined by the Proctor test (ASTM, 1984). The contractor will then examine the results of the test to determine the optimal wetness value (or range of values) at which to carry out the compacting operation. In practice, an earth fill is laid by depositing and rolling successive layers while controlling soil moisture through measured applications of water (generally by sprinkling) to the deposited material. If the soil material is too wet, it must be dried by drainage and, weather permitting, by evaporation.

Sample Problem

A Proctor compaction test was performed on medium-textured soil with the following results:

Mass wetness w:	6%	9%	12%	15%	18%	21%
Bulk density ρ_b:	1.80	1.91	1.94	1.86	1.75	1.65 g/cm^3

Plot the data and determine the optimal moisture and maximum density values. Calculate the volume wetness and degree of saturation at optimal moisture content.

An examination of the data, confirmed by the plotted graph, shows that the opitmal moisture is at or about a mass wetness w value of 12%, and the maximal density ρ_b is 1.94 g/cm^3. Accordingly, the volume wetness is

$$\theta = w\frac{\rho_b}{\rho_w} = 0.12 \times \frac{1.94\ \text{g/cm}^3}{1\ \text{g/cm}^3} = 0.23 = 23\%$$

and the degree of saturation is

$$s = \theta/f$$

where f, the porosity, can be obtained from the soil's bulk and particle densities:

$$f = 1 - (\rho_b/\rho_s)$$

Hence

$$s = \frac{\theta}{1-(\rho_b/\rho_s)} = \frac{0.23}{1-1.94/2.65} = 0.85 = 85\%$$

BOX 13.1 On the Origins of Cultivation

An important factor in the evolution of agriculture, which first took place in the Near East some 10–12 millennia ago, was the development of the tools of soil husbandry. Seeds scattered on the ground are often eaten by birds and rodents or are subject to desiccation, so their germination rate is likely to be very low. Given a limited seed stock, farmers would naturally do whatever they could to promote germination and seedling establishment. The best way to accomplish this is to insert the seeds to some shallow depth, under a protective layer of loosened soil, and to eradicate the weeds that might compete with the crop seedlings for water, nutrients, and light.

The simplest tool developed for this purpose was a paddle-shaped digging stick by which a farmer could make holes for seeds. The use of this simple device was extremely slow and laborious, however, so at some point the digging stick was modified to form the more convenient spade. This tool could be swung and made to strike the ground with greater force, not only to open the ground for seed insertion but also to loosen and pulverize the soil and eradicate weeds more effectively. In time, the spade developed a triangular blade, first made of wood but later made of stone and eventually of metal. Such a spade, initially designed to be used by one person, was later modified so that it could be pulled by a rope in order to open a continuous slit, or furrow, into which the seeds could be dropped. A second furrow could then be made alongside the first, to facilitate seed coverage. In some cases, rows were widely enough separated to permit a person to walk between the rows in order to weed the cultivated plot.

The human-pulled traction spade, or ard, gradually metamorphosed into an animal-drawn plow. The first picture of such a plow, dating to 3000 B.C.E., was found in Mesopotamia, and numerous later pictures have been found both there and in Egypt as well as in China. It was not long before these early plows were fitted with a seed funnel so that the acts of plowing and sowing could be carried out simultaneously. The same ancient implement is still used today in the Near East.

Although the advent of the plow represented a huge advance in terms of convenience and efficiency of operation, it had an important side effect. As with many other innovations, the benefits were immediate, but the full range of consequences took several generations to play out, long after the new practice became entrenched. The major environmental consequence was that plowing made the soil surface — now loosened, pulverized, and bared of weeds — much more vulnerable to accelerated erosion. In the history of civilization, contrary to the idealistic vision of the prophet Isaiah, the plowshare may have been more destructive than the sword. That is why modern efforts are directed toward reducing excessive tillage so as to prevent compaction and to promote the conservation of both soil and energy.

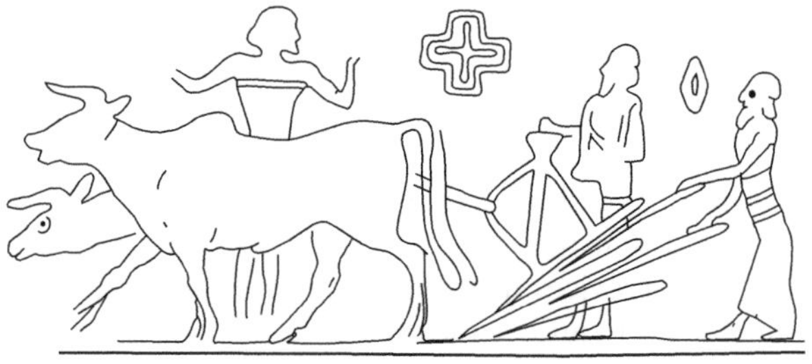

Fig. B13.1. Depiction of a seeder-plow on an ancient Mesopotamian seal.

SOIL COMPACTION IN AGRICULTURAL FIELDS

In agronomy, a soil is considered compacted when the total porosity, particularly the air-filled porosity, is so low as to restrict aeration or when the soil is so tight, and its pores so small, as to impede root penetration, infiltration, and drainage.

Soil compaction can take place in agricultural fields under the influence of man-induced mechanical forces applied at or near the soil surface. One such cause of soil compaction is trampling by livestock. However, by far the most common cause of soil compaction in modern agriculture is the effect of machinery, imposed on the soil by vehicles and soil-engaging implements.

Instances of soil compaction are prevalent in modern mechanized agriculture. The trend to reduce labor on the farm by the use of larger and heavier machinery increases the frequency of traffic over the soil and the likelihood of damage by compaction. Especially insidious is the practice of plowing with the tractor wheel running over the bottom of the open furrow, where the soil may be even more compactable than at the surface, owing to higher moisture and lower organic matter content. Furthermore, compaction in depth is much more difficult to rectify and hence longer lasting than compaction at the surface.

Especially damaging is the cultivation of clayey soils with heavy equipment when the soil is in a wet state and its strength is low. Smearing of the soil plowshares and wheels creates a "plow sole" or "plow pan." The churning and slipping action of wheels can be more important than loading per se in causing the degradation of soil structure (D. B. Davies et al., 1973). Traffic associated with harvesting in wet conditions causes deep rutting, smearing, and compaction, which in turn inhibit infiltration, drainage, and soil aeration (Swain, 1975).

As shown by Cannell (1977), roots are generally unable to enter pores narrower than their root caps; thus, if they are to grow through compacted soil they must displace soil particles to widen the pores by exerting a pressure greater than the soil's mechanical strength. Studies on the effect of mechanical impedance on root growth have been reported by, among others, Bowen et al. (1981), and Hadas et al. (1990)

Ehlers et al. (1983) conducted extensive studies of penetration resistance, bulk density, soil wetness, and root growth of oats in loessial soil. They found that root growth was inversely related to penetrometer resistance. The limiting penetrometer resistance for root growth was 3.6 MPa in the tilled surface layer and 4.6 to 5.1 MPa in the underlying soil. Dexter (1987) reported similar studies of cotton, corn, pea, and peanut roots.

Observations have shown that some roots do eventually penetrate compacted zones and that the roots of some species have a remarkable ability to enter and enlarge crevices in rocks, roads, and foundations. Where the soil is highly compact and rigid, root growth may be confined almost entirely to cracks and cleavage planes. In seeking out such cracks, plant roots are not merely passive agents; they cause differential shrinkage though preferred extraction of moisture from zones of greater penetrability.

CONTROL OF SOIL COMPACTION

The most obvious approach to the prevention of soil compaction is avoidance of all but essential pressure-inducing operations. This calls for reducing the number of operations involved in primary and secondary tillage (plowing and subsequent cultivation, respectively) and choosing the most efficient implements and the most appropriate time to effect the desirable soil condition in a single pass rather than by a repeated sequence of passes. It has long been known that excessive soil manipulation leads not only to deterioration of soil structure and to reduced productivity but also to accelerated erosion.

In recent decades, growing awareness of these hazards has resulted in the development of new field-management systems, variously called *minimum tillage, conservation tillage*, and even *zero tillage* (Wittmus et al., 1975; McGregor et al., 1975; Uri, 1999; Bradford and Peterson, 2000). Such systems reduce the number of operations, avoid unnecessary inversion of topsoil, and generally retain crop residues as protective mulch (against evaporation and erosion) over the surface. In former times, frequent cultivation of orchards and of row-cropped fields was necessary for weed control. This requirement was greatly reduced by the introduction of sprayable herbicides. However, excessive reliance on phytotoxic chemicals poses serious environmental problems.

Since random traffic by heavy machinery is a major cause of soil compaction, cultural systems have been proposed to restrict vehicular traffic to permanent, narrow lanes and to reduce the fractional area trampled by wheels to less than 10% of the land surface (Dumas et al., 1973). In row crops, seedbed preparation, involving pulverizing the top layer, can be confined to the narrow strips where planting takes place, while the inter-row zone is left in an open, cloddy condition to promote water and air penetration and to reduce erosion by water and wind. The field is thus divided into three consistent zones: (1) narrow planting strips, precision tilled; (2) narrow traffic lanes, permanently compacted; and (3) inter-row water-management beds, maintained in a cloddy condition and covered with a mulch of plant residues.

An extremely important factor is the timing of field operations in relation to soil moisture. Moist soils can be highly vulnerable to compaction. Operations that impose high pressures should, if possible, be carried out on dry soil, which is less compactible. These include traffic in general, as well as such tillage methods as subsoiling, or chiseling, which is more effective in achieving its main purpose of shattering and loosening the soil when it is dry than when it is wet.

A basic discussion of tillage in relation to soil compaction was provided by by McKyes (1985).

Part VI

THE FIELD WATER CYCLE

14. WATER ENTRY INTO SOIL

INFILTRATION

The entry of water into the soil may occur under a variety of conditions. It may take place through the entire surface uniformly or via localized furrows or crevices. It may also move upward into the soil from a source below, such as a rising water table.

When water is supplied from above to the soil surface, whether by precipitation or irrigation, it typically penetrates the surface and is absorbed into successively deeper layers of the profile. At times, however, a portion of the arriving water may fail to penetrate but instead will tend to accrue at the surface or flow over it. The penetrated water is itself later partitioned between the amount that returns to the atmosphere by direct evaporation from the soil or by the extraction and transpiration of plants and the amount that continues to seep downward and eventually recharges the groundwater reservoir.

Infiltration is the term applied to the process of water entry into the soil, generally by downward flow through all or part of the soil surface. The rate of this process, relative to the rate of water supply, determines how much water will enter the root zone and how much, if any, will run off. Hence the rate of infiltration affects not only the water economy of terrestrial plants but also the amount of overland flow and its attendant dangers of soil erosion and stream flooding. Where soil conditions, especially at the surface, limit the rate of infiltration, plants may be denied sufficient moisture while surface erosion increases. Knowledge of the infiltration process as it is affected both by the soil's properties and transient conditions and by the mode of water supply is therefore a prerequisite for efficient soil and water management.

"INFILTRATION CAPACITY," OR INFILTRABILITY

If water is sprinkled over the soil surface at a steadily increasing rate, sooner or later the supply rate will exceed the soil's finite rate of absorption, and the excess will accrue over the soil surface or run off it (Fig. 14.1). The *infiltration rate* is defined as the volume flux of water flowing into the profile per unit of soil surface area. For the special condition in which the rainfall rate exceeds the ability of the soil to absorb water, infiltration proceeds at a maximal rate, which Horton (1940) called the soil's *infiltration capacity*.

Hillel (1971) suggested the term *infiltrability* to designate the infiltration flux resulting when water at atmospheric pressure is made freely available at the soil surface. This one-word replacement avoids the extensity–intensity confusion in the term *infiltration "capacity"* and permits use of the term *infiltration "rate"* in the literal sense to represent the surface flux in any set of circumstances, whatever the rate or pressure at which the water is supplied to the soil. The infiltration rate can be expected to exceed infiltrability whenever water is ponded over the soil to a depth sufficient to cause the hydrostatic pressure at the surface to be significantly greater than atmospheric pressure. On the other hand, if water is applied slowly or at subatmospheric pressure, the actual infiltration rate may be smaller than the infiltrability.

As long as the rate of water delivery to the surface is smaller than the soil's infiltrability, water infiltrates as fast as it arrives and the process is *supply controlled* (or *flux controlled*). However, once the delivery rate exceeds the soil's infiltrability, the latter determines the actual infiltration rate, and thus the process becomes *soil controlled.* The soil may limit the rate of infiltration either at the surface or within the profile. Thus the process may be either *surface controlled* or *profile controlled.*

When a shallow layer of water is instantaneously applied and thereafter maintained over the surface of an initially unsaturated soil, the full measure of soil infiltrability comes into play from the start. Many trials of infiltration under shallow ponding have shown infiltrability to vary and generally to

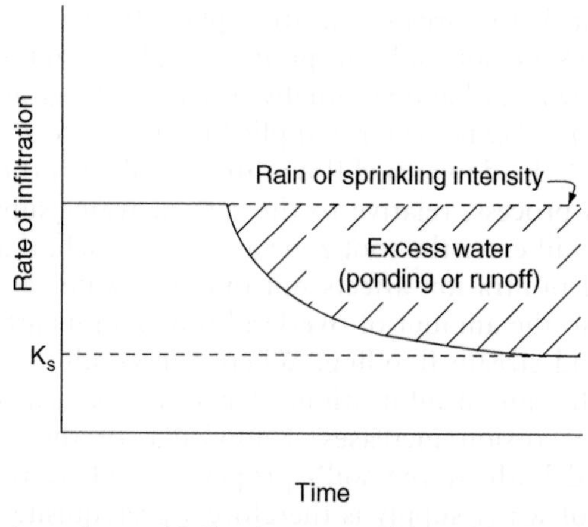

Fig. 14.1. Time dependence of infiltration rate under rainfall of constant intensity that is lower than the initial value but higher than the final value of soil infiltrability.

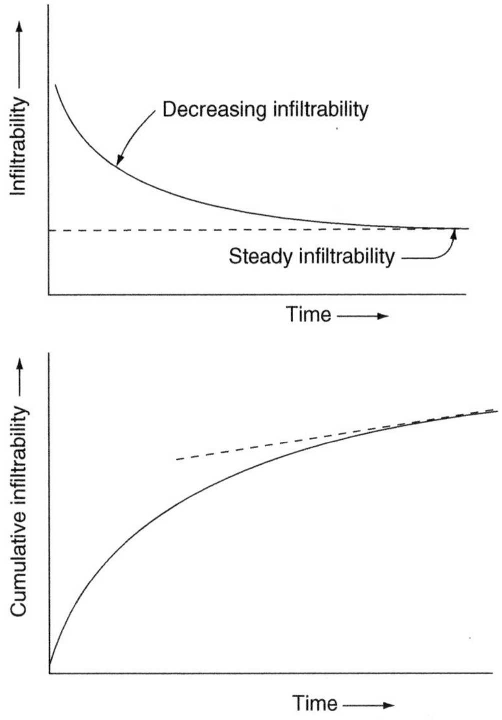

Fig. 14.2. Time dependence of infiltrability and of cumulative infiltration under shallow ponding.

decline in time. Thus, the cumulative infiltration, being the time integral of infiltration rate, typically exhibits curvilinear time dependence, with a gradually diminishing slope (Fig. 14.2).

Soil infiltrability and its variation with time are known to depend on the initial wetness and suction as well as on the texture, structure, and uniformity (or layering sequence) of the profile. In general, soil infiltrability is relatively high in the early stages of infiltration, particularly where the soil is initially quite dry, but it tends to decrease and eventually to approach asymptotically a constant rate that we call the *steady-state infiltrability* but that is often termed the *final infiltration capacity*. (The adjective *final* in this context does not signify the end of the process, but the attainment of a constant rate that can persist thenceforth.)

The decrease of infiltrability from an initially high rate can in some cases result from gradual deterioration of soil structure and the partial sealing of the profile by the formation of a surface crust. It can also result from the detachment and migration of pore-blocking particles, from swelling of clay, as well as from entrapment of air bubbles or the bulk compression of the air originally present in the soil if it is prevented from escaping during its displacement by incoming water. Primarily, however, the decline of infiltrability results from the decrease in the *matric suction gradient*, which occurs inevitably as infiltration proceeds.

If the surface of an initially dry soil is suddenly saturated (e.g., by ponding), the difference of hydraulic potential between the saturated surface and the relatively dry soil just below creates a steep matric suction gradient. As the wetted zone deepens, the same difference of potential acting over a greater distance expresses itself as a diminishing gradient. And as the wetted part of the profile deepens, the suction gradient eventually becomes vanishingly small. In a horizontal column, the infiltration rate eventually tends to zero. In downward flow into a vertical column under continuous ponding, the infiltration rate tends asymptotically to a steady, gravity-induced rate that approximates the soil's saturated hydraulic conductivity if the profile is homogeneous and structurally stable (Fig. 14.2).

PROFILE MOISTURE DISTRIBUTION DURING INFILTRATION

An examination of an initially dry, texturally uniform soil profile at any moment during infiltration under ponding generally shows the surface zone to be saturated to a depth of several millimeters or centimeters. Beneath this *saturated zone* is a less-than-saturated, lengthening zone of apparently uniform wetness, known as the *transmission zone*, beyond which occurs a *wetting zone*. In the latter zone, soil wetness increases with time at each point, but at any given time wetness decreases with depth at a steepening gradient, down to a *wetting front*. At the wetting front, the moisture gradient is so steep that there appears to be a sharp boundary between the moistened soil above and the initially dry soil beneath. The simple explanation for the existence of the wetting front is that (since the flux is the product of the gradient and the conductivity) the hydraulic conductivity of the as-yet unwetted soil is so low that water can only penetrate it when the gradient is very steep. It follows that the drier the soil is initially, the steeper must be the gradient at the wetting front.

The typical moisture profile during infiltration is shown schematically in Fig. 14.3. Repeated examinations of the moisture profile of a stable soil

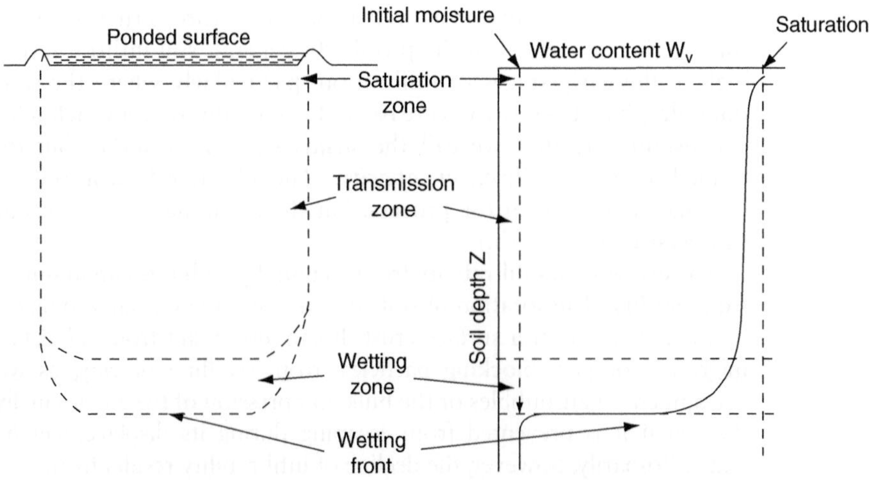

Fig. 14.3. The infiltration moisture profile. At left, a schematic section of the profile; at right, the curve of water content versus depth. The common occurrence of a saturation zone as distinct from the transmission zone may result from the structural instability of the surface zone.

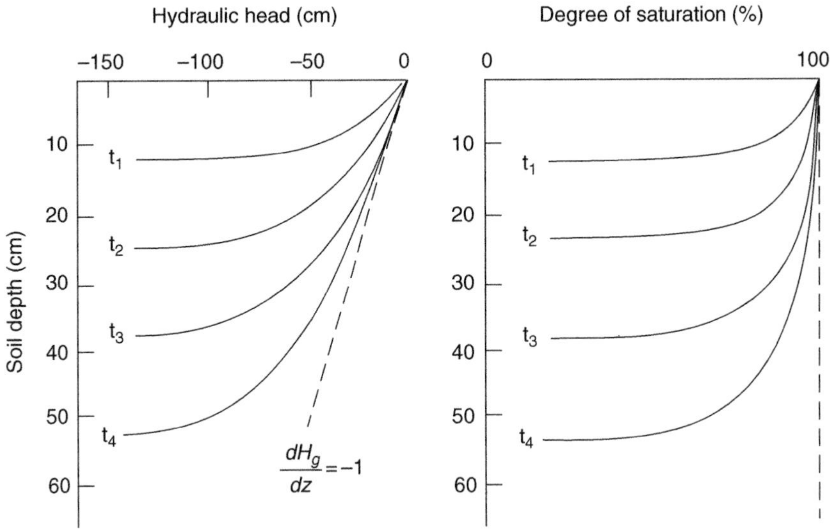

Fig. 14.4. Water-content profiles (at right) and hydraulic-head profiles (at left) at successive times (t_1, t_2, t_3, t_4) during infiltration into a uniform soil ponded at the surface. The dH_g/dz value is the gravitational head gradient. In this figure the possible existence of a saturation zone distinct from the transmission zone is disregarded.

during infiltration generally reveal that the transmission zone lengthens (deepens) continuously and that the wetting zone and the wetting front likewise move downward continuously, with the latter becoming less steep as it moves deeper into the profile. Typical families of successive moisture and hydraulic-head profiles are shown in Figure 14.4.

THE GREEN AND AMPT APPROACH

A simplistic, yet still useful, theoretical approach to the infiltration process was suggested as early as 1911 by W. H. Green and G. A. Ampt in their classic paper on flow of air and water through soils. Their theory has been found to apply particularly to infiltration into uniform, initially dry, coarse-textured soils, which exhibit a sharp wetting front. The G&A solution gives no information about the moisture profile during infiltration, but it does offer estimates of the infiltration rate and the cumulative infiltration as functions of time, $i(t)$ and $I(t)$.

The main assumptions of the G&A approach are that there exists a distinct and precisely definable wetting front during infiltration and that, although this wetting front moves continuously downward as the process proceeds, it is characterized by a constant matric suction, regardless of time and position. Furthermore, this approach assumes that, in the transmission zone behind the wetting front, the soil is uniformly wet and of constant conductivity. The wetting front is thus viewed as a plane separating a uniformly wetted infiltrated zone from an as-yet totally uninfiltrated zone. In effect, this supposes the K-ψ (hydraulic conductivity versus suction head) relation to be discontinuous, that is, to change abruptly at the suction value prevailing at the wetting front, from a high value at lower suctions to a very much lower value at higher suctions.

These assumptions simplify the flow equation, making it amenable to analytical solution. For horizontal infiltration, a Darcy-type equation can be applied directly:

$$i = dI/dt = K(H_0 - H_f)/L_f \tag{14.1}$$

where i is the flux into the soil and through the transmission zone, I is the cumulative infiltration, K is the hydraulic conductivity of the transmission zone, H_0 is the pressure head at the entry surface, H_f is the effective pressure head at the wetting front, and L_f is the distance from the surface to the wetting front (the length of the wetted zone). If the ponding depth is negligible and the surface is thus maintained at a pressure head of zero, we obtain simply

$$dI/dt = -KH_f/L_f = K(\Delta H_p/L_f) \tag{14.2}$$

Here ΔH_p is the pressure-head difference between the surface and the wetting front. This shows the infiltration rate to be proportional to the reciprocal of the distance to the wetting front.

Since a uniformly wetted zone is assumed to extend all the way to the wetting front, it follows that the cumulative infiltration I should be equal to the product of the wetting-front depth L_f and the wetness increment $\Delta\theta = \theta_t - \theta_i$ (where θ_t is the transmission-zone wetness during infiltration and θ_i is the initial profile wetness, which prevails beyond the wetting front):

$$I = L_f\,\Delta\theta \tag{14.3}$$

(In the special case where θ_t is saturation and θ_i is zero, $I = fL_t$, where f is the porosity.)

Therefore,

$$\frac{dI}{dt} = \Delta\theta\frac{dL_f}{dt} = K\frac{\Delta H_p}{L_f} = K\frac{\Delta\theta\Delta H_p}{I} \tag{14.4}$$

where dL_f/dt is the rate of advance of the wetting front. The infiltration rate is thus seen to be inversely related to the cumulative infiltration. Rearranging Eq. (14.4), we obtain

$$L_f dL_f = K\frac{\Delta H_p}{\Delta\theta}dt = \tilde{D}\,(dt) \tag{14.5}$$

where the composite term $[K(\Delta H_p/\Delta\theta)]$ can be regarded as an effective diffusivity D_{eff} for the infiltrating profile. Integration gives

$$\frac{L_f^2}{2} = K\frac{\Delta H_p}{\Delta\theta}t = \tilde{D}t \tag{14.6}$$

$$L_f = \sqrt{2Kt(\Delta H_p/\Delta\theta)} = \sqrt{2\tilde{D}t} \tag{14.7}$$

or

$$I = \Delta\theta\sqrt{2\tilde{D}t}, \quad i = \Delta\theta\sqrt{\tilde{D}/2t} \tag{14.8}$$

Thus the depth of the wetting front is proportional to $t^{1/2}$, and the infiltration rate to $t^{-1/2}$.

With gravity taken into account, the Green and Ampt approach gives

$$\frac{dI}{dt} = \Delta\theta \frac{dL_f}{dt} = K \frac{H_0 - H_f + L_f}{L_f} \tag{14.9}$$

which integrates to:

$$Kt/\Delta\theta = L_f - (H_0 - H_f) \ln[1 + L_f/(H_0 - H_f)] \tag{14.10}$$

As t increases, the second term on the right-hand side of Eq. (14.10) increases more and more slowly in relation to the increase in L_f, so at very large times the relationship becomes

$$L_f \cong Kt/\Delta\theta + \delta \qquad \text{or} \qquad I \cong Kt + \delta \tag{14.11}$$

where δ can eventually be regarded as a constant.

The Green and Ampt model was designated the *delta-function approximation* by Philip (1966a), because it corresponds exactly to the nonlinear diffusion description of infiltration in the special case where the diffusivity is wholly concentrated at the wet end of the moisture range (a so-called *Dirac delta function*). This model represents a reasonable approximation of reality in situations where the wetting front is sharp and the moisture profile differs little from a step function, for example, one-dimensional vertical systems with water supplied under positive hydrostatic pressure to soils that are coarse-textured, initially dry, and without air entrapment.

Sample Problem

The infiltration rate under shallow ponding was monitored as a function of cumulative rainfall and found to be 20 mm/hr when a total of 100 mm had infiltrated. If the eventual steady rate of infiltration is 5 mm/hr, estimate the infiltration rate at a cumulative infiltration of 200 and 400 mm. Use the Green and Ampt theory.

We refer to Eq. (14.1):

$$i = i_c + b/I$$

where i is the infiltration rate, i_c is the eventual steady rate, I is the cumulative infiltration, and b is a constant. Substituting the given values and solving for b, we get

$$20 \text{ mm/hr} = 5 \text{ mm/hr} + (b \text{ mm}^2\text{/hr})/100 \text{ mm},$$
$$b = 100(20 - 5) = 1500 \text{ mm}^2\text{/hr}$$

We use this value of b to estimate the infiltration rate at any value of cumulative infiltration.

At $I = 200$ mm: $i = 5 + 1500/200 = 12.5$ mm/hr
At $I = 400$ mm: $i = 5 + 1500/400 = 8.75$ mm/hr

Given the soil conditions described, calculate how much water can be delivered to the root zone of a crop without exceeding the soil's infiltrability if the sprinkling irrigation rate is 15 or 25 mm/hr? What is the highest steady sprinkling rate we can use if we wish to provide an irrigation of 250 mm in the shortest possible time?

At a sprinkling rate of 15 mm/hr,

$$15 \text{ mm/hr} = 5 \text{ mm/hr} + (1500 \text{ mm}^2/\text{hr})/I \text{ mm}$$
$$I = 1500/(15 - 5) = 150 \text{ mm}$$

At a sprinkling rate of 25 mm/hr,

$$I = 1500/(25 - 5) = 75 \text{ mm}$$

Note: The higher the sprinkling rate, the smaller the total amount of water that can infiltrate without exceeding the soil's infiltrability (i.e., without causing flooding or runoff).

To infiltrate 250 mm by steady sprinkling without exceeding Infiltrability, we can apply

$$i = 5 + 1500/250 = 11 \text{ mm/hr}$$

BASIC INFILTRATION THEORY

Horizontal Infiltration

We begin by combining the Darcy equation with the continuity equation to obtain the general flow equation for water in soil (see Chapters 7 and 8). For one-dimensional flow the appropriate form of the flow equation is

$$\frac{\partial \theta}{\partial t} = \frac{\partial}{\partial x}\left[K \frac{\partial H}{\partial x}\right] \tag{14.12}$$

where θ is the volumetric wetness, t is time, x is distance along the direction of flow, and H is the hydraulic head. The simplest application of this equation is in the description of lateral infiltration [termed *absorption* by Philip (1969a)] into a thin horizontal column extending in the x direction. In this case, the gravity force is zero, and water is drawn into the soil by matric suction gradients only. If the soil is texturally homogeneous and the antecedent moisture is equal throughout, the diffusivity equation can be applied:

$$\frac{\partial \theta}{\partial t} = \frac{\partial}{\partial x}\left[D(\theta) \frac{\partial \theta}{\partial x}\right] \tag{14.13}$$

This equation can be solved for the case of flow into an infinitely long column of uniform initial wetness θ_i, where the plane of entry ($x = 0$) is instantaneously brought to, and thereafter maintained at, a higher wetness θ_0. These conditions are written formally in the following way:

$$\begin{aligned} \theta &= \theta_i, \quad x \geq 0, \quad t = 0 \\ \theta &= \theta_0, \quad x = 0, \quad t > 0 \end{aligned} \tag{14.14}$$

We can now simplify Eq. (14.13) by introducing the composite variable $\lambda(\theta)$, defined as

$$\lambda(\theta) = xt^{-1/2} \tag{14.15}$$

Note that λ is a function of θ only. When substituted into Eq. (14.13) it transforms that equation from a partial to an ordinary differential equation (the only variables being θ and λ):

$$-\frac{\lambda}{2}\frac{d\theta}{d\lambda} = \frac{d}{d\lambda}\left[D(\theta)\frac{d\theta}{d\lambda}\right] \tag{14.16}$$

This mathematical sleight of hand, first introduced by Ludwig Boltzmann in 1894, has been known ever since as the *Boltzmann transformation.*

Equation (14.15) can be rewritten

$$x = \lambda(\theta)t^{1/2} \tag{14.17}$$

During infiltration into an initially dry and stable soil, the advancing wetting front is usually visible and indicates a moving zone of practically constant wetness. In this situation, $\lambda(\theta)$ can be assumed to remain constant, and therefore x (taken to be the distance to the wetting front) is proportional to $t^{1/2}$. A plot of the distance to the wetting front against $t^{1/2}$ should therefore give a straight line with a slope equal to λ.

The cumulative volume of water entered through a unit area of the entry plane ($x = 0$) in time t is given by

$$I = \int_{\theta_i}^{\theta_0} x\, d\theta \tag{14.18}$$

where I is the cumulative infiltration, θ_i is the initial volumetric wetness of the uninfiltrated soil, and θ_0 is the saturation wetness. Substitution of the Boltzmann variable [Eq. (14.15)] gives

$$I = t^{1/2}\int_{\theta_i}^{\theta_0} \lambda(\theta)\, d\theta = st^{1/2} \tag{14.19}$$

where

$$s = \int_{\theta_i}^{\theta_0} \lambda(\theta)\, d\theta = I/t^{1/2} \tag{14.20}$$

The coefficient s is a constant (a definite integral) and is called the *sorptivity* (Philip, 1969a). Note that s depends on both θ_i and θ_0. Therefore, it is defined only in relation to a fixed initial state θ_i and an imposed boundary condition θ_0. The dimensions of sorptivity are length per square root of time ($LT^{-1/2}$). From Eq. (14.19) we note that a plot of cumulative infiltration I versus $t^{1/2}$ should give a straight line with a slope equal to s.

Introducing the Boltzmann transformation into Eq. (14.13) also requires transforming the boundary conditions. Accordingly, Eq. (14.14) becomes

$$\begin{aligned} \theta &= \theta_i, & \lambda &\to \infty \\ \theta &= \theta_0, & \lambda &= 0 \end{aligned} \tag{14.21}$$

A numerical solution is generally needed because the $D(\theta)$ relationship makes Eq. (14.16) nonlinear. A quasi-analytical technique was introduced by Parlange (1971). Computer-based simulation methods were devised by de Wit and van Keulen (1972) and by Elrick and Laryea (1979).

Vertical Infiltration

Downward infiltration into an initially unsaturated soil generally occurs under the combined influence of suction and gravity gradients. As the water penetrates deeper and the wetted part of the profile lengthens, the suction gradient decreases, since the difference in pressure head (between the saturated surface and the unwetted part of the profile) divides itself over an ever-increasing distance. As this trend continues, eventually the suction gradient in the upper part of the profile becomes negligible, leaving the constant gravitational gradient as the only force moving water downward. Since the gravitational head gradient has the value of unity (the gravitational head decreasing at the rate of 1 m with each meter of vertical depth below the surface), it follows that the flux tends to approach the hydraulic conductivity as a limiting value.

In a uniform profile (without any impeding layer) under prolonged ponding, the water content of the wetted zone should, theoretically, approach saturation. In practice, however, because of air entrapment (or encapsulation in bubbles or bypassed pores), the soil-water content often does not attain total saturation but some maximal value lower than saturation, which has been called *satiation*. Total saturation is assured only when the air phase in a soil sample is evacuated (subjected to a vacuum) or replaced with a readily soluble gas (such as CO_2) and then wetted with deaired water.

Darcy's equation for vertical flow is

$$q = -K(dH/dz) = -K[d(H_p - z)/dz] \tag{14.22}$$

where q is the flux, K is hydraulic conductivity, and H is the total hydraulic head (the sum of a pressure head H_p and a gravity head $H_z = -z$). The foregoing assumes that the datum plane is at the soil surface ($z = 0$) and that z is the vertical distance from the soil surface downward (i.e., the depth). At the soil surface, $q = i$, the infiltration rate. In an unsaturated soil, H_p is negative and can be expressed as a suction head, ψ. Hence,

$$q = K(d\psi/dz) + K \tag{14.23}$$

Combining these formulations of Darcy's equation, Eqs. (14.22) and (14.23), with the continuity equation $\partial\theta/\partial t = -\partial q/\partial z$ gives the general flow equation

$$\frac{\partial\theta}{\partial t} = \frac{\partial}{\partial z}\left(K\frac{\partial H}{\partial z}\right) = -\frac{\partial}{\partial z}\left(K\frac{\partial\psi}{\partial z}\right) - \frac{\partial K}{\partial z} \tag{14.24}$$

If soil wetness θ and suction head ψ are uniquely related, we can use the chain rule to rewrite the left-hand side of Eq. (14.24) thus: $\partial\theta/\partial t = (\partial\theta/\partial\psi)(\partial\psi/\partial t)$. This transforms the equation into

$$C\frac{\partial\psi}{\partial t} = \frac{\partial}{\partial z}\left(K\frac{\partial\psi}{\partial z}\right) + \frac{\partial K}{\partial z} \tag{14.25}$$

where C $(= -\partial\theta/\partial\psi)$ is defined as the *specific* (or *differential*) *water capacity* (i.e., the change in water content in a unit volume of soil per unit change in matric potential).

Alternatively, we can transform the right-hand side of Eq. (14.24), again using the chain rule, to render $-\partial\psi/\partial z = -(\partial\psi/\partial\theta)(\partial\theta/\partial z) = (1/C)(\partial\theta/\partial z)$. We thus obtain

$$\frac{\partial\theta}{\partial t} = \frac{\partial}{\partial z}\left(\frac{K}{C}\frac{\partial\theta}{\partial z}\right) - \frac{\partial K}{\partial z} \quad \text{or} \quad \frac{\partial\theta}{\partial t} = \frac{\partial}{\partial z}\left(D\frac{\partial\theta}{\partial z}\right) - \frac{\partial K}{\partial z} \tag{14.26}$$

where D, once again, is the soil-moisture diffusivity, which we propose calling the *hydraulic diffusivity* (see Chapter 8). Equations (14.24), (14.25), and (14.26) can all be considered forms of the *Richards equation*, named for L. A. Richards, who first combined Darcy's equation for unsaturated flow with the continuity equation (L. A. Richards, 1931).

Note that the preceding three equations contain two terms on their right-hand sides; the first pertains to the role of the suction (or wetness) gradient and the second to the role of gravity. Whether the one or the other force is dominant depends on the initial and boundary conditions and on the stage of the process considered. For instance, when infiltration takes place into an initially dry soil, the suction gradients at first can be much stronger than the gravitational gradient, and the initial infiltration rate into a horizontal column approximates the infiltration rate into a vertical column. Water from a furrow will therefore tend at first to infiltrate laterally almost to the same extent as vertically (Fig. 14.5). However, when infiltration takes place into an initially wet soil, the suction gradients are weak from the start and become negligible much sooner (Fig. 14.6).

The first mathematically rigorous solution of the flow equation applied to vertical infiltration was given by Philip (1957a). More comprehensive reviews of infiltration theory were subsequently provided by the same author (Philip, 1969a; 1993). Philip's original solution pertained to the case of an infinitely

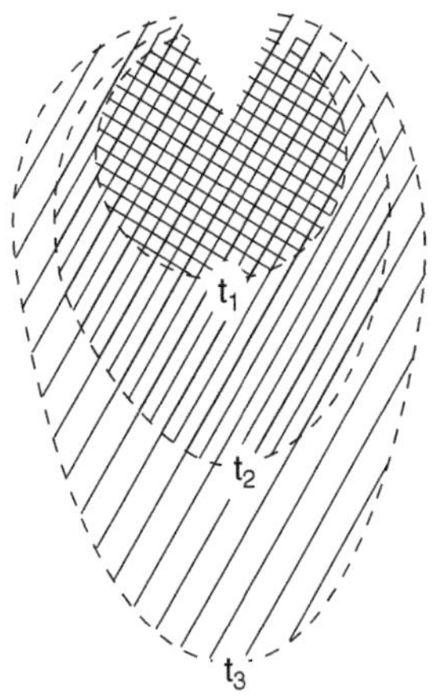

Fig. 14.5. Infiltration from an irrigation furrow into an initially dry soil. The wetting front is shown after different periods of time ($t_1 < t_2 < t_3$). At first the strong suction gradients cause infiltration to be nearly uniform in all directions. As the infiltration process continues, the suction gradients diminish and the gravitational gradient eventually predominates.

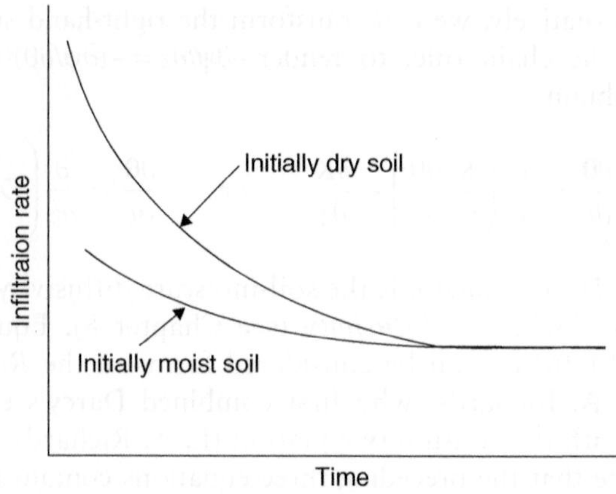

Fig. 14.6. Infiltrability as function of time in an initially dry and in an initially moist soil.

deep, uniform soil of constant initial wetness θ_i, assumed at time zero to be submerged under a thin layer of water that instantaneously increases soil wetness at the surface from its initial value to a new value θ_0 (near saturation) that is thereafter maintained constant. Mathematically, these conditions are stated as

$$t = 0, \quad z > 0, \quad \theta = \theta_i$$
$$t = 0, \quad z = 0, \quad \theta = \theta_0 \tag{14.27}$$

His solution took the form of a power series:

$$z(\theta, t) = \sum_{n=1}^{\infty} f_n(\theta)\, t^{n/2} = f_1(\theta)\, t^{1/2} + f_2(\theta)\, t + f_3(\theta)\, t^{3/2} + f_4(\theta)\, t^2 + \cdots \tag{14.28}$$

where z is the depth to any particular value of wetness θ and the coefficients $f_n(\theta)$ are calculated successively from the diffusivity and conductivity functions.

Philip's solution also describes the time dependence of cumulative infiltration I in terms of a power series:

$$I(t) = \sum_{n=1}^{\infty} j_n(\theta)\, t^{n/2} = st^{1/2} + (A_2 + K_0)t + A_3 t^{3/2} + A_4 t^2 + \cdots + A_n t^{n/2} \tag{14.29}$$

in which the coefficients $j_n(\theta)$ are calculated from $K(\theta)$ and $D(\theta)$ and the coefficient s is called the *sorptivity*. Differentiating Eq. (14.29) with respect to t we obtain the series for the infiltration rate

$$i(t) = \frac{1}{2} st^{-1/2} + (A_2 + K_0) + \frac{3}{2} A_3\, t^{1/2} + 2A_4 t + \cdots + \frac{n}{2} A_n t^{n/2-} \tag{14.30}$$

In practice, it is generally sufficient for an approximate description of infiltration to replace Eq. (14.29) and (14.30) by two-parameter equations of the type

$$I(t) = st^{1/2} + At, \qquad i(t) = \frac{1}{2} st^{-1/2} + A \tag{14.31}$$

TABLE 14.1 Steady Infiltration Rates for Different Soil Types

Soil type	Steady infiltration rate (mm/h)
Sands	>20
Sandy and silty soils	10–20
Loams	5–10
Clayey soils	1–5
Sodic clayey soils	<1

where t is not too large. In the limit, as t tends to infinity, the infiltration rate decreases monotonically to its final asymptotic value $i(\infty)$. At large times, it is possible to represent Eq. (14.31) as

$$I = st^{1/2} + Kt, \qquad i = s/2t^{1/2} + K \tag{14.32}$$

where K is the hydraulic conductivity of the soil's upper layer, which in a soil under ponding is nearly equal to the saturated conductivity K_s.

Recall that the sorptivity has been defined (according to Philip, 1969a) in terms of the horizontal infiltration equation as the ratio of the cumulative infiltration to the square root of time:

$$s = I/t^{1/2} \tag{14.33}$$

As such, it embodies in a single parameter the influence of the matric suction and conductivity on the transient flow process that follows a step-function change in surface wetness or suction. Strictly speaking, one should write $s(\theta_0, \theta_i)$ or $s(\psi_0, \psi_i)$, since s has meaning only in relation to an initial state of the medium and an imposed boundary condition. Philip also defined an *intrinsic sorptivity* — a parameter that accounts for the viscosity and surface tension of the fluid.

It should be obvious from the foregoing that the effects of ponding depth and initial wetness (Fig. 14.6) can be significant during early stages of infiltration but diminish in time and eventually tend to vanish in a very deeply wetted profile. Typical values of the "final" (steady) infiltration rate are shown in Table 14.1. In actual cases the infiltration rate can be considerably higher, particularly during the initial stages of the process and in well-aggregated or cracked soils, or lower, as in the presence of a surface crust or of an impeding layer inside the profile.

Sample Problem

A horizontal infiltration trial was conducted with a soil-filled tube having a cross-sectional area of 5×10^3 mm^2. After 15 min, cumulative infiltration totaled 1.5×10^5 mm^3. What is the expectable cumulative infiltration and infiltration rate after 1, 4, and 16 hr?

We use Philip's horizontal infiltration ("absorption") theory to calculate the sorptivity s according to Eq. (14.19),

$$s = I/t^{1/2}$$

where t is time and I is cumulative infiltration:

$$s = (1.5 \times 10^5 \text{ mm}^3/5 \times 10^3 \text{ mm}^2)/(15 \text{ min} \times 60 \text{ sec/min})^{1/2}$$
$$= 30/30 = 1 \text{ mm/sec}^{1/2}$$

We now apply this value to estimate cumulative infiltration at various times, again using Eq. (14.19): ($I = st^{1/2}$):
After 1 hr:

$$I = 1 \text{ mm/sec}^{1/2} \times (3600 \text{ sec})^{1/2} = 60 \text{ mm}$$

After 4 hr:

$$I = 1 \text{ mm/sec}^{1/2} \times (14{,}400 \text{ sec})^{1/2} = 120 \text{ mm}$$

After 16 hr:

$$I = 1 \text{ mm/sec}^{1/2} \times (57{,}600 \text{ sec})^{1/2} = 240 \text{ mm}$$

The corresponding infiltration rates are obtainable by differentiating the Eq. (14.19) with respect to time t:

$$I = dI/dt = s/2t^{1/2}$$

After 1 hr:

$$i = (1 \text{ mm/sec}^{1/2})/2(3600 \text{ sec})^{1/2} = 8.33 \times 10^{-3} \text{ mm/sec} = 30 \text{ mm/hr}$$

After 4 hr:

$$i = (1 \text{ mm/sec}^{1/2})/2(14{,}400 \text{ sec})^{1/2} = 4.166 \times 10^{-3} \text{ mm/sec} = 15 \text{ mm/hr}$$

After 16 hr:

$$i = (1 \text{ mm/sec}^{1/2})/2(57{,}600 \text{ sec})^{1/2} = 2.0833 \times 10^{-3} \text{ mm/sec} = 7.5 \text{ mm/hr}$$

Now assume that the same column is vertical. If the effective saturated hydraulic conductivity of the soil is 2×10^{-3} mm/sec, estimate the cumulative infiltration and infiltration rate values at 1, 4, and 16 hr. If the initial volumetric wetness is 0.05 (5%) and the saturation wetness is 0.45, estimate the depth of the wetting front at the same times.

We use Eq. (14.32) for cumulative infiltration in a vertical column ($I = st^{1/2} + Kt$):
After 1 hr (3600 sec):

$$I = 1 \text{ mm/sec}^{1/2} \times 60 \text{ sec}^{1/2} + 2 \times 10^{-3} \text{ mm/sec} \times 3600 \text{ sec}$$
$$= 60 \text{ mm} + 7.2 \text{ mm} = 67 \text{ mm}$$

After 4 hr (14,400 sec):

$$I = 1 \text{ mm/sec}^{1/2} \times 120 \text{ sec}^{1/2} + 2 \times 10^{-3} \text{ mm/sec} \times 14{,}400 \text{ sec}$$
$$= 120 \text{ mm} + 28.8 \text{ mm} = 148.8 \text{ mm}$$

After 16 hr (57,600 sec):

$$I = 1 \text{ mm/sec}^{1/2} \times 240 \text{ sec}^{1/2} + 2 \times 10^{-3} \text{ mm/sec} \times 57{,}600 \text{ sec}$$
$$= 240 \text{ mm} + 115.2 \text{ mm} = 355.2 \text{ mm}$$

To calculate the infiltration rate into a vertical column, we differentiate Eq. (14.32) with respect to time:

$$i = dI/dt = s/2t^{1/2} + K$$

After 1 hr:

$$i = (1 \text{ mm/sec}^{1/2})/(2 \times 60 \text{ sec}^{1/2}) + 2 \times 10^{-3} \text{ mm/sec}$$
$$= 8.33 \times 10^{-3} \text{ mm/sec} + 2 \times 10^{-3} \text{ mm/sec} = 10.33 \times 10^{-3} \text{ mm/sec}$$
$$= 37.2 \text{ mm/hr}$$

After 4 hr:

$$i = (1 \text{ mm/sec}^{1/2})/(2 \times 120 \text{ sec}^{1/2}) + 2 \times 10^{-3} \text{ mm/sec}$$
$$= 4.166 \times 10^{-3} \text{ mm/sec} + 2 \times 10^{-3} \text{ mm/sec} = 6.166 \times 10^{-3} \text{ mm/sec}$$
$$= 22.2 \text{ mm/hr}$$

After 16 hr:

$$i = (1 \text{ mm/sec}^{1/2})/(2 \times 240 \text{ sec}^{1/2}) + 2 \times 10^{-3} \text{ mm/sec}$$
$$= 2.083 \times 10^{-3} \text{ mm/sec} + 2 \times 10^{-3} \text{ mm/sec} = 4.083 \times 10^{-3} \text{ mm/sec}$$
$$= 14.7 \text{ mm/hr}$$

To calculate depth of the wetting front L_f we use a variant of Eq. (14.3) ($L_f = I/\Delta\theta$):
After 1 hr:

$$L_f = (67.2 \text{ mm})/(0.45 - 0.05) = 168 \text{ mm}$$

After 4 hr:

$$L_f = (148.8 \text{ mm}/0.4 = 372 \text{ mm}$$

After 16 hr:

$$L_f = (355.2 \text{ mm})/0.4 = 888 \text{ mm}$$

Note: The student is invited, to plot the calculated values of I, i, and L_f against time and against the square root of time. The relationships involved will become clearer in the process.

INFILTRATION INTO LAYERED PROFILES

Experimental studies have shown that the infiltration process can be greatly affected by soil-profile heterogeneity. Although in any conducting soil the matric suction and hydraulic head must be continuous throughout the profile regardless of layering sequence, the wetness and conductivity may exhibit abrupt discontinuities at interlayer boundaries.

The advent of computer-based numerical methods for solving nonlinear partial differential equations subject to complex boundary conditions made possible a flexible approach to infiltration theory. No longer is the theory shackled to the restrictive assumptions that had been necessary previously in order to formulate the process mathematically in closed form, as required for the attainment of analytical solutions.

An early demonstration that the computer can be used effectively to obtain a mathematical description of infiltration in heterogeneous soil profiles with strata of differing properties and initial wetness values was given by Hanks and Bowers (1962). In their model the initial wetness (θ_i) was allowed to vary with depth z, and different $K(\psi)$ and $\theta(\psi)$ functions were assigned to soil layers of different thicknesses.

Later, F. C. Wang and Lakshminarayana (1968) modified the formulation of infiltration to make more explicit the expression of heterogeneity, that is, the dependence of K and ψ on z and θ in the form $K = K(\theta, z)$ and $\psi = \psi(\theta, z)$. Both models were still based on the surface-ponding boundary condition. Subsequently, investigations have provided even more comprehensive descriptions of infiltration into variously constituted soil profiles (e.g., Hillel and Talpaz, 1977; Hillel, 1977; Ross, 1990).

One typical situation is that of a coarse layer of higher saturated hydraulic conductivity overlying a less conductive finer-textured layer. In such a case the infiltration rate is at first controlled by the more conductive top layer, but when the wetting front reaches and penetrates into the finer-textured sublayer, the infiltration rate can be expected to drop and tend to that of the deeper soil alone. Thus, in the long run it is the layer of lesser conductivity that controls the process. If infiltration continues for long, then positive pressure heads (a "perched water table") can develop in the top layer, just above its boundary with the impeding finer layer.

In the opposite case — infiltration into a profile with a fine-textured layer overlying a coarse-textured one — the initial infiltration rate is again determined by the upper layer. As water reaches the interface with the coarse lower layer, however, the infiltration rate typically decreases. Water at the wetting front is normally under suction, and this suction may be too high to permit ready entry into the relatively large pores of the coarse layer. The wetting front often pauses at the interface, until the pressure of the water just above the interface builds up sufficiently to enter the sublayer.

INFILTRATION INTO CRUST-TOPPED SOILS

A very important special case of a layered soil is that of a profile that develops a thin, less permeable "crust," or "seal," at the surface. Such a seal can form over a bare soil (devoid of a protective cover of vegetation or stubble) under the beating action of raindrops or as a result of the spontaneous slaking and breakdown of soil aggregates during wetting of the surface.

The impact of a raindrop on an exposed surface is related to its kinetic energy E_k:

$$E_k = mv^2/2 \qquad (14.34)$$

Here m is raindrop mass and v is its velocity. The total action of a rainstorm might be expected therefore to be a function of the sum of the kinetic energies of all the drops:

$$E_{k,tot} = \sum m_i v_i^2/2 \qquad (14.35)$$

wherein m_i, v_i are, respectively, the masses and velocities of raindrops of successive size groups.

As a raindrop of any size falls earthward through the atmosphere, it is accelerated by gravity but encounters the viscous resistance of the atmosphere.

That resistance increases as the velocity of the falling drop increases. When the air resistance equals the downward force of gravity, acceleration ceases and the drop continues to fall at a constant, "terminal" velocity that depends on its mass. The spectrum of drop sizes, hence also of terminal velocities, varies from storm to storm and from place to place.

The impact of raindrops on the soil is influenced by surface roughness, slope, incident angle, and the possible presence of standing water on the surface. It is also influenced by the structure and wetness of the exposed soil; hence the impact varies with time, being maximal at the early stages of a rainstorm, when the soil surface is at its driest state and most vulnerable to slaking. The destructive impact of rainfall on soil structure can be mitigated by the presence of a protective cover of vegetation or a mulch that can intercept the drops before they strike the soil directly.

Another important factor affecting the formation of a surface seal (crust) is the chemical composition of the infiltrating solution. Irrigation water with a high sodium adsorption ratio (SAR), coupled with a low overall concentration of salts, can induce dispersion and swelling of the clay present in the soil's surface layer, which in turn can have a strong effect on soil infiltrability. The dispersed aggregates collapse and close the interaggregate cavities, and migrating clay particles tend to lodge in soil pores beneath the surface.

The effect of sodium in the added water is strongest at the soil surface because of the mechanical agitation of the applied water and because of the absence of a confining matrix. Extreme clay dispersion and clogging can occur at SAR values as low as 10 if the solution concentration is lower than 2 mol/m^3 (Oster and Schroer, 1979). Particularly destructive to surface soil structure is the alternation of sodic irrigation water with rainfall, which results in the drastic reduction of soil infiltrability.

Surface crusts are characterized by greater bulk density, finer pores, and lower saturated conductivity than the underlying soil. Once formed, a surface crust can greatly impede water intake by the soil (Fig. 14.7), even if the crust is quite thin (say, not more than several millimeters in thickness) and the soil is otherwise highly permeable. Failure to account for the formation of a crust can result in gross overestimation of infiltration.

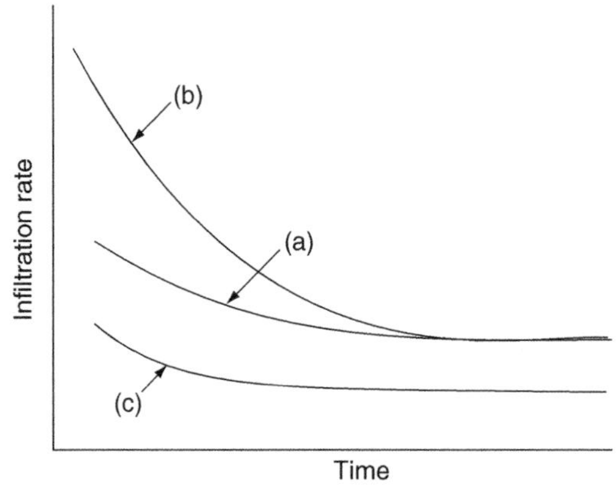

Fig. 14.7. Infiltrability as a function of time in (a) a uniform soil, (b) a soil with a more permeable upper layer, and (c) a soil with a surface crust.

An analysis of the effect of a developing surface crust on infiltration was carried out by W. M. Edwards and Larson (1970). Hillel (1964) and Hillel and Gardner (1969, 1970a) used a quasi-analytical approach to calculate fluxes during steady and transient infiltration into crust-capped profiles from knowledge of the basic hydraulic properties of the crust and of the underlying soil. Further developments of infiltration theory for crusted soils have been published by Ahuja and Swartzendruber (1992).

INSTABILITY OF WETTING FRONTS DURING INFILTRATION

An interesting and still incompletely understood phenomenon is that of *unstable flow*, particularly *wetting-front instability*, which occurs most notably in the transition of infiltrating water from a fine-textured top layer to a coarse-textured sublayer. Rather than advance as a smooth front, the percolating water may concentrate at certain points to break into the sublayer in the form of fingerlike or tonguelike protrusions. Flow through the coarse-textured layer thereafter occurs through "chimneys" or "pipes" traversing the layer rather than through the entire layer uniformly (Fig. 14.8).

Unstable flow can allow small "outlaw" streams of water to drain out of sites such as "sanitary" landfills and industrial dumps, and carry raw pollutants, including toxic products, into groundwater while evading the filtration and purification mechanisms of the soil's greater volume.

The existence of instability at the moving interface between two immiscible fluids of different densities or viscosities (such as oil and water) has long been recognized (Wooding, 1969). Interfacial instability can also occur between two miscible fluids, as in the penetration of a saline solution into a porous medium

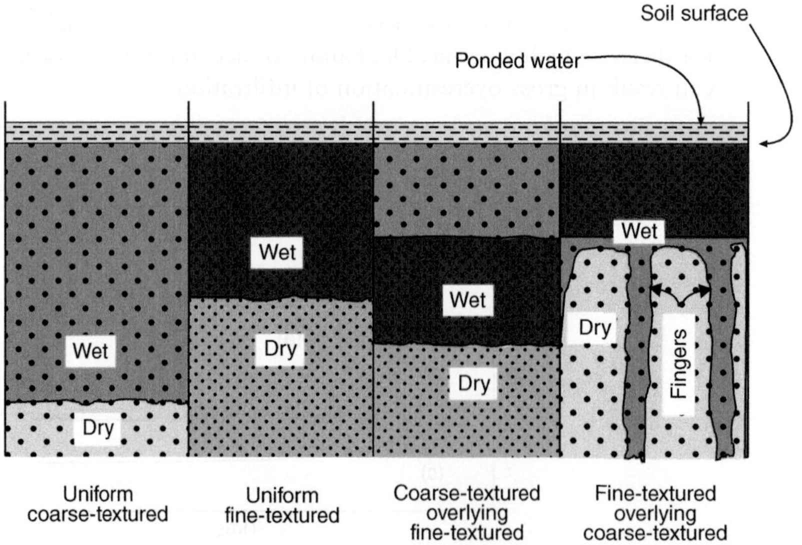

Fig. 14.8. Appearance of "fingers" during infiltration into a layered profile, with a fine-textured layer overlying a coarse-textured layer.

initially filled with fresh water. However, the occurrence of instability during the flow of water in unsaturated soil has caught the attention of soil physicists only in the last few decades.

The earliest reference to the phenomenon was apparently made by Tabuchi (1961) and by D. E. Miller and Gardner (1962). Bridge and Collis-George (1973) reported a similar occurrence in their experiments of infiltration into a two-layer profile of fine sand overlying coarse sand. Peck (1965) observed instability patterns during infiltration into an unsaturated sand where the air phase was confined so that its pressure increased as the wetting front moved downwards.

A convincing demonstration of this effect at the transition from a layer of fine-textured material into an underlying layer of coarser material (of larger pores) was published by E. D. Hill and Parlange (1972). They postulated that the formation of fingers in this case is due to an instability at an air–water interface, which is gravity driven and is more pronounced as the pore sizes of the coarse sublayer increases. In these fingers, the infiltrating water moved in saturated columns surrounded by an unsaturated cylinder of partially wetted soil. The overall flow rate through the system was governed by the number of such fingers per unit total area of interface, that is, by the fraction of the soil volume occupied by the conducting fingers. Their approach was followed by a series of laboratory tests by Glass et al. (1988, 1989).

The intriguing question is thus why any perturbations that might appear in the wetting front are in some cases flattened out by lateral diffusion so that the front remains stable, while in other cases they become self-generating fingers that destabilize the wetting front.

A general explanation was attempted by Raats (1973), who pointed out that profile layering is not a prerequisite for wetting-front instability, which may also be due to increased air pressure ahead of the wetting front or to water repellency of soil layers. Raats based his analysis on the Green and Ampt model (described earlier in this chapter) and offered the hypothesis that stability depends on the change of velocity u of the wetting front with distance L: A chance perturbation in an initially flat wetting front will tend to grow if u increases with L and will tend to disappear if u decreases with L.

In the case of a surface crust, Raats stated the following criterion for the onset of instability:

$$\Omega > (L - h_c)/K_L \tag{14.36}$$

in which Ω is the hydraulic resistance of the crust (its thickness divided by its hydraulic conductivity), L is the depth to the wetting front, h_c is the pressure head at the wetting front, and K_L is the hydraulic conductivity of the subcrust soil. Accordingly, instability is favored by a high crust resistance, a small value of $-h_c$ (i.e., a low suction, or high antecedent wetness, of the subcrust soil), and a large value of K_L. In the general cases of layered profiles, the wetting front can become unstable if

$$K_{st} < K_{sb}[(d + H_b + H_c - H_a)/L + 1] \tag{14.37}$$

where d is the depth of ponded water, H_b is barometric pressure, H_c is the capillary pressure head at the wetting front, H_a is the soil air pressure head, L is the soil depth, and subscripts t and b refer to the top and bottom layers, respectively.

The problem was subsequently reassessed by Philip (1975), who applied the methods of hydrodynamic stability analysis (C. C. Lin, 1955). He used the same Green and Ampt (delta-function) model as did Raats, but proposed a different criterion of stability, stating that "the physical attribute of the system which is fundamental to the question of stability is the water pressure gradient behind the wetting front. When this gradient assists the flow, the flow is stable; when this gradient opposes it, the flow is unstable."

In examining the implications of his criteria for particular cases, Philip suggested that flow can become unstable if (1) the wetting front reaches a stratum so water repellent that the pressure head needed to penetrate it is positive; (2) air entrapment and compression ahead of the wetting front cause a sufficient rise in air pressure there; (3) the air-entry pressure is more negative than the wetting-front moisture potential during redistribution; and (4) hydraulic conductivity increases in depth at a sufficient rate. In all the various instances of instability that Philip identified, it is essential that the flow be gravity assisted.

An analysis by Hillel and Baker (1988), later tested by R. S. Baker and Hillel (1990), emphasized the importance of the *suction of water entry* (SWE) into an initially dry sublayer. This parameter had been ignored in previous treatments of unstable flow. Because SWE determines the effective hydraulic conductivity of the sublayer, it also determines whether or not flow is accelerated with distance down the profile. When the wetting front reaches the interlayer boundary, it pauses until the suction there falls (or the pressure builds up) sufficiently to allow entry of water into the large pores of the sublayer. If at this point the conductivity of the sublayer exceeds that of the top layer, the sublayer cannot conduct throughout its entire volume, since it would then be conducting more water than it is receiving (an obvious impossibility). Hence the spatially distributed flow field with its parallel streamlines must begin to constrict at the critical depth (typically, at or just below the transition from the less permeable top layer to a more permeable sublayer), thus causing the streamlines to converge. In the case of a wide flow field, this forms spatially separated, partial-volume, flow paths (fingers).

The theory by Hillel and Baker also predicts the fractional cross-sectional area (F) of fingers that form under specified sets of conditions. In the general case;

$$F = [K_t/K_u(S_e)][(H_0 + Z_i + S_e)/Z_i] \tag{14.38}$$

Here K_t is the saturated hydraulic conductivity of the less permeable top layer, K_u is the unsaturated hydraulic conductivity of the sublayer (a function of the matric suction of water entry into the sublayer, S_e), H_0 is the hydraulic head imposed on the surface by the ponded water, and Z_i is the vertical thickness of the top layer (depth of the interlayer boundary below the surface).

In the extreme case, where the sublayer is of very coarse texture whose water entry value is effectively zero suction (i.e., a layer having such large pores or being so water repellent that it can only be penetrated by water under positive atmospheric pressure), and if the ponding depth at the surface is negligible, we get:

$$F = K_{st}/K_{sb} \tag{14.39}$$

Thus, for example, if the saturated conductivity of the sublayer is an order of magnitude (×10) greater than that of the top layer, then we may expect the fingers to occupy some 10% of the cross-sectional area of the sublayer; if it is two orders of magnitude greater, then the fingers will only occupy some 1% of the soil volume. In such conditions detecting vadose-zone fingering in the field by random sampling from above would be extremely difficult. (The hydrological term *vadose zone* refers to the entire volume of unsaturated but water-conducting porous material — including the soil and the strata below it — that is above the water table. In other words, it is the *unsaturated zone* that overlies the *saturated zone*.)

Evidence from the field reveals that unstable flow is not a mere curiosity, of interest only to theoreticians. Hendrickx et al. (1988b), who conducted a bromide tracer trial in the Netherlands, found that after 5 weeks with 120 mm rainfall, the bromide concentration in the groundwater was 13 times higher when an unstable wetting front had formed than could be expected with a stable wetting front. In Wisconsin, Kung (1990) found that water and solutes flowed through less than 10% of the soil volume at the 3.0- to 3.6-m-depth layer.

OTHER FORMS OF PREFERENTIAL FLOW

The phenomenon of fingering, described earlier, is only one of several possible forms of *preferred pathway flow*, also known as *preferential flow*, in the soil. In contrast with the mode of *distributed flow*, in which water permeates the entire porous network of the soil matrix and moves through its entire volume, preferential flow occurs via distinct pathways that constitute only a fraction (sometimes only a small fraction) of the soil's total volume.

Most readily evident are flow patterns that follow preexisting features in the soil profile, such as clay or sand lenses, cavities, and fissures. This last may be formed either physically by the shrinking and cracking of clay or biologically by burrowing animals (such as earthworms, ants, and rodents) or decaying roots. Finally, human manipulation of the land (including certain types of tillage, such as subsoiling, or chiseling) may form cracks that, in turn, constitute pathways for preferential flow.

A fairly common case is that of a clayey soil that shrinks on drying and forms cracks (or *macropores*) that begin at the surface and extend to some depth in the profile. When such a soil is subsequently subjected to surface flooding or to a heavy downpour of rain, water under positive pressure immediately penetrates the cracks and quickly fills their volume, known as *macropore storage*. Thereafter, infiltration takes place from the water-filled cracks sideways into the soil blocks between the cracks, as well as downward from the bottoms of the cracks and from the soil surface (Fig. 14.9). The process is therefore multidimensional, and its pattern depends on the geometry of the cracks, the conductivity (or sorptivity) of the soil matrix, as well as the mode and duration of wetting.

During the initial period of infiltration, the intake rate of water by cracked soils can be extremely high (Mitchell and van Genuchten, 1991) — much higher than the saturated conductivity of the soil matrix. However, if the infiltration process continues long enough, the influence of the surface cracks eventually diminishes as the wetting front penetrates deeper into the profile, well beyond the reach of the cracks, and as the cracks themselves tend to close due to the swelling of clay under prolonged imbibition.

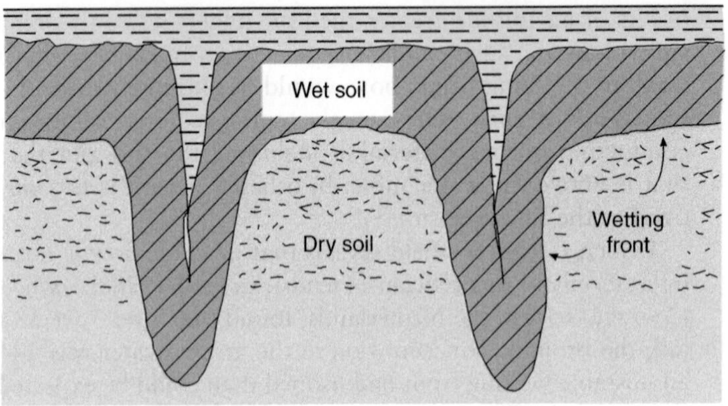

Fig. 14.9. Shape of the wetting front during ponded infiltration into a soil with open cracks.

The role of macropores in soil-water dynamics was described by Beven and Germann (1982). They pointed out that flow in the two domains (the macropores between soil blocks and the "micropores" within soil blocks) is governed by different potential gradients. They also showed that the sizes and the structures of the macropores determine the pattern of water flow. Accordingly, they classified macropores into four groups, on the basis of their morphology:

1. *Pores formed by burrowing animals* (e.g., moles, gophers, and wombats): Such pores are usually tubular and may range in diameter from 1 mm to over 50 mm.
2. *Pores formed by plants roots*: These pores are also tubular, and their sizes and extensions depend on the type of plant. After the roots have decayed, these root channels remain for a time as hollow passages. Even in the case of living roots, preferential flow may take place along the soil–root interface (Gish and Jury, 1983).
3. *Cracks and fissures*: These are formed by shrinkage of clay or by chemical weathering of the bedrock. They may also be formed by freeze–thaw cycles and such methods of cultivation as subsoiling.
4. *Natural soil "pipes"*: These can form under the erosive action of subsurface flows in highly permeable noncohesive soil materials when subjected to strong hydraulic gradients.

W. M. Edwards et al. (1988) further clarified the influences of the size, frequency, vertical and lateral distribution, and continuity of macropores on the hydrological characteristics of soil profiles. They found that worm holes may extend to 1 meter below the surface and can be quite durable.

Definitive work on infiltration processes in soils with macropores was carried out by Bouma (1981), Germann and Beven (1985), Beven and Clark (1986), and Dagan (1986). Modeling studies were reported by Hoogmoed and Bouma (1980), and Huyakorn et al. (1983, 1985), among others. Preferential flow has obvious implications regarding solute transport in the soil (Germann et al., 1984; van Genuchten et al., 1984; Berkowitz et al., 1988; Steenhuis et al., 1990).

RAIN INFILTRATION

When rain or sprinkling intensity exceeds soil infiltrability, free water appears on the soil surface. In principle, the infiltration process should then be similar to that in the case of shallow ponding. If rain intensity is less than the initial infiltrability value of the soil but greater than the final value, then at first the soil will absorb at less than its potential rate and the flow of water in the soil will occur under unsaturated conditions; however, if the rain is continued at the same intensity and as soil infiltrability declines, the soil surface will eventually become saturated. Thenceforth the process will continue as in the case of ponding infiltration. Finally, if rain intensity is at all times lower than the infiltrability, the soil will continue to absorb the water as fast as it is applied without ever attaining saturation. After a long time, as the suction gradients become negligible, the wetted profile will acquire a wetness for which the conductivity is equal to the water supply rate, and the lower this rate, the lower the degree of saturation of the infiltrating profile. This effect is illustrated in Fig. 14.10.

The process of infiltration under rain or sprinkler irrigation was studied by Youngs (1964), and by Rubin (1966). This last author, who used a numerical solution of the flow equation for conditions pertinent to this problem, recognized three modes of infiltration due to rainfall: (1) *nonponding infiltration*, involving rain not intense enough to produce ponding, (2) *preponding infiltration*, due to rain that has not yet but can eventually produce ponding, and (3) *rainpond infiltration*, characterized by the presence of ponded water.

Rainpond infiltration is usually preceded by preponding infiltration, the transition between the two being called *incipient ponding*. Thus, nonponding

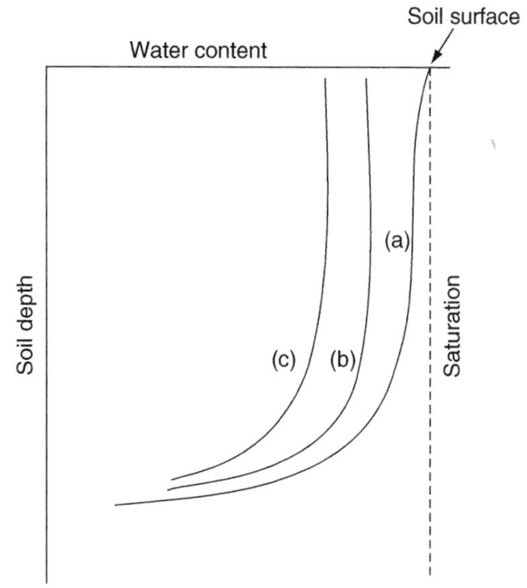

Fig. 14.10. The water-content distribution profile during infiltration (a) under ponding, (b) under sprinkling at relatively high intensity, and (c) under sprinkling at a very low intensity.

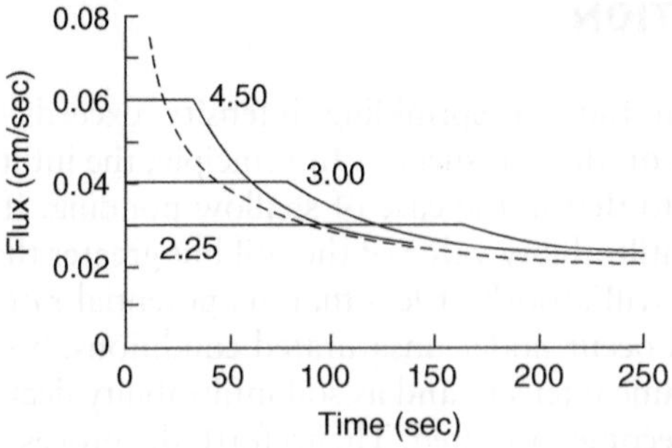

Fig. 14.11. Relation between surface flux and time during infiltration into Rehovot sand due to rainfall (solid lines) and flooding (dashed line). The numbers labeling the curves indicate the magnitude of the relative rain intensity (r/K_s). (After Rubin, 1966.)

and preponding infiltration rates are dictated by rain intensity and are therefore *supply controlled* (or *flux controlled*), whereas rainpond infiltration rate is determined by the pressure (or depth) of water above the soil surface as well as by the suction conditions and conductivity relations of the soil. Where the positive pressure of water at the soil surface is small (not substantially exceeding atmospheric pressure), rainpond infiltration, like ponding infiltration in general, is *profile controlled.*

In the analysis of rainpond or ponding infiltration, the surface boundary condition generally assumed is that of a constant pressure at the surface. On the other hand, in the analysis of nonponding and preponding infiltration, the water flux through the surface is considered to be equal either to the rainfall rate or to the soil's infiltrability, whichever is the lesser. In actual field conditions. rain intensity might increase and decrease alternately, at times exceeding the soil's saturated conductivity (and its infiltrability) and at other times dropping below it. Since periods of decreasing rain intensity involve complicated hysteresis phenomena (Hillel, 1980b), the analysis of variable-intensity rainstorms is rather difficult. However, Rubin's analysis was based on the assumption of no hysteresis.

Figure 14.11 describes infiltration rates into a sandy soil during preponding and rainpond infiltration under three rain intensities. The horizontal part of each curve corresponds to preponding infiltration, and the descending parts to rainpond infiltration periods. As pointed out by Rubin (1966), the rainpond infiltration curves are of the same general shape and approach the same limiting infiltration rate. However, they do not constitute horizontally displaced parts of a single curve and do not coincide with the infiltration rate under flooding (shown as a dashed line in Fig. 14.11).

15. SURFACE RUNOFF AND WATER EROSION

SURFACE WATER EXCESS

Whenever the rate of water supply to the soil surface exceeds the rate of infiltration, free water, called *surface water excess*, tends to accumulate over the soil surface. Where the surface is not perfectly flat and smooth, this excess water collects in depressions, forming puddles. The total volume of water thus held, per unit area is called the *surface storage capacity*. It depends on the geometric irregularities of the surface as well as on the overall slope of the land (Fig. 15.1). Only when surface storage is filled and the puddles begin to overflow can actual runoff begin. The term *surface runoff* thus represents the portion of the water supply to the surface that is neither absorbed by the soil nor retained on its surface but that runs downslope. Surface runoff typically begins as sheet flow, but as it accelerates and gains in erosive power, it eventually forms channels variously called *rills* or *gullies*. Running water tends to scour the soil surface and deepen those channels, which generally form a treelike pattern of converging branches with numerous confluences leading to larger streams.

There exists a wide spectrum of flow patterns. On the one extreme is the thin, sheetlike runoff called *overland flow*. It is likely to be the primary type in surface runoff from small natural areas or fields having little topographic relief. The next distinctive type is found in the smallest stream channels, which gather the overland flow in a continuous fashion along their length to form the lowest order of *stream flow*. As these small streams merge with one another, they form channels of higher order, which collect converging tributaries as well as continuous lateral inflows.

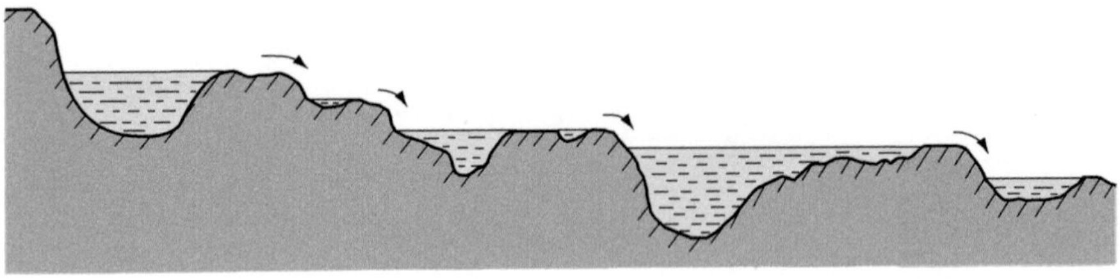

Fig. 15.1. Effect of surface roughness and slope on surface (pocket) storage of rainfall excess. Most of the water thus accumulated generally infiltrates (and some evaporates) after cessation of the rain.

OVERLAND FLOW

Overland flow is considered laminar flow in the initial condition. When the depth and velocity of the flow increase to critical values, turbulence entering the laminar flow domain will not dampen out. Vennard (1961) specified a Reynolds number of 500 as a criterion for the onset of turbulence. (Recall that the *Reynolds number* is a dimensionless index incorporating the product of the mean fluid velocity in a channel and its effective hydraulic radius, divided by its kinematic viscosity.)

To obtain a simplified formulation of overland flow, we begin with the basic equation governing the rate of flow (discharge) in a channel:

$$Q = vA \tag{15.1}$$

where Q is the flow rate, v is the average velocity, and A is the cross-sectional area of flow.

Conservation of matter (continuity) requires that, assuming no rainfall and infiltration, the change in water level h with time be equal to the negative change of flow rate with distance:

$$\partial h/\partial t = -\partial Q/\partial x \tag{15.2}$$

Taking into account the time rates of rainfall (r) and of infiltration (i), we have

$$\partial h/\partial t + \partial Q/\partial x = r - i \tag{15.3}$$

Mean velocity and cross-sectional area of flow depend on the shape and size of the channel. Since flow is constrained by shear stresses resulting from the immobility of the water in immediate contact with the sides and bottom of the channel (assuming at this stage that these boundaries are fixed in space), the hydraulic resistance of a channel is generally proportional to the area of the perimeter of contact per unit volume of flowing liquid. This contact area can be characterized in terms of the *hydraulic radius*, or *hydraulic mean depth*, of a channel, defined as the cross-sectional area divided by the wetted perimeter.

Now consider an infinitely wide channel without sides. Such a "wide-open channel" is represented by a laterally uniform sloping surface with a layer of water flowing over it, as illustrated in Fig. 15.2. Here the effective hydraulic radius is h. As an approximation, let us assume that the two forces acting on the flowing water are (1) the component of the water's weight acting in the direction of the bed slope, and (2) the shear stresses developed at

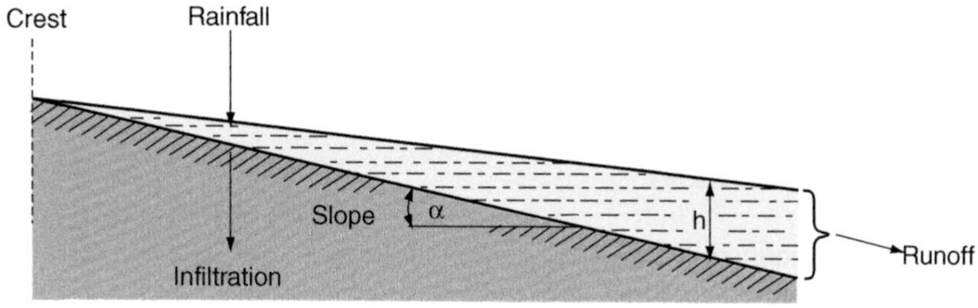

Fig. 15.2. Surface-water excess and sheet overland flow.

the solid-to-water boundary. If the velocity is more or less uniform along the slope, then the two forces must be approximately balanced, and we get

$$\tau_0 = \rho g h \alpha \qquad (15.4)$$

where ρ is the liquid's density, g is the acceleration of gravity, h is the water depth, α is the slope, and τ_0 is the bottom shear stress.

Shear stress τ_0 is known to be related to the average velocity squared:

$$\tau_0 = a\rho v^2 \qquad (15.5)$$

where a is a proportionality factor. Combining the last two equations gives

$$v = [(g/a)h\alpha]^{1/2} = Ch^{1/2}\alpha^{1/2} \qquad (15.6)$$

where C is the *Chezy coefficient*, named after the French engineer who first discovered this relationship two centuries ago. A large amount of empirical work conducted since then has shown that Chezy's coefficient C is not a constant but depends on water depth h and surface roughness n as follows:

$$C = h^{1/6}/n \qquad (15.7)$$

Combining this with Eq. (15.1) gives the equation usually attributed to Robert Manning (1891):

$$V = h^{2/3}\alpha^{1/2}/n \qquad (15.8)$$

Multiplying by h converts the left side of this equation to discharge per unit width:

$$Q = h^{5/3}\alpha^{1/2}/n \approx h^2\alpha^{1/2}/n \qquad (15.9)$$

In practice, a power of 2 for h rather than 5/3 has been found to give satisfactory results for overland flow. Measured values of the roughness coefficient n have been tabulated (e.g., Sellin, 1969) and range from 0.01 for extremely smooth channels to 0.1 for flood plains with a dense growth of timber. A value of 0.03 seems to be reasonable for soil surfaces (either bare or with a cover of short grass).

The last version of *Manning's equation* can be combined with Eq. (15.3) to yield the *kinematic wave equation*:

$$\partial h/\partial t + \beta h^{m-1} (\partial h/\partial x) = r - i \qquad (15.10)$$

where $\beta = \alpha^{1/2}/n$ and $m = 2$.

The foregoing set of equations was used by Hillel and Hornberger (1979) to model overland flow from texturally heterogeneous fields.

RUNOFF INDUCEMENT

Uncontrolled runoff is never desirable, and its control requires additional investment in most agricultural situations. In *humid regions*, where rainfall may be excessive, surface drainage is often necessary, and the land must be shaped and managed so as to direct the runoff via protected (grassed or concrete-lined) channels. In *semiarid regions*, by way of contrast, natural rainfall is on the margin of being insufficient for dryland farming, hence farmers take measures to minimize runoff so that as much of the precipitation as possible infiltrates into the soil.

However, in many *arid regions* large tracts of land are unusable for conventional rainfed farming, owing to insufficient or unstable rainfall, nonarable soils (too shallow, stony, or saline), or unsuitable topography. From the point of view of farmers (though not of ecologists, who are concerned over an area's natural biota rather than over crop production per se), rain falling on such land is almost totally lost. Often rainfall amounts per storm are insufficient either to recharge the groundwater or to support an economically viable crop. This water generally infiltrates to a shallow depth only and is quickly returned to the atmosphere either by direct evaporation or by transpiration of ephemeral vegetation. Some of this water may also appear as runoff, causing flash floods typical of the desert, although natural runoff seldom exceeds about 10% of annual precipitation (Hillel, 1982). However, when it is feasible, the possibility of controlling and increasing the amount of surface runoff obtainable from limited sections of such lands can be of great importance, particularly where water is scarce and the runoff thus obtained can augment the water supply of an inhabited region.

The art of collecting and utilizing runoff as been practiced by desert-dwelling societies since antiquity. Remnants of extensive runoff farming systems are found in the American Southwest. In desert areas of the Middle East, ancient systems of terraced stream beds were used for growing crops, whose water supply was augmented by runoff water collected from the adjoining hillsides and brought to the fields by means of sophisticated conveyance systems (Hillel, 1982). There is evidence that the builders of these systems understood the inadequacy of natural runoff and therefore perfected techniques for inducing soil crusting so as to increase runoff. Noticing that the soil had a natural tendency to crust, which was obstructed, however, by the desert's natural cover of gravel (commonly known as "desert pavement"), the ancient inhabitants of the Negev region in Israel raked the stones off the surface in order to expose the finer soil material and induce the formation of a dense crust (Tadmor et al., 1957). Even so, the ancient runoff farmers needed a water-contributing area 20 times or more larger than the area to which water was brought for crop production. Similar techniques were used to collect runoff in cisterns for subsequent use as drinking water for humans and domestic animals.

Modern technology holds the promise of more efficient runoff inducement than was possible in ancient times. Runoff can now be increased severalfold by means of mechanical treatments (stone clearing, smoothing, and compaction) as well as by a variety of chemical treatments to seal and stabilize the surface. The soil surface can be made water repellent or impermeable either by the application of mechanical barriers to water movement (such as plastic, aluminum foil, concrete, or sheet metal) or by the artificial formation of an impervious soil crust with the aid of various materials, such as sodic salts, sprayable asphalt emulsions, and silicone water-repellents (Hillel, 1982).

Several systems have been tried with respect to size and arrangement of the contributing area in relation to the water-receiving area (Shanan and Tadmor, 1976; Hillel, 1982). A small watershed can be treated in its entirety so as to provide the maximal amount of water at the outflow of the basin for conveyance to irrigated fields. On a still smaller scale, strips of land can be treated on a slope so as to contribute their share of rainfall as runoff to adjacent lower-lying "run-on" strips in which crops can he grown (Hillel, 1994). A third possible approach is to form *microwatersheds*, wherein each contributing area serves a single tree or row of plants (Shanan et al., 1970: Fairbourn and Kemper, 1971).

International development agencies are engaged in the effort to apply these ancient methods to modern farming communities in arid and semiarid regions (FAO, 1989).

SOIL EROSION BY WATER

Agricultural cultivation can have various negative effects on soils, insofar as it may set in motion processes of degradation, such as accelerated decomposition of unreplenished organic matter, depletion of nutrients, breakdown of aggregates and of structural stability, crusting, compaction, waterlogging, salination, and erosion by water and wind. Of these processes, one of the most widespread is soil erosion. It results not only in loss of soil productivity on sites where it occurs but also in the off-site deposition of sediments that may pollute surface and underground water resources as well as clog streams, reservoirs, and estuaries. The erosion process is indeed a major transport mechanism for agricultural chemicals, significant amounts of which do not remain where they are applied but spread into the larger environment. The main culprit process is erosion by water, though in arid conditions erosion by wind may be equally insidious.

Geologic erosion is a natural and inevitable process that takes place continually and universally. This slow and inexorable process is, however, greatly accelerated by human activity. Fundamentally, agriculture consists of removing an area's natural vegetation and introducing in its stead a selection of domesticated and specialized plants, called *crops*. The very act of eradicating the native vegetation and of baring and pulverizing the soil so as to make it receptive to planted crops makes the soil much more vulnerable to erosion. Most vulnerable is the soil's top layer, which is often the soil's most fertile layer (being richest in nutrient-releasing plant and animal residues). The loss of topsoil requires farmers to use greater amounts of chemical fertilizers to compensate for the reduced natural fertility, and the increased application of such chemicals itself exacerbates the hazard of groundwater and surface water pollution.

Soil erosion is a three-stage process (D. F. Post, 1996): (1) detachment of particles from the soil body that is their original domain; (2) transport of the detached soil particles by flowing water or by wind; (3) deposition of the transported particles, a stage known as *sedimentation*.

In water erosion, detachment of soil particles generally occurs under the impact of striking raindrops or by the scouring action of flowing water (whether laminar or turbulent) over the soil surface. As it runs downslope, that flow (called surface *runoff* or *overland flow*) then carries the detached particles in suspension. When the runoff water finally comes to rest in a low-lying area, it deposits its suspended load, known as *fluvial sediment*.

The topic of soil erosion has somehow been neglected by many soil physicists, whose traditional interests have focused on the physical attributes and processes of the soil directly affecting plant growth (primarily soil structure and soil moisture and their management) and, more recently, soil and water pollution. Hardly any of the texts or courses on soil physics deal with the physics of erosion per se, except tangentially or cursorily. That lacuna needs to be redressed. In the last couple of decades soil physicists have indeed expanded their sphere of interest to explore the interfaces of their discipline with such sister disciplines as meteorology, ecology, geochemistry, hydrology, and hydrogeology.

The topic of sedimentology, insofar as it also relates to soil physical processes, similarly demands greater attention from soil physicists in the effort to define the mechanisms and functional relationships involved in the process and to develop a systematic theoretical basis for understanding and controlling it.

Research on erosion has been carried out mainly by civil and agricultural engineers. Their efforts have been directed toward identifying the factors affecting erosion, quantifying soil loss rates, and designing systems to reduce those rates (Schwab et al., 1993).

BOX 15.1 Demonstrating the Significance of Soil Erosion

Two early leaders of the U.S. Soil Conservation Service, Hugh H. Bennett and Walter Clay Lowdermilk, who surveyed the global dimensions of the soil erosion problem, wrote in the *1938 Yearbook of Agriculture*: "Soil erosion is as old as farming. It began when the first heavy rain struck the first furrow turned by a crude implement of tillage in the hands of prehistoric man. It has been going on ever since, wherever man's culture of the earth has bared the soil to rain and wind."

When asked to testify on soil erosion before a committee of Congress, the two of them, without a word, placed a thick towel on the committee's polished table and poured a large cup of water onto it. The towel soaked up the water. Next they removed the wet towel and, still saying nothing to the puzzled members of the committee, poured a second cupful of water on the bare table. The water splashed and trickled off the table and onto the laps of the distinguished committee members. Every one of them then understood the dire consequences of removing the soil cover from the land.

SHEET, RILL, AND GULLY EROSION

Early investigators of water erosion made a distinction between sheet erosion, rill erosion, gully erosion, and stream-channel erosion (Fig. 15.3). Sheet erosion was defined as the uniform removal of thin layers of soil from a more or less smooth slope, carried by the distributed (rather than concentrated) flow of runoff water over the soil surface. This is an idealized concept that rarely occurs as such. In reality, the erosion process is hardly ever uniform, and very soon after it begins the sediment-carrying runoff tends to concentrate in small tills, which wend their way downslope. Being small and shallow, these rills are hardly noticeable at first and are easily obscured by cultivation. However, as the

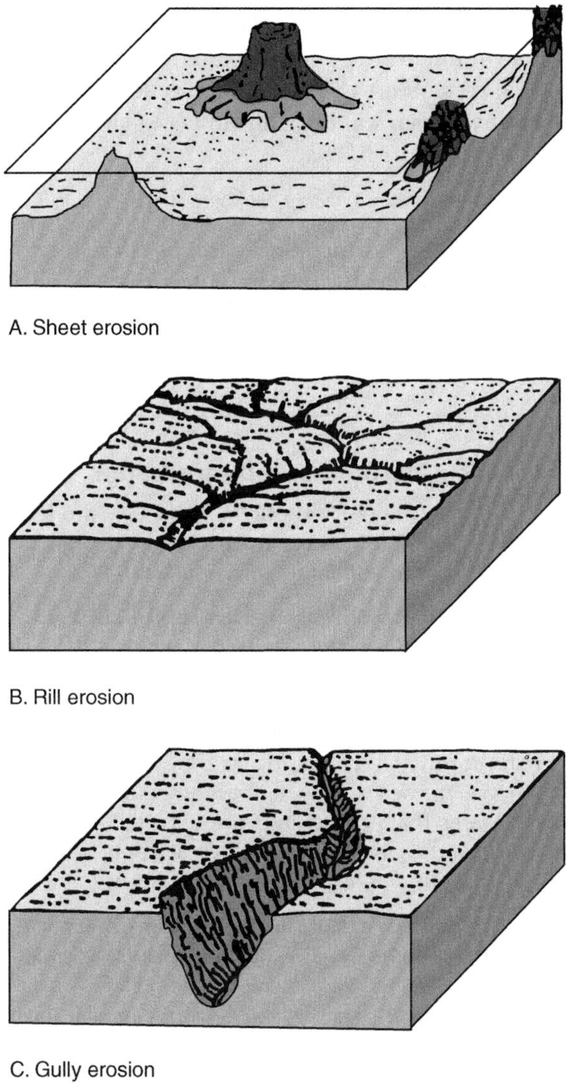

Fig. 15.3. (A) Sheet, (B) rill, and (C) gully erosion.

process continues and is repeated over the course of successive rainstorms, the rills gradually become deeper. They scour the soil to form distinct gullies, which can no longer be obscured by conventional tillage and which converge downslope to form rivulets. As gullies cut further into the soil, they may reach a tight and practically impervious subsoil. Although bottom erosion may slow down, further erosion may occur along the slumping sides of the gullies.

Watson and Laflen (1986) described splash and sheet erosion as *inter-rill erosion* and related it to the characteristics of the soil, slope, and rain as follows:

$$D_i = K_i i^2 S_f \tag{15.11}$$

where D_i is inter-rill erosion rate, K_i is inter-rill soil erodibility, i is rainfall intensity, and S_f is an empirical slope factor.

Rill erosion is the scouring and transport of soil by a concentrated flow of water (Schwab et al., 1993). The erosivity of the flowing water is a function of the hydraulic shear τ of the water flowing in the rill. The soil factors that affect rill erosion are the rill erodibility K_r and the critical shear τ_c, that is, the shear below which soil detachment is negligible (Lane et al., 1987). These variables are related as follows:

$$D_r = K_r(\tau_r - \tau_c)(1 - Q_s/T_c) \tag{15.12}$$

where D_r is the rill detachment rate, Q_s is the rate of sediment flow in the rill, and T_c is the sediment transport capacity of the rill. The hydraulic shear of the flowing water is $\tau = \rho grs$, where ρ is the density of water, g is the acceleration

BOX 15.2 Plato on Soil Degradation

The Greek philosopher Plato (427–327 B.C.E.), who founded the first Academy, in one of his Dialogues, had Critias bewail the degradation of soil that took place in Greece even two and a half millennia ago:

> What now remains of the formerly rich land is like the skeleton of a sick man, with all the fat and soft earth having wasted away and only the bare framework remaining. Formerly, many of the mountains were arable. The valleys that were full of rich soil are now marshes. Hills that were once covered with forests and produced abundant pasture now produce only food for bees. Once the land was enriched by yearly rains, for they were not lost as they are now by flowing from the bare land into the sea. The soil was deep, it absorbed and kept the water in the loamy soil, and the water that soaked into the hills fed springs and running streams everywhere. Now the abandoned shrines at spots where formerly there were springs attest that our description of the land is true.

Has the picture changed? Yes. It has gotten worse. What then took place in specific areas is now a global occurrence. The early astronauts described what they saw as they orbited the earth. They noted widespread forest fires in tropical areas and dust clouds in overgrazed semiarid areas. The island of Madagascar, like a wounded giant, appeared to be bleeding into the sea, its red soils raked off the denuded slopes by the monsoonal rains and carried away by the streams.

of gravity, r is the hydraulic radius of the till, and s is the hydraulic gradient of rill flow.

The transport capacity of rills has been expressed in terms of the following relationship (Foster and Highfill, 1983):

$$T_c = B\tau^{3/2} \tag{15.13}$$

where T_c is the transport capacity per unit width, B is the transport coefficient based on soil and water conditions, and τ is the hydraulic shear in the rill channel.

EROSIVITY OF RAINFALL

When a rainstorm strikes a bare soil surface, quantities of soil material are splashed into the air, and many particles are splashed repeatedly. The amount of soil splashed, measured by means of collecting pans placed over the soil, was found to be 50–90 times greater than the runoff losses. A heavy rainstorm may splash as much as 200 tons of soil (Schwab et al., 1993). Although most of the detached sediment is not immediately lost from the field (most of it being trapped by depressions in the surface), it has the effect of clogging the surface pores and reducing the soil's infiltrability, thereby inducing even greater runoff and greater soil erodibility. Raindrop impact on bare soil was depicted in high-speed photographs by Mihara (1951).

Particle detachment and splash obviously depend on rainfall intensity, the size distribution of drops and their terminal velocities, the direction and steepness of slope, wind, soil conditions (texture, looseness, size and stability of aggregates, roughness of surface), and the possible presence of impediments to splash, such as vegetation, litter, and gravel. Splashed particles may move more than 0.6 m in height and more than 1.5 m laterally on level surfaces. On sloping ground, the splashed material moves preferentially. Surface roughness is also an important factor. Where the surface is pitted, the depressions may fill with water during high-intensity rainfall to form puddles, which then absorb some of the energy of the rain and much of the splashed material, thus reducing the soil loss from the field as a whole.

Hudson (1981) defined the following attributes of rain pertaining to erosion:

1. The *intensity* of a rain is the amount of water falling on a unit area in a unit of time, generally expressed as millimeters per hour. It usually varies during the course of a rainstorm. The time pattern of rain intensity also differs from storm to storm, from place to place, and from season to season.
2. The *duration* of a rain is the length of time from the start of a rainstorm to its ending. This is often arbitrarily determined, because rainstorm events typically exhibit alternating periods of stronger and weaker rain intensity as well as respite periods during which only a drizzle occurs. Is the squall that follows a drizzly respite part of the same rainstorm, or is it a new one?
3. The *energy* of a rainstorm is the summation of the kinetic energies of all raindrops falling on a unit area. As such, it is a function of the size

distribution of the raindrops and their respective terminal velocities. Measurements of raindrop sizes (reported by Hudson, 1981) have shown that they range from a fraction of a millimeter to about 5 mm in diameter. Raindrops larger than 5 mm generally tend to disintegrate because of air turbulence. A positive relationship has been found between rain intensity and median drop size, but this relationship is not consistent for different types of rains in different regions.

A falling raindrop is subject to gravitational acceleration. As the raindrop gains speed, however, the resistance of the air also increases, until it becomes equal to the accelerating force. Thereafter, the drop ceases to accelerate but continues to fall at a constant speed, called the *terminal velocity*. It is analogous to the phenomenon described in Chapter 3 in the section on the settling of particles suspended in an aqueous medium.

The terminal velocity of each drop depends on its size, as shown in Fig. 15.4. Note that the largest raindrops, having a diameter of about 5 mm, fall at a terminal velocity of about 9 m/sec (roughly equivalent to the speed limit in the streets of a city). The abrupt collision of such drops with the soil has the impact and effect of a miniature bomb: It detaches and splashes soil particles and creates a minicrater at the surface, as shown in Fig. 15.5.

The erosive power of rainfall is related to the kinetic energy of the raindrops (which is proportional to the product of the mass and velocity squared, i.e., $Ek = mv^2$) and to their momentum (equal to the product of the mass and velocity, $M = mv$). If the size distribution of raindrops is known, their terminal velocities can be determined. The energy and momentum of the rainstorm as a whole can then be calculated by summation. Attempts have also been made to measure the momentum or kinetic energy of a rainstorm directly by various devices (Hudson, 1981; Romkens et al., 2002).

According to Hudson (1981), the threshold level of intensity at which a rain becomes erosive is about 25 mm/h. In this respect, an important difference exists between temperate and tropical regions. In temperate regions,

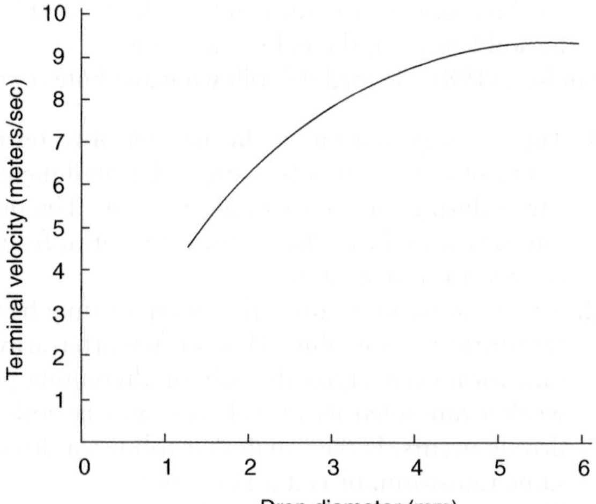

Fig. 15.4. Terminal velocity of raindrops as a function of size. (From Laws, 1940).

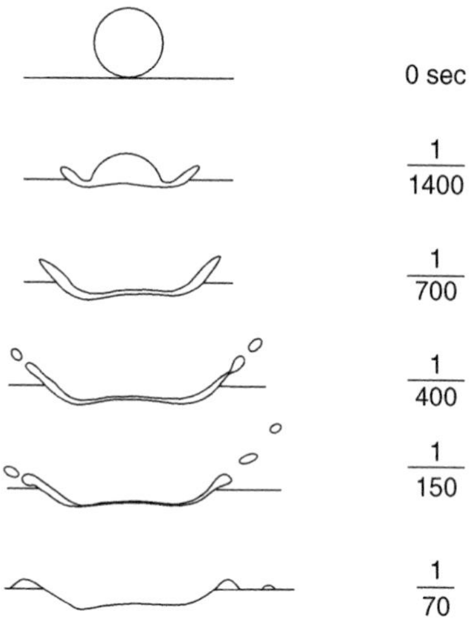

Fig. 15.5. Impact of a raindrop on an erodible soil surface.

only about 5% of the rain falls at intensities great enough to be erosive, whereas in tropical regions as much as 40% of the rain contributes to soil erosion. Moreover, the total amount of rain per season is likely to be greater in the humid tropics, so altogether those regions are more prone to severe erosion.

ERODIBILITY OF SOIL

The actual amount of erosion depends as much on the erodibility of the soil as it does on the erosivity of the rain. The *erodibility* of a soil (i.e., its vulnerability to the erosivity of rain or of running water) depends on many factors. Prominent among these factors are the texture and structure of the soil. For example, a weakly aggregated soil is more erodible than a stably aggregated one, and a dispersed clay is much more erodible than a flocculated clay. Similarly, a soil pulverized by excessive tillage or trampling is more erodible than an untilled and mulch-covered soil. Other relevant factors are the slope and the type and density of the vegetative cover. Soil erodibility indexes have been described in numerous publications (e.g., Hudson, 1981; Schwab et al., 1993).

Apart from the widespread occurrence of surface erosion, there is an entirely different and even more catastrophic form of erosion that takes place in certain topographic and climatic conditions. It is the slumping of rain-saturated soil from steep slopes, where shallow soil is underlain by smooth bedrock. Such landslides may result in the massive loss of the entire soil cover of a hillside. The author has observed these occurrences in such places as New Zealand and Nepal, especially where the slope had been denuded of its original forest cover and where roadbeds that have been cut into the bottom of the slope destabilize the soil uphill. In Nepal, where the landscape is particularly steep and geologically unstable, the volume of soil detached during such events may be large enough to temporarily block the flow of an entire stream.

THE "UNIVERSAL SOIL LOSS EQUATION"

The following empirical equation is known as the universal soil loss equation (USLE):

$$A = RKLSCP \tag{15.14}$$

Here, A is the estimated average annual soil loss, in metric tons per hectare; R is the rainfall erosivity factor (depending on its intensity, quantity, and duration); K is the soil erodibility factor (a measure of the soil's susceptibility to erosion, affected by soil texture, organic matter content, structure and its stability, and permeability); LS is the topographic factor (combining the effects of slope length and steepness); C is the surface cover factor (depending on whether the soil is vegetated, mulched, or bare); and P is the management factor (relating the soil loss with the given management practices to the losses that would occur with up-and-down slope cultivation).

The advantage of the universal soil loss equation is its simplicity. Its disadvantage is that it may be too simplistic. Being a strictly empirical equation, the USLE can best be applied in conditions that are basically similar to those where the various factors have been calibrated experimentally in the field. Any attempt to apply the equation away from the context in which it was devised and calibrated may be misleading (to apply it in Africa, say, where the soils and the rainfall regimen are very different from those in the United State), unless there is supporting experimental data obtained locally.

The erosion process is not a steady, orderly, easily predictable process. Much of it takes place episodically, in rare but violent events. A single torrential rainstorm striking the soil just when its surface happens to be bare and pulverized (hence most vulnerable) may cause more soil loss in a few hours than a whole season's "normal" rainfall over a fully vegetated field.

The USLE was originally proposed in 1965 for estimating sheet and rill erosion from cultivated fields in the United State east of the Rocky Mountains (Wischmeier and Smith, 1978). It has since been applied throughout the United State and in many other countries, for rangelands and forestlands as well as cultivated lands. Various revisions have been made to the original formulation (El Swaify et al., 1985; Lal, 1994; D. F. Post, 1996). However, the equation remains largely empirical, and the topic of soil erosion yet awaits the development of a more fundamental and comprehensive mechanistic approach.

CONTROL OF SOIL EROSION

Effective control of erosion by water consists of minimizing the impact of raindrops and the velocity of running water on the soil surface. This includes enhancing infiltrability and surface storage, improving soil structure, protecting the surface by a cover crop or a mulch (to prevent raindrops from striking the exposed soil), minimizing cultivation and performing it on the contour rather than up and down the slope, and avoiding both compaction and excessive pulverization (Romkens et al., 2002). An ancient and still common practice of soil conservation is the shaping of sloping land by means of terraces or

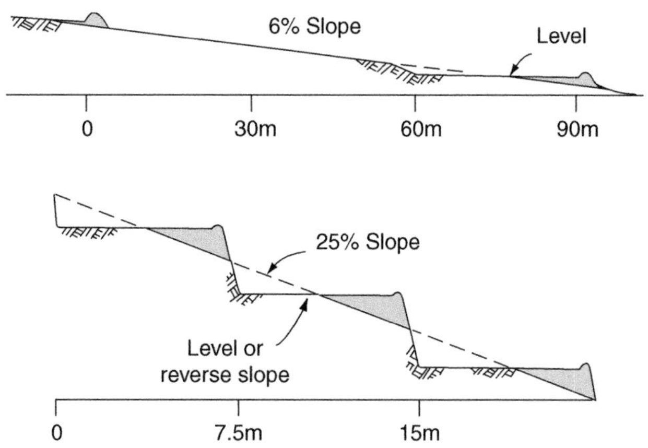

Fig. 15.6. Terrace shapes on slopes of different steepness.

contour strips so as to reduce the inclination of the surface and the length of slope segment, thereby checking the downhill acceleration of running water (Schwab et al., 1993; Hillel, 1994). Figure 15.6 illustrates different types of modern terraces.

16. REDISTRIBUTION AND RETENTION OF SOIL MOISTURE

SOIL WATER STORAGE

When rain or irrigation ceases and free water on the surface disappears (by infiltration, runoff, or evaporation), the process of infiltration comes to an end. Downward water movement within the soil, however, does not cease immediately and may in fact persist for a long time as soil moisture percolates within the profile. The upper soil layers, wetted to near saturation during infiltration, do not retain their full water content, because part of the water moves down into the lower layers under the influence of gravity and possibly also of suction gradients. In the presence of a high water table, or if the profile is initially saturated throughout, this postinfiltration movement is herein termed *internal drainage*. In the absence of groundwater, or where the water table is too deep to affect the relevant depth zone, and if the profile is not initially wetted to saturation throughout its depth, this movement is called *redistribution*. Its effect is to redistribute water in the soil by increasing the wetness of successively deeper layers at the expense of the infiltration-wetted upper layers of the profile.

In some cases, the rate of redistribution diminishes rapidly, becoming imperceptible after several days. Thereafter the initially wetted part of the soil appears to retain its moisture, unless this moisture evaporates from the soil directly or is taken up by plants. In other cases, redistribution may continue at an appreciable, though diminishing, rate for many days or weeks.

The importance of the redistribution process should be self-evident, because it determines the amount of water retained at various times by the different-depth zones in the soil profile and hence can affect the water economy of plants. The rate and duration of downward flow during redistribution determine the effective *soil-water storage*. This property is vitally important, particularly in relatively dry regions, where water supply is infrequent and plants must rely for long periods on the unreplenished reservoir of water within the rooting zone. Even in relatively humid regions, where the water supply by precipitation would seem to be sufficient, inadequate soil-moisture storage may deprive growing plants of a major portion of the water supply and can cause crop failure. Finally, the redistribution process is important because it often determines how much water will flow through the root zone (rather than be retained within it) and hence how much leaching of solutes and recharge of groundwater will take place. As we shall show, soil-water storage is generally not a fixed quantity or static property but a dynamic process, determined by the time-variable rates of inflow to, and outflow from, the relevant soil volume.

INTERNAL DRAINAGE IN DEEPLY WETTED PROFILES

We have already made a distinction between the postinfiltration movement of soil water in cases where a groundwater table is present at some shallow depth (i.e., at a depth not exceeding a few meters) and that in cases where groundwater is either nonexistent or too deep to affect the state and movement of soil moisture in the root zone.

At the groundwater table (also called the *phreatic surface*), whatever its depth, soil water is at atmospheric pressure. Beneath this level the hydrostatic pressure exceeds atmospheric pressure; above this level soil water is generally under suction. Internal drainage in the presence of a water table tends eventually toward a state of equilibrium. At equilibrium, the suction head at each point corresponds to its height above the free-water level. As equilibrium is approached, the downward drainage flux declines while the hydraulic gradient decreases, both approaching zero in time. (That is, of course, provided there is no further addition of water by another episode of infiltration or abstraction of water by evapotranspiration — processes that would prevent the attainment of static equilibrium.)

At equilibrium, if attained, soil wetness would increase in depth to a value of saturation just above the water table, a depth distribution that would mirror the soil-moisture characteristic curve. Phenomena involving falling water tables and groundwater drainage will be treated in our next chapter. In the present section, however, we deal with the internal drainage of profiles initially wetted to near saturation throughout their depth. We shall treat this process as if it were unaffected by any water table, which, if it exists, is assumed to lie too far below the zone of interest to be of any direct consequence.

In the hypothetical case of a deeply and uniformly wetted profile, suction gradients would hardly exist. Internal drainage should occur under the influence of gravity alone. If so, downward flow through any arbitrary plane (say, the bottom of the root zone) should be equal to the hydraulic conductivity and should diminish in time as the conductivity diminishes due to reduction of water content in the infiltration-wetted soil above the plane considered.

Recall the one-dimensional (vertical) form of Darcy's equation:

$$q = -K(\theta)\frac{\partial(-\psi - z)}{\partial z} \tag{16.1}$$

where q is downward flux, $K(\theta)$ is hydraulic conductivity (a function of wetness θ), z is depth, and ψ is matric suction. With $\partial\psi/\partial z$ assumed to be zero, we have simply

$$q = K(\theta) \tag{16.2}$$

If we happen to know the functional dependence of K on θ, we can formulate q as an explicit function of θ. For instance, if the function is exponential, say,

$$K(\theta) = ae^{b\theta} \tag{16.3}$$

where a and b are constants, we get

$$q = ae^{b\theta} \qquad \text{or} \qquad \ln q = A + b\theta \tag{16.4}$$

where $A = \ln$ a. Thus, even a small decrease in θ can result in a steep (logarithmic) decrease of the flux q.

A simple approach to the internal drainage process is possible if the soil profile can be assumed to drain uniformly, that is, if the soil is equally wet and its wetness diminishes at an equal rate throughout the draining profile. Observations have shown this to be a reasonable approximation in many (but certainly not all) cases. In the absence of any flow through the upper soil surface, the flux q_b, through any plane at depth z_b, must equal the rate of decrease of total water content W, where $W = \theta z_b$ (the product of the wetness θ and depth z_b):

$$q_b = K(\theta) = -dW/dt = -z_b(d\theta/dt) \tag{16.5}$$

This equation can be used to measure the depth-averaged hydraulic conductivity as a function of wetness in a field plot that is under an impervious surface (say, a sheet of plastic material). Note that in this case the downward flux increases in proportion to depth. Note also that it diminishes in time, as does the rate of decrease of soil wetness, in accordance with the functional decrease of conductivity with the remaining soil wetness. Once the characteristic $K(\theta)$ function is established for a given soil, it becomes possible to predict the time dependence of θ and of drainage rate q for any depth.

An alternative approach is to assume that the downward drainage flux is proportional to the total amount of water remaining in the draining profile. Thus

$$-dW/dt = \lambda W \tag{16.6}$$

where, as before, $-dW/dt$ is the time rate of decrease of profile water content (storage) W and λ is a proportionality constant. Equation (16.6) can be integrated by separation of variables:

$$\int (dW/W) = -\lambda\int dt \tag{16.7}$$

to give

$$\ln W = -\lambda t + c \tag{16.8}$$

wherein c is the integration constant. The last equation can be rewritten as

$$W = e^c / e^{\lambda t} \tag{16.9}$$

At $t = 0$ (the beginning of the process), $W = e^c$, a constant that we can designate W_i (initial water content of the profile). Thus

$$W = W_i e^{-\lambda t} \qquad \text{or} \qquad W/W_i = e^{-\lambda t} \tag{16.10}$$

which is the well-known equation used to describe exponential decay processes, such as radioactive decay. The proportionality constant λ is called the *decay constant*, and it can be interpreted as the fraction of the remaining soil-water content that drains per unit time. To characterize the stability of soil-moisture storage, we can use the concept of *half-life* ($t_{1/2}$), which is the time required for the cumulative drainage of half the initial amount of water present in the soil (i.e., the value of t in the last equation at which $W = W_i/2$). Thus, at $t = t_{1/2}$, $e^{\lambda t} = 2$. Hence,

$$t_{1/2} = \ln 2/\lambda = 0.693/\lambda \tag{16.11}$$

If Eq. (16.6) is not by itself a good description of the internal drainage process, as W tends rather unrealistically to approach zero in time, the equation can perhaps be made more realistic by providing for soil-water content to approach asymptotically some finite residual value W_r rather than zero. As an example, let us assume that W_r corresponds to the product of the soil depth considered, Z_b, and the equilibrium value of soil wetness at a matric suction value of, say, 33 kPa (one-third of a bar). The value of W_r obviously depends on soil texture, as do the values of W_i, λ, and indeed $t_{1/2}$.

For a more fundamental approach to the problem of predicting the dependence of flux and wetness on time, we must go to the general flow equation, which can be stated as

$$\frac{\partial \theta}{\partial t} = \frac{\partial}{\partial z}\left[K(\theta)\frac{\partial \psi}{\partial z}\right] - \frac{\partial K(\theta)}{\partial z} \tag{16.12}$$

This equation can be solved subject to the following conditions:

$$\begin{array}{llll} t = 0, & z \geq 0, & \theta = \theta_s & \text{(initially saturated profile)} \\ t > 0, & z = 0, & q = 0 & \text{(no further inflow or outflow through the surface)} \\ t > 0, & z = z_b, & q = K(\theta_b) & \text{(flux equals conductivity at } z_b\text{)} \end{array} \tag{16.13}$$

Results obtained by numerical solution of the equation for the conditions shown (Hillel and van Bavel, 1976) are illustrated for soils of different textures in Figs. 16.1–16.3, which show the pattern of decreased soil-moisture storage during internal drainage. The sandy soil, although it contains less water at saturation, is seen to drain much more rapidly at first. Thus, it loses half again as

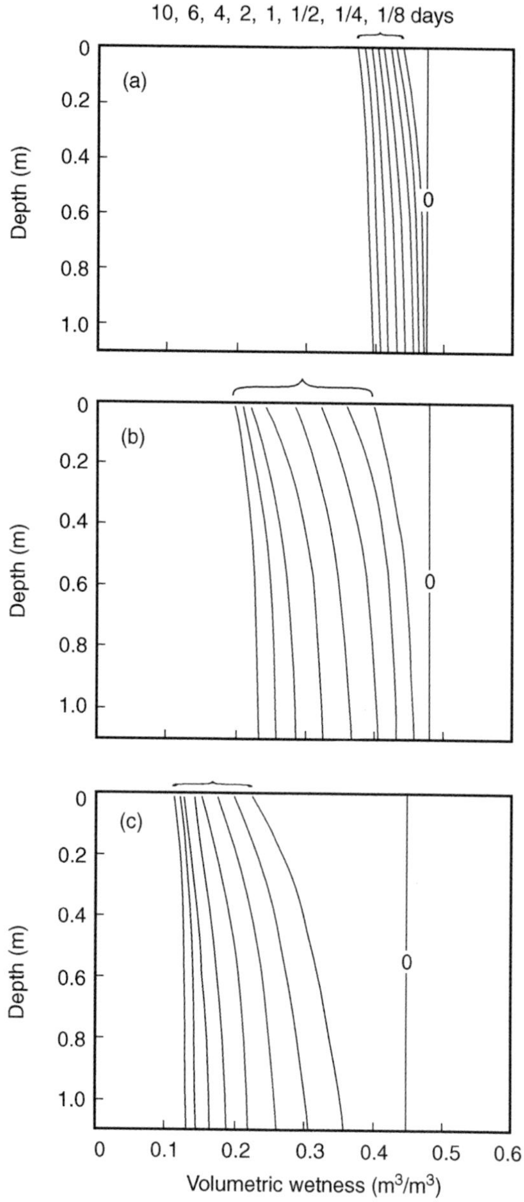

Fig. 16.1. The changing soil-moisture distribution during drainage from initially saturated uniform profiles of (a) clay, (b) loam, and (c) sand. The numbers indicate duration of the process (days). (After Hillel and van Bavel, 1976.)

much water as the loam and nearly five times as much water as the clay during the first two days. Thereafter these differences diminish and are eventually reversed as further drainage from the sand slows down to a very low rate, while drainage from the loam, and even more so from the clay, persists at an appreciable rate for many more days.

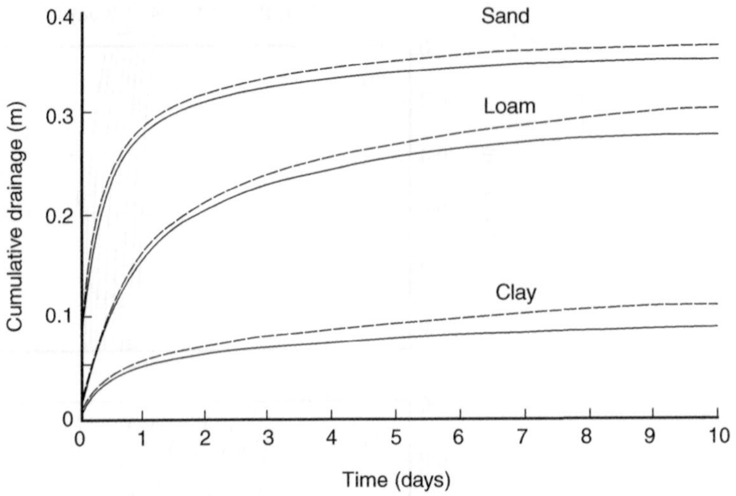

Fig. 16.2. Cumulative gravity drainage from initially saturated uniform profiles (1.16 m deep) of sand, loam, and clay. Dashed lines, drainage alone; solid lines, simultaneous drainage and evaporation. (After Hillel and van Bavel, 1976.)

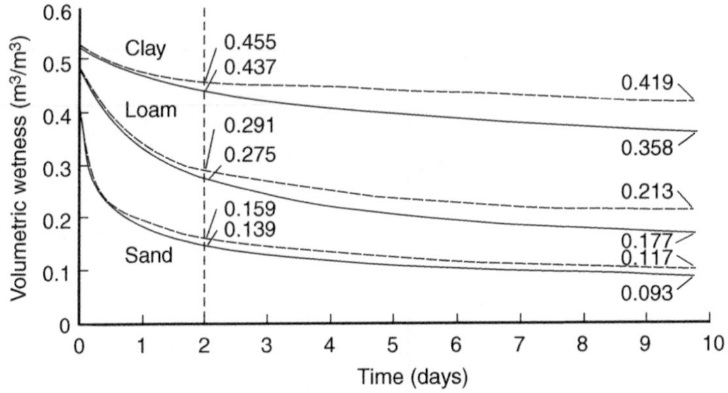

Fig. 16.3. Volumetric wetness at depth 0.41 m as function of time in initially saturated uniform profiles of sand, loam, and clay. Dashed lines, drainage without evaporation; soild lines, simultaneous drainage and evaporation. (After Hillel and van Bavel, 1976.)

Sample Problem

Assume gravity drainage out the bottom of a uniformly wetted soil profile, 1 m deep, with an initial volume wetness of 45%. Further assume the hydraulic conductivity K to be an exponential function of the soil's volume wetness θ, namely, $K = ae^{b\theta}$, where $a = 1.5 \times 10^{-5}$ mm/day and $b = 35$. Estimate the remaining soil wetness when the drainage flux has diminished to 1/2, then to 1/10, and then to 1/100 of the evapotranspiration (ET) rate, which is 5 mm/day.

With gravity alone (no suction gradients), vertical drainage occurs at a rate q, equal to the hydraulic conductivity, which is itself a function of the soil's wetness at the depth considered (e.g., at bottom of the root zone):

$$q = K(\theta) = ae^{b\theta} = 1.5 \times 10^{-5}e^{35\theta}$$

We can now calculate the flux at different assigned wetness values to obtain a plot of log q versus θ, as follows:

When $\theta = 0.45$, $q = (1.5 \times 10^{-5})e^{35\times0.45} = 103.8$ mm/day
When $\theta = 0.40$, $q = (1.5 \times 10^{-5})e^{35\times0.40} = 18$ mm/day
When $\theta = 0.35$, $q = (1.5 \times 10^{-5})e^{35\times0.35} = 3.15$ mm/day
When $\theta = 0.30$, $q = (1.5 \times 10^{-5})e^{35\times0.30} = 0.54$ mm/day
When $\theta = 0.25$, $q = (1.5 \times 10^{-5})e^{35\times0.25} = 0.095$ mm/day
When $\theta = 0.20$, $q = (1.5 \times 10^{-5})e^{35\times0.20} = 0.017$ mm/day

From the q-θ plot we can determine the θ values at which $q = 2.5$ mm/day (ET/2), 0.5 mm/day (ET/10), and 5×10^{-2} mm/day. The θ values turn out to be approximately 43, 30, and 23.2%, respectively. Which of the q values is to be considered negligible is a matter of individual decision.

Alternatively, we can calculate the θ remaining at any flux value, as follows:

$$q = K(\theta) = ae^{b\theta}$$

Hence $\ln q = \ln a + b\theta$. Therefore $\theta = (\ln q - \ln a)/b$.

The actual plot of q versus θ is left as an optional exercise.

Implicit in the foregoing discussion was the supposition that the soil profile is texturally or structurally uniform in depth. Quite a different situation can prevail when the profile is layered. An impeding layer inside or below the root zone can greatly inhibit internal drainage. In many cases such a layer can be harmful to crops, for it might cause retention of excessive moisture, and perhaps of salts, in the root zone. In some cases, however, as in rapidly draining sands, it might be desirable to form an artificial barrier at some depth to promote greater retention of moisture within reach of plant roots.

REDISTRIBUTION IN PARTIALLY WETTED PROFILES

Perhaps more typical than the condition described in the preceding section (namely, that of a soil that is deeply and uniformly wetted throughout its profile) are cases in which the end of the infiltration process consists of a wetted zone in the upper part of the profile and an unwetted zone beneath. The postinfiltration movement of water in such a profile can truly be called *redistribution*, because the relatively dry deeper layer (beyond the infiltration wetting front) draws water from the upper one so that soil moisture redistributes itself between the layers. The time-variable rate of redistribution depends not only on the hydraulic properties of the conducting soil but also on the initial wetting depth as well as on the relative dryness of the bottom layers. When the initial wetting depth is small and the underlying soil is relatively dry, strong suction gradients augment the gravitational gradient in causing a rapid rate of redistribution. On the other hand, when the initial wetting depth is considerable and the underlying soil itself is fairly moist, the suction gradients tend to be small, so redistribution occurs primarily under the influence of gravity, as in the case of internal drainage described earlier.

At first the decrease of soil wetness in the initially wetted zone can be expected to occur more rapidly during the redistribution of moisture in profiles that had wetted to shallow depth than in the internal drainage of profiles that had been wetted deeply. Sooner or later, however, the redistribution process "spends itself out," so to speak, and the flux slows down for two reasons: (1) the suction gradients between the wetter and the drier zones diminish as the former loses, and the latter gains, moisture; (2) as the initially wetted zone quickly desorbs, its hydraulic conductivity falls correspondingly. With both gradient and conductivity decreasing simultaneously, the flux also falls rapidly. The rate of advance of the wetting front slows down, and this front, which was relatively sharp during infiltration, gradually flattens out and dissipates during redistribution. This trend is illustrated in Fig. 16.4. The figure shows that the initially wetted upper zone drains monotonically, though at a decreasing rate. On the other hand, the sublayer at first wets up but eventually begins also to drain.

The time dependence of soil wetness in the upper zone is illustrated in Fig. 16.5 for a sandy soil, in which the unsaturated conductivity falls off rapidly with increasing suction, and for a clayey soil, in which the decrease of conductivity is more gradual and hence redistribution tends to persist longer.

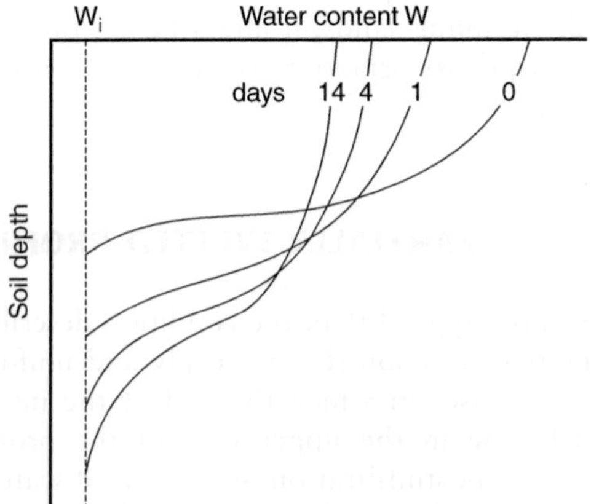

Fig. 16.4. The changing moisture profile in medium-textured soil during redistribution following a partial irrigation. The successive moisture profiles shown are for 0, 1, 4, and 14 days after the irrigation. W_i, preirrigation (antecedent) soil wetness.

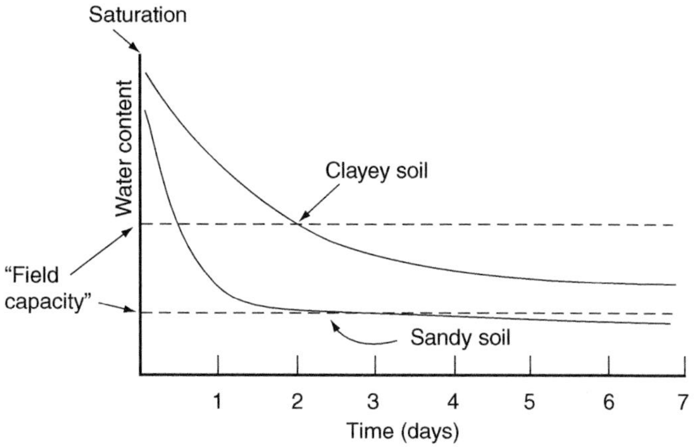

Fig. 16.5. The monotonic decrease of soil wetness with time in the initially wetted zone of a clayey and a sandy soil during redistribution. The dashed lines show the wetness values remaining in each soil after two days.

HYSTERETIC PHENOMENA IN REDISTRIBUTION

Unlike the internal drainage process, in which a deeply wetted soil desorbs monotonically, the redistribution process that takes place in a partially wetted profile necessarily involves hysteresis; the upper part of the profile desorbs while the lower part absorbs water from it. It is the nature of hysteresis that even if the two sections were to attain equilibrium with the same matric potential, the desorbed section of the soil would be wetter than the section having undergone sorption. Moreover, an intermediate zone, just below the infiltration-wetted zone, first absorbs water and then begins to desorb as the redistribution process progresses deeper into the profile. This pattern of sorption and desorption is portrayed schematically in Fig. 16.6.

We recall from Chapter 8 that the relation between wetness and suction is not unique but depends on the history of wetting and drying that took place at each point in the soil. This relationship, when plotted, exhibits two limiting curves that apply when wetting or drying starts from the extreme conditions of dryness or saturation, respectively. Between the wetting and drying branches (the "primary soil-moisture characteristic curves") there lies an infinite number of possible "scanning curves," which describe wetting or drying between different intermediate values of wetness. Such scanning curves tend to be "flatter" than the primary desorption curve; hence a greater increase in suction is needed to effect a unit drop in wetness during desorption after partial wetting than in desorption after complete wetting.

If hysteresis were nonexistent, we would expect a draining vertical column under the influence of gravity to tend toward an equilibrium such that soil wetness would increase with depth as the suction forces increase with elevation (see Chapter 8). With hysteresis, however, more of the moisture is retained in the infiltration-wetted part of the profile, where it might remain available for subsequent plant use, rather than flow downward beyond the reach of

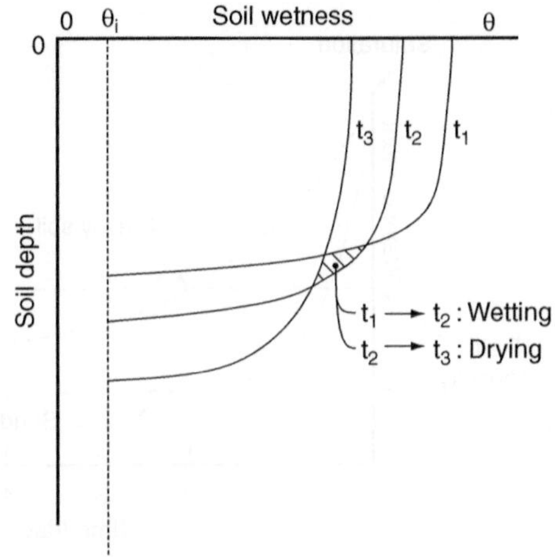

Fig. 16.6. Water content profiles at the end of infiltration (t_1) and at two later points in time as a result of redistribution.

plant roots. Evidence for this was provided by Rubin (1967), as shown in Fig. 16.7. The role of hysteresis can be especially important in semiarid regions, where limited rainfall may wet the soil only to shallow depths, so retaining moisture in the root zone is likely to be critical to native plants as well as to crops.

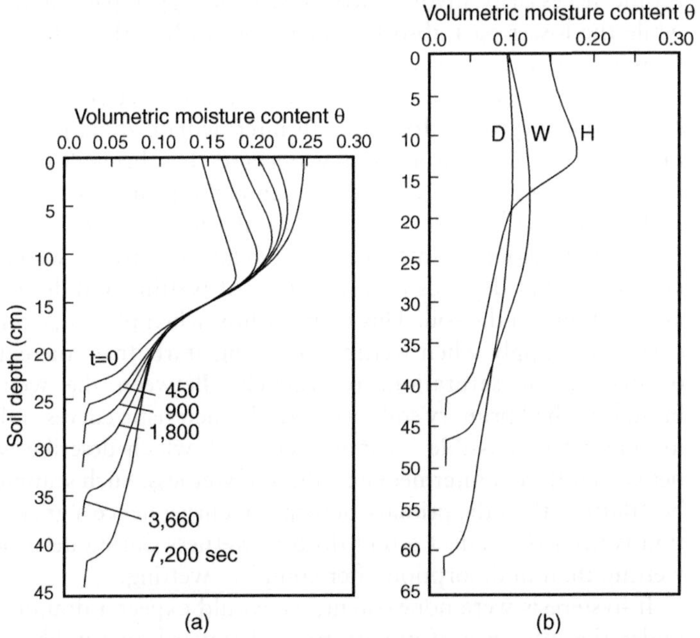

Fig. 16.7. Calculated soil-moisture profiles after redistribution in a sandy soil, illustrating the effect of hysteresis on retention of moisture in the upper layers. In the right-hand figure, the hysteretic redistribution profile (H) is compared with the nonhysteretic profile based on the sorption characteristic curve (W) and with the nonhysteretic profile based on the desorption characteristic curve (D). (After Rubin, 1967.)

The steep reduction in hydraulic conductivity that takes place in the draining part of the profile, the overall decrease in matric suction gradients, and the effect of hysteresis all act in concert to promote moisture retention in the upper part of the profile. If we did not know of these synergistic effects, we might never understand how the soil, being in effect a "bottomless barrel," nevertheless manages to retain enough water in its upper part long enough to support the growth of vegetation even in arid regions, where water supply is meager and infrequent.

ANALYSIS OF REDISTRIBUTION PROCESSES

Once again we return to the general equation for flow in a vertical profile, Eq. (16.12):

$$\partial\theta/\partial t = (-\partial/\partial z)[K(\partial\psi/\partial z) + K]$$

In the case of redistribution involving hysteresis, this equation is written

$$(\partial\theta/\partial\psi)_{\mathrm{h}}(\partial\psi/\partial t) = (-\partial/\partial z)[K_{\mathrm{h}}(\psi)(\partial\psi/\partial z)] - \partial K_{\mathrm{h}}(\psi)/\partial z \qquad (16.14)$$

In the foregoing, θ is volumetric wetness, t is time, K is conductivity, z is depth, ψ is suction head, and the subscript h indicates a hysteretic function. After the cessation of infiltration and in the absence of evaporation, flux through the soil surface is zero. Therefore the hydraulic gradient at the surface must also be zero. Conservation of matter requires that

$$\int_{z=0}^{z=\infty} \theta \, dz = \text{constant} \qquad (16.15)$$

for all time, provided no sinks are present (e.g., no extraction of water by plant roots).

When redistribution begins, the upper portion of the profile, which was wetted to near saturation during the preceding infiltration process, begins to desorb monotonically. Below a certain depth, however, the soil first wets during redistribution and then begins to drain, and the value of wetness at which this turnabout takes place diminishes with depth. Each point in the soil thus follows a different scanning curve, and the conductivity and water-capacity functions vary with position.

To solve Eq. (16.12), one dependent variable (namely, ψ) can be eliminated:

$$\partial\theta/\partial t = (-\partial/\partial z)[(K/c)(\partial\theta/\partial z) + K] \qquad (16.16)$$

The solution can then be sought, subject to the following initial and boundary conditions:

$$\begin{aligned} t &= 0, & z &> 0, & \theta &= \theta(z) \\ t &> 0, & z &= 0, & q &= (K/c)(\partial\theta/\partial z) + K = 0 \\ t &> 0, & z &= \infty, & \theta &= \theta_{\mathrm{i}} \end{aligned} \qquad (16.17)$$

where q is the flux, $\theta(z)$ is a function of soil depth describing the initial post-infiltration moisture profile, θ_{i} is the soil's preinfiltration wetness, c is the differential water capacity $d\theta/d\psi$, and the other variables are as previously defined. The ratio K/c is the hydraulic diffusivity D.

Equation (16.16) has been solved numerically by means of an implicit difference equation. To take account of hysteresis, Rubin (1967) used empirical equations for the dependence of ψ on θ in primary wetting, in primary drying, and in scanning from wetting to drying, as well for the relation of K to θ. (Empirical equations cannot be expected to pertain to any but the particular soil considered; the results of the analysis, however, are thought to be valid in principle, provided the basic assumptions are met in the real situation, at least approximately.)

Rubin's findings (Fig. 16.7) indicate that the hysteretic moisture profile is not bounded by the two possible nonhysteretic profiles (one assuming the desorbing branch of the soil-moisture characteristic, the other assuming the sorbing branch). The hysteretic redistribution was shown to be clearly slower than the nonhysteretic one. These results demonstrate the importance of hysteresis in redistribution processes, particularly in coarse-textured soils, which often exhibit more pronounced hysteresis than do fine-textured soils.

A different approach to the analysis of redistribution was taken by Gardner, Hillel, and Benyamini (1970a, b). Equation (16.16) can be solved analytically by separation of variables (W. R. Gardner, 1962b) in the special case where the empirical relation $D = C\theta^n$ applies (C and n being constants). In this procedure, it is assumed that the solution is of the form $\theta = T(t)Z(z)$, where T is a function of t alone and Z is a function of z alone. It is further assumed that $K = B\theta^m$ (with B and m constants). Their analysis suggests that both where matric suction forces predominate and where, alternatively, the gravity force predominates, the time dependence of soil-water content obeys a relation of the type

$$W, \theta_{\text{ave}} \quad \text{or} \quad \int \theta \, dz \approx a(b + t)^{-c} \tag{16.18}$$

where W is the total water content and θ_{ave} is the mean wetness of the initially wetted zone at time t during redistribution. This theory accords with experimentally measured patterns of redistribution in a sandy loam soil irrigated with different quantities of water (Figs. 16.8 and 16.9). The constants in Eq. (16.18) were shown to be related to the soil's hydraulic conductivity and diffusivity. Of these, the value of b could be neglected after a day or so, so the simplified equation could be cast into logarithmic form:

$$\log W = \log a - c \log t \tag{16.19}$$

The values of a and c can be obtained from a graph of log W versus log t, provided the data indicate a straight line, as shown in Fig. 16.9. Eq. (16.18) can also be differentiated with respect to time to yield the rate of decrease of water content in the initially wetted zone ($-dW/dt$), equal to the flux through the initial (end-of-infiltration) wetting front:

$$-dW/dt = ac/(b + t)^{c+1} \tag{16.20}$$

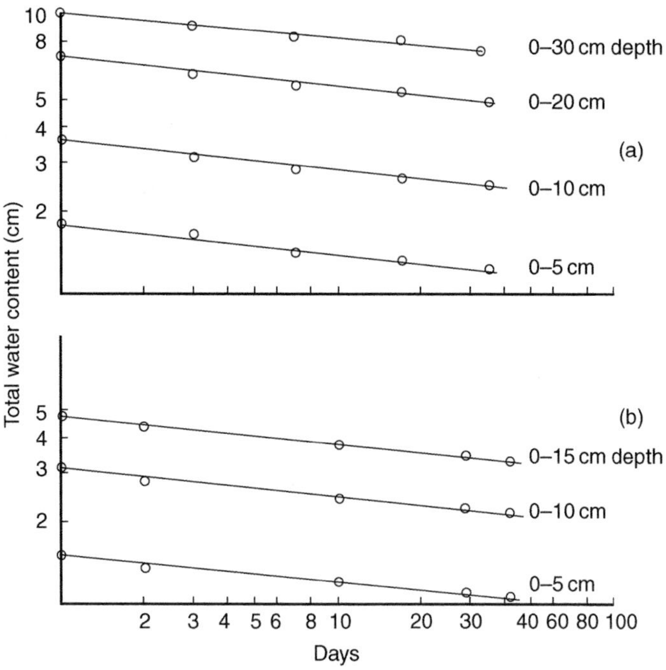

Fig. 16.8. Total amount of water retained in various depth layers within the initially wetted zone of a fine sandy loam during redistribution following irrigations of (a) 100 and (b) 50 mm of water. (After Gardner, Hillel, and Benyamini, 1970a, b.)

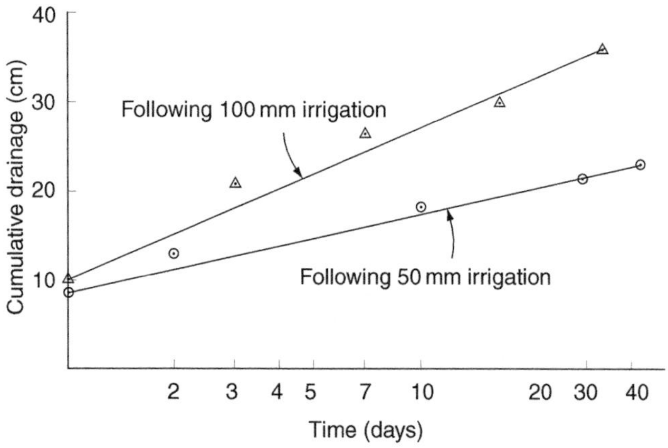

Fig. 16.9. Cumulative downward flow through the initial (end of infiltration) wetting front during redistribution after irrigations of 50 and 100 mm of water. (After Gardner, Hillel, and Benyamini, 1970a, b.)

Sample Problem

Use the equation of L. A. Richards et al. (1956) to estimate the water content of a 1-m-deep profile at the end of 1, 2, 3, 4, 5, 6, and 7 days of redistribution if the initial profile water content is 500 mm, factor $a = 40$, and exponent $b = 0.8$. Estimate the time necessary for the downward flux at a depth of 1 m to diminish to 10 mm/day.

Using Eq. (16.21),

$$dW/dt = -at^{-b},$$

we get

$$dW = -at^{-b}\,dt$$

Integrating, we obtain

$$W = -[a/(-b+1)]t^{-b+1} + c$$

where c, the constant of integration, equals the profile water content at $t = 0$. We can now calculate the water content at various times:

At zero time,	$W = 500 - [40/(-0.8+1)] \times 0^{0.2} = 500$ mm
After 1 day,	$W = 500 - (40/0.2) \times 1^{0.2} = 300$ mm
After 2 days,	$W = 500 - (40/0.2) \times 20^{0.2} = 270$ mm
After 3 days,	$W = 500 - (40/0.2) \times 30^{0.2} = 251$ mm
After 4 days,	$W = 500 - (40/0.2) \times 40^{0.2} = 236$ mm
After 5 days,	$W = 500 - (40/0.2) \times 50^{0.2} = 224$ mm
After 6 days,	$W = 500 - (40/0.2) \times 60^{0.2} = 214$ mm
After 7 days,	$W = 500 - (40/0.2) \times 70^{0.2} = 205$ mm

To estimate the time needed for the flux to diminish to any specified value q_n we again use Eq. (16.22) $[t_n = (a/q_n)^{1/b}]$, hence

$$t = (40/10)^{1.25} = 5.66 \text{ days}$$

THE CONCEPT OF *FIELD CAPACITY*

Early observations that the rate of water-content change during internal drainage and redistribution decreases in time were construed to mean that the flow rate becomes negligible within a few days or even that flow ceases entirely. The presumed water content at which internal drainage allegedly ceases, termed the *field capacity*, had for a long time been accepted almost universally as an actual physical property, characteristic and constant for each soil.

Though the field capacity concept originally derived from rather crude data of water content in the field, some workers sought to explain it in terms of a static equilibrium or the conjectured (but never proven) occurrence of an abrupt physical discontinuity in "capillary water." It was commonly assumed that the application of a certain quantity of water to a unit depth of soil will fill the "deficit to field capacity," and that any quantity of water delivered to the soil will not penetrate beyond the depth to which that quantity can "replenish" the field capacity. It became a common practice to calculate the amount of irrigation to be applied at any particular time on the basis of the deficit to the field capacity of the soil depth to be wetted, as if each soil layer were a bucket that must be filled before it can overflow to a lower bucket.

In recent decades, with the development of theory and with more precise experimental techniques for the in situ determination of soil wetness and of unsaturated flow, the field capacity concept has been recognized as arbitrary. The redistribution process is in fact continuous and exhibits no abrupt "breaks" or static levels. Although its rate decreases constantly, in the absence of a water table the process continues and equilibrium is approached, if at all, only after very long periods. The soils for which the field capacity concept is most applicable are coarse-textured ones, in which internal drainage is initially most rapid but soon slows down owing to the relatively steep decrease of hydraulic conductivity with increasing matric suction (see Chapter 8). In medium- or fine-textured soils, however, redistribution can persist at an appreciable rate for many days.

As an example, we can cite the case of a loessial silt loam in the Negev region of Israel, in which the changes in water content (shown in Table 16.1) were observed in the 0.6- to 0.9-m-depth zone following a wetting to a depth exceeding 1.50 m. The plot was covered with a paper mulch and a layer of dry soil to prevent evaporation. The data show that water loss continued (albeit at a diminishing rate) for over five months.

An irrigation farmer accustomed to frequent irrigations is interested mainly in the short-run storage of moisture in the soil. For such a farmer, the field capacity of the loessial soil described can be taken at about 18%. By way of contrast, a dryland farmer is likely to be interested in accumulating and storing soil moisture from one season to the next or even from one year to the next. The latter farmer, obviously, cannot assume that the same soil retains a water content of 18% but only 14% or even less. The difference, 4% by mass, represents about one-third of the amount of soil moisture classically considered available for crop use!

Various laboratory methods have been proposed for the estimation of field capacity. These include equilibration of presaturated samples in the laboratory with a centrifugal force 1000 times the gravity force (the so-called

TABLE 16.1 Retention of Moisture (% by Mass) in a Loessial Soil (0.6- to 0.9-m Depth) Following a Deep Wetting[a]

End of infiltration	29.0
After 1 day	20.2
2 days	18.7
7 days	17.5
30 days	15.9
60 days	14.7
156 days	13.6

[a] After Hillel (1971).

"moisture equivalent") or with a matric suction value of 10 or 33 kPa (1/10 or 1/3 bar). Although the results of such tests may correlate with measurements of soil-moisture storage in the field in certain circumstances, it is a fundamental mistake to expect criteria of this sort to apply universally, since they are solely static in nature while the process they purport to represent is highly dynamic.

Notwithstanding the fundamental impossibility of fixing unequivocally any unique point in time at which internal drainage or redistribution ceases, there remains a universal need for a simple criterion to characterize the ability of soils to retain moisture (i.e., the upper limit of soil-water content that can be depended on, more or less, in the field). The rapid slowing of internal drainage or redistribution is an important property of soils, responsible for retaining moisture (albeit temporarily) for the vital needs of plants during periods between rains or irrigations.

PHYSICALLY BASED CONCEPTS OF MOISTURE STORAGE

Granted that the traditional definition of field capacity is subjective, what criterion of moisture storage can be used in its stead? In the first place, since no laboratory system yet devised is capable of duplicating soil-water dynamics in situ, it should be obvious that profile moisture storage must be measured directly in the field. Second, the field determination itself must be made reproducible by standardizing a consistent procedure. Such vague specifications as "wet the soil to the depth of interest" and "allow the soil to drain for approximately 2 days" are not good enough. Wetting depth is extremely important, and the preferred depth is the maximal one — considerably beyond the "depth of interest." The internal drainage of a profile initially wetted to the entire depth of the root zone is a much more reproducible and reliable criterion than redistribution in a profile wetted to some unspecified partial depth without regard to antecedent conditions.

Third, the measurement of soil-moisture content and depth distribution should be made repeatedly rather than only once at an arbitrary time such as 2 days. Periodically repeated measurements, preferably by a nondestructive method (neutron gauging or time-domain reflectometry), will provide information on the dynamic pattern of internal drainage and allow evaluation of whether any single value of soil moisture at any specifiable time is distinct enough to be designated as the field capacity (Romano and Santini, 2002). To make this judgment objectively, one must decide at the outset what drainage rate is low enough to be considered "negligible." A likely criterion might be 10% of the mean daily potential evapotranspiration (PET). Thus, if PET is 5 mm/day, one might specify a drainage flux of 0.5 mm/day through the bottom of the root zone as small enough to be disregarded. For different soils, the characteristic time may vary from one day to a few weeks, depending on soil hydraulic properties and on the wetness or flux criterion used.

A still better approach is to characterize, for each field to be considered, the internal drainage process (starting from well-defined initial and boundary conditions) as a complete function of time. One way to do this is to specify the half-life of stored moisture, as explained earlier in this chapter. Another way is to fit the measured function to an empirical equation such as

$$q = -dW/dt = at^{-b} \tag{16.21}$$

It would then be up to the user to calculate what value of soil moisture gives the best estimate of soil-moisture storage to suit his or her purposes. Assuming, for the sake of argument, that the data obtained in the field accords with the preceding equation, then, for whatever limiting value of flux q_n one wishes to regard as negligible, one gets

$$t_n = (a/q_n)^{1/b} \tag{16.22}$$

where t_n is the time period required for the drainage rate to fall to a negligible value. Constants a and b must be calculated from the internal drainage curve as measured in the field.

17. GROUNDWATER DRAINAGE AND POLLUTION

BASIC CONCEPTS OF GROUNDWATER HYDROLOGY

Of the earth's total amount of fresh (nonsaline) water, some 75% exists in the frozen state in polar ice caps and glaciers, while less than 2% is in *surface waters*, such as lakes and streams, and a relatively minute amount is contained in the generally unsaturated soil. That leaves nearly 22% of our planet's fresh water that permeates porous rocks and sediments, generally at some depth below the ground surface. It is this amount that we call *groundwater*.

In many regions, groundwater constitutes an important source of fresh water for domestic, agricultural, or industrial use. Since an adequate water supply is a prerequisite for the development of settlements and industries, including the agricultural industry, and since injudicious exploitation of groundwater can deplete the *groundwater reservoir* and diminish its quality, it is important to acquire and disseminate knowledge pertaining to the behavior of groundwater and to methods of managing it. Extraction of groundwater in excess of the *annual recharge* (including the natural percolation of precipitation water, seepage from reservoirs and streams, and artificial injection through wells and by surface ponding) will cause depletion, whereas an excess of recharge over extraction will cause a buildup of groundwater.

An *aquifer* (literally, a "carrier of water") is a porous geological formation that accumulates and transmits water in sufficient quantities to serve as a source of supply for human use. Aquifers vary greatly in their characteristics. Some consist of unconsolidated coarse sediments, such as sand and gravel, and some are pervious bedrock, such as sandstone, conglomerate, limestone, or even fissured volcanic formations. A formation that, in contrast with an aquifer, neither contains nor transmits significant amounts of water is called an *aquifuge*. An example is a formation of tight, impermeable granite. Between the two extremes, aquifer and aquifuge, hydrologists recognize an intermediate type of formation that contains water but, owing to low permeability, cannot transmit enough to serve as a source of supply (e.g., clay layers and shales). Such a formation is called an *aquiclude*.

A distinction is made between two major types of aquifers: confined and unconfined. A *confined* aquifer occurs when a permeable stratum is overlain by an impermeable stratum (i.e., when an aquifer is "confined" by an aquiclude). A confined aquifer is generally recharged not through the aquiclude but from areas (possibly some distance away) where the permeable formation outcrops, as shown in Fig. 17.1. Often, the water contained in a confined aquifer is under pressure, called *artesian pressure* (after the district of Artois in France), which may be sufficient to bring water up to ground surface whenever a well is drilled through the aquiclude.

The equilibrium height to which water rises in an unpumped well is called the *piezometric level* (from the Greek word meaning "to press", hence a *piezometer* is a device to measure pressure). The static level of water in a well that penetrates a confined aquifer indicates the hydrostatic pressure of water in that aquifer (M. H. Young, 2002). Not all rock formations can serve as aquifers. Layers that are sufficiently permeable to transmit water vertically to or from a confined aquifer but not permeable enough to transport water lat-

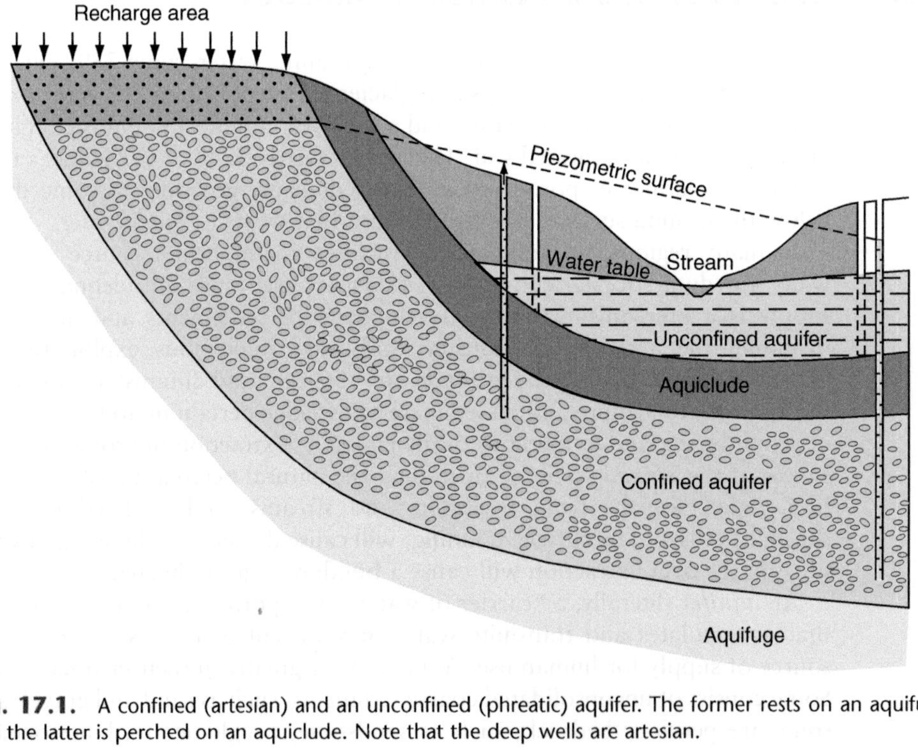

Fig. 17.1. A confined (artesian) and an unconfined (phreatic) aquifer. The former rests on an aquifuge, and the latter is perched on an aquiclude. Note that the deep wells are artesian.

erally are called *aquitards* (Bouwer, 1978). An aquifer bound by one or two aquitards is called a *leaky* or *semiconfined* aquifer.

An unconfined aquifer (also called a *phreatic* aquifer, from the Greek *phrear*, "a well") is one in which the water level is free to fluctuate up and down, to rise and fall periodically, and thus to seek an equilibrium level. Phreatic aquifers may reach levels close to the ground surface, and at times and places even rise above it. Typically, however, the soil remains unsaturated to some depth (called the *unsaturated zone* or *vadose zone*, or the *zone of aeration*), below which a saturated condition prevails in what is known as the *zone of saturation*.

For more fundamental expositions of groundwater theory, the reader is referred to publications by Freeze and Cherry (1979), Ward and Robinson (1990), and Bras (1990). Problems of groundwater pollution were described by Fried (1976) and by Pepper et al. (1996).

CONFINED GROUNDWATER FLOW

We now describe a number of simple cases illustrating flow phenomena in confined aquifers. The simplest example is that of flow through a horizontal layer of uniform thickness and properties, sandwiched between impervious layers. As illustrated in Fig. 17.2, flow takes place from the channel on the left to the channel on the right. If the water level in each of the two channels is maintained at a constant level, the flow will take place at a steady state, that is, a state in which the flux and gradient are time invariant and the flux is space invariant as well.

The process can be formulated as follows. Begin with the Darcy equation (with q being flux, K conductivity, H hydraulic head, and x horizontal distance):

$$q = -K(dH/dx) \tag{17.1}$$

Now separate differential variables:

$$dH = -(q/K)\,dx$$

Next, integrate

$$\int_{H_1}^{H_2} dH = -(q/K)\int_0^x dx \tag{17.2}$$

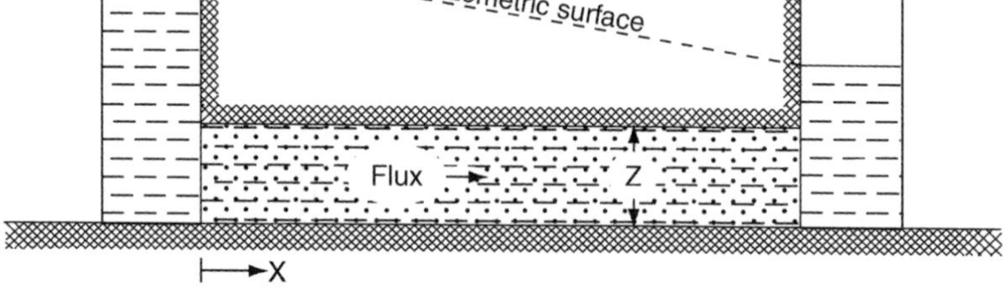

Fig. 17.2. Flow in a horizontal confined aquifer of uniform thickness and hydraulic properties.

to obtain

$$H_2 - H_1 = -(q/K)x \qquad \text{or} \qquad H_2 = H_1 - (q/K)x \tag{17.3}$$

which indicates that, as long as the flux and conductivity are constant, the hydraulic head diminishes linearly with distance along the direction of flow.

To calculate the *discharge* Q, that is, the total volume of flow per unit time per unit thickness (perpendicular to the plane of the drawing), we multiply the flux by the height of the aquifer Z:

$$Q = qZ = -KZ(dH/dx) = -\tau(dH/dx) \tag{17.4}$$

Here $\tau = KZ$ is a composite parameter called the *transmissivity* that characterizes the amount of water obtainable from an aquifer under a unit hydraulic gradient. In evaluating the potential usefulness of an aquifer as a source of water, the transmissivity rather than the conductivity is the parameter of interest, since a less permeable aquifer can deliver more water than a highly permeable one if the dimensions of the former are sufficiently larger than those of the latter.

Now let us consider the slightly more complex case of steady radial flow toward a single well. Suppose that the initial piezometric head at the well was H_i and that this head was uniform for an infinite distance in the region surrounding the well, as shown in Fig. 17.3. As we begin to pump water out of the well, we notice the piezometric head at, and in the vicinity of, the well. A hydraulic-head gradient is thereby created, and water flows radially inward toward the well from the surroundings, creating a *cone of depression* in the piezometric head. If we continue pumping water at a steady rate, and if the head remains constant at some distance from the well, the cone of depression will stabilize at the exact configuration that will supply the well with the amount of water withdrawn per unit time. Henceforth the depression of piezometric head (from the original level H_i) at each point, called the *drawdown*, will be a function of radial distance and of discharge, but not of time.

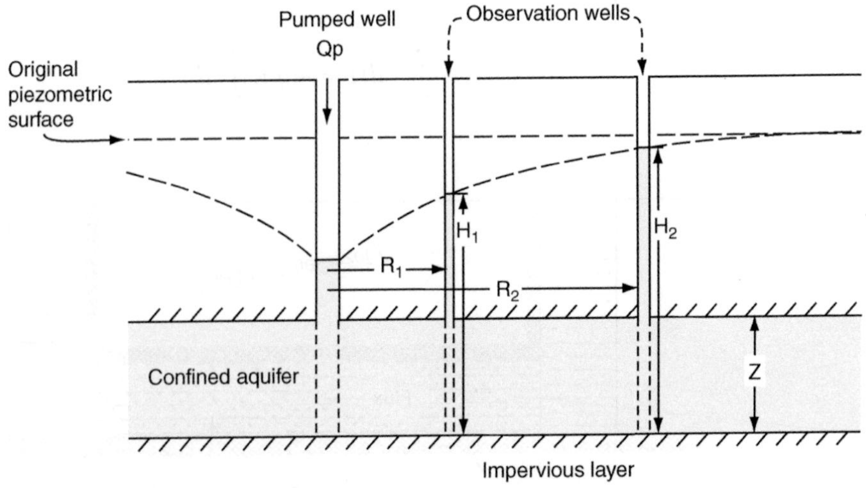

Fig. 17.3. Cone of depression formed during radial low to a pumped well.

To formulate this process in mathematical terms, we cast Darcy's law into the radial form (with r, the radius, as the space coordinate):

$$q = -K(dH/dr) \tag{17.5}$$

To obtain the discharge through any concentric cylindrical surface at any radial distance r from the well, we multiply the flux q by the area through which flow takes place, namely, $2\pi rZ$ (where $2\pi r$ is the circumference of the circle and Z, once again, is the height, or thickness, of the aquifer):

$$Q = -2\pi rZK\ (dH/dr) \tag{17.6}$$

At steady state, the discharge through all concentric cylindrical surfaces must equal the discharge from the well, that is, the pumping rate $-Q_p$ (volume withdrawn per unit time), so that in Eq. (17.6) $Q = -Q_p$. We can now separate the differential variables:

$$dH = (Q_p/2\pi ZK)\ (dr/r)$$

and integrate between any arbitrary radial distances r_1 and r_2:

$$\int_{H_1}^{H_2} dH = (Q_p/2\pi ZK)\int_{r_1}^{r_2} dr/r \tag{17.7}$$

to obtain

$$H_2 - H_1 = (Q_p/2\pi ZK)\ \ln(r_2/r_1) \tag{17.8}$$

If we wish to measure the aquifer transmissivity $\tau = ZK$, we can drill two observation wells at radial distances r_1 and r_2 from the pumping well, begin pumping at a steady rate, and wait until the piezometric levels in the two observation wells stabilize. Then, knowing the values of r_1, r_2, H_1, H_2, and Q_p, we can calculate the value of τ from the well-known *Thiem equation*:

$$\tau = ZK = [Q_p/2\pi(H_2 - H_1)]\ \ln(r_2/r_1) \tag{17.9}$$

Equation (17.9) presupposes steady-state conditions. To obtain a solution pertaining to the transient-state, falling-head phase of the process, we must first consider the change in aquifer storage resulting from a change in piezometric head.

In an unconfined aquifer, a fall in piezometric head means a fall in the water table and consequent desaturation of a corresponding depth increment in the soil, so the effective height (or thickness) of the aquifer is visibly reduced. But a confined aquifer remains saturated even while the piezometric head declines, so it is difficult to see where the water comes from.

A decline of piezometric head implies a decrease of hydrostatic pressure, which causes decompression (expansion) of water. The volume of water remaining in the aquifer can thus remain constant, although some water is withdrawn. Strictly speaking, however, the volume does not remain constant, since the overburden (i.e., the weight of the strata overlying the confined aquifer), formerly supported in part by the hydrostatic pressure in the aquifer, causes an incremental compression of the aquifer (possibly resulting in surface subsidence) as the hydrostatic pressure is partially relieved. Consequently, a decline in piezometric head even in a confined aquifer releases water, and the volume released per unit area per unit decline in piezometric level is generally called *storativity*.

Sample Problem

Two confined aquifers, geometrically similar to the one depicted in Fig. 17.2, were compared, as follows:

	Aquifer A	*Aquifer B*
Horizontal distance between channels	200 m	90 m
Hydraulic-head difference between channels	10 m	15 m
Discharge per unit length of channel	$0.02\ m^3/m\ hr$	$0.05\ m^3/m\ hr$
Vertical thickness of conducting layer	8 m	6 m

Calculate the transmissivity and hydraulic conductivity values.

The system is described by a form of Eq. (17.4):

$$Q = KZ(\Delta H/\Delta x) = \tau(\Delta H/\Delta x)$$

where Q is discharge, K is hydraulic conductivity, Z is vertical thickness of the aquifer, ΔH is hydraulic-head difference, Δx is distance ($\Delta H/\Delta x$ being the hydraulic gradient), and τ is transmissivity (= KZ).

To calculate transmissivity, we set

$$\tau = Q/(\Delta H/\Delta x)$$

and to calculate conductivity we set

$$K = Q/Z(\Delta H/\Delta x) = \tau/Z$$

For Aquifer A,

$$\tau = (0.02\ m^3/m\ hr)/(10\ m/200\ m) = 0.4\ m^2/hr$$
$$K = (0.4\ m^2/hr)/8\ m = 0.05\ m/hr = 1.39 \times 10^{-5}\ m/sec$$

For aquifer B,

$$\tau = (0.08\ m^3/m\ hr)/(15\ m/90\ m) = 0.48\ m^2/hr$$
$$K = (0.48\ m^2/hr)/6\ m = 0.08\ m/hr = 2.22 \times 10^{-5}\ m/sec$$

Sample Problem

Consider a confined aquifer of the type depicted in Fig. 17.3, with a single pumping well and two piezometers located at 50-m intervals along a radial line from the well. Water is pumped out continuously at the rate of $100\ m^3/hr$. After the "cone of depression" (drawdown) stabilizes and steady-state conditions are established, the hydraulic-head difference between the two piezometers is 2 m. Calculate the transmissivity of the aquifer. If its thickness is 60 m, calculate the aquifer's hydraulic conductivity.

This steady-state radial-flow case is described by Eq. (17.9), known as the *Thiem equation*:

$$\tau = ZK = [Q_p/2\pi(H_2 - H_1)] \ln(r_2/r_1)$$

where τ is transmissivity, Z is aquifer thickness, K is hydraulic conductivity, Q_p is pump discharge, and H_1 and H_2 are hydraulic-head values at the piezometers located at radial distances r_1 and r_2, respectively, from the well. Substituting the appropriate values, we can write:

$$\tau = 60 \text{ m} \times K = [(100 \text{ m}^3/\text{hr})/(6.28 \times 2 \text{ m})] \ln(100 \text{ m}/50 \text{ m}) = 5.52 \text{ m}^2/\text{hr}$$
$$K = (5.52 \text{ m}^2/\text{hr})/60 \text{ m} = 2.55 \times 10^{-5} \text{ m/sec}$$

UNCONFINED GROUNDWATER FLOW

Whereas water in unsaturated soil is affected by suction gradients and its movement is subject to variations in conductivity resulting from changes in soil wetness, groundwater (by definition) is always under positive hydrostatic pressure and hence saturates the soil. Thus, no suction gradients and no variations in wetness or conductivity normally occur below the water table. The hydraulic conductivity is maximal and fairly constant in time, though it may vary in space and direction. (We are disregarding here possible effects due to overburden pressures and swelling phenomena as well as to flocculation–dispersion phenomena.)

Despite the differences between the saturated and unsaturated zones, the two are not independent realms but parts of a continuous flow system. Groundwater is recharged by percolation through the unsaturated zone, and the position of its surface (the water table) is determined by the relative rates of recharge versus discharge. Reciprocally, the position of the water table affects the moisture profile and flow conditions above it. One problem encountered in attempting to distinguish between the unsaturated and saturated zones is that the boundary between them may not be at the water table but at some elevation above it, corresponding to the upper limit of the capillary fringe (at which the suction is equal the soil's air-entry value). This boundary may be diffuse and hardly definable, especially when affected by hysteresis.

Water may seep into or out of the saturated zone via the surface or by lateral flow via the sides of a channel or a porous drainage tube. The groundwater table is hardly ever entirely level and may exhibit steep gradients in the drawdown regions near drainage channels, tubes, and wells. Where the topography of the land is variable, as well as where the inflow from precipitation or from stream seepage varies areally, the water table can change in depth and may in places and at times intersect the soil surface and emerge as free water ("outflow").

If the depth of the water table remains constant, the indication is that the rate of inflow to the groundwater and the rate of outflow are equal. In other words, where there is downward seepage out of the unsaturated zone, this must be offset by downward or horizontal outflow of the groundwater if the water table is to remain stationary. On the other hand, a rise or fall of the

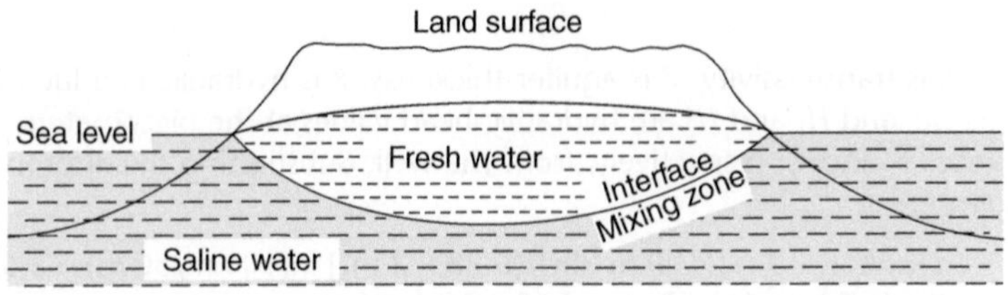

Fig. 17.4. Idealized representation of a Ghyben–Herzberg lens of fresh water floating on saline water under an island or narrow peninsula.

water table indicates a net recharge or discharge of groundwater, respectively. Such vertical displacements of the water table can occur periodically, as under a seasonally fluctuating regimen of rainfall or irrigation. The rise and fall of the water table can also be affected by barometric pressure changes, though generally to a minor degree.

Groundwater flow can be geometrically complex where the profile is heterogeneous or anisotropic or where sources and sinks of water are distributed unevenly. If the profile above the water table consists of a sequence of layers such that a highly conductive one overlies one of low conductivity, then it is possible for the flow rate into the top layer to exceed the transmission rate through the lower layer. In such circumstances, the accumulation of water over the interlayer boundary can result in the development of a "perched" water table with positive hydrostatic pressures. If infiltration persists at a relatively high rate, the two bodies of water will eventually merge. After infiltration ceases, the perched water table tends gradually to dissipate by evapotranspiration and by downward seepage into the primary water table.

In some cases, groundwaters of different density (caused by temperature or salinity differences) may come into contact. An example is the occurrence along seacoasts of a body of fresh water (called a *Ghyben–Herzberg lens*; see Fig. 17.4) over a body of saline water. At the interface of the two bodies, miscible displacement phenomena can be observed (Bear, 1969).

BOX 17.1 Quicksand

Shallow groundwater may, at certain places and times, exhibit upward pressure gradients. Where the overlying material consists of fine, smooth, rounded sand grains with little tendency to mutual adherence, such upwelling of water may cause the grains to be buoyed up so that they are, in effect, practically weightless (a process called *elutriation*). In such conditions, the sand readily yields to pressure applied from the top, and tends to "swallow" heavy objects placed on the surface. Such was frequently the fate of outlaws in the old Wild West movies. They could have avoided that fate had they studied soil physics.

ANALYSIS OF FALLING WATER TABLE

Truly steady-state flow processes are rare in unconfined aquifers (as elsewhere). More typical are transient-state processes, which involve a change in water-table height. Transient flow under such conditions has been described in terms of the *specific yield* concept, generally defined as the volume of water extracted from the groundwater per unit area when the water table is lowered a unit distance. In other words, it is the ratio of drainage flux to rate of fall of the water table. The assumption that there exists anything like a fixed value of *drainable porosity* and that this fraction of the soil volume drains instantaneously as the water table descends is a gross approximation. In fact, the volume of water drained increases gradually with the increasing suction that accompanies the progressive descent of the water table.

According to Luthin (1966), the drainable porosity f_d is not a constant but a function of the negative capillary pressure h (i.e., the matric suction head) and can be written as $f_d(h)$. As the water table drops from h_1 to h_2, the volume of water V_w drained out of a unit column will be

$$V_w = \int_{h_2}^{h_1} f_d(h)\, dh \tag{17.10}$$

The function $f_d(h)$ is related to the soil moisture characteristic, which is difficult to formulate as an exact function. However, it is sometimes possible to write an approximate expression for the relation of drainable porosity, specific yield, and retention for different soils to capillary pressure (or suction) and still end up with a reasonable prediction of the amount of water that drains out of a profile when the water table is lowered. The simplest equation to use is that of a straight line, $f_d = ah$, where a is the slope of the line. The quantity of water drained is now given by

$$V_w = \int_{h_2}^{h_1} ah\, dh = \frac{a}{2}(h_1^2 - h_2^2) \tag{17.11}$$

The time-dependent vertical drainage of a soil column following a drop of the water table has been studied by Ward and Robinson (1990), who considered the interrelationships among porosity, specific yield, and retention for different soils. Gardner (1962a) assumed a constant mean diffusivity and a linear relation between hydraulic head and soil wetness. His equation is

$$V_w/V_{w\infty} = 1 - (8/\pi^2)\exp(-D_m \pi^2 t/4L^2) \tag{17.12}$$

where V_w is the volume of water per unit area removed during time t, $V_{w\infty}$ is the total drainage after infinite time, D_m is the weighted mean diffusivity of the soil, and L is the column length (height).

The inherent complexity of these relationships makes it difficult to devise comprehensive and exact theories. Hence researchers have tended to devise approximate theories. The applicability of such theories depends on the validity of the assumptions made in their derivation. Some approximate solutions that provide reasonably accurate predictions of drainage outflow do not give an accurate description of the changing moisture and suction profiles above the water table (Fig. 17.5). More exact solutions of nonsteady drainage problems can be obtained by numerical techniques (e.g., Neuman, 1975; Sinai et al., 1987). The role of encapsulated air in the process of drainage was considered by Collis-George and Yates (1990).

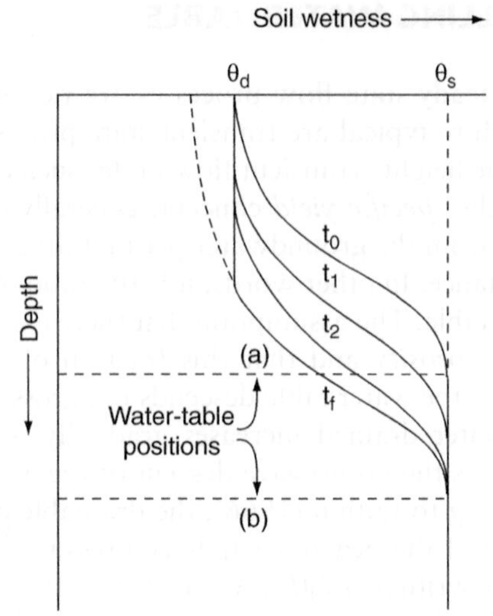

Fig. 17.5. Idealized succession of soil moisture profiles following a rapid drop of the water table from position (a) to (b), where t_0 and t_f represent the equilibrium moisture profiles at time zero and at final completion of vertical drainages, respectively; t_1 and t_2 represent intermediate times during vertical drainage; and θ_s indicates saturation and θ_d the presumed wetness after drainages; that is, $\theta_s - \theta_f$ = "drainable" porosity. Note that in reality θ_d is generally not a constant, but itself depends on proximity to the water table (as shown by the dashed line extension of the t_f profile), on siol-profile characteristics, and on the fact that true equilibrium is practically never attained. Moreover, the soil-moisture profile above a fluctuating water table also depends on hysteresis and, if the water table is close to he soil surface, on plant activity as well.

Sample Problem

An initially saturated vertical soil column is drained by dropping the water table abruptly to a level 1 m below its original height, as depicted in Figure 17.6. Plot the fractional amount of drainable water removed as a function of time if the weighted mean diffusivity of the draining soil is 10^{-6} m^2/sec. Use Eq. (17.12).

Following Gardner (1962a), the volume of water V_w drained per unit area, as a fraction of the total drainable water $V_{w\infty}$ (i.e., the cumulative volume per unit area that drains after infinite time), is a function of time t, weighted mean diffusivity D_m, and column length L:

$$V_w/V_{w\infty} = 1 - (8/\pi^2)\exp(-D_m\pi^2 t/4L^2)$$

We can now substitute the appropriate values of D_m and L and assign successive values of t to obtain actual solutions, as follows:

At $t = 1$ hr = 3600 sec,

$$V_w/V_{w\infty} = 1 - 8/(9.87 \times e^{\varepsilon})$$

where $\varepsilon = 10^{-6} \times 9.87 \times 3600/(4 \times 1^2) = 8.88 \times 10^{-3}$. Hence:

$$V_w/V_{w\infty} = 0.197 = 19.7\%$$

At $t = 4$ hr = 14,400 sec,

$$\varepsilon = 10^{-6} \times 9.87 \times 14{,}400/(4 \times 1^2) = 3.55 \times 10^2$$
$$V_w/V_{w\infty} = 1 - 8/(9.87 \times e^{0.0355}) = 0.218 = 21.8\%$$

At $t = 1$ day = 86,400 sec,

$$\varepsilon = 10^{-6} \times 9.87 \times 86{,}400/(4 \times 1^2) = 0.213$$
$$V_w/V_{w\infty} = 1 - 8/(9.87 \times e^{0.213}) = 0.345 = 34.5\%$$

At $t = 2$ days = 172,800 sec,

$$\varepsilon = 10^{-6} \times 9.87 \times 172{,}800/(4 \times 1^2) = 0.426$$
$$V_w/V_{w\infty} = 1 - 8/(9.87 \times e^{0.426}) = 0.471 = 47.1\%$$

At $t = 4$ days = 345,600 sec,

$$\varepsilon = 0.852,$$
$$V_w/V_{w\infty} = 1 - 8/(9.87 \times e^{0.852}) = 0.654 = 65.4\%$$

At $t = 8$ days = 691,200 sec,

$$\varepsilon = 1.704$$
$$V_w/V_{w\infty} = 1 - 8/(9.87 \times e^{1.704}) = 0.853 = 85.3\%$$

The actual plotting of these data is left as an exercise for students, who may also wish to compare Gardner's theory with one or more of the other theories extant.

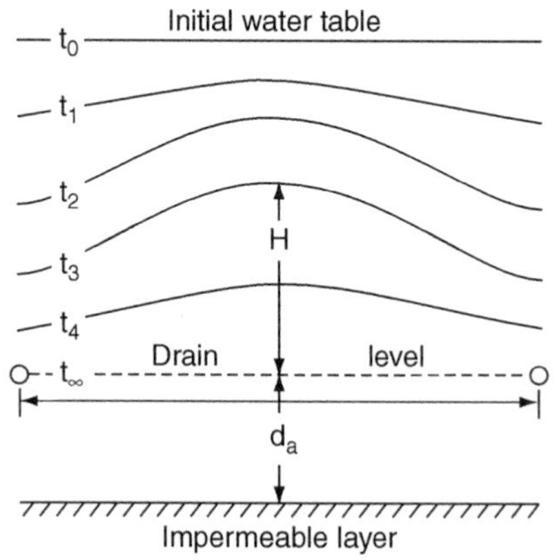

Fig. 17.6. Idealized succession of water-table positions during drainage of an initially saturated profile.

EQUATIONS FOR FLOW OF UNCONFINED GROUNDWATER

As stated in Chapters 7 and 8, Darcy's law alone is sufficient to describe only steady-flow processes. For unsteady flow, Darcy's law must be combined with the mass conservation law to obtain the general flow equation. For homogeneous isotropic media, the equation takes the form (de Wiest, 1969)

$$\frac{\partial \theta}{\partial t} = -\left(\frac{\partial q_x}{\partial x} + \frac{\partial q_y}{\partial y} + \frac{\partial q_z}{\partial z}\right)$$

$$\frac{\partial \theta}{\partial t} = K\left(\frac{\partial^2 H}{\partial x^2} + \frac{\partial^2 H}{\partial y^2} + \frac{\partial^2 H}{\partial z^2}\right) \tag{17.13}$$

where q_x, q_y, and q_z are the fluxes in the x, y, and z directions, respectively, θ is wetness, t is time, K is conductivity, and H is hydraulic head. In a saturated, stable medium, there is no change of wetness (water content) with time, and we obtain

$$K_s\left(\frac{\partial^2 H}{\partial x^2} + \frac{\partial^2 H}{\partial y^2} + \frac{\partial^2 H}{\partial z^2}\right) = 0 \tag{17.14}$$

where K_s is the saturated conductivity. Since K_s is not zero, it follows that

$$\frac{\partial^2 H}{\partial x^2} + \frac{\partial^2 H}{\partial y^2} + \frac{\partial^2 H}{\partial z^2} = 0 \tag{17.15}$$

This is known as the *Laplace equation*. The expression $\partial^2/\partial x^2 + \partial^2/\partial y^2 + \partial^2/\partial z^2$, or in vector notation ∇^2, is known as the *Laplacian operator*. Accordingly, we can write Laplace's equation $\nabla^2 H = 0$. If, instead of using Cartesian coordinates (x, y, z), we cast Eq. (17.15) into cylindrical coordinates (r, α, z), we obtain

$$\frac{1}{r}\frac{\partial}{\partial r}\left(r\frac{\partial H}{\partial r}\right) + \frac{1}{r^2}\frac{\partial^2 H}{\partial \alpha^2} + \frac{\partial^2 H}{\partial z^2} = 0 \tag{17.16}$$

Laplace's equation also applies to systems other than fluid flow in porous media, that is, to flow of heat in solids and of electricity in electrical conductors. Solutions for boundary conditions appropriate to such systems, some of which are also applicable to soil-water flow, are given by Carslaw and Jaeger (1959).

The direct analytical solution of Laplace's equation for conditions pertinent to groundwater flow is generally not possible. Therefore, it is often necessary to resort to approximate or indirect methods of analysis. Where flow is restricted to two dimensions, the equation becomes

$$\partial^2 H/\partial x^2 + \partial^2 H/\partial y^2 = 0 \tag{17.17}$$

which is more readily soluble (Childs, 1969; Domenico, 1972).

In the solution of problems relating to unconfined groundwater flow toward a shallow sink (a drainage tube or ditch), it is often convenient to employ the

Dupuit–Forchheimer assumptions (Kirkham, 1967) that, in a system of gravity flow toward a shallow sink, all the flow is horizontal and that the velocity at each point is proportional to the slope of the water table but independent of depth. Though these assumptions are obviously not correct in the strict sense and can in some cases lead to anomalous results, they often provide feasible solutions in a form simpler than obtainable by rigorous analysis.

DRAINAGE OF SHALLOW GROUNDWATER

The term *drainage* can be used in a general sense to denote outflow of water from soil. More specifically, it can serve to describe the artificial removal of excess water or the set of management practices designed to prevent the occurrence of excess water (E. Farr and Henderson, 1986). The removal of free water tending to accumulate over the soil surface by appropriately shaping the land is termed *surface drainage* and is outside the scope of our present discussion. Finally, *groundwater drainage* refers to the outflow or artificial removal of excess water from within the soil, generally by lowering the water table or by preventing its rise.

Soil saturation per se is not necessarily harmful to plants. The roots of many plants can thrive in water, provided it is free of toxic substances and contains sufficient oxygen to allow normal respiration. Plant roots must respire constantly, and most terrestrial plants are unable to transfer the required flux of oxygen from their canopies to their roots. The problem is that water in a saturated soil seldom can provide sufficient oxygen for root respiration. Excess water in the soil tends to block soil pores and thus retard aeration and in effect strangulate the roots.

In *waterlogged soils*, gas exchange with the atmosphere is restricted to the surface zone of the soil, while within the profile proper, oxygen may be almost totally absent and carbon dioxide may accumulate. Under anaerobic conditions, various substances are reduced from their normally oxidized states. Toxic concentrations of ferrous, sulfide, and manganous ions can develop. These, in combination with products of the anaerobic decomposition of organic matter (e.g., methane) can greatly inhibit plant growth. At the same time, nitrification is prevented, and various plant and root diseases (especially fungal) are more prevalent.

The occurrence of a high water table may not always be clearly evident at the very surface, which may be deceptively dry even while the soil is completely waterlogged just below the surface zone. Where the effective rooting depth is thus restricted, plants may suffer not only from lack of oxygen in the soil but also from lack of nutrients. If the water table drops periodically, shallow-rooted plants growing in waterlogged soils may even, paradoxically, suffer from occasional lack of water, especially when the transpirational demand is very high.

High moisture conditions at or near the soil surface make the soil susceptible to compaction by animal and machinery traffic. Necessary operations (e.g., tillage, planting, spraying, and harvesting) are thwarted by *poor trafficability* (i.e., the ability of the ground to support vehicular traffic and to provide the necessary traction for locomotion). Tractors are bogged down and cultivation tools are clogged by the soft, sticky, wet soil. Furthermore, the surface zone of a wet soil does not warm up readily at springtime, owing to greater thermal inertia and downward conduction and to loss of latent heat via the higher evaporation rate. Consequently, germination and early seedling growth are retarded.

In a warm climate, the evaporation rate and, hence, the hazard of salinity are likely to be greater than in a cool climate. The process of evaporation inevitably results in the deposition of salts at or near the soil surface, and these salts can be removed and prevented from accumulating only if the water table remains deep enough to permit leaching without subsequent resalination through capillary rise of the groundwater (see Chapter 9). Irrigated lands, even in arid regions, frequently require drainage. In fact, irrigation without drainage is, more often than not, unsustainable (Hillel, 2000).

In large areas, therefore, drainage is required for the long-term maintenance of soil productivity. Irrigated agriculture cannot long be sustained in many arid regions unless drainage is provided for salinity control as well as for effective soil aeration. On the other hand, the presence of a shallow water table in the soil profile (provided it is not too shallow) can in certain circumstances be beneficial. Where precipitation or irrigation water is scarce, the availability of groundwater within reach of the roots can supplement the water requirements of crops. However, to obtain any lasting benefit from the presence of a water table in the soil, its level and fluctuation must be controlled.

DRAINAGE DESIGN EQUATIONS

Various equations, empirically or theoretically based, have been proposed for the purpose of determining the desirable depths and spacings of drain pipes or ditches in different soil and groundwater conditions. Since field conditions are often complex and highly variable, these equations are generally based on assumptions that idealize and simplify the flow system. The available equations are therefore approximations that should not be applied blindly. Rather, the assumptions must be examined in the light of all information obtainable concerning the circumstances at hand.

One of the most widely applied equations is that of Hooghoudt (1937), designed to predict the height of the water table that will prevail under a given rainfall or irrigation regime when the conductivity of the soil and the depth and horizontal spacing of the drains are known. This equation, like others of its type, oversimplifies the real field situation, for it disregards additional factors affecting groundwater movement, such as soil layering and the variable rate of evapotranspiration. It is based on the following tacit assumptions:

1. The soil is uniform and of constant hydraulic conductivity.
2. The drains are parallel and equally spaced.
3. The hydraulic gradient at each point beneath the water table is equal to the slope of the water table above that point.
4. Darcy's law applies.
5. An impervious layer exists at a finite depth below the drain.
6. The supply of water from above is at a constant flux q.

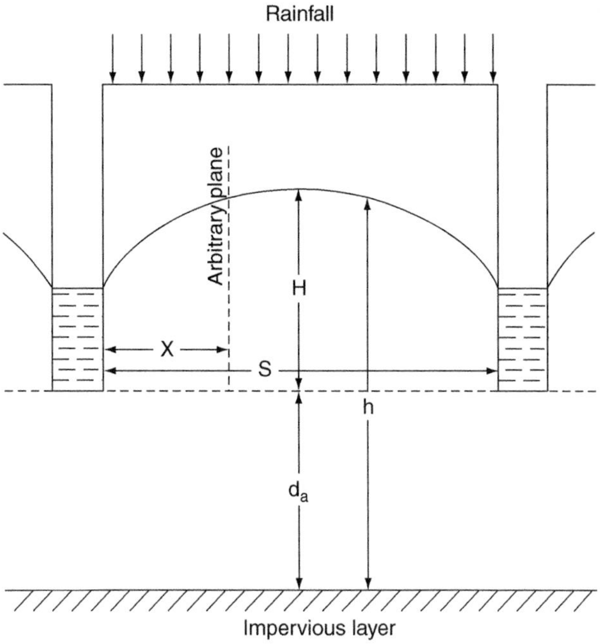

Fig. 17.7. Model used in derivation of Hooghoudt's equation.

To derive the Hooghoudt equation, let us examine flow in a profile section of a field having a width of one unit and bounded on its sides by two adjacent drains (tubes or ditches) a distance S apart (Fig. 17.7). Assuming symmetry, we can draw a vertical midplane between the drains, which will divide flow toward one drain from flow to the other. Now let us consider flow toward one of the drains through any arbitrary vertical plane located a distance x from that drain. The quantity of water passing through this plane per unit time must be equal to the percolation flux q multiplied by the width from the arbitrary plane to the midplane between the drains. This width is $(S/2 - x)$. Accordingly, the horizontal flow per unit time through the arbitrary plane is

$$Q = -q\left(\frac{1}{2}S - x\right) \tag{17.18}$$

The value of Q can also be obtained from Darcy's law. If we assume the effective gradient to be equal to the slope of the water table (dh/dx) at the arbitrary vertical plane, we get

$$Q = -Kh\ (dh/dx) \tag{17.19}$$

where K is the hydraulic conductivity and h is the height of the water table above an impervious layer that is assumed to form the "floor" of the flow system. Now we can equate the two equations:

$$q\left(\frac{1}{2}S - x\right) = Kh(dh/dx) \quad \text{or} \quad \frac{1}{2}qS\,dx - qx\,dx = Kh\,dh \tag{17.20}$$

which can be integrated to yield

$$qSx - qx^2 = Kh^2 \tag{17.21}$$

Assuming that at $x = 0$ (i.e., at the drain), $h = d_a$ (the height of the drain above the impervious floor), while at $x = \frac{1}{2}S$ (the midplane), $h = H + d_a$ (where H is the maximal height of the water table above the drains), we obtain Hooghoudt's equation for the elliptical shape of the water table between drains:

$$S^2 = (4\ KH/q)\ (2d_a + H) \tag{17.22}$$

This equation has been widely used for determining the desirable spacing and depth of drains needed to maintain the water table below a certain level. That level, as well as the average infiltration flux and hydraulic conductivity, must be known a priori. A depth must also be known, or assumed, for the impervious layer.

In the event that H is negligible compared to d_a, we can write

$$Q = -Kd_a\ (dH/dx)$$

Equating this with (17.18), we obtain

$$q\left(\frac{1}{2}S - x\right) = Kd_a\left(dH/dx\right),\quad \frac{1}{2}qS\,dx - qx\,dx = Kd_a\,dH$$

Integration yields

$$qSx - qx^2 = 2KHd_a + \text{constant} \tag{17.23}$$

Since $H = 0$, at $x = 0$, the constant of integration is also $= 0$. Therefore

$$H = qx(S - x)/2Kd_a \tag{17.24}$$

At the midpoint between drains, where $x = \frac{1}{2}S$, the maximum height H_{max} of the water-table mound

$$H_{max} = qS^2/8Kd_a \tag{17.25}$$

The height of the rise of the water table between drains is related directly to the recharging flux q and the square of the distance S between drains and inversely to the soil's hydraulic conductivity K.

An equation describing the water-table drop at the midpoint between drains following an abrupt water-table drop at the drains is the *Glover equation* (Smedema and Rycroft, 1983):

$$S^2 = (\pi^2 K h_{i,\,ave}\ t/f_d)\ \ln(4H_i/\pi H) \tag{17.26}$$

Here $h_{i,ave}$ is the average initial depth of the water-bearing stratum, f_d is the assumed drainable porosity, H_i is the initial height of the midpoint water table above the drains, and H is the height at any time t.

The ranges of depth and spacing generally used for the placement of drains in field practice are shown in Table 17.1.

In Holland, the country with the most experience in drainage, common criteria for drainage are to provide for the removal of about 7 mm/day and to prevent a water-table rise above 0.5 m from the soil surface. In more arid

TABLE 17.1 Prevalent Depths and Spacings of Drainage Tubes in Various Soil Types

Soil type	Hydraulic conductivity (m/day)	Spacing of drains (m)	Depth of drains (m)
Clay	1.5	10–20	1–1.5
Clay loam	1.5–5	15–25	1–1.5
Loam	5–20	20–35	1–1.5
Fine, sandy loam	20–65	30–40	1–1.5
Sandy loam	65–125	30–70	1–2
Peat	125–250	30–100	1–2

regions, because of the greater evaporation rate and groundwater salinity, the water table must generally be kept much deeper. In the Imperial Valley of California, the drain depth ranges from 1.5 to 3 m and the water-table depth midway between drains is about 1.20 m. For medium- and fine-textured soils the depth should be greater still where the salinity risk is high. Since there is a practical limit to the depth of drain placement, it is the density of drain spacing that must be increased under such circumstances. Setting parallel drainage lines closer together is an alternative way to ensure that their drawdown curves will be deeper at the midpoint between adjacent lines.

Sample Problem

Use the Hooghoudt equation to compare the necessary drain spacings S for two soils with hydraulic conductivity K values of 10^{-6} and 10^{-7} m/sec. Assume the allowable maximum water-table mound H_{max} to be 1 m above the drains, which are 2 m (d_a) above an impervious stratum. Total rainfall is 1200 mm and total evapotranspiration 1000 mm during a 6-month growing season.

Recall Eq. (17.25):

$$H_{max} = qS^2/8Kd_a$$

which we can solve for S:

$$S = (8Kd_aH_{max}/q)^{0.5}$$

where q is average percolation flux. Substituting the given values, we have:

For Soil A,

$$S = [(8 \times 10^{-6}\text{m/sec} \times 2\text{ m} \times 1\text{ m})/(0.2\text{ m}/15{,}760{,}000\text{ sec})]^{0.5}$$

in which the number of seconds by which we divide the denominator represents a 6-month period. Hence,

$$S = 35.5\text{ m}$$

For Soil B,

$$S = [(8 \times 10^{-7}\text{ m/sec} \times 2\text{ m} \times 1\text{ m})/(0.2\text{ m}/15{,}760{,}000\text{ sec})]^{0.5} = 11.23\text{ m}$$

GROUNDWATER POLLUTION

As already stated, the excess of infiltration over evapotranspiration normally flows downward, eventually reaching the water table and contributing to groundwater recharge (Gee and Hillel, 1988). If the infiltrated water contains solutes that are not extracted by plants, reacted or retained in the soil, or volatilized into the atmosphere (or if the percolating water picks up additional solutes en route through the soil), such solutes may contribute to groundwater pollution.

Nitrates

Prominent among the soluble components that may lead to groundwater pollution are the nitrates. They are, as a rule, highly soluble and are not retained by the soil's exchange complex. Nitrates may form in situ by the decomposition of organic matter in well-aerated soils or may be added to the soil as fertilizers. Nitrates may also be removed from the soil by plants (for which nitrogen in soluble form constitutes an essential nutrient), or they may decompose microbially in a process known as *denitrification*, which reduces the nitrogen to a gaseous form that may escape the soil by diffusion to the free atmosphere.

If soil-formed or soil-borne nitrogen migrates to bodies of water (streams and lakes), it may cause the *eutrophication* (i.e., the nutrient enrichment) of water bodies, leading to the proliferation of algae and the oxygen deprivation of numerous species of fish. Finally, if nitrates accumulate in groundwater, they may render the water unsuitable for drinking. Excessive concentrations of nitrates in the water given to babies (mixed into their formula) have been blamed for methemoglobinemia (the so-called "blue baby" syndrome), an abnormality of the blood that impairs its ability to convey oxygen.

In well-aerated soils, nitrogen-containing compounds such as plant residues and animal manures are oxidized to form nitrates. Human household wastes applied to the soil (as in septic tanks, leaching fields, and "sanitary" landfills) further contribute nitrates. The liberal and often excessive application of nitrogenous fertilizers in agriculture also adds to the nitrate pollution of groundwater. Additional important sources are cattle and poultry feedlots, typically located in the vicinity of large cities. Such centers are based on importing feed from extensive and sometimes distant areas, and hence thay tend to concentrate the resulting wastes in population centers, just where groundwater constitutes a major source of drinking water.

Although nitrates move readily through the soil and subsoil, the full effect of their migration on groundwater quality may not be felt for quite a while. That is because of the large volume of dilution within the groundwater aquifer. Moreover, in many cases the water table is quite deep, so the time needed to reach it is long.

The amounts of nitrates already en route to the water table may be such as to cause significant pollution eventually. Well below the root zone, nitrates are no longer subject to degradation and remain essentially stable as they migrate toward the water table. It may take years or even decades to arrive, but eventually they will. The disconcerting prospect is that even if further contributions of nitrates to deep percolation are avoided (as they should be), the process of nitrate addition to groundwater will inevitably continue for some time, due to prior sources that are beyond reach and hence beyond remedy.

As an example, let us assume that the excess of infiltration over evapotranspiration is, say, 100 mm per year. Averaged over time, that would then dictate the average flux leaving the root zone and moving more or less steadily through the vadose zone toward the water table. The average velocity of that movement can be estimated by dividing the net average flux (q_{ave}) by the volumetric wetness θ of the vadose zone:

$$v_{ave} = q_{ave}/\theta \quad (17.27)$$

Nitrates are subject to various processes in the biologically active zone of the soil, including uptake by roots and denitrification by microorganisms. The latter process occurs in ill-drained soils. Even in apparently well-drained soils, however, the interiors of dense clods may remain anoxic long enough to produce conditions favorable to biochemical reduction, especially denitrification. However, denitrification requires the presence of carbon as an energy source. In its absence, denitrification ceases, even though the soil may be water-logged.

Pesticides

Agricultural pesticides comprise a large and varied group of compounds, applied in the field in a constant effort to control infestations by so-called "pests" — namely, weeds, insects, nematodes, fungi, etc. An agricultural field is, after all, an artificially delineated tract where farmers generally try to eradicate the preexisting community of plants and animals in order to grow domesticated plants not capable of surviving in the natural environment. Since most fields remain open areas not isolated from their surroundings, nature's variety of plants and animals, oblivious to being tagged pests, continually try to reinvade their stolen domain. Hence the farmers fight against them is a never-ending one.

Many of the modern pesticides are synthetic compounds that are *xenobiotic* (foreign to natural ecosystems) and hence resistant to chemical and biological degradation and highly persistent. Their toxicity is usually not confined to the target pests, so pesticide applications in the field generally leave residues that may be highly mobile. They may enter the food chain and produce unintended and unexpected (often harmful) effects on nontarget organisms. If soluble, such residues are carried to surface bodies of water as well as to groundwater.

In recent years, awareness has grown of the hazards posed to the larger environment by the wanton use of persistent broad-spectrum pesticides. Consequently, efforts are being made to produce pesticides that are more specific and that break down rapidly after serving their purpose. Additional efforts are being made to use biological control agents (natural enemies of the pest organisms) instead of, on in conjunction with, the chemical pesticides. This practice, known as *integrated pest management*, has been gaining increasing acceptance. Even so, chemical pesticides are still in widespread and perhaps excessive use and pose a threat of groundwater pollution.

Petroleum Products

Numerous petroleum products are used in industrial countries as fuels (for machines of all sorts and for generating electricity), as raw materials for synthesizing plastics, and as lubricants. Such products may be applied to the soil either deliberately or inadvertently. Thousands of underground fuel tanks, for instance, leak part of their contents (e.g., gasoline and fuel oil) into the subsoil, from whence such materials may migrate to and pollute the groundwater below. Although they contain volatile as well as soluble components, as a class these products are known as *non-aqueous-phase liquids* (NAPLs).

In large part, NAPLs are immiscible in water. The modes of their movement and interactions with the water phase in the soil's pores are complex processes that constitute an active area of research at present (Dracos, 1987; Corapciouglu and Baehr, 1987). As water and NAPLs move simultaneously, their interface seldom constitutes a neat plane, and since water is generally the wetting liquid and is often (though not always) of lower viscosity, it tends to displace and bypass masses of NAPLs, which thereby remain trapped in the soil as discontinuous pockets or globules. Efforts have been made to remediate NAPL-contaminated soils and groundwater aquifers by venting the medium so as to drive off the volatile fractions of light NAPLs (e.g., gasoline), by introducing microbes to decompose the contaminants, by emulsifying the heavier residues in water with the aid of detergents, and by filtering the contaminated water prior to its use R. S. Baker, 1998.

Other Toxic Chemicals

The toxic wastes of innumerable industrial processes are disposed of in the ground, whether legally or illegally, openly or surreptitiously. The eventual fate of such compounds depends on their nature, on the type of soil in which they are placed, on the prevailing climate, and — of course — on the soil-water regime. Of particular concern are the so-called "heavy metals" (e.g., copper, lead, zinc, cadmium, and mercury), which may be present in compound or in elemental form. There are several sources for these metals, including lead in gasoline, paints, and various industrial effluents. The ions of such metals may be retained in the soil's exchange complex, but since such retention is not permanent the metals may leach into lakes, rivers, and groundwater, especially under the influence of acid rain. Among the metals with high toxicity are the cations of mercury (Hg^{2+}), lead (Pb^{2+}), arsenic, beryllium, boron, cadmium, chromium, copper, nickel, manganese, selenium, silver, tin, and zinc (R. M. Miller, 1996; Gerba, 1996).

BOX 17.2 On the Role of Macropores

Where macropores exist, a dual-flow regime may occur, consisting of (1) episodic, rapid spurts of water and water-borne materials via the macropores to the deeper layers of the subsoil or directly to groundwater; and (2) slow, continuous (though at a variable rate) conduction of water and solutes through the soil matrix, driven by hydraulic gradients that are variously directed (not always vertically downward). The first type of flow is discontinuous in time and may bypass the greater volume of the soil matrix. The second type of flow is much more persistent. In this version of the "hare and tortoise" race, the one may predominate over the other, or vice versa, depending on circumstances.

The pattern of groundwater recharge may thus be bimodal, with a sharp wave of short duration soon after the start of a rainfall or irrigation event, representing flow through the open macropores, followed later by a flat wave with a slow rise and even slower descent, representing flow through the micropores of the soil matrix. And just as the rates and quantities of the two flows differ, so may differ the transports of different solutes, including potential pollutants. Recently introduced ("exotic") solutes may be conveyed from the surface by the rapid spurts, whereas the solutes present deeper inside the profile, including those generated within the soil as mineralized products of organic matter, are conveyed in the second wave.

18. EVAPORATION FROM BARE SOIL AND WIND EROSION

EVAPORATION PROCESSES

Evaporation in the field can take place from plant canopies, from the soil surface, or, more rarely, from a free-water surface (F. E. Jones, 1991). Evaporation from plants, called *transpiration*, is the principal mechanism of soil-water transfer to the atmosphere when the soil surface is covered with vegetation. Soil-water uptake and transpiration by plants is, however, the subject of our next chapter. When the surface is at least partly bare, evaporation can take place from the soil directly. Since these two interdependent processes are generally difficult to separate, they are commonly lumped together and treated as if they were a single process, called *evapotranspiration*. Some scientists, however, object to the latter term, believing it to be both cumbersome and unnecessary; they refer to all processes of vapor transfer to the atmosphere — from soil and plants alike — as evaporation (Monteith, 1973).

In the absence of vegetation and when the soil surface is subject to radiation and wind effects, evaporation occurs entirely from the soil. This process is the subject of our present chapter. It is a process that, if uncontrolled, can involve very considerable losses of water in both irrigated and unirrigated agriculture. Under annual field crops, the soil surface may remain largely bare throughout the periods of tillage, planting, germination, and early seedling growth, periods in which evaporation can deplete the moisture of the surface soil and thus hamper the growth of young plants during their most vulnerable stage. Rapid drying of a seedbed can thwart germination and thus doom an entire crop from the outset. The problem can also be acute in young orchards, where the soil surface is often kept bare continuously for several years, as well as in dryland farming in arid zones, where the land is regularly fallowed for several months to collect and conserve rainwater from one season to the next.

Evaporation of soil moisture involves not only loss of water but also the danger of soil salination. This danger is greatest in arid areas, where annual rainfall is low, irrigation water is brackish, and the groundwater table is high.

PHYSICAL CONDITIONS

Three conditions are necessary for evaporation to occur and persist. First, there must be a continual supply of heat to meet the latent heat requirement (about 2.5×10^6 J/kg, or 590 cal/g of water evaporated at 15°C). This heat can come from the body itself, thus causing it to cool, or — as is more common — it can come from the outside in the form of radiated or advected energy. Second, the vapor pressure in the atmosphere over the evaporating body must remain lower than the vapor pressure at the surface of that body (i.e., there must be a vapor pressure gradient between the body and the atmosphere), and the vapor must be transported away by diffusion or convection or both. These two conditions — supply of energy and removal of vapor — are generally external to the evaporating body and are influenced by meteorological factors, such as air temperature, humidity, wind velocity, and radiation, which together determine the *atmospheric evaporativity* (the maximal flux at which the atmosphere can vaporize water from a free-water surface).

The third condition for evaporation to be sustained is that there be a continual supply of water from or through the interior of the body to the site of evaporation. This condition depends on the content and potential of water in the body as well as on its conductive properties, which together determine the maximal rate at which water can be transmitted to the evaporation site (usually, the surface). Accordingly, the actual evaporation rate is determined either by the *evaporativity* of the atmosphere or by the soil's own ability to deliver water (sometimes called the *evaporability* of soil moisture), whichever is the lesser (and hence the limiting factor).

If the top layer of soil is initially quite wet, as it typically is at the end of an infiltration episode, the process of evaporation will generally reduce soil wetness and thus increase matric suction at the surface. This, in turn, will cause soil moisture to be drawn upward from the layers below, provided they are sufficiently moist.

Among the sets of conditions under which evaporation may take place are the following:

1. A shallow groundwater table may be present within the soil, or it may be absent (or too deep to affect evaporation). Where a groundwater table occurs close to the surface, continual flow may take place from the saturated zone beneath through the unsaturated soil to the surface. If this flow is more or less steady, continued evaporation can occur without materially changing the

soil-moisture content (though nonvolatile solutes my accumulate at the surface). In the absence of shallow groundwater, however, the loss of water at the surface and the resulting upward flow of water in the profile will necessarily be a transient-state process tending gradually to dry out the soil.

2. The soil profile may be uniform. Alternatively, soil properties may change gradually in various directions, or the profile may consist of distinct layers differing in texture or structure.

3. The profile may be shallow, resting on an impervious boundary (e.g., bedrock), or it may be deep. If the bottom is deep enough to remain unaffected by processes at the surface, the profile is called *semi-infinite*.

4. The flow pattern may be one-dimensional (vertical), or it may be two- or three-dimensional, as in the presence of vertical cracks forming secondary evaporation planes.

5. Conditions may be nearly isothermal or strongly nonisothermal. In the latter case, temperature gradients and the conduction of heat and vapor through the system may interact with liquid water flow.

6. External environmental conditions may remain constant or fluctuate. The fluctuations may be regular (diurnal or seasonal) or highly irregular (e.g., spells of cool or warm weather).

7. Soil-moisture flow may be governed by evaporation alone or by both evaporation (at the top of the profile) and internal drainage (or redistribution) down below.

8. The soil may be stable or unstable. For instance, the surface zone may be compacted denser under traffic or raindrop impact and it may become infused with salt.

9. The surface may or may not be covered by a layer of mulch (e.g., plant residues) differing from the soil in hydraulic, thermal, and diffusive properties.

10. Finally, the evaporation process may be continuous over an extended period or may be interrupted by sporadic episodes of rewetting (e.g., intermittent rainfall or irrigation).

To be studied systematically, each of these circumstances, as well as others not listed but perhaps equally relevant, must be formulated in terms of a specific set of initial and boundary conditions. We now describe a few of the circumstances under which evaporation of soil moisture may occur. We begin with a description of capillary rise, which is often a precursor and contributor to the process of salination, especially in the presence of a high water table.

CAPILLARY RISE FROM A WATER TABLE

The rise of water in the soil from a free-water surface (i.e., a water table) has been termed *capillary rise*. This term derives from the *capillary model*, which regards the soil as analogous to a bundle of capillary tubes, predominantly wide in the case of a sandy soil and narrow in the case of a clay soil. Accordingly, the equation relating the equilibrium height of capillary rise h_c to the radii of the pores is

$$h_c = (2\gamma \cos \alpha)/r\rho_w g \qquad (18.1)$$

where γ is the surface tension, r is the capillary radius, ρ_w is the water density, g is the gravitational acceleration, and α is the wetting angle, normally (though not always justifiably) taken as zero.

This equation predicts that water will rise higher, though typically less rapidly, in a clay than in a sand. The reason is that the former soil type contains narrower pores. However, soil pores are not individual capillary tubes of uniform or constant radius, and hence the height of capillary rise will differ in different pores. Above the water table, matric suction will generally increase with height. Consequently, the number of water-filled pores, and hence wetness, will diminish in each soil as a function of height. The rate of capillary rise, that is, the flux, generally decreases with time as the soil is wetted to greater height and as equilibrium is approached.

The wetting of an initially uniformly dry soil by upward capillary flow, illustrated in Fig. 18.1, is a rare occurrence in the field. In its initial stages, this process is similar to infiltration, except that it takes place in the opposite direction, that is, against the direction of gravity. At later stages of the process, the flux tends not to a constant value, as in downward infiltration, but to zero. The reason is that the direction of the gravitational gradient is opposite to the direction of the matric suction gradient, and when the latter (which is large at first but decreases with time) approaches the magnitude of the former, the overall hydraulic gradient approaches zero.

Such an ideal state of static equilibrium between the gravitational head and the suction head is the exception rather than the rule under field conditions. In general, the condition of soil water is not static but dynamic — constantly in a state of flux rather than at rest. Where a water table is present, soil water generally does not attain equilibrium, even in the absence of vegetation, since the soil surface is subject to solar radiation and the evaporative demand of the ambi-

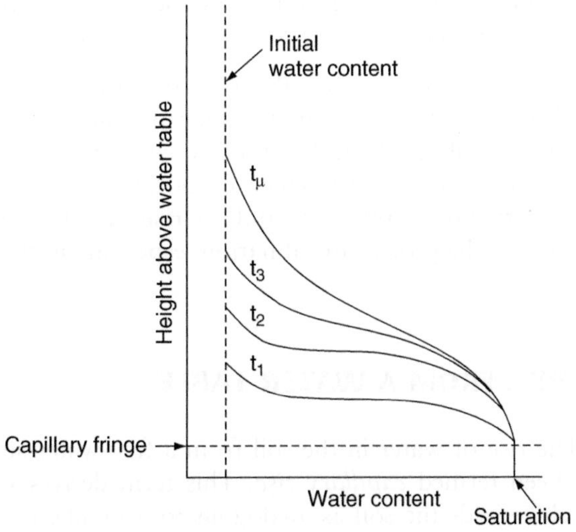

Fig. 18.1. The upward infiltration of water from a water table into a dry soil: water content distribution curves (moisture profiles) for various times ($t_1 < t_2 < t_3 < t_\infty$), where t_∞ is the profile after an infinitely long time (equilibrium). Note that the equilibrium curve is in effect the wetting branch of the soil-moisture characteristic. Note also that what is at first a sharp wetting front, representing the limit of the upward advancing water, gradually becomes diffuse and ends up as a smooth curve characteristic of the particular soil's pore size distribution.

ent atmosphere. However, if soil and external conditions are constant — that is, if the soil is of stable structure, the water table is stationary, and atmospheric evaporativity also remains constant (at least approximately) — then, in time, a steady-state flow situation can develop from water table to atmosphere via the soil. To be sure, we must hasten to qualify this statement by noting that in the field the flow regime will at best be a quasi-steady-state flow, since diurnal fluctuations and other perturbations will prevent attainment of truly stable flow conditions. Nevertheless, the representation of this process as a steady-state flow may be a useful approximation from the analytical point of view.

STEADY EVAPORATION IN THE PRESENCE OF A WATER TABLE

Solutions of the flow equation for the steady-state upward flow of water from a water table through the soil profile to an evaporation zone at the soil surface were given by several workers, including W. R. Gardner (1958), Ripple et al. (1972), and Hillel (1977).

The equation describing steady upward flow is

$$q = K(\psi)\,(d\psi/dz - 1) \tag{18.2}$$

or

$$q = D(\theta)\,(d\theta/dz) - K(\psi) \tag{18.3}$$

where q is flux (equal to the evaporation rate under steady-state conditions), ψ is suction head, K is hydraulic conductivity, D is hydraulic diffusivity, θ is volumetric wetness, and z is height above the water table. The equation shows that the flow stops ($q = 0$) when the suction profile is at equilibrium ($d\psi/dz = 1$). Another form of Eq. (18.2) is

$$q/K(\psi) + 1 = d\psi/dz \tag{18.4}$$

Integration should give the relation between depth and suction or wetness:

$$z = \int \frac{d\psi}{1 + q/K(\psi)} = \int \frac{K(\psi)}{K(\psi) + q}\,d\psi \tag{18.5}$$

or

$$z = \int \frac{D(\theta)}{K(\theta) + q}\,d\theta \tag{18.6}$$

In order to perform the integration in Eq. (18.5), we must know the functional relation between K and ψ, that is, $K(\psi)$. Similarly, the functions $D(\theta)$ and $K(\theta)$ must be known if Eq. (18.6) is to be integrated. An empirical equation for $K(\psi)$, given by W. R. Gardner (1958), is

$$K(\psi) = a(\psi^n + b)^{-1} \tag{18.7}$$

where parameters a, b, n are constants that must be determined for each soil. In this formulation, the suction head ψ is expressed in terms of centimeters. Accordingly, Eq. (18.2) becomes

$$e = q = a(d\psi/dz - 1)/(\psi^n + b) \tag{18.8}$$

where e is the evaporation rate.

With Eq. (18.7), Eq. (18.5) can be used to obtain suction distributions with height for different fluxes as well as fluxes for different surface-suction values. The theoretical solution is shown graphically in Fig. 18.2 for a fine sandy loam soil with an n value of 3. The curves show that the steady rate of capillary rise and evaporation depends on the depth of the water table and on the suction at the soil surface. This suction is dictated largely by the external conditions: The greater the atmospheric evaporativity, the greater the suction at the soil surface on which the atmosphere is acting. However, increasing the suction at the soil surface, even to the extent of making it infinite, can increase the flux through the soil only up to an asymptotic maximal rate, which depends on the depth of the water table. Even the driest and most evaporative atmosphere cannot steadily extract water from the surface any faster than the soil profile can transmit from the water table to that surface. The fact that the soil profile can limit the rate of evaporation is a remarkable and useful feature of the unsaturated flow system. The maximal transmitting ability of the profile depends on the hydraulic conductivity of the soil in relation to the suction, $K(\psi)$.

Disregarding constant b of Eq. (18.7), W. R. Gardner (1958) obtained the function

$$q_{\max} = Aa/d^n \tag{18.9}$$

where d is the depth of the water table below the soil surface and a and n are constants from Eq. (18.7), A is a constant that depends on n, and $q_{\max}$ is the limiting (maximal) rate at which the soil can transmit water from the water table to the evaporation zone at the surface.

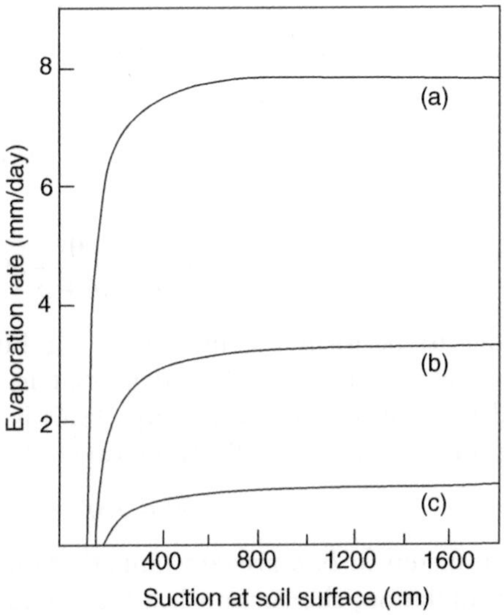

Fig. 18.2. Steady upward flow and evaporation from a water table as a function of the suction prevailing at the soil surface, with the water table at a depth of: (a) 90 cm, (b) 120 cm, (c) 180 cm. The soil is a fine sandy loam, with $n = 3$. (After W. R. Gardner, 1958.)

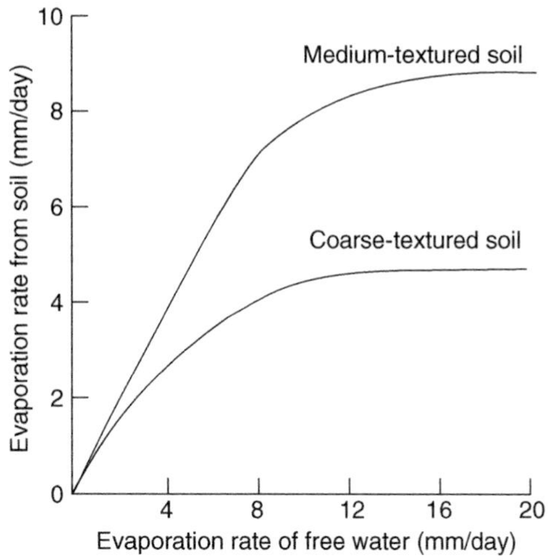

Fig. 18.3. Theoretical relation between evaporation rate from coarse- and medium-textured soils (water-table depth 60 cm) and evaporation rate from free-water surface. (After W. R. Gardner, 1958.)

We can now see how the actual steady evaporation rate is determined either by the external evaporativity or by the water-transmitting properties of the soil, depending on which of the two is lower and therefore limiting. Where the water table is near the surface, the suction at the soil surface is low and the evaporation rate is dictated by external conditions. However, as the water table deepens and as the suction at the soil surface increases, the evaporation rate approaches a limiting value regardless of how high external evaporativity may be.

Equation (18.9) suggests that the maximal evaporation rate decreases with water-table depth most steeply in coarse-textured soils (in which parameter n is greater because the conductivity falls off more steeply with increasing suction than in clayey soils). Nevertheless, a sandy loam soil can still evaporate water at an appreciable rate (Fig. 18.2), even when the water table is as deep as 1.80 m. Figure 18.3 illustrates the effect of texture on the evaporation rate.

A flexible treatment of steady-state evaporation, based on numerical methods of solution, was developed by Ripple et al. (1972). Their results are illustrated in Figs. 18.4 and 18.5. Their procedure makes it possible to estimate the steady-state evaporation from bare soils (including layered ones) with a high water table. The required data include soil-moisture characteristic curves, water-table depth, and standard records of air temperature, air humidity, and wind speed. The theory takes into account both the relevant atmospheric factors and the soil's capability to transmit water in liquid and vapor forms. The possible effects of thermally induced water transfer (except in the vapor phase) and of salt accumulation at the surface are additional factors yet to be fully taken into account. More recent solutions for steady-state evaporation from a shallow water table were given by Warrick (1988).

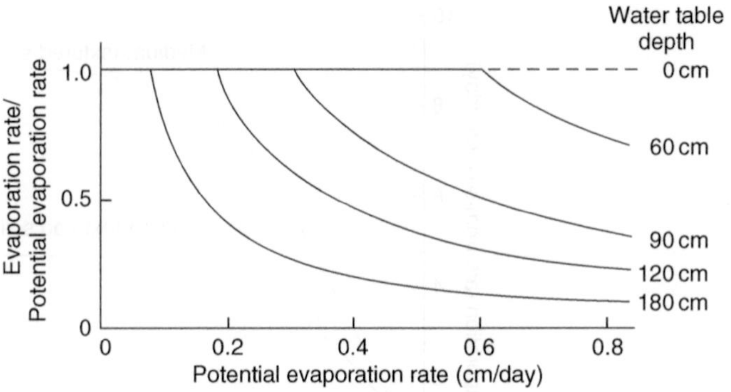

Fig. 18.4. Dependence of relative evaporation rates e/e_{pot} on the potential evaporation rate for a clay soil. Numbers at curves indicate the depth to the water table, in centimeters. (After Ripple et al., 1972.)

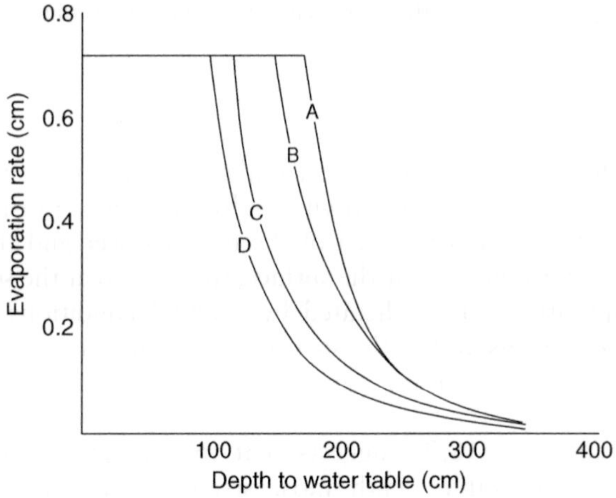

Fig. 18.5. Influence of layering on the relation between the evaporation rate and the depth to the water table. Limiting curves of soil-water evaporation are shown for a homogeneous soil (A), a two-layer soil with upper-layer thickness of either 3 cm (B) or 10 cm (C), and a three-layer soil with the thickness of the intermediate and uppermost layers equal to 10 cm each (D). (After Ripple et al., 1972.)

Sample Problem

The potential rate of evaporation from a saturated uniform soil with a high water table is 9 mm per day. Prior studies have shown that the soil's behavior conforms to Eq. (18.9). The parameters of this equation are cast in terms of centimeters, with the composite coefficient $Aa = 4.88\ \text{cm}^2/\text{sec}$ and the exponential constant $n = 3$. Estimate the threshold depth beyond which the water table must be lowered if evaporation is to be reduced and the water-table depth at which the evaporation rate will fall to 10% of the potential rate. Assume steady-state conditions. Plot the expectable daily evaporation rate as a function of water-table depth.

We begin with Eq. (18.9), which, according to W. R. Gardner (1958) estimates the maximal evaporation rate possible from a soil with a high water table:

$$q_{max} = Aa/d^n$$

where d is depth of the water table below the soil surface. Accordingly, the maximal depth of the water table still capable of supplying the surface with the steady flux needed to sustain the potential evaporation rate is

$$d = (Aa/q_{max})^{1/n} = (4.88\ \text{cm}^2/\text{sec})/[(0.9\ \text{cm/day})/(86{,}400\ \text{sec/day})]^{1/n} = 77.7\ \text{cm}$$

Note that any water-table position higher than −77.7 cm would seem to allow a flux greater than the climatically determined potential rate; hence for a shallow water table the actual evaporation rate is flux controlled at the surface. For deeper water-table conditions, however, it becomes profile controlled.

To calculate the water-table depth for which q will be 10% of the potential rate, we can write

$$d_{0.1} = \left[\frac{4.88}{0.1 \times 0.9/86{,}400}\right]^{1/3} = 167.3\ \text{cm}$$

To plot the q_{max} versus d curve, we obtain the following data:

From $d = 0$ to $d = 77.7$ cm, $q_{max} = 0.90$ cm/day
At $d = 80$ cm, $q_{max} = 4.88/80^3 = 0.82$ cm/day
At $d = 100$ cm, $q_{max} = 4.88/100^3 = 0.42$ cm/day
At $d = 150$ cm, $q_{max} = 4.88/150^3 = 0.12$ cm/day
At $d = 200$ cm, $q_{max} = 4.88/200^3 = 0.05$ cm/day
At $d = 250$ cm, $q_{max} = 4.88/250^3 = 0.027$ cm/day

The actual plotting of these data is left to the enterprising student.

EVAPORATION IN THE ABSENCE OF A WATER TABLE

Steady evaporation from soils is not a widespread occurrence, since, even where a high water table exists, neither its depth nor external conditions can remain constant for very long. More commonly, soil-moisture evaporation occurs under unsteady conditions and results in a net loss of water from the soil; that is, it results in *drying*. The process of drying involves considerable losses of soil moisture, especially in arid regions, where these losses can amount to 50% or more of total precipitation. In this section, we consider the drying of initially wetted soil profiles that do not contain a water table anywhere near enough to the surface to affect the evaporation process appreciably.

We begin once again by assuming that external conditions, and hence atmospheric evaporativity, are constant. Under such conditions, the soil-drying process has been observed to occur in three recognizable stages:

1. An initial, so-called *constant-rate stage*, which occurs early in the process, while the soil is wet and conductive enough to supply water to the site of evaporation at a rate commensurate with the evaporative demand imposed by the atmosphere. During this stage, the evaporation rate is limited by, and hence also controlled by, external meteorological conditions (i.e., radiation, wind, air temperature and humidity, etc.) rather than by the properties of the soil profile. As such, this stage, being *weather controlled*, is analogous to *the flux-controlled stage* of infiltration (in contrast with the *profile-controlled stage*; see discussion of rain infiltration in Chapter 14). The evaporation rate during this stage might also be influenced by soil surface conditions, including surface reflectivity and the possible presence of a mulch, insofar as these can modify the effect of the meteorological factors acting on the soil. In a dry climate, this stage of evaporation is generally brief and may last only a few hours to a few days. (*Note*: the term *constant-rate stage* of evaporation really makes sense only in connection with laboratory experiments under constant evaporativity, whereas in natural conditions evaporativity varies continually, due to changes in the intensity of incoming radiation during the day–night cycle.)

2. An intermediate *falling-rate stage*, during which the evaporation rate falls progressively below the potential rate (the atmospheric evaporativity). At this stage, the evaporation rate is limited or dictated by the rate at which the gradually drying soil profile can deliver moisture toward the evaporation zone. Hence it can also be called the *profile-controlled stage*. This stage may persist for a much longer period than the first stage.

3. A residual *slow-rate stage*, which is established eventually and which may persist at a nearly steady rate for many days, weeks, or even months. This stage apparently comes about after the surface zone has become so desiccated that further conduction of liquid water through it effectively ceases. Water transmission through the desiccated layer thereafter occurs primarily by the slow process of vapor diffusion, and it is affected by the vapor diffusivity of the dried surface zone and by the adsorptive forces acting over molecular distances at the particle surfaces. This stage is called the *vapor-diffusion stage* and can be important where the surface layer becomes quickly desiccated (e.g., if the surface consists of a loose assemblage of clods).

Whereas the transition from the first to the second stage is generally a sharp one, the second stage generally blends into the third stage so gradually that the last two cannot be separated easily. A qualitative explanation for the occurrence of these stages follows.

During the initial stage, the soil surface gradually dries and soil moisture is drawn upward in response to steepening evaporation-induced gradients. The rate of evaporation can remain nearly constant as long as the moisture gradients toward the surface compensate for the decreasing hydraulic conductivity (resulting from the diminishing water content of the surface zone). We can restate this in terms of Darcy's law, $q = K(d\psi/dz)$, by noting that the flux q remains constant because the gradient $d\psi/dz$ increases sufficiently to offset the decrease of K. Sooner or later, however, the soil surface approaches equilibrium with the overlying atmosphere (i.e., becomes more or less *air dry*). From this moment on, the suction gradients toward the surface cannot increase any more and, in fact, must begin to decrease as the

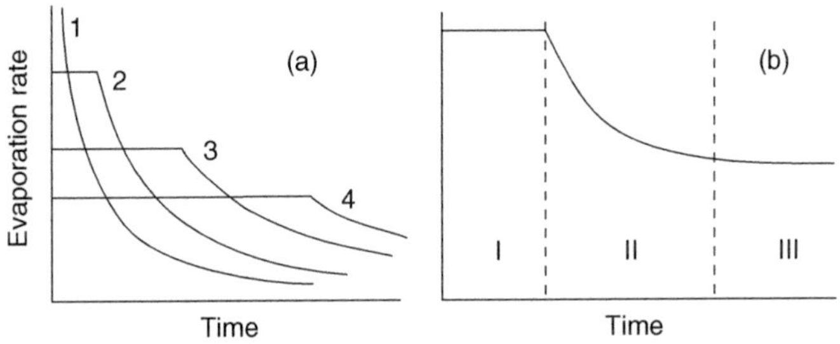

Fig. 18.6. (a) Evaporation rate versus time under different evaporativities (curves 1–4 in order of decreasing initial evaporation rate); and (b) relation of relative evaporation rate (actual rate as a fraction of the potential rate) versus time, indicating the three stages of the drying process.

soil in depth loses more and more moisture. Since, as the evaporation process continues, both the gradients and the conductivities at each depth near the surface tend to diminish simultaneously, it follows that the flux toward the surface and the evaporation rate inevitably diminish as well. As shown in Fig. 18.6, the end of the first (i.e., the beginning of the second) stage of drying can occur rather abruptly. The pattern of evaporation under different evaporative conditions is also shown in terms of cumulative evaporation in Fig. 18.7.

The tendency of the moisture gradients toward the soil surface to become steeper during the first stage of the process, as the surface becomes progressively drier, and their tendency to become less and less steep during the second stage, after the surface has dried to its final "air-dry" value, is illustrated in Fig. 18.8. Continuation of the evaporation process for a prolonged period is sometimes accompanied by the downward movement into the profile of a "drying front" (van Keulen and Hillel, 1974) and the development of a distinct desiccated zone (Hillel, 1977), through which water can move from the still-moist underlying layers only by vapor diffusion.

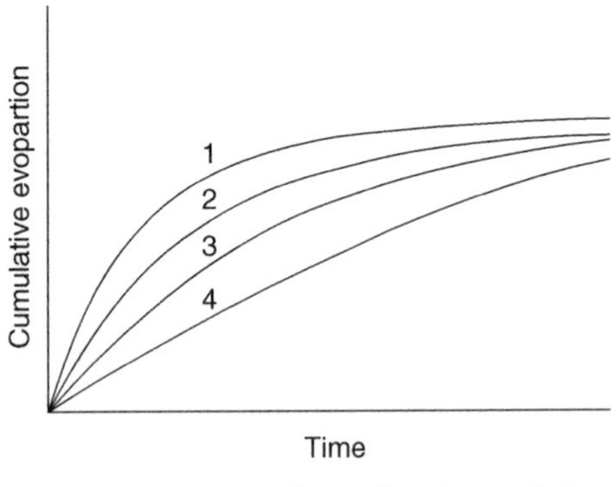

Fig. 18.7. Relation of cumulative evaporation to time (curves 1–4 are in order of decreasing initial evaporation rate).

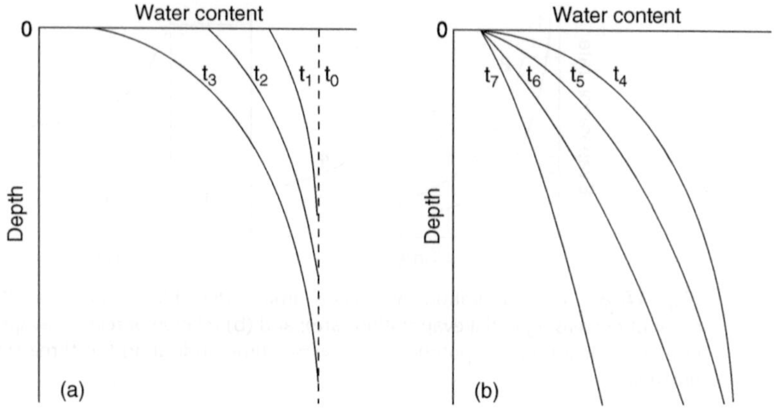

Fig. 18.8. The changing moisture profile in a drying soil: (a) The first stage, during which the gradients toward the surface become steeper until the surface becomes air dry. (b) The second stage, in which the moisture gradients decrease as the deeper layers lose moisture by continued upward movement.

The length of time the initial stage of drying lasts depends on the intensity of the meteorological factors that determine atmospheric evaporativity as well as on the depth and the conductive properties of the soil itself. Under similar external conditions, the first stage of drying is sustained longer in a clayey than in a sandy soil, since clayey soils retain higher wetness and conductivity values because suction develops in the upper zone.

For a semi-infinite soil column subject to infinite initial evaporativity at the surface, neglecting gravity, W. R. Gardner and Hillel (1962) provided the following solution for the cumulative evaporation E, indicating a linear dependence on the square root of time:

$$E = 2(\theta_i - \theta_f)\sqrt{Dt/\pi} \tag{18.10}$$

The evaporative flux e, which is the time derivative of E, is thus inversely proportional to the square root of time:

$$e = dE/dt = (\theta_i - \theta_f)\sqrt{D/\pi t} \tag{18.11}$$

W. R. Gardner and Hillel (1962) used a different formulation to predict the rate of evaporation from finite-length columns during the decreasing-rate stage of drying:

$$e = -dW/dt = D(\theta_{ave})W\pi^2/4L^2 \tag{18.12}$$

where θ_{ave} is the average volumetric wetness obtained by dividing the total water content of the soil W by the depth of wetting L and $D(\theta)$ is the known diffusivity function. The cumulative evaporation can be obtained by integrating Eq. (18.12) with respect to time.

The use of computer-based simulation techniques in the analysis of drying processes under various initial and boundary conditions was demonstrated by Hanks and Gardner (1965), Hillel (1977), and Campbell (1985).

Sample Problem

Calculate the 10-day course of evaporation and evaporation rate during the drying process of an infinitely deep, initially saturated, uniform column of soil subjected to an infinitely high evaporativity. Assume the initial volume wetness θ_i to be 48%, the air-dry value θ_f to be 3%, and the weighted mean diffusivity to be 10^4 mm²/day.

The problem posed can be solved approximately by use of Eq. (18.10):

$$E = 2(\theta_i - \theta_f)(Dt/\pi)^{1/2}$$

where E is cumulative evaporation. We now solve successively for days 1, 2, 3, etc., to obtain the time course of cumulative evaporation. Thus:

After day 1, $E = 2(0.48 - 0.03)(10{,}000 \times 1/3.1416)^{1/2} = 50.8$ mm
After day 2, $E = 2 \times 0.45(10{,}000 \times 2/3.1416)^{1/2} = 71.8$ mm
After day 3, $E = 2 \times 0.45(10{,}000 \times 3/3.1416)^{1/2} = 88.0$ mm
After day 4, $E = 2 \times 0.45(10{,}000 \times 4/3.1416)^{1/2} = 101.6$ mm
After day 5, $E = 2 \times 0.45(10{,}000 \times 5/3.1416)^{1/2} = 113.6$ mm
After day 6, $E = 2 \times 0.45(10{,}000 \times 6/3.1416)^{1/2} = 124.4$ mm
After day 7, $E = 2 \times 0.45(10{,}000 \times 7/3.1416)^{1/2} = 134.4$ mm
After day 8, $E = 2 \times 0.45(10{,}000 \times 8/3.1416)^{1/2} = 143.7$ mm
After day 9, $E = 2 \times 0.45(10{,}000 \times 9/3.1416)^{1/2} = 152.4$ mm
After day 10, $E = 2 \times 0.45(10{,}000 \times 10/3.1416)^{1/2} = 160.6$ mm

To estimate the mean evaporation for each day, we calculate the midday rate using Eq. (18.11):

$$e = dE/dt = (\theta_i - \theta_f)\sqrt{D/\pi t}$$

Midday 1, $e = 0.45(10{,}000/0.5\pi)^{1/2} = 35.9$ mm/day
Midday 2, $e = 0.45(10{,}000/1.5\pi)^{1/2} = 20.7$ mm/day
Midday 3, $e = 0.45(10{,}000/2.5\pi)^{1/2} = 16.0$ mm/day
Midday 4, $e = 0.45(10{,}000/3.5\pi)^{1/2} = 13.6$ mm/day
Midday 5, $e = 0.45(10{,}000/4.5\pi)^{1/2} = 12.0$ mm/day
Midday 6, $e = 0.45(10{,}000/5.5\pi)^{1/2} = 10.8$ mm/day
Midday 7, $e = 0.45(10{,}000/6.5\pi)^{1/2} = 10.0$ mm/day
Midday 8, $e = 0.45(10{,}000/7.5\pi)^{1/2} = 9.3$ mm/day
Midday 9, $e = 0.45(10{,}000/8.5\pi)^{1/2} = 8.7$ mm/day
Midday 10, $e = 0.45(10{,}000/9.5\pi)^{1/2} = 8.2$ mm/day

Note: The sum of daily rates thus estimated does not equal cumulative evaporation for the 10-day period, since the midday rates calculated for the first three days (when the rate descends steeply as a function of time) underestimate the effective mean rates. Moreover, representing a continuously drying system in terms of a single "weighted mean diffusivity" is itself an oversimplification. More realistic treatment, however, requires more complex mathematics.

THE "DRYING-FRONT" PHENOMENON

An interesting phenomenon is the possible development of what might be called a "drying-front," which moves downward into the soil as the drying process progresses. Such a front may be discerned from the inflection of the

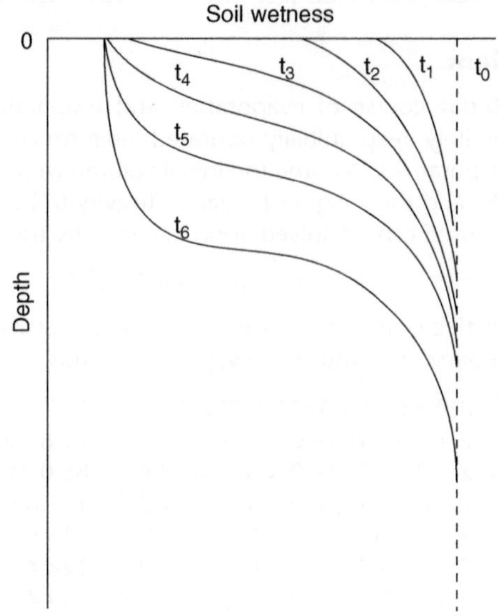

Fig. 18.9. Development of a drying front and its movement into the soil during the course of soil-moisture evaporation (hypothetical). Note that t_1, t_2, t_3, etc., represent the soil-moisture profile at (initially uniform) progressively later stages of the drying process.

soil-moisture profile. This curve (Fig. 18.9) is usually convex along its entire length at first, but may eventually become concave in its upper part. It appears that this inflection results from the soil's diffusivity versus wetness function.

The hydraulic diffusivity for liquid water is known to decrease exponentially with a decrease in the soil's volumetric wetness. However, though the liquid diffusivity falls, the vapor diffusivity rises as the soil dries, so the overall diffusivity often indicates a minimum at some low value of wetness beyond which further drying results in an increase, rather than further decrease, of diffusivity. This phenomenon was pointed out by Philip (1974; see Fig. 18.10). Its possible effect on the drying process and the shape of the moisture profile in the surface zone was analyzed by van Keulen and Hillel (1974). Their work showed that a

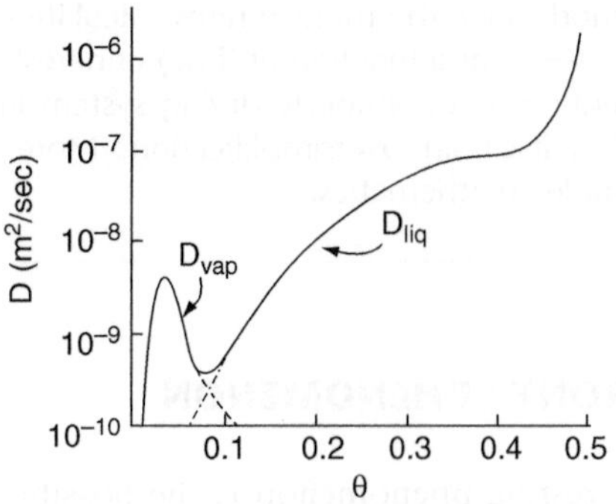

Fig. 18.10. Relation between moisture diffusivity D and moisture content θ for Yolo light clay. The contribution of the vapor phase to D becomes dominant at $\theta < 0.06$. (After Philip, 1974.)

"hooked" diffusivity function (i.e., D rising with decreasing θ in the dry range) does indeed result in an inflected soil-moisture profile and in the development of what might be called a "drying front."

Sample Problem

Consider the drying of a uniform soil wetted to a depth of 1 m, at which an impervious horizon exists, allowing no drainage. The initial volume wetness is 24%, and the initial diffusivity is 4×10^4 mm²/day. Assume a diffusivity function of the type $D = a \exp b\theta$, where $a = 12$ and $b = 20$. Estimate the evaporation rate as it varies during the first 10 days of the drying process, under an evaporativity of 12 mm/day.

The case described can be represented by the finite-column model of Gardner and Hillel

$$e = -dW/dt = D(\theta_m)W\pi^2/4L^2$$

where e is the evaporation rate (mm/day), W is the total profile water content (mm), t is time (days), D is diffusivity (mm²/day, a function of mean wetness θ_m), and L is the length of the wetted profile (mm). The mean wetness θ_m is related to total profile water content W by

$$\theta_m = W/L \quad \text{or} \quad W = L\theta_m$$

Substituting the appropriate values into our equation, we calculate the first day's evaporation rate:

$$e_1 = [(4 \times 10^4 \text{ mm}^2/\text{day}) \times (0.24 \times 10^3 \text{ mm}) \times 9.87]/4 \times 10^6 \text{ mm}^2 = 23.7 \text{ mm/day}$$

This is much greater than the evaporativity, which is "only" 12 mm/day. Hence we assume that the evaporation rate is at first controlled by, and equal to, the evaporativity. So we deduct 12 mm from the water content of the profile and proceed to calculate the second day's evaporation rate.

Our total profile water content W is now 240 minus 12 mm (the first day's evaporation) = 228 mm, and our new $\theta = 228/1000 = 0.228$ (22.8%). The new value of $D = 12e^{20\times0.223} = 2.38 \times 10^4$ mm²/day. The second day's evaporation rate is

$$e_2 = [(2.38 \times 10^4 \text{ mm}^2/\text{day}) \times 228 \text{ mm} \times 9.87]/4 \times 10^6 \text{ mm}^2 = 13.4 \text{ mm}$$

This is still higher than the potential evaporation rate (the evaporativity), hence we once again deduct 12 mm from the water content and proceed to calculate the third day's evaporation rate, updating our variables as follows: $W = 228 - 12$ mm = 216 mm, $\theta = 216/1000 = 0.216$, and $D = 12e^{20\times0.216} = 9 \times 10^3$ mm²/day.

$$e_3 = [(9 \times 10^3 \text{ mm}^2/\text{day}) \times 216 \text{ mm} \times 9.87]/4 \times 10^6 \text{ mm}^2 = 4.8 \text{ mm}$$

This evaporation rate is less than the evaporativity, so we conclude that the first stage of the process has ended and the falling-rate, profile-controlled phase has begun.

Toward the fourth day's events we calculate that

$$W = 216 - 4.8 = 211.2 \text{ mm}$$

$$\theta = 211.2/1000 = 0.2112$$

$$D = 12e^{20\times0.2112} = 8.2 \times 10^3 \text{ mm}^2/\text{day}$$

$$e_4 = [(8.2 \times 10^3 \text{ mm}^2/\text{day}) \times 211.2 \text{ mm} \times 9.87]/4 \times 10^6 \text{ mm}^2 = 4.3 \text{ mm}$$

Updating: $W = 211.2 - 4.3 = 206.9$ mm, $\theta = 0.207$, $D = 12e^{20\times0.207} = 7.5 \times 10^3$ mm^2/day.

$$e_5 = [(7.5 \times 10^3 \text{ mm}^2/\text{day}) \times 206.9 \text{ mm} \times 9.87]/4 \times 10^6 \text{ mm}^2 = 3.8 \text{ mm}$$

Now $W = 206.9 - 3.8 = 203.1$, $\theta = 0.2031$, $D = 12e^{20\times0.2031} = 7.0 \times 10^3$ mm^2/day.

$$e_6 = [(7.0 \times 10^3 \text{ mm}^2/\text{day}) \times 203.1 \text{ mm} \times 9.87]/4 \times 10^6 \text{ mm}^2 = 3.5 \text{ mm}$$

Now $W = 203.1 - 3.5 = 199.6$, $\theta = 0.1996$, $D = 12e^{20\times0.1996} = 6.5 \times 10^3$ mm^2/day.

$$e_7 = [(6.5 \times 10^3 \text{ mm}^2/\text{day}) \times 199.6 \text{ mm} \times 9.87]/4 \times 10^6 \text{ mm}^2 = 3.2 \text{ mm}$$

Now $W = 199.6 - 3.2 = 196.4$, $\theta = 0.1964$, $D = 12e^{20\times0.1964} = 6.1 \times 10^3$ mm^2/day.

$$e_8 = [(6.1 \times 10^3 \text{ mm}^2/\text{day}) \times 196.4 \text{ mm} \times 9.87]/4 \times 10^6 \text{ mm}^2 = 3.0 \text{ mm}$$

Now $W = 196.4 - 3.0 = 193.4$, $\theta = 0.1934$, $D = 12e^{20\times0.1934} = 5.7 \times 10^3$ mm^2/day.

$$e_9 = [(5.7 \times 10^3 \text{ mm}^2/\text{day}) \times 193.4 \text{ mm} \times 9.87]/4 \times 10^6 \text{ mm}^2 = 2.7 \text{ mm}$$

Now $W = 193.4 - 2.7 = 190.7$, $\theta = 0.1907$, $D = 12e^{20\times0.1907} = 5.4 \times 10^3$ mm^2/day.

$$e_{10} = [(5.4 \times 10^3 \text{ mm}^2/\text{day}) \times 190.7 \text{ mm} \times 9.87]/4 \times 10^6 \text{ mm}^2 = 2.5 \text{ mm}$$

The total 10-day cumulative evaporation is thus estimated to be

$$e_{\text{total}} = 12 + 12 + 4.8 + 4.3 + 3.8 + 3.5 + 3.2 + 3.0 + 2.7 + 2.5 \approx 51.8 \text{ mm}$$

This amount of evaporation is, interestingly, only one-third of the amount we calculated in the previous problem for the same period from an infinitely deep, initially saturated column of soil subject to infinite evaporativity. Here, evaporation is limited by the finiteness of both the evaporativity and the column itself.

DIURNAL FLUCTUATIONS OF TOP LAYER MOISTURE AND HYSTERESIS EFFECTS

The foregoing treatments of the evaporation process were based on the precept that the soil surface is subjected to a constant, meteorologically induced, *evaporativity*, which dictates the *potential evaporation rate*. In nature, however, evaporativity is not constant but variable. It fluctuates diurnally and varies from day to day, so it may become difficult to discern the stages defined in our preceding sections. The resulting course of evaporation may not be described accurately by a simplistic theory based on the assumption of constant evaporativity.

Detailed experimental observations by R. D. Jackson (1973) and R. D. Jackson et al. (1973) showed that the surface-zone soil moisture content fluctuates in accord with the diurnal fluctuation of evaporativity. The soil surface dries during daytime and rewets during nighttime by sorption from the moister layers beneath (see Fig. 18.11). This pattern was found in a layer of soil several centimeters thick. The amplitude of the diurnal fluctuation of moisture tends to diminish with depth and time, and the daily maxima and minima exhibit an increasing phase lag at greater depths. Because of the high evaporativity in midday, the surface evidently desiccates sooner than it

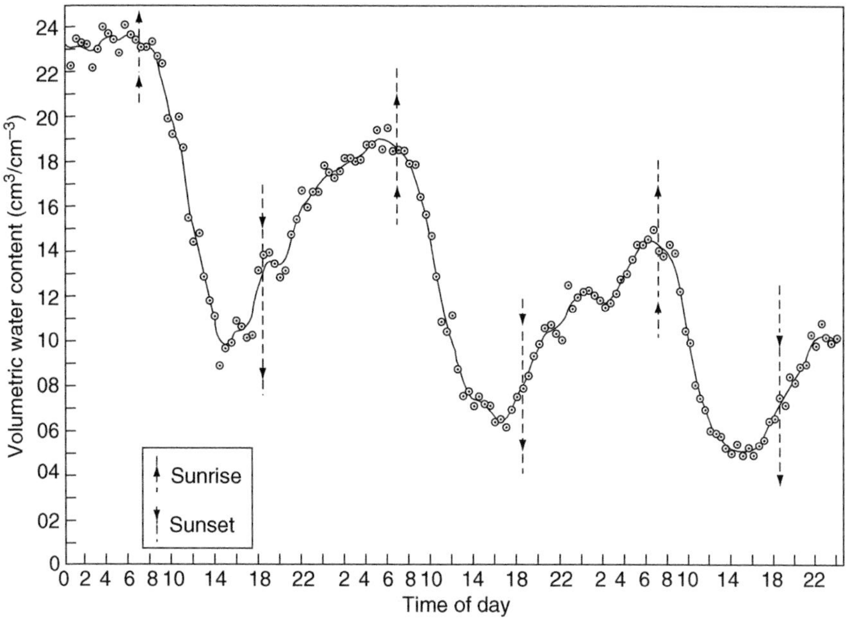

Fig. 18.11. Time course of changing volumetric wetness at the surface (0–5 mm) of a loam soil during drying, 5–7 days after irrigation (After R. D. Jackson, 1973.)

would if the evaporativity were to remain constant at its average value. However, Jackson et al. were unable on the basis of their evidence to resolve the question as to whether the early formation of a dry surface layer might serve to reduce cumulative evaporation below that which would occur had evaporativity remained steady. They noted, in any case, that the concept of three stages of drying appears to have little meaning under natural field conditions.

Subsequently, Hillel (1975, 1976a, 1977) developed a dynamic simulation model capable of monitoring the evaporation process continuously through repeated cycles of rising and falling evaporativity. The model was used in an attempt to clarify the extent to which the diurnal pattern of evaporativity may influence the overall quantity of evaporation and the resulting soil-moisture distribution in space and time. Of particular interest is the fluctuating wetness of the near-surface zone, where germination takes place. Computations carried out for a 10-day period accord with experimental findings that the diurnal cycle of evaporativity causes nighttime resorption of moisture from below and hence a somewhat higher average wetness in the soil surface zone than would occur otherwise.

The alternating desorption and resorption of moisture by the soil surface zone inevitably involves the phenomenon of *hysteresis*, defined as the dependence of the equilibrium state of soil moisture (namely, the relation between wetness and suction) on the direction of the antecedent process (whether sorption or desorption). Several investigators (e.g., Rubin, 1966; Bresler et al., 1969; Vachaud and Thony, 1971) investigated the effect of hysteresis on soil-water dynamics.

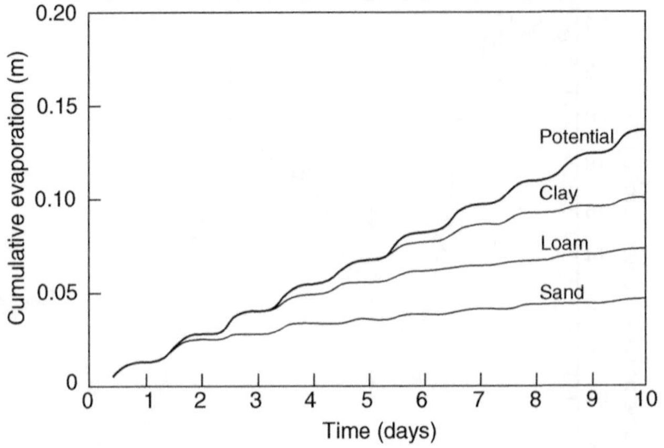

Fig. 18.12. Cumulative evaporation during simultaneous drainage and evaporation from initially saturated uniform profiles of sand, loam, and clay. (After Hillel and van Bavel, 1976.)

A priori reasoning leads us to suppose that the hysteresis effect might tend to retard evaporation in a regime of fluctuating evaporativity. After its strong daytime desiccation, the surface zone of the soil draws moisture from below during the night, so the top layer is in the process of sorption while the underlying donor layer is in the process of desorption. In principle, the hysteresis phenomenon causes a sorbing zone of soil to approach potential equilibrium with a desorbing zone of the same soil while the former is at a lower moisture content and hence at a lower value of hydraulic conductivity. It would seem to follow that the hysteresis effect can contribute to the self-arresting tendency of the evaporation process.

A test of this hypothesis was carried out by Hillel (1975), who modified his earlier simulation model to account for hysteresis. He reported a systematic reduction of cumulative evaporation as the range of hysteresis (i.e., the displacement between the primary desorption and sorption curves) increased. At its greatest, the reduction of evaporation due to hysteresis amounted to 33% of the nonhysteretic evaporation. A similar study of evaporation from soils of various textures (Hillel and van Bavel, 1976) showed that differences in hydraulic properties can also influence cumulative evaporation. Over time, coarse-textured (sandy) soils tend to evaporate the least, whereas fine-textured (clayey) soils can sustain the first stage of drying longer and hence evaporate the most, under both steady and cyclic evaporativity (Fig. 18.12).

NONISOTHERMAL EVAPORATION

The discussion so far has dealt with isothermal conditions only. However, a more complete treatment of evaporation should not neglect the effect of temperature gradients and heat flow. The heat required to vaporize water must be transported to the evaporation site, which requires a nonzero temperature gradient. Fritton et al. (1970) has indicated that the isothermal-flow equation can approximate cumulative evaporation for both wind and radiation treatments, but this equation cannot portray the soil-water distribution pattern in the case of nonisothermal flow. They concluded that where temperature gradients are significant, analysis based on simultaneous heat and mass transfer gives better fit with experimental data than the isothermal equation.

Rose (1968) attempted to establish the magnitude of vapor versus liquid water movement during evaporation under nonisothermal conditions. The effect of warming the soil is to lower the suction and to raise the vapor pressure of soil water. Hence the effect of a thermal gradient is to induce flow and distillation from warmer to cooler regions. When the soil surface is warmed by radiation, this effect would tend to counter the tendency to upward flow of water in response to evaporation-induced moisture gradients. For this reason the evaporation rate might be lower when the surface is dried by radiation (which warms the soil) than when it is dried by wind (which cools the soil), even when the two modes of energy input are of comparable evaporation potential (Hanks et al., 1967).

A comprehensive model of nonisothermal evaporation was developed by van Bavel and Hillel (1975, 1976). Their work was based on the use of actual weather data (radiation, air temperature, humidity, and wind speed) and soil physical properties as inputs. The model permits the calculation of both potential and actual evaporation as a resultant of the interaction between weather and soil factors, rather than its imposition as a forcing function. This model also predicts the partitioning of radiant energy into sensible and latent heat and the resulting pattern of soil temperature. A more elaborate model of spatially variable evaporation from bare soil was offered by Evett et al. (1994).

The frequently made assumption that potential evaporation from a bare soil surface is approximately equal to the evaporation from a free-water surface ignores the many differences (e.g., in heat capacity, thermal conductivity, color, albedo, and roughness) between a soil and a body of water. All these factors affect the energy balance of the surface and hence the amount of energy "allocated" to evaporation. Almost from the start, the emittance and long-wave radiance decline with time as the soil loses water by drainage and surface drying. Concurrently, the albedo rises and less shortwave energy is retained. Further, as the soil drains and dries, its thermal conductivity and heat capacity decrease, affecting the soil heat flux. Surface temperature and air stability (affected by the buoyancy of the lower air as it is heated by the ground) might also vary.

EVAPORATION FROM IRREGULAR SURFACES AND SHRINKAGE CRACKS

Thus far we have discussed the drying of soils as if it were entirely a one-dimensional process, that is, vertically downward from a smooth horizontal surface plane. In many cases, the surface is far from smooth and horizontal: The land as a whole may slope in various directions and at various degrees of inclination. And the surface may exhibit the sort of geometrically complex microrelief (ridges and furrows, large and small clods) that would cause the drying process to be uneven and three-dimensional, at least in its early stages.

Differences in temperature and drying pattern between soil surfaces of differing roughness characteristics (e.g., different directions and elevations of ridges) have long been recognized in principle and even utilized in practice in attempts to enhance germination and crop-stand establishment in unfavorable climates (J. E. Adams, 1970; Hillel and Rawitz, 1972). As examples we might mention the practice of planting on ridge tops or on sun-facing ridge slopes in cool and wet climates and the opposite practice of planting in furrows or depressions in warm and dry climates. However, a fundamental definition of the role of surface configuration in evaporation and drying is difficult to achieve, because it includes not only three-dimensionality in the soil realm but also complex effects in the above-ground realm involving radiation exchange, air motion, and vapor-transfer processes.

A particular example of multidimensional evaporation is that of a soil tending to form vertical fissures, or *cracks*, during the process of drying. Especially prone to cracking are the soils known as *grumusols*, or *vertisols*, which have a high content of montmorillonitic clay and occur typically in an environment of alternating wet and dry seasons. They appear in every continent and are estimated to cover about 40 million hectares in all. These soils are characterized by an extreme tendency to expand on wetting and to contract on drying.

The cracks that form in a grumusol may be as wide as 0.1 m and as deep as 1 m and assume a characteristic polygonal pattern. Such cracks can desiccate and tear plant roots and dry the soil to an extreme degree down to considerable depths. The wide cracks permit the entry of turbulent air, which tends to sweep out water vapor from the deeper reaches of the soil profile (J. E. Adams and Hanks, 1964). The exposed vertical sides of the cracks then become secondary evaporation planes, which increase the effective evaporating surface threefold or fourfold over the evaporating surface of a noncracking soil. Lateral

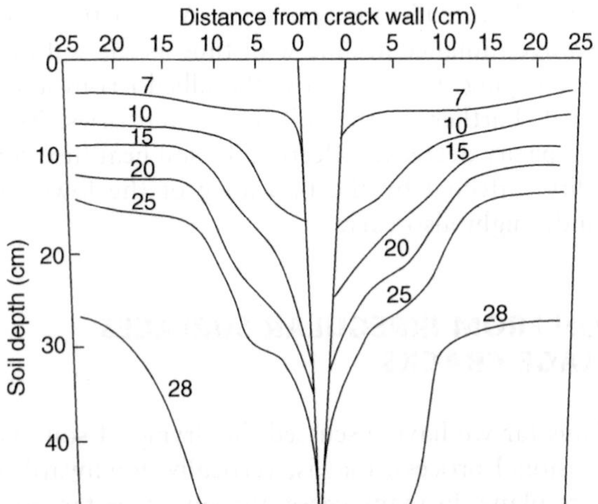

Fig. 18.13. Moisture distribution (mass wetness, %) in the vicinity of a vertisol's shrinkage crack after a summer growing season. (After Ritchie and Adams, 1974.)

movement of moisture toward crack walls can take place from a distance exceeding 10 cm. Consequently, cracks can increase evaporation by 12–30% (Ritchie and Adams, 1974).

The process of crack formation may be self-generating, in the sense that once a crack begins at the surface, preferential drying and shrinkage along its bottom and sides can cause the crack to extend into the soil. However, in the field the cracking pattern appears to remain consistent through repeated cycles of wetting and drying. A thorough wetting causes the soil blocks between the cracks to expand and close the cracks; subsequent drying opens them up again. After the soil surface becomes air dry, evaporation from shrinkage cracks becomes the dominant component of the drying process, as deeper zones in the profile become the main. A two-dimensional representation of soil-moisture distribution as it develops in the vicinity of a shrinkage crack is shown in Fig. 18.13.

REDUCTION OF EVAPORATION FROM BARE SOILS

In principle, evaporation flux from the soil surface can be modified in three alternative or complementary ways: (1) by controlling energy supply to the site of evaporation (e.g., modifying the albedo through color or structure of the soil surface or through shading of the surface); (2) by reducing the potential gradient, or the force driving water upward through the profile (e.g., lowering the water table, if present, or warming the surface so as to set up a downward-acting thermal gradient); or (3) by decreasing the conductivity or diffusivity of the profile, particularly of the surface zone (e.g., tillage, soil conditioning, incorporation of organic matter into the topsoil).

The choice of means for reducing evaporation depends on the stage of the process one wishes to regulate: whether it is *the first stage*, in which the effect of meteorological factors acting on the soil surface dominates the process, or *the second stage*, in which the internal water supply to the surface, determined by the transmitting properties of the profile, becomes the rate-limiting factor. Methods influencing the first stage are not necessarily effective during the second stage.

Covering or mulching the surface with vapor barriers or with reflective materials can reduce the intensity with which external factors, such as radiation and wind, act on the surface (Hanks, 1992). Thus, surface treatments can retard evaporation during the initial stage of drying. A similar effect may result from application of materials that lower the vapor pressure of water (F. E. Jones, 1991). Retardation of evaporation during the first stage can provide the plants with a greater opportunity to utilize the moisture of the uppermost soil layers, an effect that can be vital during the germination and establishment phases of plant growth.

During the second stage of the drying process, the effect of surface treatments is likely to be only slight, and reduction of the evaporation rate and of eventual water loss will depend on decreasing the diffusivity or conductivity of the soil profile in depth. Deep tillage, for instance, by possibly increasing the range of variation of diffusivity with changing water content, may reduce the rate at which the soil can transmit water toward the surface during the second stage of the drying process. However, the evaporation rate is usually much lower during the second stage than during the first, so the investment needed to affect evaporation at this stage might not be worthwhile, especially if the actual reduction achieved is small.

An irrigation regime having a high irrigation frequency may cause the soil surface to remain wet and the first stage of evaporation to persist, resulting in a maximum rate of water loss. Water loss by evaporation from a single deep irrigation is generally smaller than from several shallow ones with the same total amount of water. However, water losses due to percolation are likely to be greater from deep irrigations than from shallow ones. New water application methods, such as drip (or trickle) irrigation, which concentrates the water in a small fraction of the area while maintaining the greater part of the surface in a dry state, are likely to reduce the direct evaporation of soil moisture significantly (Hillel, 1997).

BOX 18.1 Why Soils Wet Quickly but Dry Slowly

Significantly, infiltration is an inherently faster process than evaporation: A few hours of infiltration can charge up a soil profile with an amount of water that many weeks of evaporation will not fully extract. Why is it so?

Are the forces driving infiltration stronger than those driving evaporation? Not generally. Although infiltration is aided by gravity and evaporation generally proceeds against gravity, the other components of the potential gradient generally predominate. In the case of infiltration, the matric potential gradient between the wetted soil surface and the relatively dry layers below can be strong enough (initially, at least) to increase the infiltration rate several-fold over the rate that would occur if gravity were the only force in action. However strong the downward matric potential gradients might be within the soil profile (say, 10 or 20 bars per meter), the potential gradients between the soil and the dry atmosphere are likely to be much greater. The moisture potential of the warm, dry air over a field in an arid region may be equivalent to a suction of hundreds of bars. So if the forces involved do not account for the difference between the rates of infiltration and evaporation, what does?

One answer is that the soil's surface zone is, in effect, a preferential valve. Its hydraulic conductivity is intrinsically higher during infiltration than during evaporation. In infiltration, the surface zone is wetted to saturation or near saturation, hence its conductivity is maximal. In effect, the surface becomes a wide-open "funnel," as receptive to water as its saturated conductivity will allow. During evaporation, in contrast, the surface dries out rather quickly, thereby constituting, in effect, a narrowing bottleneck to the transmission of water.

Another effect results from the diurnal fluctuation of atmospheric evaporativity. In daytime the surface is desiccated; at night it is partially rewetted by the upward movement of moisture from deeper soil layers. As the surface is alternately dried and rewetted, hysteresis comes into play (Hillel, 1977). It has the effect of reducing the rewetting of the surface, so it further suppresses evaporation. By thus enhancing moisture conservation and availability, the soil's physical attributes and processes serve to promote plant growth. This is only one example of the vital role played by soil physical properties in sustaining life on earth.

SOIL EROSION BY WIND

Bare soil surfaces, especially when dry and loose, are vulnerable to erosion by wind. The wind erosion process consists of three distinct phases: detachment and lifting of particles, transportation, and deposition. As in the case of water erosion, a distinction can also be made between the *erosivity* of the fluid medium (in this case, flowing air) and the *erodibility* of the soil surface.

The detachment and uplifting of particles result from the effects of wind gusts and air turbulence in the lower layer of the atmosphere on the soil surface. In general, the minimal wind speed (measured at a height of 0.3 m above ground surface) needed to initiate wind erosion is about 5 m/sec (Schwab et al., 1993). The initiation of wind erosion is strongly affected by the microtopography of the soil surface, especially by the presence of protruding hillocks or hummocks. It is also enhanced by the midday heating of a dry soil surface, which creates conditions of aerodynamic instability as the air that is warmed by the soil becomes buoyant and tends to rise. In desert areas, the combination of this buoyancy and the turbulent wind gusts striking the ground promotes the uplift of soil particles and the appearance of spiral whirls known as dust devils, which may rise to a height of scores of meters.

In wind erosion, air gusts pick up masses of loose particles from the surface (an action called *deflation*) and then convey this material (called *aeolian dust*) over some distance before depositing it. That distance can be great indeed. The circulation of some wind-blown dust has been shown to be global. Dust particles originating in the Sahara have even been found in Hawaii!

The proneness of the soil to wind erosion depends on surface dryness, roughness, the sizes of particles (texture), and whether the particles are loose or bound in aggregates.

Three types of soil movement by wind can be recognized: surface creep, saltation, and suspension. *Surface creep* is the rolling or sweeping motion of relatively large particles (diameters of 0.5 mm or more) that occurs within a few centimeters over the ground surface (Fig. 8.14). *Saltation* is the jerky, kangaroo-like motion of finer sand particles (0.1–0.5 mm) that follow distinct trajectories (Fig. 8.15). In this form of motion, particles are lifted and transported over a distance of some meters and then — as they strike the ground obliquely — they cause the bouncing of another particle, in a repeated

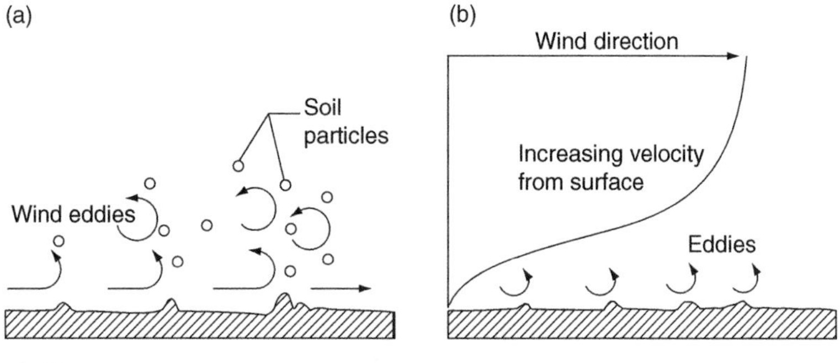

Fig. 18.14. Wind sweeping of dust particles over a rough, dry surface: (a) eddies picking up loose particles; (b) vertical profile of wind over the surface. (After Hudson, 1981.)

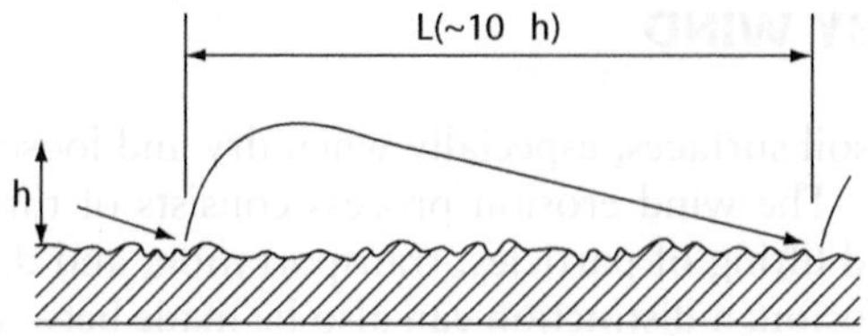

Fig. 18.15. The bouncing (saltation) of sand particles. (After Hudson, 1981.)

sequence of hop-skip-bounce events. Most saltation takes place within a half meter of the soil surface. Still finer particles (0.02–0.1 mm in diameter), once lifted off the surface, remain *suspended* in the air for extended periods. They are often carried aloft and may be transported by the upper air currents to distances of hundreds of kilometers.

The quantity and distance of windborne soil movement depends on the array of particle sizes, the wind speed, and the topography of the land. The capacity of the air to transport particles generally varies as the cube of the wind speed above a threshold value) and as the square root of the particle diameters (Schwab et al., 1993).

Deposition of sediment takes place where the wind abates so that the gravitational force exceeds the aerodynamic lift. Wind speed diminishes especially on the lee side of clumps of vegetation or other physical barriers (e.g., fences). Raindrops also bring down dust from the air.

Factors affecting wind erosion have been quantified experimentally in several locations and have been formulated in empirical regression equations (Woodruff and Siddoway, 1965). The parameters of these equations have been tabulated (Soil Conservation Service, 1988). Computer-based models have been devised to predict wind erosion on the basis of known meteorological and soil conditions, with the latter including tillage effects on soil roughness and erodibility (Hagen, 1991).

Various measures have been devised to control wind erosion. Among these are the maintenance of a protective sod or mulch over the soil surface, the increase of surface cloddiness and roughness by means of appropriate tillage, the stabilization of surface-zone soil aggregates, and the use of mechanical or vegetative windbreaks.

BOX 18.2 The North American Dust Bowl of the 1930s

A spectacular example of wind erosion occurred in the 1930s in America's Southern Great Plains region. Known as the Dust Bowl, it resulted from the introduction of the plow into the vast, semiarid, wind-swept grasslands of that region. During the first three decades of the 20th century, the region enjoyed a wet period. The prairie was green, and when plowed and planted it produced bountiful crops of wheat. But rainfall there is inherently unstable. For example, in Oklahoma's Cimarron County, the annual rainfall averaged about 500 mm between 1900 and 1930 and as much as 700 mm between 1914 and 1923. Then in 1934 and the following few years it was below 350 mm.

During the boom decades from 1900 to 1930, farmers by the thousands settled the southern Great Plains. They broke the sod, planted their seeds in the soft soil, watched the rain-satiated wheat come up, and then harvested its bumper crop. In the Texas Panhandle alone, the area under wheat had expanded from 35,000 to 800,000 hectares between 1909 and 1929. This enormous change in the land's surface was made possible by the mass production of tractors. Those tractors later became the leading Dust Bowl villains. They were "snubnosed monsters," wrote John Steinbeck in *The Grapes of Wrath*, "raising the dust and sticking their snouts into it."

Eventually the copious rains ceased, to be replaced by an inexorable drought. The year 1934 was particularly hot and dry, and the land lay bare and parched under the merciless sun, its excessively pulverized soil easy prey to the whipping winds. The winds swept up the loose soil as if it were talcum powder and rolled it along in billowing red-brown clouds that eclipsed the sun and obliterated fences and covered houses and choked animals and people. And so a great exodus began — 40% of the people of Cimarron County, for example, abandoned their homes and moved out of the region.

The effects of the Dust Bowl were not confined to that region. The fine dust wafted thousands of meters aloft and drifted across the entire continent. On May 12, 1934, the *New York Times* reported that the city "was obscured in a half-light . . . and much of the dust seemed to have lodged itself in the eyes and throats of weeping and coughing New Yorkers." Washington, D.C., was overhung with a thick cloud of dust denser than ever seen before. The entire eastern seaboard of the United States was blanketed in a heavy fog composed of millions of tons of the rich topsoil swept up into the continental jet stream. Even ships far out in the Atlantic found themselves showered with Great Plains dust.

The great American Dust Bowl of the 1930s is not merely a thing of the past. It is being repeated on an equally vast scale in such regions as the Sahel in sub-Saharan Africa. The process by which a semiarid region is denuded of vegetation and its soil destabilized to the point of uncontrolled erosion and degradation is often called *desertification* (Dregne, 1994; Hillel and Rosenzweig, 2002).

Part VII

SOIL–PLANT–WATER RELATIONS

19. PLANT UPTAKE OF SOIL MOISTURE

WATER REQUIREMENTS OF PLANTS

Nature, despite its celebrated laws of conservation, can in some ways be exceedingly wasteful, or so it appears from our own partisan viewpoint. One of the most glaring examples is the way it requires plants to draw quantities of water from the soil far in excess of their essential metabolic needs. In dry climates, plants growing in the field may consume hundreds of tons of water for each ton of vegetative growth. That is to say, the plants must inevitably transmit to an unquenchably thirsty atmosphere most (often well over 90%) of the water they extract from the soil.

The release of water vapor by plants to the atmosphere, a process called *transpiration*, is not an essential or an active physiological function of plants but by and large a passive response to the atmospheric environment. In principle, plants can thrive in an atmosphere practically saturated with vapor and hence requiring very little transpiration. Rather than by plant growth per se, transpiration is caused by the vapor pressure gradient between the normally water-saturated tissue of the leaves and the dry atmosphere. In other words, it is exacted from the plants by the *evaporative demand* of the climate in which they live.

In a sense, the plant in the field can be compared to a wick in an old-fashioned kerosene lamp. Such a wick, its bottom dipped into a reservoir of fuel while its top is subject to the burning fire that consumes the fuel, must constantly transmit the liquid from bottom to top under the influence of physical forces imposed on the passive wick by the conditions prevailing at its two ends. Similarly, the plant has its roots in the soil-water reservoir while its leaves are subject to the radiation of the sun and the sweeping action of the wind, which require it to transpire practically unceasingly. This analogy, to be sure, is a gross

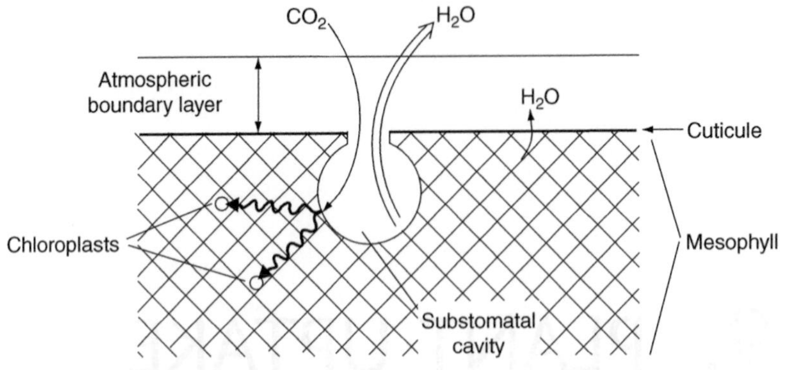

Fig. 19.1. Schematic representation of transpiration through the stomate and the cuticle and of the diffusion of CO_2 into the stomate and through the mesophyll to the chloroplasts. (After Rose, 1966.)

oversimplification, since plants are not entirely passive and in fact are able at times to limit the rate of transpiration by shutting the stomates of their leaves (Fig. 19.1). (The *stomates* are the small openings in the leaves through which gaseous exchange takes place between the interior of the leaf and the external atmosphere.) However, for this limitation of transpiration most plants pay, sooner or later, in reduced growth potential, since the same stomates that transpire water also serve to absorb the carbon dioxide needed in photosynthesis. Reduced transpiration often results in warming of the plants and hence in increased respiration and further reduction of net photosynthesis.

To grow successfully, a plant must achieve a water economy such that the demand made on it for water is balanced by the supply available to it. The problem is that the evaporative demand by the atmosphere is practically continuous, whereas rainfall occurs only occasionally and irregularly. To survive during dry spells between rains, the plant must rely on the diminishing reserves of water contained in the pores of the soil, which itself loses water by direct evaporation and internal drainage. How efficient is the soil as a water reservoir for plants? How do plants draw water from the soil, and to what limit can they thrive while soil moisture diminishes? How is the rate of transpiration determined by the interaction of plant, soil, and meteorological factors? These and related questions are the topics of this chapter.

BOX 19.1 Plants Adapted to Different Environments

Plants that inhabit water-saturated domains are called *hydrophytes*, or aquatic plants. Plants adapted to drawing water from shallow water tables are called *phreatophytes*. In contrast, plants that can grow in arid regions by surviving long periods of thirst and then recovering quickly when water is supplied are called *xerophytes*, or desert plants. Such plants may exhibit special features, called *xeromorphic* (such as succulent tissues, thickened epidermis and a waxy surface cuticle, recessed stomates, and reduced leaf area), that are designed to store water and minimize its loss.

Some plants are especially adapted to growing in a saline environment and are called *halophytes*. Plants not so adapted to saline conditions are called *glycophytes*. It takes an extreme degree of adaptability (or masochism?) for certain plants, called *xero-halophytes*, to grow in an environment that is both dry and saline at the same time.

Finally, there are plants that grow best in moist but aerated soils, generally in semi-humid to semiarid climates. Such intermediate plants are called *mesophytes*. Most crop plants belong in this category. Mesophytes control their water economy by developing extensive root systems and optimizing the ratio of roots to shoots and by regulating the aperture of their stomates to curtail transpiration during periods of water shortage. The latter effect, however, necessarily entails restriction of photosynthesis. Moreover, curtailment of transpiration reduces the evaporative cooling of plants, so they tend to warm up under the sun's radiation, and as they warm up their rate of respiration rises. In effect, therefore, a thirsty plant consumes its own reserves and further reduces its growth potential. For these reasons, thirsty plants are generally less productive than plants that are well endowed with water throughout their growing season.

THE SOIL–PLANT–ATMOSPHERE CONTINUUM

Current approaches to the issue of soil-water extraction and utilization by plants are based on recognition that the field, with all its components — soil, plant, and ambient atmosphere taken together — constitutes a physically integrated, dynamic system in which various flow processes occur interdependently like links in a chain. This unified system has been called the *soil–plant–atmosphere continuum* (SPAC, for short) by J. R. Philip (1966b). The universal principle operating consistently throughout the system is that water flow always takes place spontaneously from regions where its potential energy state is higher to where it is lower. Former generations of soil physicists, plant physiologists, and meteorologists, each group working separately in what it considered to be its separate and exclusive domain, tended to obscure this principle by expressing the energy state of water in different terms (e.g., the "tension" of soil water, the "diffusion pressure deficit" of plant water, and the "vapor pressure" or "relative humidity" of atmospheric water) and hence failed to communicate readily across the self-imposed boundaries of their respective disciplines.

As we have come to understand, the various terms used to characterize the state of water in different parts of the soil–plant–atmosphere system are merely alternative expressions of the energy level, or *potential*,of water. Moreover, the very occurrence of differences, or gradients, of this potential between adjacent locations in the system constitutes the force inducing flow within and between the soil, the plant, and the atmosphere. This principle applies even though different components of the overall potential gradient are effective to varying degrees in motivating flow in different parts of the soil–plant–atmosphere system.

To describe the interlinked processes of water transport throughout the SPAC, we must evaluate the pertinent components of the energy potential of water and their effective gradients as they vary in space and time. As an approximation, the flow rate through each segment of the system can be taken to be proportional directly to the potential gradient and inversely to the segment's resistance. The flow path includes liquid water movement in the soil toward the roots, liquid and perhaps vapor movement across the root-to-soil contact zone, absorption into the roots and across their membranes to the vascular tubes of the xylem, transfer through the xylem up the stem and branches to the leaves, evaporation in the intercellular spaces within the leaves, vapor diffusion through the substomatal cavities and out the stomatal apertures to the boundary air layer in direct contact with the leaf surface, and through it, finally, to the turbulent atmosphere, which carries away the water extracted by the plant from the soil.

BASIC ASPECTS OF PLANT–WATER RELATIONS

Close observation of how higher plants are structured reveals much about how they function in their terrestrial environment. It is intriguing to compare the shape of such plants to the shape of familiar higher animals. The characteristic feature of warm-blooded animal bodies, in contrast with plants, is their minimal area of external surface exposure. Apart from a few protruding organs needed for mobility and sensory perception, such animals are rather compact and bulky in appearance. Not so is the structure of higher plants, whose vital functions require them to maximize rather than minimize surface exposure, both above and below ground. The aerial canopies of plants frequently exceed the area of covered ground many times over. Such a large surface helps the plant to intercept and collect sunlight and carbon dioxide, two resources that are normally diffuse rather than concentrated.

Even more striking is the shape of roots, which proliferate and ramify throughout a large volume of soil while exposing an enormous surface area. A single annual plant can develop a root system with a total length of several hundred kilometers and with a total surface area of several hundred square meters. (Estimates of total length and surface area of roots are even larger if root hairs are taken into account.) The need for such exposure becomes apparent if we consider the primary function of roots, which is to gather water and nutrients continuously from a medium that often holds only a meager supply of water per unit volume and that generally contains soluble nutrients only in very dilute concentrations. And while the atmosphere is a well-stirred and thoroughly mixed fluid, the soil solution is a sluggish and unstirred fluid that moves toward the roots at a grudgingly slow pace, so the roots have no choice but to move toward it. Indeed, roots forage constantly through as large a soil volume as they can, in a constant quest for water and nutrients. Their movement and growth, involving proliferation in the soil region where they are present and extension into ever new regions, are affected by a host of factors additional to moisture and nutrient, including temperature, aeration, mechanical resistance, the possible presence of toxic substances, and the primary roots' own *geotropism* (i.e., their preference for a vertically downward direction of growth).

Green plants are the earth's only true *autotrophs*, able to create new living matter from inorganic raw materials. We refer specifically to the synthesis of basic sugars (subsequently elaborated into more complex compounds) by the combination of atmospheric carbon dioxide and soil-derived water, accompanied by the conversion of solar radiation into chemical energy in the process of *photosynthesis*, usually described by the following deceptively simple formulas:

$$6CO_2 + 6H_2O + \text{sunlight energy} \rightarrow C_6H_{12}O_6 + 6O_2$$
$$\underset{\text{glucose}}{nC_6H_{12}O_6} \rightarrow \underset{\text{starch}}{(C_6H_{10}O_5)_n} + nH_2O \qquad (19.1)$$

We, along with the entire animal kingdom, owe our lives to this process, which not only produces our food but also releases into the atmosphere the elemental oxygen we need for our respiration. Plants also respire, and the process of respiration represents a reversal of photosynthesis, in the sense that some photosynthetic products are reoxidized and decomposed to yield the original constituents (water, carbon dioxide, and released energy). Thus

$$C_6H_{12}O_6 + 6O_2 \rightarrow 6CO_2 + 6H_2O + \text{thermal energy} \qquad (19.2)$$

In examining these formulas, we note immediately the central role of water as a major metabolic agent in the life of the plant, that is, as a source of hydrogen atoms for the reduction of carbon dioxide in photosynthesis and as a product of respiration. Water is also the solvent and hence conveyor of transportable ions and compounds into, within, and out of the plant. It is, in fact, a major structural component of plants, often constituting 90% or more of their total "fresh" mass. Much of this water occurs in cell vacuoles under positive pressure, which keeps the cells turgid and gives rigidity to the plant as a whole.

Only a small fraction of the water normally absorbed by plants growing in dry weather is used in photosynthesis, while most (often over 98%) is lost as vapor in the process of *transpiration*. This process is impelled by the exposure to the atmosphere of a large area of moist cell surfaces, necessary to facilitate absorption of carbon dioxide and oxygen. Mesophytes are extremely sensitive, and vulnerable, to lack of sufficient water to replace the amount lost in transpiration. Water deficits impair plant growth and, if extended in duration, can be fatal.

WATER RELATION OF PLANT CELLS AND TISSUES

Water plays a crucial role in the processes of plant growth, which involve cell division and cell expansion. The latter process occurs as each pair of divided cells imbibes water. The resulting internal pressure, called *turgor*, stretches the new cells' elastic walls, which thicken through the deposition of newly synthesized material.

The cell wall, consisting mainly of cellulose and other polysaccharides, and the protoplasmic material within, called *cytoplasm*, are generally permeable to water molecules diffusing or flowing into and out of the cell. Since the cell wall contains fairly large interstices, it does not restrict solute movement. However, inside the cell wall (Fig. 19.2) and surrounding the cytoplasm is a lipoprotein "cell membrane" called the *plasmalemma*, which is selectively permeable and

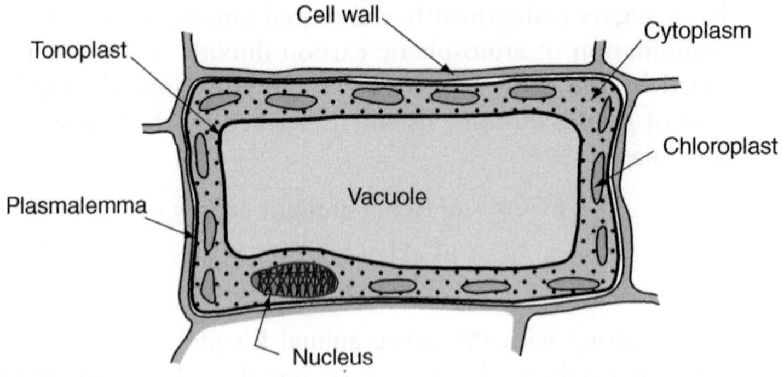

Fig. 19.2. The internal structure of a mature mesophyll (leaf) cell, showing the solute-confining membranes.

thus helps regulate what enters and leaves a plant cell. Most solutes are prevented from diffusing freely in and out of the cell's internal solution phase, generally contained in a central "cavity" called a *vacuole*. In a mature cell, the vacuole often occupies the greater part of the cell's volume. It is surrounded by a "vacuolar membrane" called the *tonoplast*.

Although some solutes are selectively conveyed through the membranes by a complex process called *active transport*, by and large the cell acts as a small *osmometer*, drawing water by osmosis from the surrounding aqueous phase into the generally more concentrated solution in the vacuole. The *osmotic potential* of cell water, due to the sugars and salts dissolved in it, is negative with respect to that of pure water (generally taken as the zero reference) and may range, in terms of pressure units, from −0.5 to −5 Mpa (−5 to −50 bar).

The expansive pressure caused by osmosis is countered by the elastic cell walls, which are tensed as they are stretched and hence tend to press inward. This contractile tendency can be expressed in terms of a positive pressure potential. The overall potential of water inside the cell ϕ depends on the osmotic potential of the cell solution ϕ_0 and on the effective potential resulting from the pressure exerted by the walls ϕ_p, called the *turgor pressure*. If these were the only components of the cell's water potential, we could write (Kramer and Boyer, 1995; Boyer,1995):

$$\phi = \phi_0 + \phi_p \tag{19.3}$$

If we immerse a cell in pure water at atmospheric pressure, the cell will imbibe water until its affinity for water is counterbalanced by the cell wall's "turgor" pressure. The cell will then be fully hydrated and turgid. If a plant is "thirsty" (or, according to a currently prevalent expression, if it is subject to "moisture stress"), its cells will not be fully turgid and its *relative hydration* (i.e., its water content relative to that when it is fully turgid at a water potential of zero) will be less than unity. The functional relationship between the relative hydration of plant cells and their water potential is as important in plant physiology as the relationship between soil wetness and soil-water potential is in soil physics.

If we immerse a cell not in pure water but in a solution more concentrated than the cell's own solution, outward osmosis will take place and the cell will dehydrate. In extreme cases, the cell may dehydrate to the point of *plasmolysis*, at which point the effective turgor pressure falls to zero. At this point the cytoplasm may actually separate from the walls and collapse inward.

An interesting illustration of the importance of turgor pressure changes is the *stomatal control mechanism*. The opening (aperture) or closure of stomates is affected by light intensity and atmospheric CO_2 concentration (i.e., stomates tend to close in the dark and at high CO_2 concentrations). Stomatal aperture is also affected by temperature and the plant's own internal rhythms. Most of all, however, stomates are affected by tissue hydration and, more specifically, by the turgidity of the paired *guard cells* that constitute what is in effect a sensitive valve. Increasing hydration and turgidity causes these cells to bulge outward, crescent-like, thus dilating the space between them. Low turgidity causes them to collapse against each other, thus shutting off the diffusive outlet for water vapor (also serving as the inlet for carbon dioxide). A transverse section and surface views of a leaf with open and shut stomates are shown in Fig. 19.3. Methods of measuring plant water potential were described by Boyer (1995).

In the absence of transpiration, plants absorb water from the soil by means of the osmotic mechanism, and plant water generally exhibits a positive hydrostatic pressure. Evidence of this positive pressure is found in the tendency of some plants to exude water during the night (a phenomenon called *guttation)* and in the tendency of excised stems or roots to ooze out water (an occurrence often attributed to "root pressure"). However, when transpiration takes place, the pressure of water in the leaves diminishes and generally falls below atmospheric pressure. The resulting tension, or suction, induces upward mass flow from roots

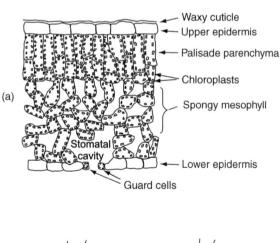

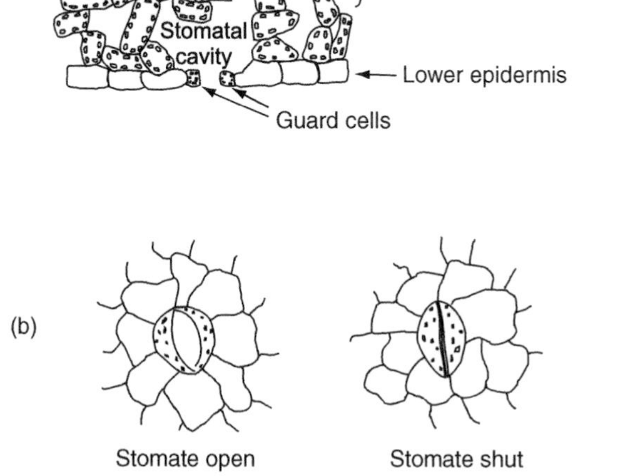

Fig. 19.3. A leaf section: (a) transverse view and (b) surface view, showing an open and a closed stomate.

to leaves through the tubelike capillary vessels of the *xylem*. It is remarkable that these capillary vessels can maintain the cohesive continuity of liquid water columns even under a tension of hundreds of kilopascals, as needed to draw water from relatively dry soils and transport it to the tops of trees scores of meters high. The rate at which water is extracted from the soil and transmitted to the transpiring leaves obviously depends not only on the magnitude of the potential or pressure gradients but also on the hydraulic properties of the conveyance system, that is, on the conductance of the stems and on the conductance as well as absorptive properties of the roots.

STRUCTURE AND FUNCTION OF ROOTS

The anatomy of roots is designed, as it were, to perform a number of essential functions, including absorption and conveyance of water and nutrients from the soil and anchorage of the plant's superstructure. A longitudinal section of a growing root reveals several sections, as shown in Fig. 19.4a. The tip of the root is shielded by a *root cap*, which is often of mucilaginous consistency and may thus help to lubricate the path of the advancing root. Following the root cap is a zone of rapid cell division called the *apical meristem*. Next is a region of *cell elongation*, where the initially isodiametric cells elongate in the direction of the root axis, thus pushing the root tip forward and causing the root to extend farther into the soil. Behind the region of cell elongation is that of *cell differentiation*. where different groups of cells develop specific characteristics and assume specialized functions.

The innermost cells of the root become *vascular* tissue, while the outermost cells develop thin radial protrusions known as *root hairs*. The latter have the effect of greatly increasing the surface area of contact over which the root can interact with the soil and absorb nutrients. Farther back, the older section of

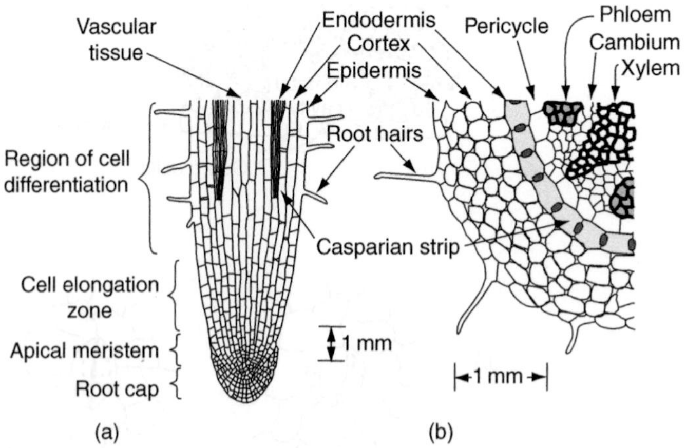

Fig. 19.4. The structure of a growing root: (a) longitudinal section through the root cap and the zones of cell division, elongation, and differentiation; and (b) cross section, showing the concentric arrangement of the epidermis, cortex, endodermis, pericycle, and vascular bundles. (After Nobel, 1974.)

the root becomes somewhat hardened and less absorbent, and, as the root converges with other roots, it becomes thicker and more stemlike.

A cross-sectional view of a typical root (some distance away from the root tip), illustrated in Fig. 19.4b, shows a central vascular core, called the *stele*, surrounded by three concentric, sleevelike layers of cells: the *endodermis*, the *cortex*, and the *epidermis*.

The vascular system of the roots connects with that in the stems and leaves. It consists of (1) *xylem vessels*, conveying water and mineral nutrients from the soil, along with organic compounds metabolized in the roots, to the upper parts of the plant, and (2) *phloem vessels*, serving to translocate the organic compounds (such as carbohydrates) that are synthesized in the leaves and to distribute these throughout the plant. The two types of vessels are illustrated in Fig. 19.5. Xylem vessels are hollow cells, practically devoid of protoplasm, arranged end to end and joined by *perforation plates* to form continuous tubes. The aqueous solution drawn from the soil by the absorbing sections of the roots, typically with a concentration of the order of 10^{-2} *M* (Epstein, 1977; Nobel, 1970), flows up through the root system and then through the stem and petiole and leaf veins into the leaves, with motion occurring in the direction of decreasing hydrostatic pressure. In contrast with the xylem, the conducting cells in the phloem, called *sieve cells*, contain cytoplasm and remain metabolically active. The tissue between the xylem and the phloem is the *cambium*,whose cells divide and differentiate to form xylem and phloem. Between the cambium and the endodermis is a layer called the *pericycle*,whose cells can divide and lead to the formation of lateral or branch roots.

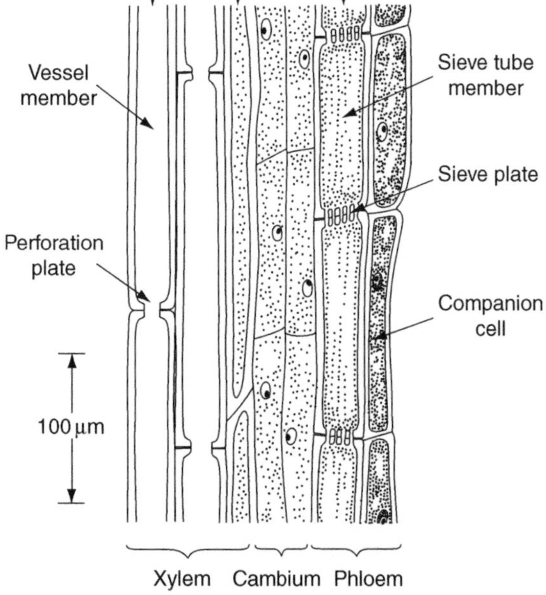

Fig. 19.5. Longitudinal section through xylem and phloem vessels, showing their internal structures. (After Nobel, 1974.)

BOX 19.2 The Plight of Plants in Arid Regions

Perhaps we can identify more fully with the plight of arid-zone plants if we can imagine ourselves to be living under a government that taxes away 99% of our income while requiring us to keep our reserves in a bank that is embezzled daily. The onerous regime governing plant life is the dry climate, which requires each plant to transpire such an exorbitant share of the moisture it obtains by its roots. The bank on which plants depend is the soil, which loses moisture continuously by direct upward evaporation from the surface as well as by downward seepage below the root zone.

Plants live simultaneously in two realms, the conditions in each of which change constantly and not necessarily in coordination. The canopy of the plant lives in the open atmosphere, subject to varying sunlight, temperature, wind, and humidity (as well as to competition from adjacent plants and to predation by herbivorous animals). The roots of the plant live in the dark, tight, cool soil, which provides anchorage as well as moisture and nutrients. How the plant can sense both realms and respond optimally to their changing conditions is source of wonderment. It is certainly amazing that plants can do so without a brain or a nervous system, and without access to a computer. We, who have access to computers, can barely begin to understand how green plants function in the real environment: how they sustain themselves and, indeed, how they sustain us.

The *endodermis is* a layer of cells whose radial walls are impregnated with a waterproof waxy material forming a shield termed the *casparian strip*. To cross this barrier, water and solutes must either flow along the cell walls or enter and flow across the cytoplasm. Outside the endodermis are several layers of cells known as the *cortex*, a "loose" tissue with numerous intercellular air spaces through which gases (e.g., O_2 and CO_2) can diffuse. Finally, surrounding the entire root and serving as its "skin" is the *epidermis,*where the root hairs originate.

VARIATION OF WATER POTENTIAL AND FLUX IN THE SOIL–PLANT SYSTEM

To characterize the soil–plant–atmosphere continuum physically, one must evaluate the pertinent components of the energy potential of water and the effective potential gradients as they vary from one segment to another along the entire path of water movement. The total potential difference between soil moisture and atmospheric humidity can amount to tens of megapascals (hundreds of bars), as illustrated in Fig. 19.6, and in an arid climate can even exceed 100 MPa. Of this total, the potential drop in the soil toward the roots may vary from less than 100 kPa to several hundred kilopascals (i.e., from a fraction of a bar to several bars). Except where the soil is fairly dry, the potential drop in the roots (from cortex to xylem) is likely to be somewhat greater than that in the soil. The potential drop in the xylem from roots to leaves will generally not exceed a few hundred kilopascals. Altogether, therefore, the summed potential drop in the soil and plant will be of the order of,

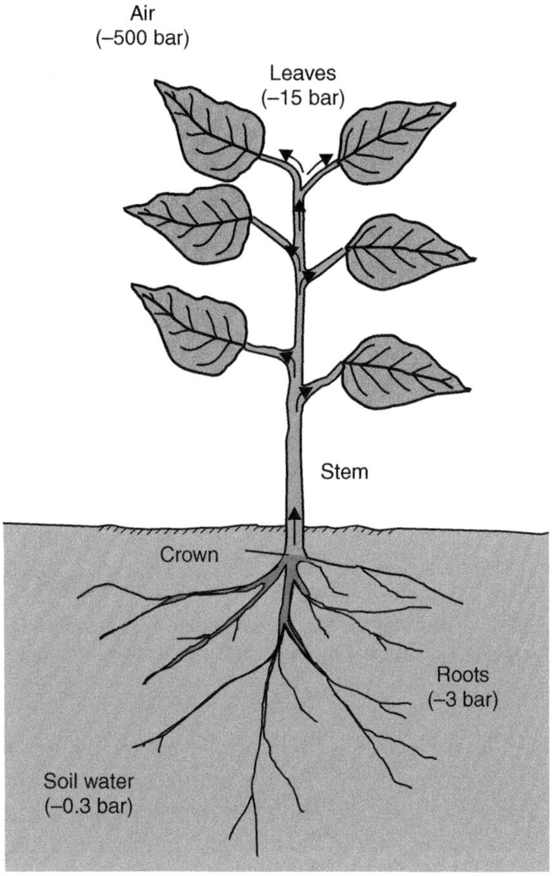

Fig. 19.6. Variation of water potential along the transpiration stream.

say, 1–3 MPa. Clearly, the major portion of the overall potential difference in the SPAC occurs between the leaves and the atmosphere.

Figure 19.7 describes the distribution of water potentials in the SPAC. This figure is not drawn to scale and is meant only to illustrate general relationships. Curve 1 is for a low value of water suction in the soil (AB) and hence also at the root surfaces (B). In the mesophyl cells (DE), water suction ("negative" potential) is below the critical value at which the leaves may lose their turgidity, hence the plant is able to transport water from the soil to the atmosphere without wilting. Area E represents the substomatal cavity. In curve 2, soil-water suction is equally low but the transpiration rate is higher, and water suction in the leaf mesophyl approaches the critical wilting values (say, 2–3 MPa ≈ 20–30 bar). Curve 3 is for when soil-water suction is relatively high but the transpiration rate is low. Curve 4, finally, indicates the extreme condition when soil-water suction and transpiration rate are both high. In this case, leaf-water suction is likely to exceed the critical value, causing the plant to wilt.

Soil-water suction increases as soil wetness decreases. Consequently, the plant-water suction required to extract water from the soil must increase correspondingly. The soil will deliver water to the root as long as the water

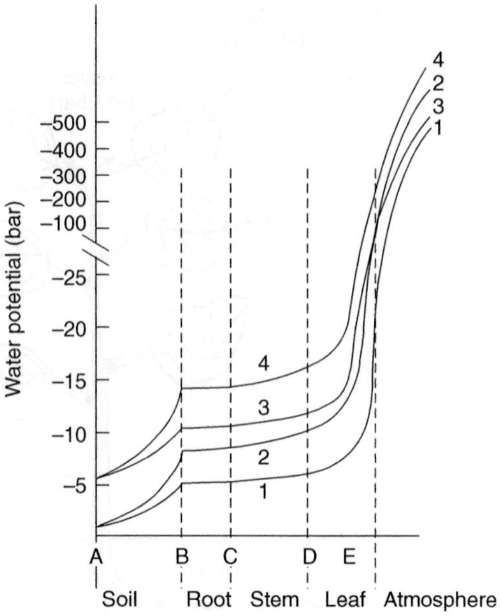

Fig. 19.7. The distribution of potentials in the soil–plant–atmosphere continuum under different conditions of soil moisture and atmospheric evaporativity.

suction in the latter is maintained greater than in the former. However, as a root extracts water from the soil in contact with it, the suction in the contact zone may increase and tend to equal the root suction. Water uptake may then cease unless additional water can move in from the farther reaches of the soil as a result of the suction gradients that form in the periphery of the roots. In order for this additional water to become available to the plant, not only must it be at a suction lower than the root-water suction, but it must also move toward and into the root at a rate sufficient to compensate the plant for its own continuing loss of water to the atmosphere in the process of transpiration. The dependence of suction on distance from the root during water uptake was calculated by W. R. Gardner (1960a) for several soils and was found to be quite flat until the initial suction approached 15 bar. This is illustrated in Fig. 19.8.

As long as the plant does not wilt, and as long as the influx of radiation and heat to the canopy results in change of phase only, it is possible to assume steady-state flow through the plant. This means that the transpiration rate is equal to the plant transport and that both are equal to the soil-water uptake rate:

$$q = -\Delta\phi_s/R_s = -\Delta\phi_{s-r}/R_{s-r} = -\Delta\phi_p/R_p = -\Delta\phi_{p-a}/R_{p-a} \qquad (19.4)$$

where $\Delta\phi_s$ is the potential drop in the soil toward the roots, $\Delta\phi_{s-r}$ is the potential drop between the soil and root xylem, $\Delta\phi_p$ is the potential drop in the plant to the leaves, and $\Delta\phi_{p-a}$ is the potential drop between the leaves and the atmosphere. As mentioned, the magnitudes of these potential increments are of the order of 1 MPa, 1 MPa, 1 MPa, and 50 MPa, respectively, and the corresponding resistances are proportional. It follows that the resistance R_{p-a} between the leaves

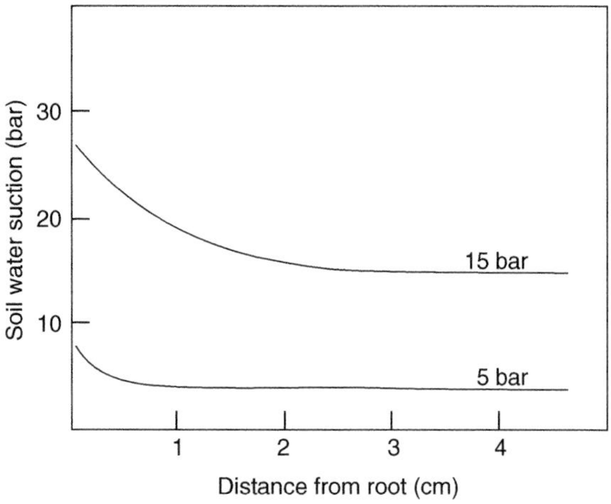

Fig. 19.8. Relation of soil-water suction to distance from the root axis in a sandy soil with an uptake rate of 0.1 cm^3 per centimeter of root per day. The two curves are for two different levels of soil-water suction at a distance form the root (After Gardner, 1960a.)

and the atmosphere might be 50 or more times greater than the resistances in the plant and the soil. When the stomates close, as they tend to do at noontime during a hot day, the resistance $R_{p\text{–}a}$ becomes even greater and may result in a reduced rate of transpiration.

The suction difference between soil and root needed to maintain a steady flow rate depends, as stated, on the conductivity K and the flow rate q. When soil-water suction is low and conductivity high, the suction in the root need not differ greatly from the mean suction in the soil. When soil-water suction increases and soil conductivity decreases, the suction difference (or gradient) needed to maintain the same flow rate must increase correspondingly. As long as the transpiration rate required of the plant is not too high, and as long as the hydraulic conductivity of the soil is adequate and the density of the roots is sufficient, the plant can extract water from the soil at the rate needed to maintain normal physiological activity. However, the moment the rate of extraction drops below the rate of transpiration (either because of a high evaporative demand by the atmosphere and/or because of low soil conductivity and/or because the root system is too sparse), the plant experiences a net loss of water; if it cannot adjust its root-water suction or its root density so as to increase the rate of soil-water uptake, the plant may suffer loss of turgor and be unable to grow normally. If this situation persists, the plant wilts. It follows that as atmospheric evaporativity rises and soil conductivity diminishes, the average soil-water suction at which wilting occurs will also tend to diminish.

Denmead and Shaw (1962) were evidently the first to present experimental confirmation of the effect of dynamic conditions on water uptake and transpiration. They measured the transpiration rates of corn plants grown in containers and placed in the field under different conditions of irrigation and atmospheric evaporativity. Under an evaporativity of 3–4 mm/day, the actual transpiration rate began to fall below the potential

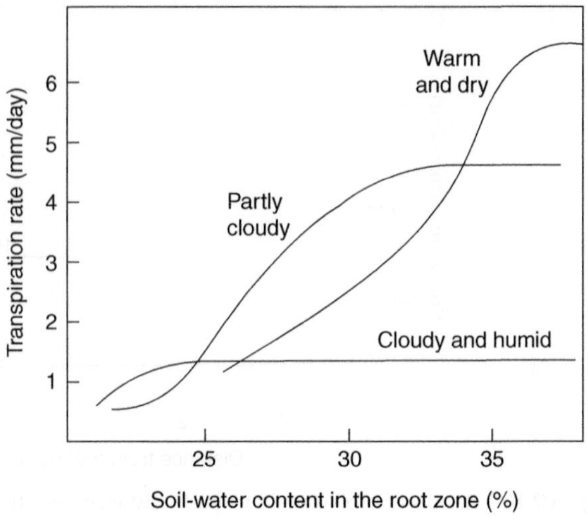

Fig. 19.9. Relation of actual transpiration rate of soil-water content, under different meteorological conditions. (After Denmead and Shaw, 1962.)

rate at an average soil-water suction of about 200 kPa (2 bar). Under more extreme meteorological conditions, with an evaporativity of 6–7 mm/day, this drop in transpiration rate began at a soil-water suction value of 30 kPa (0.3 bar). On the other hand, when potential evaporativity was very low (1.4 mm/day), no drop in transpiration rate was noticed until average soil-water suction exceeded 1.2 MPa (12 bar). The volumetric water contents at which the transpiration rates fell varied between 23%, under the lowest evaporativity, and 34%, under the highest evaporativity measured. This is illustrated in Figs. 19.9 and 19.10.

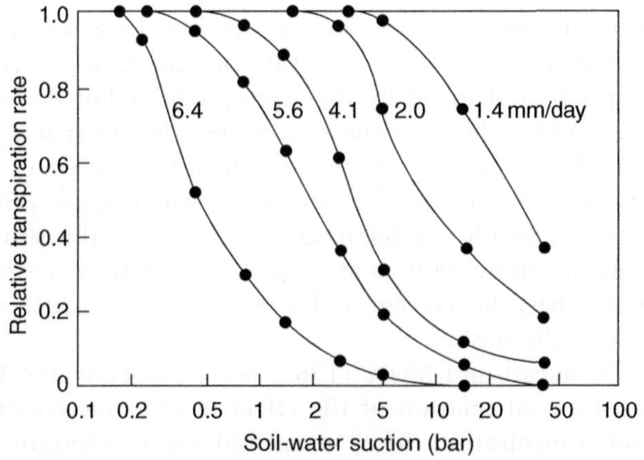

Fig. 19.10. The relation of relative transpiration rate to average soil-water suction, under different meteorological conditions. The numbers represent different rates of potential evapotranspiration. (After Denmead and Shaw, 1962.)

MATHEMATICAL MODELING OF WATER UPTAKE BY ROOTS

The roots of a plant function as an elaborate pumping and conveyance system reaching into, probing, and drawing water from various soil depths in a manner that is somehow "programmed" to maximize the plant's chances for survival and self-perpetuation. Soil-moisture content and potential are seldom uniform throughout the root zone, and neither is the density or hydraulic resistance of the roots. How the root system of a plant senses the root zone as a whole and integrates its response so as to utilize soil moisture to best advantage is a phenomenon still imperfectly understood. One classical view was that the root system adjusts its water withdrawal pattern so as to maintain the total soil-moisture potential constant throughout the root zone. On the other hand, an often-observed pattern of water withdrawal is such that the top layer is depleted first and the zone of maximal extraction moves gradually into the deeper layers.

Since the soil usually extends in depth considerably below the zone of root activity, it is of interest to establish how the pattern of soil-water extraction by roots relates to the pattern of water flow into, within, through, and below the root zone. Some drainage through the root zone is generally considered necessary to prevent deleterious accumulation of salts, particularly in arid zones; yet excessive drainage might involve unnecessary loss of nutrients as well as of water. If a groundwater table is present at a shallow depth, it may contribute to the supply of crop-water requirements by upward capillary flow, but it might also infuse the root zone with salts. Considerable upward flow into the root zone from underlying moist layers might also be possible even in the absence of a water table, depending on soil-moisture characteristics and on the irrigation regime. In fact, the opposite processes of downward drainage and upward capillary flow can occur in a sequential or alternating pattern at varying rates so that the net outflow or inflow of water for the root zone as a whole can only be determined by integrating the fluxes taking place through the bottom of that zone continuously over a period of time.

Numerous models have been developed in recent decades, with the aim of achieving a quantitative formulation and prediction of the sequential and simultaneous processes involved in the uptake of soil moisture by plants. These models vary widely in aim, structure, and level of detail. Some are based on empirical and others on mechanistic approaches to the complex of soil–plant–water relations.

Early modelers of soil-moisture extraction by roots differed widely in their quantitative assessment of the relative importance of the root-versus soil-resistance terms at various stages of the extraction process. Conflicting approaches assumed either the one term to predominate over the other or vice versa. Recent theoretical and experimental studies have suggested that both resistance terms can be important. Root resistance probably predominates in situations of low soil-moisture tension (i.e., high soil hydraulic conductivity), and the soil's hydraulic resistance gains importance as the extraction process continues and causes progressive depletion of soil moisture.

In principle, two alternative approaches can be taken to modeling the uptake of soil water by roots. The first is to consider the convergent radial flow of soil water toward and into a representative individual root, taken to be a cylindrical sink uniform along its length (i.e., of constant and definable thickness and absorptive properties). The root system as a whole can then be described as a set of such individual roots, assumed to be regularly spaced in the soil at definable distances that may vary within the soil profile. This approach, called the *microscopic-scale* or *single-root* model, usually involves casting the flow equation in cylindrical coordinates and solving it for the distribution of potentials, water contents, and fluxes from the root outward. The solution can be attempted either by analytical means, which usually require rather restrictive assumptions, or by numerical means.

The alternative approach is to regard the root system in its entirety as a diffuse sink that permeates each depth layer of soil uniformly, but at a density that differs from layer to layer. This approach, termed the *macroscopic-scale* or *root system* model, disregards the flow patterns toward individual roots and thus avoids the geometric complication involved in analyzing the distribution of fluxes and potential gradients on a microscale. The major shortcoming of the macroscopic approach is that it is based on the gross spatial averaging of the matric and osmotic potentials and takes no account of the increase in suction and in concentration of salts in the soil at the immediate periphery of each absorbing root.

EFFECT OF ROOT GROWTH ON SOIL-WATER UPTAKE

One of the main shortcomings of the early models of water uptake by roots was the omission of root growth as a process affecting the pattern of moisture extraction and movement in the soil profile and possibly enhancing the ability of plants to avoid stress due to moisture deficit. A plant with growing roots can reach continuously into moist regions of the soil rather than depend entirely on the dwindling reserves of moisture and the increasing hydraulic resistance of a fixed root zone. The process of root growth, if rapid enough, can reduce the effect of the localized drawdown of both matric and osmotic potential around each root as well as expand the effective volume of soil tapped by the root system as a whole.

In principle, and insofar as it relates to water uptake, we can consider overall root growth as consisting of several sequential or concurrent processes, including proliferation, extension, senescence, and death. As used in the present context, the term *proliferation* applies to the localized increase of rooting density (i.e., by branching) within each layer without any increase in the volume of the root zone as a whole. Root *extension* is the additional process by which roots from any layer invade an adjacent (generally an underlying) layer and increase its rooting density. The process of *senescence* involves suberization and the gradual reduction of root permeability. With further aging, the older roots eventually become totally inactive and, to all intents and purposes, can be considered dead.

The phenomenon of root extension is characteristic of a stand of young plants, such as an annual crop in its early stages of growth. The rate of root-system extension tends to diminish as plant stands mature. The vertical extent of root-system penetration is often limited by such factors as the lack of aeration and the mechanical impedance of the deeper layers in the soil profile.

The possible effects of root growth processes on plant-water relations were studied by Hillel and Talpaz (1976), who considered the time course of plant-water potential at the *crown* of the roots (where all the roots converge and the stem emerges from the soil with a single value of water potential). Without root extension in depth, this simulation suggests that root proliferation alone may add relatively little to the period of time the plant can maintain the potential transpiration rate without experiencing stress. On the other hand, root systems capable of extending themselves deeper into the soil profile can prolong that time span significantly by tapping into a greater store of moisture. The predicted soil-moisture profiles resulting from extraction by various root systems are shown in Fig. 19.11.

Subsequently, Huck and Hillel (1983) developed a more comprehensive model of root growth and water uptake. In addition to soil hydraulics, their model accounted for such plant physiological functions as photosynthesis, respiration, and transpiration. They pointed out that since a terrestrial plant lives in two realms, the atmosphere and the soil, in each of which conditions vary continuously and not necessarily in concert, we must consider that a plant responds to its above-ground and below-ground environments conjunctively. In order to thrive, each plant must develop both a root system extensive enough to supply sufficient water and nutrients and a canopy elaborate enough to synthesize the carbohydrates needed for growth and reproduction, and it must do so in optimal fashion. Insufficient canopy growth will limit photosynthesis and, hence, constrain root growth, just as insufficient root growth will limit water supply and, eventually, canopy growth.

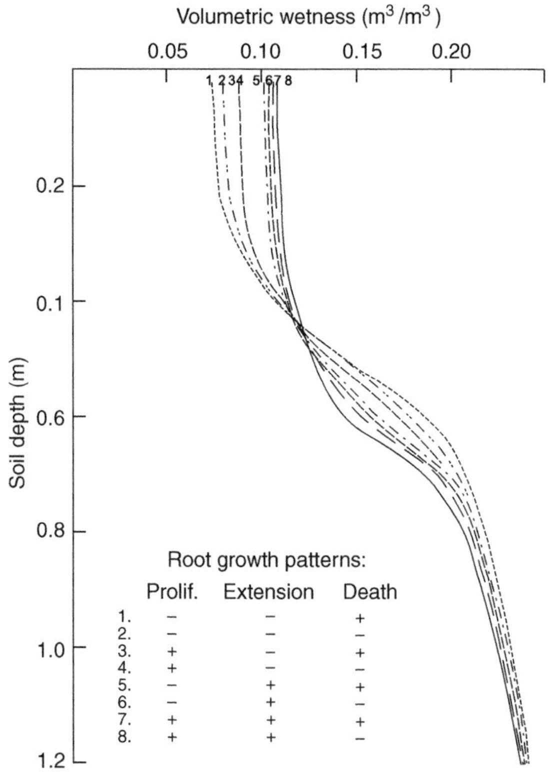

Fig. 19.11. Soil-moisture profiles at the end of a 10-day simulation of moisture extraction by root systems with various growth patterns. (After Hillel, 1977.)

The simultaneous processes of photosynthesis, transpiration, canopy and root growth, as well as respiration are seen to be mutually dependent. Therefore, the hydrologic, edaphic, agronomic, and climatic viewpoints need to be reconciled to provide the linkage between root activity and canopy growth. Because of the complex nature of the interactions and feedback mechanisms involved, it seems impossible to deal with soil–plant–water relations except in the context of a comprehensive model. To be useful, however, a model must be neither too complex and incomprehensible nor too simplistic and arbitrary, and insofar as possible it should be based on the quantitative formulation of the major processes in terms of their basic mechanisms.

We now briefly describe the principles underlying the model of Huck and Hillel (1983). The first process to be considered is the flow of carbon through the plant system (Fig. 19.12). Carbon enters the plant in the course of photosynthesis, which depends on the absorption of a fraction of solar radiation

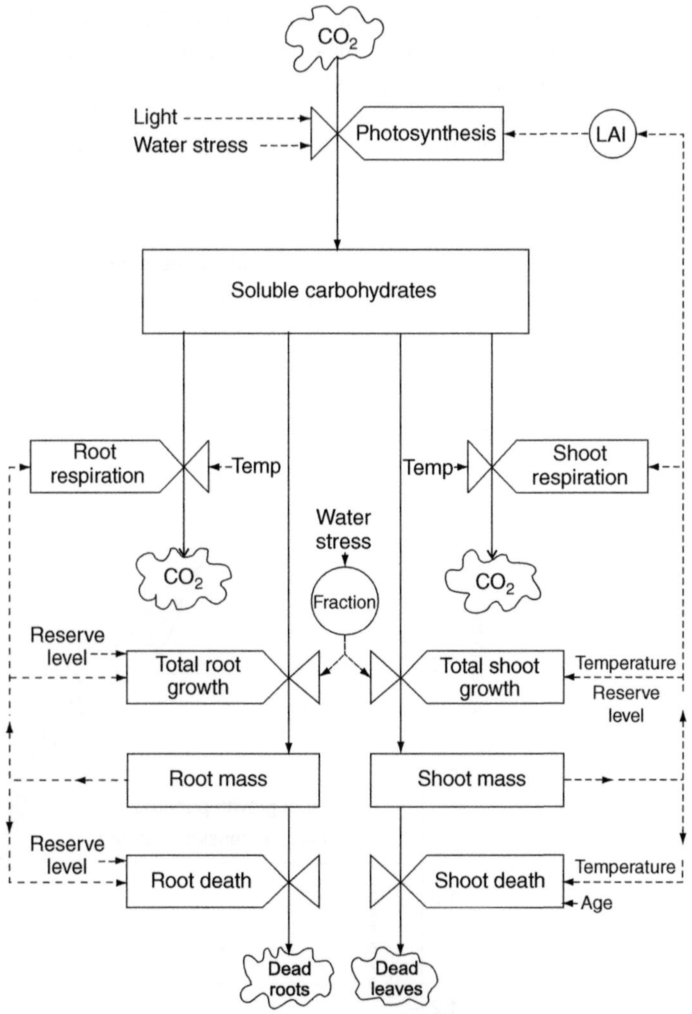

Fig. 19.12. The cycle of carbon in a living plant. (After Huck and Hillel, 1983.)

known as the *photosynthetically active radiation*. The rate of the process also depends on leaf area, although the relationship is not linear, since, as the canopy becomes denser, some shading of lower leaves by higher ones takes place (de Wit, 1965). The relationship between leaf area and biomass should also be known. The rate of photosynthesis is further constrained by leaf-water content, which, in turn, affects the stomatal control mechanism and, thus, the rate of CO_2 absorption by the leaves (Lommen et al., 1971).

Disposition of the products of photosynthesis, namely, the sugars and readily hydrolyzable starches, together termed *soluble carbohydrates*, is itself a complex process. Before any of this material can be incorporated into new structural matter (growth), the respiratory requirements of both shoots and roots must be satisfied (Penning de Vries, 1975). *Maintenance respiration* depends on both the mass and temperature of the respective tissues. Beyond maintenance, actual growth is accompanied by additional respiratory activity (termed *growth respiration*) to generate the chemical energy needed for the synthesis of new tissue.

The partitioning of the energy reserves available for growth between roots and shoots is affected by the relative values of water potential in different parts of the plant. As long as the plant is fully hydrated, the model assumes that soluble carbohydrates are transformed preferentially into the structural carbohydrates of new branches and leaves (i.e., into shoot growth). With greater leaf area, an increased capacity for photosynthesis and a greater water demand normally result. As the water requirement of a growing plant increases, however, it is the shoots (specifically, the leaves) that first experience water stress, because they are directly exposed to the atmosphere's evaporative demand. On the other hand, the roots, being nearer the source of the water, maintain a more hydrated state. Because they are better able to maintain turgor, the roots gradually become the preferred sink for the plant's available (soluble) carbohydrates. As the canopy begins to experience water stress, the fraction of soluble carbohydrates allocated to shoot growth diminishes, while the fraction allocated to root growth increases, thus enhancing the water-drawing power of the plant and, hence, its chances for survival. The preferential allocation of resources to roots or to shoots is thus seen to shift back and forth, depending on transient conditions.

Differences among species in the pattern of carbohydrate partitioning between roots and shoots evidently affect their adaptability to different conditions. Plants adapted to arid environments tend to develop smaller canopies and relatively deeper root systems, whereas plants conditioned to grow in more humid environments tend to develop more extensive canopies and relatively shallower root systems.

The second set of processes to be considered by the model is water movement through the soil–plant complex (Fig. 19.13). The water-storage capacity of plant tissue is small compared to the volume of water transpired so nearly all the flow is directly related to transpiration, which, in turn, is strongly linked to photosynthesis (because both flows involve the exchange and transmission of gases through the same stomatal passages). At full canopy cover, the rate of transpiration is a function of the rates of energy supply and vapor removal, as long as the leaves are fully hydrated and their stomata are open. This potential evaporation rate is affected by radiation, temperature,

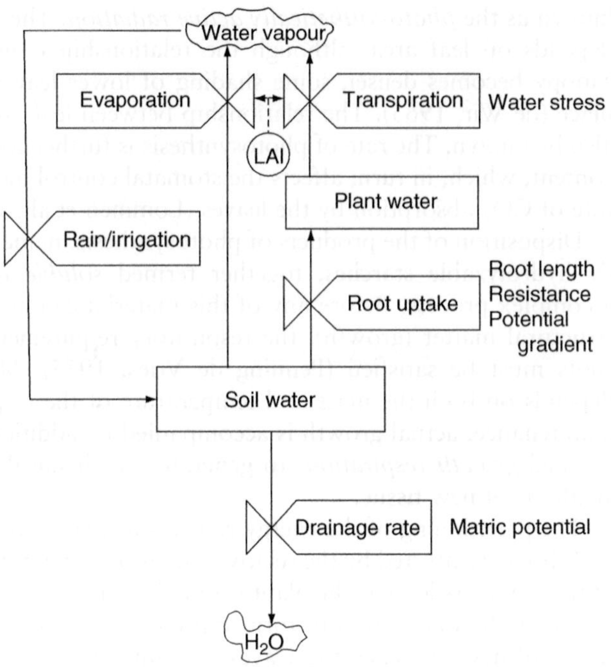

Fig. 19.13. The cycle of water in soil–plant–atmosphere continuum. (After Huck and Hillel, 1983.)

humidity, wind, and soil-heat flows (de Wit and Goudriaan, 1978). When leaf-water deficits develop and the stomata close, thus increasing canopy resistance to water outflow, transpiration is reduced.

The spatial pattern of water uptake from various regions of the soil depends on root distribution as well as on the distribution of moisture within the soil profile. Both are time variable. If the soil is uniformly wet, the water uptake pattern is mainly influenced by root distribution and root resistance to water flow (Belmans et al., 1979). When moisture has been removed from the soil regions initially having the greatest concentration of roots, the "center of gravity" of water uptake by roots must shift to wetter soil layers. Because of their immature vascular system, new roots must depend on the availability of water in the immediate vicinity, hence they grow faster in soil regions containing greater moisture reserves. Root proliferation is therefore greatly reduced in any given soil layer as soil moisture is reduced. Furthermore, fresh roots cannot invade or extend into and through a new soil layer unless parent roots are present in an adjacent layer. Thus, root growth will appear to "track" regions of moist soil.

The Huck and Hillel model has been further elaborated and tested by Hoogenboom et al. (1987). This model incorporates real-world climatic data as well measurable plant and soil characteristics.

20. WATER BALANCE AND ENERGY BALANCE IN THE FIELD

WATER AND ENERGY TRANSPORT IN THE SOIL

Any attempt to control the availability of soil moisture to plants must be based on quantitative knowledge of the dynamic balance of water in the soil. The *field water balance*, like a financial statement of income and expenditures, is an account of all quantities of water added to, subtracted from, and stored within a given volume of soil during a specified period of time. The various soil-water flow processes that we have attempted to describe in earlier chapters of this book as separate phenomena (e.g., infiltration, redistribution, drainage, evaporation, uptake of water by plants) are in fact strongly interdependent, because they occur sequentially or simultaneously.

The water balance is based on the *law of conservation of mass*, which states that matter can be neither created nor destroyed but can only change from one state or location to another. Since no significant amounts of water are normally decomposed, or recomposed, in the soil, the water content of a soil body of finite volume cannot increase appreciably without addition from the outside (as by infiltration or capillary rise), nor can it diminish unless transported to the atmosphere (by evapotranspiration) or to deeper zones (by drainage).

The field water balance is intimately associated with the *energy balance*, since it involves processes that require energy. The energy balance is an expression of the classical *law of conservation of energy*, which states that, in a given system, energy can be absorbed from, or released to, the surroundings and that along the way it can change form, but it cannot be created or destroyed.

The content of water in the soil affects the way the energy flux reaching the field is partitioned and utilized. Likewise, the energy flux affects the state and movement of water. The water balance and energy balance are inextricably linked, since they are involved in the same processes within the same environment. A physical description of the soil–plant–atmosphere system must be based on an understanding of both balances together. In particular, the evaporation process, which is often the largest component of the water balance (because most of the water added to the field is normally evaporated or transpired from it) is also likely to be the principal consumer of energy in the field. Thus, evapotranspiration depends, in a combined way, on the simultaneous supply of water and energy (Baker and Norman, 2002).

WATER BALANCE OF THE ROOT ZONE

In its simplest form, the water balance merely states that any change that occurs in the water content ΔW of a given body of soil during a specified period must equal the difference between the amount of water added to that body, W_{in}, and the amount of water withdrawn from it, W_{out}, during the same period:

$$\Delta W = W_{in} - W_{out} \tag{20.1}$$

When gains exceed losses, the water content change is positive; and conversely, when losses exceed gains, ΔW is negative. The soil volume of interest is thus treated as a "bank account" or a storage reservoir.

To itemize the accretions and depletions from the soil storage reservoir, one must consider the disposition of rain or irrigation reaching a unit area of soil surface during a particular period of time. Rain or irrigation water applied to the land may in some cases infiltrate the soil as fast as it arrives. In other cases, some of the water may pond over the surface. Depending on the slope and microrelief, a portion of this surface water may exit from the area as overland flow (*surface runoff*), while the remainder will be stored temporarily as puddles in surface depressions. Some of the latter evaporates, and the rest eventually infiltrates the soil after cessation of the rain. Of the water infiltrated, some evaporates directly from the soil surface, some is taken up by plants for growth or transpiration, some may drain downward beyond the root zone, and the remainder accumulates within the root zone and adds to soil-moisture content. Additional water may reach the defined soil volume either by runoff from a higher area or by upward flow from a water table or from wet layers present at some depth.

The pertinent volume or depth of soil for which the water balance is computed is determined arbitrarily. Thus, in principle, a water balance can be computed for a small sample of soil or for an entire watershed. From an agricultural or plant ecological point of view, it is generally most appropriate to consider the water balance of the root zone per unit area of field.

The root-zone water balance is usually expressed in integral form:

$$\text{Change in storage} = \text{Gains} - \text{Losses}$$

$$(\Delta S + \Delta V) = (P + I + U) - (R + D + E + T_r) \tag{20.2}$$

where ΔS is the change in root-zone soil-moisture storage, ΔV is the increment of water incorporated in vegetative biomass, P is precipitation, I is irrigation, U is upward capillary flow into the root zone, R is runoff, D is downward drainage out of the root zone, E is direct evaporation from the soil surface, and T is transpiration by plants. All quantities are expressed in terms of volume of water per unit land area (equivalent depth units) during the period considered.

The time rate of change of soil-moisture storage can be written as follows (assuming the rate of change of plant-water content to be relatively unimportant):

$$dS/dt = (p + i + u) - (r + d + e + t_r) \tag{20.3}$$

Here each of the lowercase letters represents the instantaneous time rate of change of the corresponding integral quantity in Eq. (20.2). The change in root-zone soil-moisture storage can be obtained by integrating the change in soil wetness over depth and time:

$$S = \int_0^z \int_{t_1}^{t_2} (\partial\theta/\partial t)\, dz\, dt \tag{20.4}$$

where θ is the volumetric soil wetness, measurable by sampling or by means of a neutron meter or TDR (see Chapter 6). Note that t is time, whereas t_r is transpiration rate in our notation.

The largest composite term in the "losses" part of Eq. (20.2) is generally the evapotranspiration $E + T$. It is convenient at this point to refer to the concept of *potential evapotranspiration* (designated E_{tp}), representing the climatic "demand" for water. Potential evapotranspiration from a well-watered field depends primarily on the energy supplied to the surface by solar radiation, which is a climatic characteristic of each location (depending on latitude, season, slope, aspect, cloudiness, etc.); it varies greatly from hour to hour but varies little from year to year. The value of E_{tp} depends secondarily on atmospheric advection, which is related to the size and orientation of the field and the nature of its upwind "fetch" or surrounding area. Potential evapotranspiration also depends upon surface roughness and soil thermal properties, which vary in time (van Bavel and Hillel, 1976). As a first approximation and working hypothesis, however, it is often assumed that E_{tp} depends entirely on the external climatic inputs and is independent of the properties of the field itself.

Actual evapotranspiration E_{ta} is generally a fraction of E_{tp}, depending on the degree and density of plant coverage of the surface as well as on soil moisture and root distribution. E_{ta} from a well-watered stand of a close-growing crop will generally approach E_{tp} during the active growing stage, but it may fall below it during the early-growth stage, prior to full canopy coverage, and again toward the end of the growing season as the matured plants begin to dry out (Hillel and Guron, 1973; Hillel, 1987a, 1997). For the entire season, E_{ta} may total 60–80% of E_{tp}, depending on water supply: The drier the soil moisture regime, the lower the actual evapotranspiration.

Another important, indeed essential, item of the field water balance is the internal drainage D out the bottom of the root zone. The function of drainage, in principle, is to release excess water that might otherwise restrict aeration and to leach excess salts. In the absence of adequate drainage (natural or — where

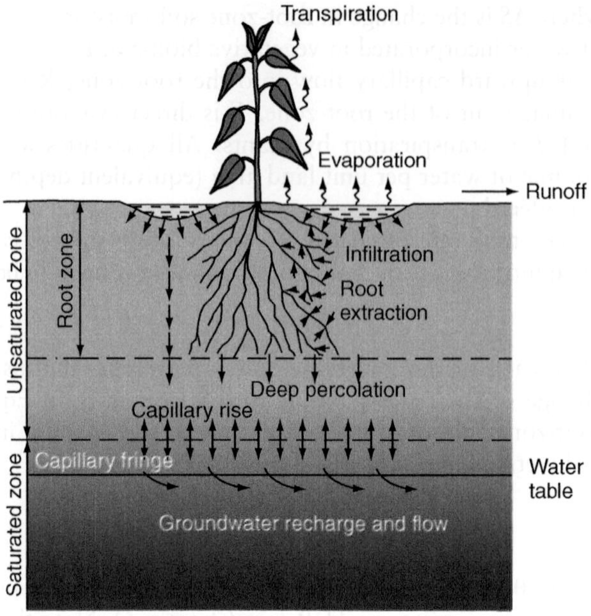

Fig. 20.1. The water balance of a root zone.

necessary — artificial), waterlogging as well as salt accumulation in the root zone are particular hazards of arid-zone farming.

The various processes included in the water balance of a hypothetical rooting zone are shown in Fig. 20.1. In this depiction, only vertical flows are considered within the soil (though lateral flows are shown to occur over the surface and below the water table). In a larger sense, any soil unit of interest constitutes a part of the overall hydrologic cycle, illustrated in Fig. 20.2.

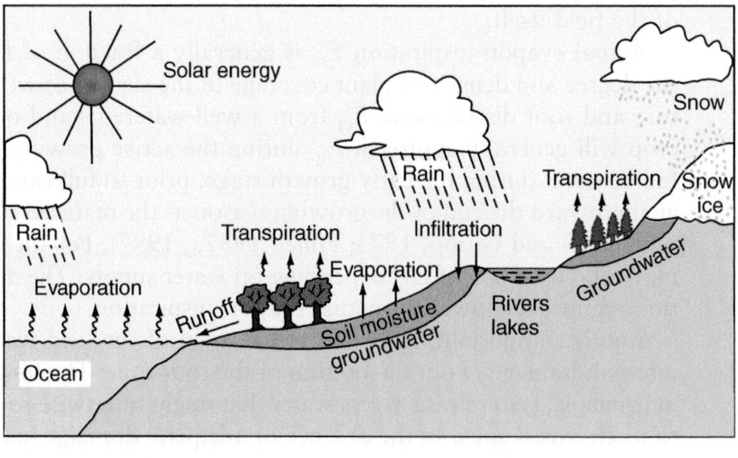

Fig. 20.2. The hydrologic cycle.

EVALUATION OF THE WATER BALANCE

Simple and readily understandable though the field water balance may seem in principle, it is still rather difficult to measure in practice. A single equation can be solved only if it has no more than one unknown. Often the largest component of the field water balance and the one most difficult to measure directly is the evapotranspiration $E + T$, also designated E_t. To obtain E_t from the water balance, we must have accurate measurements of all the other terms in the equation. It is relatively easy to measure the amount of water added to the field by rain and irrigation ($P + I$), though areal nonuniformities must be taken into account. The amount of runoff R generally is (or at least should be) small in agricultural fields, particularly in irrigated fields, so it is often considered negligible in comparison with the major components of the water balance.

Over an extended period such as a year or a growing season, the change in water content of the root zone is likely to be small in relation to the total water balance. In this case, the sum of rain and irrigation is approximately equal to the sum of evapotranspiration E_t and deep percolation D. For shorter periods, the change in soil-water storage ΔS can be relatively significant and must be measured. This measurement can be made by sampling periodically or by use of specialized instruments. Of the various instruments available, the *neutron meter* and *time-domain reflectometry (TDR)* permit repeated, undisturbed, in situ determinations of volumetric wetness.

During dry spells, without rain or irrigation, $W_{in} = 0$, so the sum of D and E_t now equals the reduction in root-zone water storage ΔS:

$$-\Delta S = D + E_t \tag{20.5}$$

Measurement of root-zone or subsoil water content by itself cannot tell us the rate and direction of soil-water movement. Even if the water content at a given depth below the root zone remains constant, we cannot conclude that the water there is immobile, since it might be moving steadily through that depth. Tensiometric measurements can help to establish the directions and magnitudes of the hydraulic gradients through the profile. Knowing the gradients, we can then assess the fluxes, provided of course that we also know the hydraulic conductivity as a function of suction or wetness for the particular soil.

The most direct method for measuring the field water balance is by use of *lysimeters*. These are generally large containers of soil, set in the field to represent the prevailing soil and climatic conditions and allowing more accurate measurement of physical processes than can be carried out in the open field. Some Iysimeters are equipped with a weighing device and a drainage system, which together permit continuous measurement of both evapotranspiration and percolation. Lysimeters may not provide a reliable measurement of the field water balance, however, when the soil or above-ground conditions of the Iysimeter differ markedly from those of the field itself.

RADIATION EXCHANGE IN THE FIELD

The term *radiation* refers to the emission of energy in the form of photons or electromagnetic waves from all bodies above 0 K. *Solar radiation* received on the earth's surface is the major component of its energy balance. Green plants are able to convert a part of the received radiation into chemical energy. They do this in the process of *photosynthesis*, on which all life on earth ultimately depends. Therefore, it is appropriate to begin a discussion of the energy balance with an account of the radiation balance. (An introductory description of the radiation and energy balances for a bare soil was given in Chapter 12.)

Solar radiation reaches the outer surface of the atmosphere at a practically constant flux (the *solar constant*) of nearly 1.4 joules per second per square meter perpendicular to the incident radiation (i.e., 1.353 kW/m^2, about 2 cal/min cm^2). Nearly all of this radiation is of the wavelength range of 0.3–3 μm (3000–30,000 Å), and about half of this radiation consists of visible light (i.e., 0.4–0.7 μm in wavelength). The solar radiation corresponds approximately to the emission spectrum of a blackbody at a temperature of 6000 K. The earth, too, emits radiation, but since its surface temperature is about 300 K, this *terrestrial radiation* is of much lower intensity and longer wavelength than solar radiation (i.e., in the wavelength range of 3–50 μm). Between the two radiation spectra, the sun's and the earth's, there is little overlap, so it is customary to refer to the first as *short-wave radiation* and to the second as *long-wave radiation.*

In passage through the atmosphere, solar radiation changes both its flux and spectral composition. About one-third of it, on the average, is reflected back to space (this reflection can become as high as 80% when the sky is completely overcast with clouds). In addition, the atmosphere absorbs and scatters a part of the radiation, so only about half of the original flux density of solar radiation finally reaches the ground. (In arid regions, where the cloud cover is sparse, the actual radiation received at the soil surface can exceed 70% of the "external" radiation; whereas in humid regions this fraction may be less than 50%.) A part of the reflected and scattered radiation also reaches the ground and is called *sky radiation.* The total of direct solar and sky radiations is termed *global radiation.*

Albedo is the reflectivity coefficient of the surface to short-wave radiation. It depends on the color, roughness, and inclination of the surface and is of the order of 5–10% for water, 10–30% for a vegetated area, 15–40% for a bare soil, and up to 90% for fresh snow.

In addition to these fluxes of incoming and reflected shortwave radiation, there occurs an exchange of long-wave (heat) radiation. The earth's surface, which converts incoming short-wave radiation to sensible heat, emits long-wave (infrared) radiation. At the same time the atmosphere also absorbs and emits long-wave radiation, part of which reaches the surface. The difference between these outgoing and incoming fluxes is called the *net long-wave radiation flux.* During the day, the net long-wave radiation may be a small fraction of the total radiation balance, but during the night, in the absence of direct solar radiation, the heat exchange between the land surface and the atmosphere dominates the radiation balance.

The overall difference between total incoming and total outgoing radiation (including both the short-wave and long-wave components) is termed *net radiation*, and it expresses the rate of radiant energy absorption by the field:

$$J_n = (J_s\downarrow - J_s\uparrow) + (J_l\downarrow - J_l\uparrow) \tag{20.6}$$

where J_n is the net radiation, $J_s\downarrow$ is the incoming flux of short-wave radiation from sun and sky, $J_s\uparrow$ is the short-wave radiation reflected by the surface, $J_l\downarrow$ the flux of long-wave radiation incoming from the sky, and $J_l\uparrow$ is the long-wave radiation reflected and emitted by the surface. At night, the short-wave fluxes are negligible, and since the long-wave radiation emitted by the surface generally exceeds that received from the sky, the nighttime net radiation flux is negative.

The flux of reflected short-wave radiation is equal to the product of the incoming short-wave flux and the reflectivity coefficient (the albedo α):

$$J_s\uparrow = aJ_s\downarrow$$

Therefore,

$$J_n = J_s\downarrow(1-\alpha) - J_l \tag{20.7}$$

where J_l is the net flux of long-wave radiation, which is given a negative sign. (Since the surface of the earth is usually warmer than the atmosphere, there is generally a net loss of thermal radiation from the surface.) As a rough average, J_n is typically of the order of 55–70% of $J_s\downarrow$ (Tanner and Lemon, 1962).

BOX 20.1 The Atmospheric Greenhouse Effect

The earth's surface tends naturally to maintain an energy balance such that the total amount of radiant energy received is equal to the total amount released. Since the flux of emitted infrared radiation depends on the surface temperature, that temperature adjusts itself over time so that the balance is maintained. The radiation balance is a self-adjusting system: If more energy comes in than goes out, the surface becomes warmer so that more energy is then emitted, and vice versa.

Recall that the rate of energy emission is proportional to the fourth power of the temperature. If the surface temperature rises from, say, 10°C (283 K) to 30°C (303 K), it will increase its rate of heat release by $(303/283)^4$, or 31%.

If the earth were without a gaseous atmosphere, calculations show, its mean surface temperature at equilibrium would be some 33°C lower than it actually is. The presence of the atmosphere causes the surface to be warm enough to promote the profusion of life as we know it. That effect of the atmosphere is called the *natural greenhouse effect*. The name is based on the perception that the effect of the atmosphere is analogous to the effect of a glass-covered greenhouse.

The interior of a greenhouse is typically warmer than the external environment, primarily because of the optical properties of its glass walls. Clear glass is nearly transparent to light, hence it admits the incoming solar radiation practically without hindrance. Inside the greenhouse, that energy is converted to heat. However, the same glass that admits more than 90% of the incoming visible light, absorbs some 90% of infrared radiation at wavelengths greater than 2000 nm. So the release of heat by infrared radiation from the interior of the greenhouse is partially blocked by the glass. As the interior gets hotter, its flux of emitted radiation intensifies and its wavelength shortens, until the fraction transmitted by the glass compensates for the incoming radiation. A new energy balance is then established, at a higher interior temperature.

The atmosphere is not exactly like the walls of a glasshouse. It is fluid and turbulent rather than rigid, so it mixes and exchanges heat and vapor with the environment in a way that glass cannot. But the greenhouse analogy is apt in one important respect: The atmosphere is preferentially permeable to short-wave radiation. Except for suspended dust particles, which absorb solar radiation, and clouds, which reflect it, the clear atmosphere is largely transparent to incoming light. Not so to outgoing heat. Although the major gases (nitrogen and oxygen, plus argon) that make up well over 99% of the atmosphere transmit infrared radiation readily, some of the gases present in the atmosphere in small concentrations tend to absorb infrared radiation. These radiatively active gases are known as the *greenhouse gases.*

The most important natural greenhouse gas, being relatively abundant and ubiquitous (though in variable concentrations) is *water vapor.* The second most important is *carbon dioxide,* which is continually cycled throughout the biosphere via such processes as photosynthesis, respiration, and decomposition of organic matter. The natural greenhouse effect, as far as it goes, is largely beneficent: It helps to create conditions that promote life on earth.

Ever since the beginning of the Industrial Revolution, little over two centuries ago, human activity has resulted in a significant infusion of carbon dioxide to the atmosphere, mainly by the burning of fossil fuels (coal and petroleum). Additional releases of carbon dioxide have resulted from the clearing of forests (first the temperate-zone forests and — increasingly — the tropical forests) as well as from the cultivation of virgin lands (which hastens the decomposition of soil organic matter). All these anthropogenic processes have already raised the concentration of CO_2 from its original level of about 270 parts per million by volume to some 370 ppmv.

Other greenhouse gases whose concentrations are also increasing due to human activity are: *chlorofluorocarbons* (CFCs), which are synthetic gases used as refrigerants, foaming agents, and propellants for spray cans; *methane* (CH_4), which is emitted from marshes, rice paddies, and the fermentative digestion of cattle; and *nitrous oxide* (N_2O), which results from the burning of fossil fuels and biomass and from fertilizer use.

Altogether, the human-enhanced greenhouse effect poses the danger of elevating the average temperature of the earth's surface excessively — perhaps by as much as 4°C by the second half of the 21st century — Although CO_2 enrichment per se may spur photosynthesis and improve water-use efficiency, many scientists are apprehensive that the enhanced greenhouse effect will eventually do more harm than good. The result might be not only a warmer earth and more extreme climate, but also a greater evaporative demand and, in some regions, a drier soil.

Reviews and analyses of the potential changes due to the enhanced greenhouse effect have been published by (among many others) Leggett (1990), Gates (1993), and Rosenzweig and Hillel (1998).

TOTAL ENERGY BALANCE

Having balanced the gains and losses of radiation at the surface to obtain the net radiation, we next consider the transformations of this energy.

Part of the net radiation received by the field is transformed into heat, which warms the soil, plants, and the atmosphere. Another part is taken up by the plants in their metabolic processes (e.g., photosynthesis). Finally, a major part is generally absorbed as latent heat in the twin processes of evaporation and transpiration. Thus,

$$J_n = LE + A + S + M \tag{20.8}$$

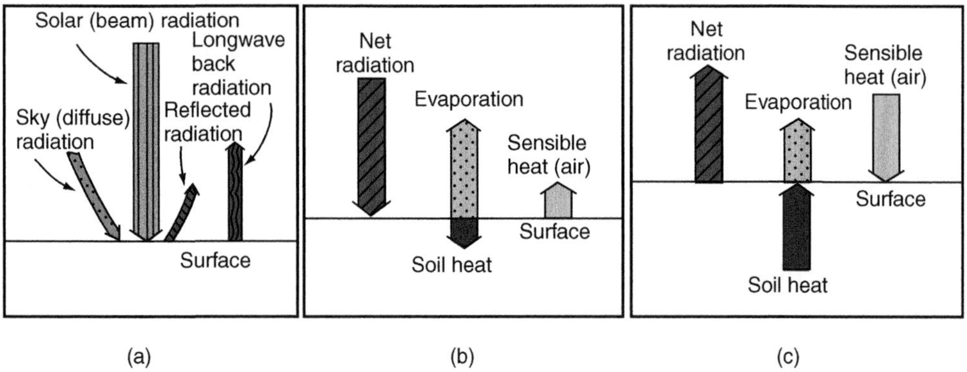

Fig. 20.3. Schematic representation of (a) the radiation balance, (b) the daytime energy balance, and (c) the nighttime energy balance. [Net radiation = (solar radiation + sky radiation) − (reflected radiation + back radiation).] Note that the daytime net radiation during the growing season is much greater than at night. (After Tanner, 1968.)

where LE is the rate of energy utilized in evapotranspiration (a product of the rate of water evaporation E and the latent heat of vaporization L), A is the energy flux that goes into heating the air (called *sensible heat*), S is the rate at which heat is stored in the soil, water, and vegetation, and M represents miscellaneous other energy terms, such as photosynthesis and respiration. The energy balance is illustrated in Fig. 20.3.

Where the vegetation is short (e.g., grass or field crops), the storage of heat in the vegetative biomass is negligible compared with storage in the soil. (The situation might be different, of course, in the case of a dense forest with massive trees.) The heat stored in the soil under sparse vegetation may be a fairly large portion of the net radiation at any particular time during the day, but the net storage over a 24-hr period is usually small, for the nighttime losses of soil heat tend to negate the daytime gains. For this reason, mean soil temperature generally does not change appreciably from day to day. The daily soil-storage term has been variously reported to be of the order of 5–15% of J_n (Evett, 2002), depending on season. In spring and summer this term is positive, but it becomes negative in autumn. Measurements of carbon dioxide exchange over active crops in the natural environment, however, have revealed that photosynthesis may in some cases account for as much as 5% of the daily net radiation where there is a large mass of active vegetation, particularly under low-light conditions. In general, though, the term M in Eq. (20.8) is much less than that.

Increasing the fraction of sunlight that can be utilized productively by plants is a major challenge to crop breeders and agronomists. One constraint is the concentration of carbon dioxide in the atmosphere, which is only about 370 parts per million. Growing plants in an atmosphere artificially enriched with CO_2 can improve the utilization of sunlight, but such enrichment is generally practical only in enclosed greenhouses.

Overall, the amounts of energy stored in soil and vegetation and fixed photochemically account for a rather small portion of the total daily net radiation, with the greater portion going into latent and sensible heat. The proportionate allocation between these terms depends on the availability of water for evaporation, but in most agriculturally productive fields the latent heat predominates over the sensible-heat term.

Sample Problem

The total incoming global radiation J_s (sun and sky) received by a particular field on a given day is 2.1×10^7 J/m^2 (500 cal/cm^2, or langleys). The albedo α is 15%. The net outgoing long-wave radiation balance J_l amounts to 4.19×10^5 J/m^2 (10 cal/cm^2). The sensible heat transfer to the air A is 5.03×10^5 J/m^2 (12 cal/cm^2), the net heat flow into the soil S is 2.514×10^5 J/m^2 (6 cal/cm^2), and the metabolic uptake of energy M is 3.35×10^5 J/m^2 (8 cal/cm^2).

Calculate the net radiation, the amount of energy available for latent-heat transfer (evapotranspiration), and the day's evapotranspiration in millimeters of water. On the following day, the sensible-heat transfer is reversed and evapotranspiration totals 7.5 mm. If everything else remains the same, calculate the amount of advected energy taken up by the field.

To calculate the net radiation J_n, we write the radiation balance, Eq. (20.7):

$$J_n = J_s{\downarrow}(1-\alpha) - J_l = 2.1 \times 10^7(1-0.15) - 4.19 \times 10^5 = 1.74 \times 10^7 \text{ J/m}^2$$
$$= 500(1-0.15) - 10 = 415 \text{ cal/cm}^2$$

The latent-heat term LE can be calculated from the overall energy balance, Eq. (20.8), when all other terms are known:

$$J_n = LE + A + S + M \qquad \text{or} \qquad LE = J_n - A - S - M$$

Using the given values, we have

$$LE = (174 - 5.04 - 2.51 - 3.35) \times 10^5 = 1.63 \times 10^7 \text{ J/m}^2$$
$$= 415 - 12 - 6 - 8 = 389 \text{ cal/cm}^2$$

Since roughly 2.43×10^6 J are required at the prevailing temperatures to vaporize 1 L (= 1 mm over 1 m^2), the amount of evaporation is

$$LE/L = (16.3/2.43) \times 10^6 = 6.7 \text{ mm}$$

Alternatively, since 580 cal are required to vaporize 1 g $\approx$ 1 cm^3 of water, the amount of evaporation is

$$LE/L = 389 \text{ cal/cm}^2/580 \text{ cal/cm}^3 = 0.67 \text{ cm} = 6.7 \text{ mm}$$

On the following day, with a positive influx of sensible heat by advection, evapotranspiration amounts to 7.5 mm, and hence the latent-heat term is

$$LE = (7.5 \times 10^{-3} \text{ m})(2.43 \times 10^9 \text{ J/m}^3) = 1.82 \times 10^7 \text{ J/m}^2$$
$$= 0.75 \text{ cm} \times 580 \text{ cal/cm}^3 = 435 \text{ cal/cm}^2$$

The energy balance is therefore

$$\text{income} = \text{disposal:} \quad J_n + A = LE + S + M$$

and the advected energy is

$$A = LE + S + M - J_n = (182 + 2.51 + 3.35 - 174) \times 10^6 \text{ J/m}^2$$
$$= 1.4 \times 10^6 \text{ J/m}^2 = 435 + 6 + 8 - 415 = 34 \text{ cal/cm}^2$$

TRANSPORT OF HEAT AND VAPOR TO THE ATMOSPHERE

The transport of sensible heat and latent heat (carried by vapor) from the soil–plant complex to the atmosphere occurs through several layers. The first is the *laminar boundary layer*, a relatively quiescent (though not stagnant) body of air in immediate contact with the surface of the evaporating body. Through this layer, which may be only about 1 mm thick, transport occurs by diffusion. Beyond this layer, turbulent transport becomes predominant in the *turbulent boundary layer*.

The sensible heat flux A is proportional to the product of the temperature gradient dT/dz and the turbulent transfer coefficient for heat k_a:

$$A = -c_p \rho_a k_a \,(dT/dz) \tag{20.9}$$

where c_p is the specific heat capacity of air at constant pressure, ρ_a is the density of air, T is temperature, and z is height above the surface.

The rate of latent-heat transfer by water vapor from the field to atmosphere, LE, is similarly proportional to the product of the vapor-pressure gradient and the appropriate turbulent-transfer coefficient for vapor.

If we assume that the transfer coefficients for heat and water vapor are equal, then the ratio between the transports of sensible heat and latent heat is

$$\beta = A/LE \approx \xi_c \,(\Delta T/\Delta e) \tag{20.10}$$

Here $\Delta T/\Delta e$ is the ratio of temperature gradient to vapor-pressure gradient in the atmosphere above a field and ξ_c is the *psychrometric constant* (66 Pa/°C ≈ 0.66 mbar/°C).

The ratio β, called the *Bowen ratio*, depends mainly on the interactive temperature and moisture regimes of the field. When the field is wet, the relative humidity gradients between its surface and the atmosphere tend to be large; much of the energy received is taken up as latent heat in the process of evaporation, so the temperature gradients tend to be small. Thus, β is rather small when the evaporation rate is high. When the field is dry, on the other hand, the relative humidity gradients toward the atmosphere are generally small and much of the received energy goes to warming the surface, so the temperature gradients tend to be steep and the Bowen ratio becomes large. In a recently irrigated field, β may be smaller than 0.2, while in a dry field in which the plants are under a water stress (with stomatal resistance coming into play), the surface may warm up and a much greater share of the incoming energy will be transferred to the atmosphere directly as sensible heat. Under extremely arid conditions, in fact, LE may tend to zero and β to infinity. With advection, sensible heat may be transferred from the air to the field rather than vice versa, and the Bowen ratio can become negative.

Transfer through the turbulent atmospheric boundary layer takes place primarily by means of *eddies*, which are irregular, swirling microcurrents of air, whipped up by the wind. Eddies of varying size, duration, and velocity fluctuate up and down at varying frequency, carrying both heat and vapor. It is reasonable to assume that momentum and carbon dioxide, as well as vapor and heat, are carried by the same eddies. While the instantaneous gradients and vertical fluxes of heat and vapor will generally fluctuate, when a sufficiently long averaging period is allowed (say, 15–60 min), the fluxes exhibit a stable statistical relationship over a uniform field.

This is not the case at a low level over an inhomogeneous surface of spotty vegetation and partially exposed soil. Under such conditions, cool, moist packets of air may rise from the vegetated spots and warm, dry air may rise from the dry soil surface, with the latter rising more rapidly owing to buoyancy. An index of the relative importance of buoyancy (thermal) versus frictional forces in producing turbulence is the *Richardson number*, $R_i = g(dT/dz)/[T(du/dz)^2]$, where dT/dz is the temperature gradient and g is the acceleration of gravity. The air profile tends to be stable when R_i is positive and unstable (buoyant) when R_i is negative (W. D. Sellers, 1965).

Using the Bowen ratio, we can write the latent-heat and sensible-heat fluxes as follows (recalling that $J_n = S + A + LE$ and that $\beta = A/LE$):

$$LE = (J_n - S)/(1 + \beta) \tag{20.11}$$

$$A = \beta(J_n - S)/(1 + \beta) \tag{20.12}$$

Thus, LE can be obtained indirectly from micrometeorological measurements in the field (i.e., J_n, S, and β) without necessitating measurements of soil-water fluxes or plant activity. The usefulness of the Bowen ratio in evapotranspiration studies was analyzed by Angus and Watts (1984).

The diurnal variation of the energy balance is illustrated in Fig. 20.4. Both the diurnal and the annual patterns of the energy balance components differ for different conditions of soil, vegetation, and climate (Sharma, 1984; Baker and Norman, 2002).

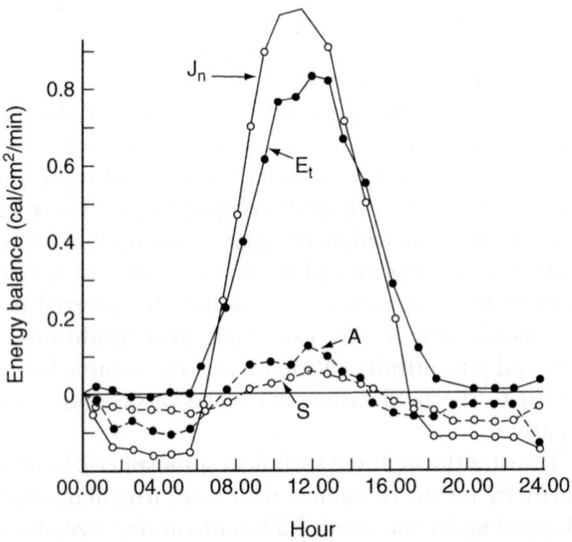

Fig. 20.4. The diurnal variation of net radiation J_n and of energy utilization by evapotranspiration E, sensible heating of the atmosphere A, and heating of the soil S under alfalfa-brome hay on Plainfield sand. Wisconsin, 4 September 1957. (After Tanner, 1960.)

ADVECTION

The equations given for the energy balance apply to extensive uniform areas in which all fluxes of energy and vapor are vertical or nearly so. However, any small field differing markedly from the neighboring areas is subject to lateral effects and may exchange energy and matter in one form or another with its surroundings. Specifically, winds sweeping over a small field may transport heat and vapor into or out of it. This phenomenon, called *advection*, can be especially important in arid regions, where small irrigated fields are often surrounded by an expanse of dry land. Under such conditions, the warm and dry incoming air can introduce energy in the form of sensible heat (in addition to the energy of net radiation), which is transformed into latent heat of vaporization as it is absorbed by the crop (Graham and King, 1961; Rosenberg, 1974).

The extraction of sensible heat from a warm mass of air flowing *over* the top of a field, and the conversion of this heat to latent heat of evaporation, is called the *oasis effect*. The passage of warm air *through* the vegetative cover has been called the *clothesline effect*. A common sight in arid regions is the poor growth of the plants near the windward edge of a field, where penetration of warm, dry wind contributes energy for evapotranspiration. Where advective heat inflow is large, evapotranspiration from rough and "open" vegetation (e.g., widely spaced row crops or trees) can greatly exceed that from smooth and close vegetation (e.g., mowed grass).

The effects of advection are likely to be small in very large and uniform fields but very considerable in small plots that differ significantly from their surroundings. With advection, latent heat "consumption" can be larger than net radiation. Hence, values of evapotranspiration and of irrigation requirements obtained from small experimental plots are not typically representative of large fields, unless these plots are "guarded" in the upwind direction by an expanse, or *fetch*, of vegetation of similar roughness characteristics and subject to a similar water regime. It should be obvious from the preceding that a small patch of vegetation, particularly if it consists of a spaced stand of shrubs or trees, can at times evaporate water in excess of the evaporation from a free-water surface such as a lake, a pond, or a pan.

Advection is not confined to small fields. It can also occur on a "macrometeorological" scale. A case in point is the periodic invasion of semihumid regions along the Mediterranean littoral by searing desert winds, variously called *sirocco* (from the Arabic word *sharkiyeh*, meaning "easterly") or *khamsin* (meaning "fifty," because it occurs most commonly during the transition seasons of spring and autumn, each said to last 50 days).

POTENTIAL EVAPOTRANSPIRATION: THE PENMAN EQUATION

The concept of *potential evapotranspiration* is an attempt to characterize the micrometeorological environment of a field in terms of an evaporative power, or demand, that is, in terms of the maximal evaporation rate the atmosphere is capable of exacting from a field of given surface properties. The concept probably derives from the common observation that when a porous body initially permeated with water is exposed and allowed to dry gradually in the open air, progressively longer increments of time are generally required to remove equal increments of water. The evaporation rate obviously depends both on the environment and on the state of wetness of the object itself. Intuitively, therefore, one might suppose that there ought to be a definable evaporation rate for the special case in which the object is maintained perpetually in as wet a state as possible and that this evaporation rate should depend solely on the meteorological environment.

Penman (1956) defined potential evapotranspiration as "the amount of water transpired in unit time by a short green crop, completely shading the ground, of uniform height and never short of water." As such, it is a useful standard of reference for the comparison of different regions and of different measured evapotranspiration values within a given region.

To provide the highest possible yields, many agricultural crops must receive sufficient rain or irrigation to prevent water from becoming a limiting factor. Knowledge of the potential evapotranspiration can therefore serve as a basis for planning the irrigation regime. In general, the actual evapotranspiration E_{tp} from various crops will not equal the potential value E_{tp}, but in the case of a close-growing crop the maintenance of optimal soil-moisture conditions for maximal yields will generally result in E_{ta} being nearly equal to (or a nearly constant fraction of) E_{tp}, at least during the active growth phase of the crop.

Various empirical approaches have been proposed for the estimation of potential evapotranspiration (e.g., Thornthwaite, 1948; Blaney and Criddle, 1950). The method proposed by Penman (1948) is physically based and hence inherently more meaningful. His equation, based on a combination of the energy balance and aerodynamic transport considerations, is a major contribution in the field of agricultural and environmental physics.

The Dalton equation for evaporation from a saturated surface is

$$LE = (e_s - e)f(u) \tag{20.13}$$

where u is the wind speed above the surface, e_s is the saturation vapor pressure (or the absolute humidity) at the temperature of the surface, and e is the vapor pressure of the air above the surface at an elevation sufficient so that e is unaffected by e_s. For the present, we shall disregard the fact that the shape of the function $f(u)$ should depend on the roughness of surface and on the stability (buoyancy) of the air over the surface.

The rate C at which a surface loses (or gains) heat by convection can be written in a similar way:

$$C = (T_0 - T_z)\xi f(u) \tag{20.14}$$

where T_0 and T_z are temperatures at the surface and at height z, respectively, and ξ is called the psychrometric constant because it appears in the equation relating vapor pressure to the readings of wet bulb and dry bulb thermometers.

If e_a is the saturated vapor pressure of the air, then

$$LE_a = (e_a - e)f(u) \tag{20.15}$$

And

$$E_a/E = 1 - (e_s - e_a)/(e_s - e) \tag{20.16}$$

is obtained by dividing Eq. (20.15) by Eq. (20.13) and rearranging the terms.

Penman assumed that $S = 0$ (i.e., the soil heat flux is negligible), so that he could write Eq. (20.11) as

$$J_n/LE = 1 + \beta \tag{20.17}$$

Since the Bowen ratio can be written

$$\beta = \xi(T_s - T_a)/(e_s - e) \tag{20.18}$$

then

$$\frac{J_n}{LE} = 1 + \xi\left(\frac{e_s - e_a}{e_s - e}\right)\Bigg/\left(\frac{e_s - e_a}{T_s - T_a}\right) \tag{20.19}$$

In Eq. (20.18) we can write

$$(e_s - e_a)/(T_s - T_a) = (\Delta e/\Delta T)T = T_a = \Delta$$

where Δ is the slope of the curve of saturated vapor pressure vs. temperature.

Now we can rewrite Eq. (20.18):

$$\frac{J_n}{LE} = 1 + \frac{\xi}{\Delta}\left[\frac{e_s - e_a}{e_s - e}\right] \tag{20.20}$$

Since $(e_s - e_a)/(e_s - e) = 1 - E_a/E$ from Eq. (20.16), algebraic rearrangement gives

$$LE = \frac{(\Delta/\xi)J_n + LE_a}{\Delta/\xi + 1} \tag{20.21}$$

Equation (20.20) is the *Penman equation*, where

$$LE_a = 0.35(e_a - e)(0.5 + u_2/100)\ (\text{mm/day})$$

e_a = saturated vapor pressure at mean air temperature (mm Hg), e = mean vapor pressure in air, and u_2 = mean wind speed in miles per day at 2 m above the ground. This equation permits calculation of the potential evapotranspiration rate from measurements of the net radiation and of the temperature, vapor pressure, and wind velocity taken at one level above the field.

The Penman formulation, based on the simultaneous solution of the heat and mass balance equations, avoided the necessity of determining the value of the surface temperature T_s. (That determination was difficult to make routinely and was subject to substantial errors before precise radiation thermometers became available.) Penman's original formulation also disregarded the possible fluctuations in the direction and magnitude of the soil-heat flux term. Moreover, it makes no provision for surface roughness, air instability (buoyancy) effects, and advection.

To correct for the differences between potential evapotranspiration from rough surfaces and potential evaporation from smooth water E_0, Penman used the following empirical factors determined in southern England:

E_0(bare soil)/E_0(water) = 0.9

E_0(turf)/E_0(water) = 0.6 in winter, ranging to 0.8 in summer

It bears emphasis that representing potential evapotranspiration purely as an externally imposed "forcing function" is a gross approximation. In fact, the field participates, as it were, in determining its evapotranspiration rate even when it is well endowed with water, through the effect of its varying albedo, thermal capacity and conductivity, aerodynamic roughness, etc. The oft-stated principle that all well-watered fields in a given location, regardless of specific characteristics, are subject to the same potential evapotranspiration is only approximately correct (van Bavel and Hillel, 1976).

The Penman formulation was modified by van Bavel (1966) to allow for short-term variations in soil-heat flux and for differences among various surfaces. His method for predicting potential evapotranspiration requires the additional measurements of net radiation and soil-heat flux. A roughness height parameter is used to characterize the aerodynamic properties of the surface, that is, to take account of the fact that, all other things being equal, potential evapotranspiration from a corn field should exceed that from a lawn, which, in turn, should normally be greater than that from a smooth bare soil.

The latent heat flux at potential evapotranspiration LE_0 is given by

$$LE_0 = \frac{(\Delta/\xi)(J_n - S) + k_v d_a}{\Delta/\xi + 1} \tag{20.22}$$

where Δ is the slope of the curve of saturation vapor pressure versus temperature at mean air temperature, ξ is the psychrometric constant, J_n is net radiation, S is soil-heat flux, d_a is the vapor-pressure deficit at elevation Z_a (namely, $e_s - e_a$), and k_v is the transfer coefficient for water vapor (a function of wind speed and surface roughness). For the dependence of k_v on mean wind speed u_2, Penman (1948) suggested empirically $k_v = 20(1 + u_2/100) = 20 + u_2/5$, where e_s and e_a are given in millimeters of mercury.

Further improvements of physically based predictions of evapotranspiration can result from the inclusion of air-stability or buoyancy effects, inasmuch as vapor transfer is enhanced whenever the thermal structure of the air becomes unstable (Szeicz et al., 1973). The advent of remote-sensing infrared thermometry has made possible continuous monitoring of surface temperature and hence also allows a better estimation of the vapor pressure at the surface.

The ratio Δ/ξ and the saturation vapor pressure e_s at various temperatures are available in standard tables in many texts on physical meteorology and on environmental physics. A summary is given in Table 20.1.

TABLE 20.1 The ratio Δ/ξ and the saturation vapor pressure e_s at various temperatures

T (°C)	10	15	20	25	30	35
Δ/ξ	1.23	1.64	2.14	2.78	3.57	4.53
e_s (mm Hg)	9.20	12.707	17.53	23.75	31.82	42.18
e_s (mbar)	12.27	17.04	23.37	31.67	42.43	56.24

Sample Problem

On a given day in early spring the daily net radiation J_n is 350 cal/cm^2, the mean air temperature T_a at standard height (2 m) is 15°C, the mean vapor pressure e_a at that height is 8 mm Hg, and the mean wind speed u_2 is 15 miles/day. On another day in late spring the net radiation is 420 cal/cm^2, mean air temperature is 20°C, mean vapor pressure is 9 mm Hg, and mean wind speed is 20 miles/day. Finally, on a day in summer J_n is 500 cal/cm^2, T_a is 25°C, e_a is 10 mm Hg, and u_2 is 25 miles/day. Estimate the potential evapotranspiration using Eq. (20.22). Assume the net soil-heat flux S to be zero in all cases. *Note*: Use the units in which Eq. (20.22) was originally cast.

Potential evapotranspiration LE_0 is given by

$$LE_0 = \frac{(\Delta/\xi)(J_n - S) + k_v d_a}{(\Delta/\xi) + 1}$$

where d_a is the mean vapor pressure deficit $(e_s - e_a)$ at standard height. Recall that Δ/ξ and e_s at several temperatures are tabulated in Table 20.1, and assume that $k_v = 20 + u_2/5$ (Penman, 1948). Accordingly,

For the early spring day:

$$LE_0 = [1.64 \times (350 - 0) + (20 + 15/5)(12.78 - 8)]/(1.64 + 1) = 259 \text{ cal/cm}^2 \text{ day}$$

For the late spring day:

$$LE_0 = \frac{2.14 \times (420 - 0) + (20 + 20/5)(17.53 - 9)}{2.14 + 1} = 351 \text{ cal/cm}^2 \text{ day}$$

For the summer day:

$$LE_0 = \frac{2.78 \times (500 - 0) + (20 + 25/5)(23.75 - 10)}{2.78 + 1} = 459 \text{ cal/cm}^2 \text{day}$$

Remembering that approximately 580 cal are required to vaporize 1 g of water, and assuming a water density of 1 g/cm^3, we can use 58 cal/cm^2 day as the latent-heat flux equivalent to the evaporation of 1 mm of water per day. Hence the values of potential evapotranspiration are estimated to be:

259/58 = 4.5 mm for the early spring day
351/58 = 6.1 mm for the late spring day
459/58 = 7.9 mm for the summer day

CANOPY RESISTANCE: THE PENMAN–MONTEITH EQUATION

Monteith (1980) provided a review and extension of the Penman equation. An essential condition for the validity of the Penman analysis is that the latent-heat and sensible-heat exchange must both take place from a surface at the same temperature. When the air in contact with the evaporating surface is vapor saturated, we can expect the evaporation rate to be maximal (i.e., to occur at the full potential rate). However, the Penman approach can be valid even if the air in contact with the surface is not saturated, provided it has a constant relative humidity, which can be related to the free energy of the evaporating water.

Monteith (1973, 1980) introduced the *transfer-resistance* concept to account for heat and vapor transfers from less-than-saturated surfaces. The resistance to vapor transfer r_v is defined by the relation

$$LE = \rho c(e_0 - e_a)/\xi r_v \tag{20.23}$$

where e_0 is the vapor pressure of air in contact with the surface, e_a is the vapor pressure of air at an arbitrary height (usually within 2 m of the surface to ensure that it is related to the state of the surface and to the turbulent transfer of vapor from the surface), and ρc is the volumetric specific heat of air. The corresponding resistance for heat transfer r_H is therefore given by

$$C = \rho c(T_0 - T_a)/r_H \tag{20.24}$$

Equations (20.13) and (20.14) can now be regarded as special cases of Eqns. (20.23) and (20.24), with $r_v = r_H = \rho c/\xi f(u)$. Eliminating surface temperature, Monteith (1980) then gives the evaporation rate in a form similar to the original Penman equation:

$$\lambda E = \frac{\Delta H + \rho c\,[e_s(T_a) - e_a]/r_H}{\Delta + \gamma(r_v/r_H)} \tag{20.25}$$

The relative humidity h at the evaporating surface is related to the water potential ψ of that surface by $h = \exp(-\psi/RT)$. Equation (20.25) then becomes:

$$\lambda E = \frac{\Delta hH + \rho c[he_s(T_a) - e_a]/r_H}{\Delta h + \gamma(r_v/r_H)} \tag{20.26}$$

Equation (20.25) is known as the *Penman–Monteith equation*. It has been used to specify the physiological control of evaporation from a crop (affected by the stomatal control mechanism) in terms of a *canopy resistance* r_c. It has also been used to predict the maximum rate of evapotranspiration from a crop as a function of a minimum value of that resistance, with r_v assumed to be the sum of the resistance to momentum transfer r_m and the canopy resistance r_c in series.

SPATIAL VARIATION OF PROPERTIES AND PROCESSES

Pedologists have long recognized that soils exhibit great differences from one location to another, and they have attributed these differences to the varied combinations of factors governing soil formation. They have tended to classify soils and draw maps with distinct boundaries between differing soils. Within each classification, however, the prevalent assumption was that the soil is more or less uniform. Any section of land, it would therefore seem, could be characterized precisely by a limited number of sampling points. Another convenient presumption was that the major physical processes in the soil take place in a single dimension, that is, vertically up or down.

This idealized picture of areally homogeneous landscape units in which soil properties and processes are exactly determinable is, alas, unrealistic. Spatial as well as temporal variability, to whatever degree, is an inevitable attribute of every section of the landscape (be it a catchment, an agricultural field, or a desert biome). Nielsen et al. (1973) were among the first to point out the importance of spatial variability and to seek ways to characterize it quantitatively.

The causes of variability can be both natural and anthropogenic. *Natural variability* results from the inherently heterogeneous nature of the original geological formation as well as of the relief, thermal and water regimes, and biotic factors that had prevailed during the long course of soil development. *Anthropogenic variability*, superposed on natural variability, results over time from the uneven treatment or management of the area by such practices as land-shaping, compaction, tillage, irrigation, drainage, cropping, and applications of chemicals.

Different types of soil properties are recognized. *Point properties* are those that can be measured at each point within the soil. Examples are temperature, pressure (liquid, gaseous, or intergranular), and chemical composition. *Volume properties*, on the other hand, have meaning only with reference to a specified (or an implied) volume. They include such properties as bulk density and porosity. When volume properties are measured, the results may depend on the size of the volume measured. Large volumes encompass and may obscure microscale irregularities.

A distinction should also be made between *static* and *dynamic soil properties*. Static properties are attributes of the soil that remain regardless of process. Examples are particle size and pore size distribution, wetness, and organic matter content. Dynamic properties only come into play in the response of the soil to applied forces that induce processes. Examples are infiltrability (the rate at which the soil absorbs water applied to its surface at atmospheric pressure) and compactibility (the tendency of a soil body to compress when subject to mechanical stresses). Dynamic properties may be related functionally to static properties (as, for instance, hydraulic conductivity is dependent on pore size distribution), but the relation is likely to be nonlinear (e.g., the conductance of each class of pores is related to the radius squared).

Static properties typically (though not invariably) exhibit bell-shaped ("normal") distribution curves. Such curves are obtained when the results of measurements made at numerous locations in a study area (in which the values of a property are randomly distributed) are plotted against the number of measurements yielding similar values. For this type of distribution, the *mean* value (average of all values) is nearly the same as the *median* (the middle value in the array) and the *mode* (the most frequently occurring value). The height versus width of the distribution curve characterizes the degree of variability. The commonly used *coefficient of variation* is the standard deviation expressed as a percentage of the arithmetic mean. When this index is applied to soil physical properties, it is usually found to be smaller for static than for dynamic properties (Mulla and Mc Bratney, 2000).

In addition to greater variation, dynamic properties tend to exhibit skewed rather than symmetrical frequency distributions. (In a *skewed* distribution, the mean, median, and mode generally differ.) That is often because small deviations from the mean values of static properties may generate large deviations from the mean values of related dynamic properties (Horowitz and Hillel, 1983). To give an example: The presence of macropores may increase porosity by no more than 10%, whereas the infiltrability may be increased by 100% or more. A way of handling skewed distributions is to try to "normalize" them by means of various mathematical devices. A common approach is to plot the logarithms of the measured values, to test whether a more-or-less linear *log-normal* curve is thereby obtained.

A further distinction is to be made between the spatial variability of soil properties and that of soil processes. Certain processes depend on certain properties (e.g., the infiltration rate depends on the hydraulic conductivity). However, processes may be additionally affected by the nonuniformity of the forcing inputs and outputs involved. Thus, runoff depends not only on the profile's hydraulic conductivity but also on the time-variable and space-variable intensity of rainfall or on the depth-variable application of irrigation.

In some cases, the relative uniformity of inputs and outputs may mask the nonuniformity of soil properties (e.g., when low-intensity rains are absorbed entirely by a soil of space-variable conductivity, or when atmospheric evaporativity extracts water uniformly from a variably wetted soil), so the processes tend to be less variable than the properties. In other cases, variable inputs combine with the soil's variable properties to magnify the resultant variability of the processes. Consider, for instance, the hypothetical case of a localized rainstorm of high intensity impacting a section of a heterogenous catchment. If the rain strikes the part of the catchment that is sandy, the water is likely to percolate straight down to the water table beneath, with minimum runoff and maximum leaching. If, however, the rainstorm takes place over a clayey section of the catchment, the likely results would be maximal runoff and minimal percolation.

In the field, soil properties may vary randomly or systematically. If variation in a section of land is random, there is equal probability that a given value of a property will be found regardless of where the measurement is made. If, on the other hand, the variation is systematic (i.e., if the study area has a distinct structure or spatial pattern), the likelihood of obtaining similar results from several measurements becomes smaller as the sites of measurement are farther apart. Statistical procedures for testing the spatial structure of a study area include *autocorrelation* and *variograms*.

Techniques for assessing probable values of properties in any location between measured sites are based on geostatistics, a methodology developed by mining engineers (Journel and Huijbregts, 1978). A statistical procedure called *kriging* yields more realistic estimates than the older method of linear interpolation, because it considers the spatial trend of a property based on the array of values surrounding the point of interest (Warrick and Nielsen, 1980; Webster and Oliver, 1990; Warrick, 1998; van Es, 2002).

Recognizing the existence of spatial variability implies the acceptance of uncertainty. We can never determine a property or quantify a process exactly. All that capricious nature allows us to do is to assess the probability that the property or process of interest lies within a specifiable range of values. A single mean value is not sufficient to characterize an area, since two areas may have the same mean values of a property but behave quite differently if the spatial distributions of the property differ. Knowing the spatial configuration of an agricultural field's properties and processes can provide guidance on how to adjust irrigation, fertilization, and pest-control treatments to accord with the variability of the field, rather than apply these practices uniformly (and in many cases, wastefully, with damage to the environment), as has long been the common practice. This methodology is known as *precision agriculture.* Rehabilitation of disturbed lands will similarly benefit from better understanding of their spatial characteristics.

21. IRRIGATION AND WATER-USE EFFICIENCY

THE ROLE OF IRRIGATION

Water constitutes the single most important constraint to increasing food production in our hungry world. So tenuous and delicate is the balance between the incessant demand for water by crops and its sporadic supply by precipitation that even short-term dry spells often reduce production significantly, and prolonged droughts can cause total crop failure and mass starvation. *Irrigation* is the practice of supplying water artificially to permit farming in arid regions and to offset drought in semiarid or semihumid regions. As such, it can play a key role in feeding an expanding population. Even in areas where total rainfall is ample, it may be unevenly distributed during the year so that only with irrigation is stable multiple cropping possible. In fact, the potential productivity of irrigated land generally exceeds that of unirrigated ("rain-fed") land, due both to increased yields per crop and to the possibility of multiple cropping.

By some estimates, irrigated agriculture produces one-third of the world's crop on one-sixth of the world's cropland. The efficiency of irrigation helps to relieve pressure on the earth's natural ecosystems, including its rainforests. Yet, in too many places, the sustainability of irrigation is threatened by processes of degradation. The proper management of irrigation, based on a fundamental understanding of the processes involved, is therefore of vital importance.

The practice of irrigation consists of introducing water into the part of the soil profile that serves as the root zone, for the subsequent use of a crop. A well-managed irrigation system optimizes the spatial and temporal supply of water,

BOX 21.1 The Early History of Irrigation

The decisive process in the origin of civilization was the development of dependable food-producing systems. That process involved a sequence of steps, starting with an initially extensive hunting–gathering economy that tended to become increasingly intensive and culminating in a complete revolution in human society and its management of the environment. A crucial step in the process was the selection of favorable wild plants in their natural habitats, their domestication and transformation into crops, and their propagation in selected areas.

That process first took place in the Near East some 10,000 to 12,000 years ago, in the arc of rain-fed uplands and foothills fringing the Fertile Crescent on the west, north, and east. Prominent among the native plant resources of that region were species of the Graminea and Leguminosa families (the former including the progenitors of wheat, barley, oats, and rye; the latter including lentils, peas, chickpeas, and vetch). The nutritious seeds of such plants could be collected and stored to provide food for several months. Most native plants scatter their seeds as soon as they mature and are therefore difficult to harvest efficiently. A few anomalous plants, however, due to chance mutations, retain their seeds. The preferential selection of such seeds, and their propagation in favorable plots of land, constituted the beginnings of agriculture, providing the early farmers with crops that could be harvested more uniformly and dependably than could the wild plants.

Eventually, farming was extended from the relatively humid centers of its origin toward the river valleys of the Jordan, the Tigris–Euphrates, the Nile, and the Indus. Because the climate of these river valleys is generally arid, a new type of agriculture, based primarily on irrigation, came into being. The early farmers of those river valleys, around 5000 B.C.E., relied at first on natural irrigation by the unregulated seasonal floods to water the banks or floodplains of the rivers. As soon as the flood withdrew, they could cast their seeds in the mud. At times, however, the flood did not last long enough to wet the soil thoroughly. So the riparian farmers learned to build dikes around their plots, thus creating basins in which a desired depth of water could be impounded until it soaked into the ground and wetted the soil fully. The basins also retained the nutrient-rich silt and prevented it from running off with the receding floodwaters. Channels were dug to convey river water to the farther reaches of the plain that might not receive the water otherwise. By these means, irrigation farmers strove to gain greater control over the water supply to their crops.

As the population of the river valleys grew, the necessity arose to intensify production still further. Instead of just one crop per year, the farmers of Mesopotamia and Egypt, for example, could possibly grow two, three, or even four, given the year-round warmth and the abundant sunshine of the local climate. To do so, they needed to draw water at will from the rivers or from shallow wells dug to the water table. At first they drew water manually in ceramic or animal-skin containers, which they then carried using shoulder yokes. In time, a new technology was invented: mechanical water-lifting devices. The simplest of these is the device known today in Arabic as the *shadoof* — a wooden pole used as a lever, with the long arm serving to raise bucketfuls of water and the short arm counterweighted with a mass of mud or a stone. A more sophisticated device, invented many centuries later in Egypt and attributed to Archimedes, is the so-called *tamboor*, which consists of an inclined tube containing a tight-fitting spiral fin. Both of these devices are human powered. A more elaborate water-lifting device is the animal-powered *saqiya* waterwheel. All of these devices were used in the Near East until quite recently but have of late been replaced by modern motorized pumps.

not necessarily to obtain the highest yields per unit of land or per unit amount of water, but to maximize the benefit-to-cost ratio. Because of the economic considerations involved, which are necessarily specific to each location and set of circumstances, some aspects of irrigation management efficiency extend beyond the scope of this book on soil physics. The present chapter addresses some of the essential physical and environmental factors involved in irrigation.

CLASSICAL CONCEPTS OF SOIL-WATER AVAILABILITY TO PLANTS

The concept of *soil-water availability*, while never clearly defined in physical terms, has long excited controversy among adherents of different schools of thought. Veihmeyer and Hendrickson (1955) claimed that soil water is equally available throughout a definable range of soil wetness, from an upper limit (*field capacity*) to a lower limit (the *permanent wilting point*), both of which are characteristic and constant for any given soil. They postulated that plant functions remain unaffected by any decrease in soil wetness until the permanent wilting point is reached, at which point plant activity is curtailed abruptly. The simplicity of this conceptual model, though based on arbitrary limits, helped to popularize it among irrigators.

We have already pointed out (Chapter 16) that the field capacity concept, though useful in some cases, lacks a universal physical basis. The wilting point, if defined simply as the value of soil wetness of the root zone at the time plants wilt, is not easy to recognize. Root-zone soil moisture may be reasonably uniform for plants grown in restricted pots, but it is generally far from uniform in the field. Moreover, wilting is often a temporary phenomenon, which may occur in midday even when the soil is quite wet. The *permanent wilting percentage* (Hendrickson and Veihmeyer, 1950) has been defined as the root-zone soil wetness at which wilted plants can no longer recover turgidity even when placed in a saturated atmosphere for 12 hours. This is still an arbitrary criterion, since plant-water potential may not reach equilibrium with the average soil-moisture potential in such a short time. In any case, plant response depends as much on the evaporative demand (its variation and peak intensity) as on soil wetness (which itself is a variable function of space and time).

The *equal availability* hypothesis enjoyed wide popularity for some decades, in the 1900s'. Later investigators produced evidence that soil-water availability decreases with decreasing soil wetness and that a plant may suffer progressive water stress and reduction of growth considerably before the wilting point is reached. Still other investigators, seeking to compromise between the opposing views, attempted to divide the so-called "available range" of soil wetness into "readily available" and "decreasingly available" ranges, and searched for a "critical point" somewhere between field capacity and wilting point as an additional criterion of soil-water availability. These different hypotheses, in vogue through most of the 20th century, are represented graphically in Fig. 21.1.

None of these hypotheses was based on a comprehensive theoretical framework that could encompass the array of factors likely to influence the water regime of the soil–plant–atmosphere system as a whole. Rather, they tended to draw general conclusions from a limited set of experiments conducted under specific and sometimes poorly defined conditions.

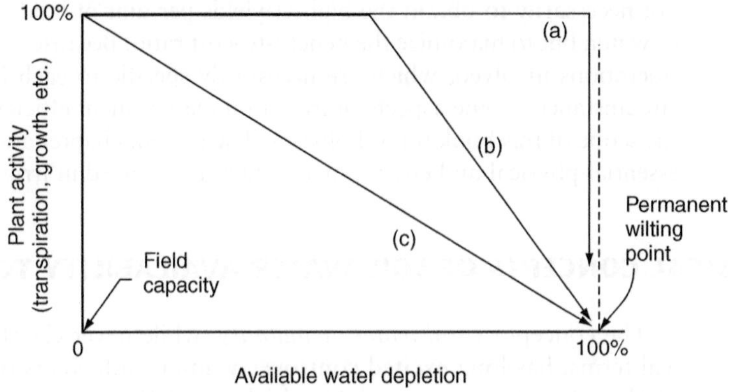

Fig. 21.1. Three classical hypotheses regarding the availability of soil water to plants: (a) equal availability from field capacity to wilting point, (b) equal availability from field capacity to a "critical moisture" beyond which availability decreases, and (c) availability decreases gradually as soil-moisture content decreases.

TRADITIONAL PRINCIPLES OF IRRIGATION MANAGEMENT

In practice, the old hypothesis of equal availability of water to crops within a fixed range of soil moisture resulted in a regimen of repeated, infrequent, but massive applications of water intended to fill the soil reservoir to its "field capacity." Each irrigation was followed by a prolonged period of depletion to nearly the permanent wilting point before another irrigation was to be applied to replenish the "deficit" to field capacity. The traditional irrigation cycle thus consisted of a brief episode of soil saturation, followed by an extended time (many days and sometimes weeks) of soil moisture extraction by the crop. In this manner the root zone was subjected to alternating periods of excessive wetness (with consequent disruption of soil aeration) and then of excessive desiccation (to the detriment of roots, especially in the surface layer). Practical limitations on the frequency of irrigation by the conventional methods of water application made it difficult to try alternative regimens based on continuous maintenance of a more nearly optimal level of soil moisture in the root zone.

Many irrigation projects were designed a priori to supply water to each farm unit on a fixed and infrequent schedule rather than to make water continuously available on demand. The traditional mode of irrigation seemed to make good economic sense because many furrow, flood, or portable sprinkler systems have a fixed cost associated with each application of water. With such systems, it is desirable to minimize the number of irrigations per season by increasing the interval of time between successive irrigations. For example, if the cost of portable sprinkler tubes is a dominant consideration, it obviously pays to make maximal use of such tubes by minimizing the amount required per unit area irrigated and by rotating the available tubing from site to site so as to cover the greatest overall area possible before having to return again to the same site for reirrigation.

The classical questions involved in irrigation management are *when* to irrigate and *how much* water to apply at each irrigation. To the first question, the traditional reply has been: Irrigate when the available moisture is nearly depleted, as determined by sampling or neutron gauging or tensiometry at some presumably representative depth or depths within the root zone, or by observing indications of incipient moisture stress in the plants. To the second question, the traditional reply was: Apply sufficient water to bring up the moisture reserve of the soil root zone to field capacity, plus a "leaching fraction" of, say, 10–20% for salinity control. In the last third of the 20th century, however, newer concepts of irrigation management evolved.

MODERN PRINCIPLES OF IRRIGATION MANAGEMENT

Ever since the decade of the 1960s, evidence has been accumulating that crops may respond with a pronounced increase in yield when irrigation is provided in sufficient quantity and frequency to prevent the occurrence of moisture stress at any time during the growing season. Such an effect can be produced by maintaining soil-moisture content in the root zone at a high level and soil-moisture suction at a low level while taking care to avoid excessive wetting and impairment of soil aeration as well as leaching out of nutrients. More precise control of soil moisture must also be aimed at preventing excessive percolation and the consequent rise of groundwater, which in turn poses the danger of salination.

The desired conditions were difficult to achieve by the traditional surface and sprinkler irrigation methods prevalent until a few decades ago. The advent of newer irrigation methods (including permanent installations, called "solid set," of low-intensity sprinklers, subirrigation by means of porous tubes, and — especially — the so-called "microirrigation" techniques of drip, trickle, bubbler, or microsprayer irrigation) has made it possible to establish and maintain soil-moisture conditions at a more nearly optimal level than before. These newer methods now make possible the optimization of soil moisture, salinity, fertility, and aeration simultaneously. Since these new irrigation systems are capable of delivering water to the soil in small quantities as often as desirable with no appreciable additional cost for the extra number of irrigations, the economic constraints on high-frequency irrigation have been lifted (Rawlins and Raats, 1975).

As the frequency of irrigation increases, the infiltration period becomes a more important part of the irrigation cycle. Changing the irrigation cycle from an extraction-dominated to an infiltration-dominated process brings into play a different set of relationships governing water flow. For example, with daily (rather than weekly or monthly) applications of water, the pulses of moisture resulting from intermittent irrigation are damped out within a few centimeters of the surface, so flow below this point is essentially steady and gravity driven.

By adjusting the rate and quantity of application in accordance with such measurable variables as potential evapotranspiration, soil hydraulic conductivity, and the soil solution's salt and nutrient concentration, it is possible to control the level of suction prevailing within the root zone as well as the through-flow (drainage) rate and thus also the leaching fraction. Control of the crop's soil environment thus passes into the hands of the irrigation manager more completely than ever before. Properly managed, the new systems hold the promise of saving water even while improving growing conditions and increasing yields.

Since a high-frequency irrigation system can be adjusted to supply water at very nearly the exact rate required by the crop, the irrigator need no longer depend on the soil's own capacity to store water. The consequences of this fact are far reaching: New lands, until recently considered unsuited for irrigation, can now be brought into production. One outstanding example is the case of coarse sands and gravels, in which the storage of soil moisture is minimal and in which the surface conveyance and application of water (e.g., in ditches or furrows) would involve inordinate losses by excess and nonuniform seepage. Such soils can now be irrigated quite readily, even on sloping ground and with hardly any leveling, by means of a drip irrigation system, for instance.

Theoretically, with high-frequency irrigation the irrigator need no longer worry about whether soil moisture is depleted or when plants begin to suffer stress. Such situations are avoided entirely. To the old question "When to irrigate?" the modern irrigationist answers, "As often as practicable; if possible, daily." To the question "How much water to give?" the answer is, "Enough to meet the evapotranspirational demand and to prevent salination of the root zone." Potential evaporation can be measured or calculated, and soil salinity can be monitored by sampling the soil or the soil solution or by using salinity sensors (Oster and Willardson, 1971).

ADVANTAGES AND LIMITATIONS OF NEW IRRIGATION METHODS

Of particular interest is the method of *drip* (or *trickle*) irrigation, along with its many variants, such as microsprayer, or "spitter," irrigation. Collectively called *microirrigation*, these methods have gained recognition and are being introduced in many irrigated areas. The idea of applying water slowly, literally drop by drop, at a rate that is continuously absorbed by the soil's root zone at specific points, is not an entirely new notion. However, what has finally made it practical is the rather recent development of low-cost plastic tubing and variously designed emitter fittings. System assemblies are now available that are capable of maintaining sufficient pressure in thin lateral tubes to ensure uniform discharge throughout the field as well as a controlled rate of drip, trickle, or spray discharge through the narrow-orifice emitters with a minimum of clogging (Figs. 21.2 and 21.3).

The application system has been supplemented by ancillary equipment, including filters, fertilizer injectors, and timing or metering valves (enabling the irrigator to predetermine the schedule and quantity of each irrigation or sequence of irrigations). Field trials in varied locations have resulted in increased yields of both orchard and field crops, perennial as well as annual, especially in adverse conditions of soil, water, and climate. Drip irrigation has also been found suitable for greenhouses and gardens and lends itself readily to labor-saving automation.

The justifiable enthusiasm for the new methods, however, carries certain dangers. Hasty adoption of microirrigation without enough care in adaptation to local crops and to soil and weather conditions can result in disappointment. The method offers many potential advantages, yet it is no panacea. Inefficiency,

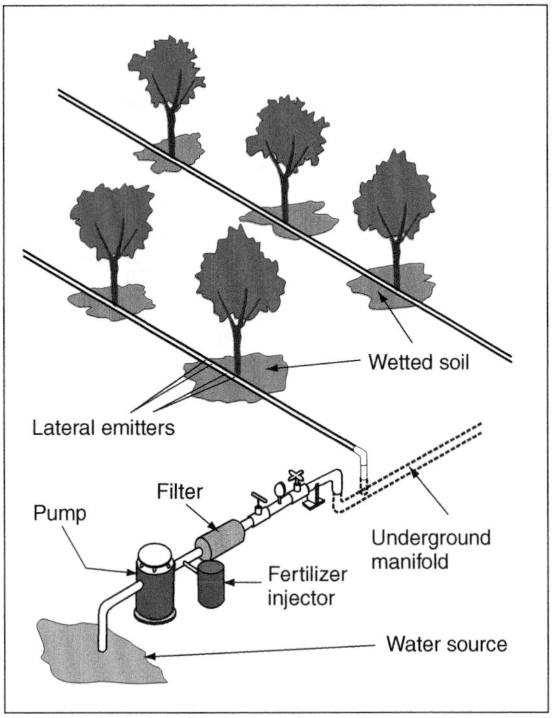

Fig. 21.2. Partial-area wetting around orchard trees under drip irrigation.

in fact, is just as easy to achieve in the operation of a drip or trickle system as it is in the operation of conventional systems. Let us therefore list some of the possible advantages of drip irrigation while mentioning some of its limitations.

One advantage, already mentioned, is the possibility of obtaining favorable moisture conditions even in problematic soils, such as coarse sands and clays, which are ill suited to conventional ways of irrigation (Bucks et al., 1982; Hillel, 1997). Another is the possibility of delivering water uniformly to plants in a field of variable elevation, slope, wind velocity and direction, soil texture, and infiltrability. Still another potential advantage is the ability to maintain the soil at a highly moist yet unsaturated condition so that soil air remains a continuous phase capable of exchanging gases with the atmosphere. High moisture reduces the soil's mechanical resistance to root penetration and development.

Where salinity is a hazard, as when the irrigation water is brackish, the continuous supply of water ensures that the osmotic pressure of the soil solution will remain low near the water source. Moreover, drip irrigation, because it is applied underneath the plant canopy, avoids the hazard of leaf scorch and reduces the incidence of fungal diseases, both of which may occur with sprinkler irrigation. Since drip irrigation wets the soil only in the immediate vicinity of each emitter, the greater part of the surface (particularly the inter-row areas) remains dry and hence less prone to weed infestation and soil compaction. In summation, properly managed microirrigation seems to offer the best opportunity at present to optimize the water, nutrient, and air regimes in the root zone (Rawitz and Hillel, 1974; Bucks et al., 1982; Hillel, 1987a, 1997).

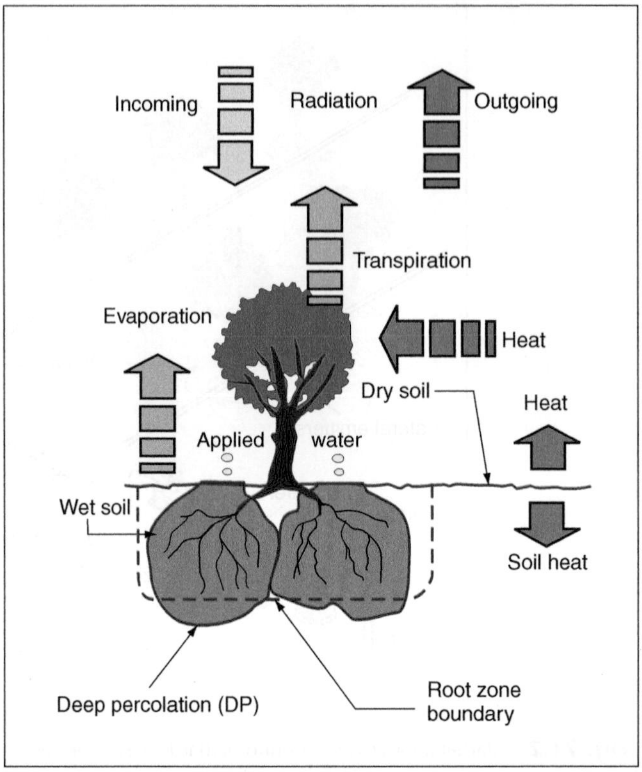

Fig. 21.3. Radiation and water balances on a plant under localized irrigation.

The fact that drip irrigation and other forms of localized microirrigation wet only a small fraction of the soil volume can also become a problem. While it is a proven fact that even large trees can grow on less than 50% of what is generally considered to be the normal root zone of a field (provided enough water and nutrients are supplied within this restricted volume), the crop becomes extremely sensitive and vulnerable to even a slight disruption of the irrigation system or schedule. If the system does not operate perfectly and continuously, crop failure can result because the soil-moisture reservoir available to the plants is extremely small.

WATER-USE EFFICIENCY AND WATER CONSERVATION

Any concept of efficiency is a measure of the output obtainable from a given input. Irrigation or water-use efficiency can be defined in different ways, however, depending on the nature of the inputs and outputs considered. For example, one can define as an economic criterion of efficiency the financial return in relation to the money invested in the installation and operation of a water supply and delivery system. The problem is that costs and prices fluctuate from year to year and vary widely from place to place and thus may not be universally comparable. Perhaps a more objective criterion for the relative merit of an irrigation systems is an agronomic one — namely, the added yield (resulting from irrigation) per unit of land area or, better yet, per unit amount of water applied. Often, the criterion of efficiency depends on one's point of view.

A widely applicable expression of efficiency is the *agronomic* or *crop water-use efficiency*, which has long been defined as the amount of vegetative dry matter produced per unit volume of water consumed by the crop. Because most of the water taken up by plants in the field is transpired (in arid regions, as much as 99%), the plant water-use efficiency is in effect the reciprocal of the so-called *transpiration ratio* defined as the mass of water transpired per unit mass of dry matter produced.

What we shall refer to as *technical efficiency* is what irrigation engineers call *irrigation efficiency*. It is generally defined as the net amount of water added to the root zone divided by the amount of water taken from some source. As such, this criterion of efficiency can be applied to large regional projects or to individual farms or to specific fields. In each case, the difference between the net amount of water added to the root zone and the amount withdrawn from the source represents the loss incurred in conveyance and distribution.

In practice, many (perhaps most) irrigation projects still operate in an inherently inefficient way. In many of the surface irrigation schemes, one or a few farms may be allocated large flows representing the entire discharge of a lateral canal for a specified period of time. Where water is delivered to the consumer only at fixed times and where charges may be assessed per delivery regardless of the actual amount used, customers tend to take as much water as they can while they can. This often results in overirrigation, which not only wastes water but also causes project-wide and perhaps even region-wide problems connected with the disposal of return flow, waterlogging of soils, leaching of nutrients, and excessive elevation of the water table requiring expensive drainage to rectify. Although it is difficult to arrive at reliable statistics, it has been estimated that the average irrigation efficiency in such schemes is less than 50%. Since it is a proven fact that, with proper management, irrigation efficiencies of 80–90% can actually be achieved, there is obviously room, and need, for much improvement.

The volume of water "consumed" in the field results from evapotranspiration rather than transpiration alone. It thus includes the amount of water evaporated directly from the soil surface without being taken up by the plants. In addition, evapotranspiration often includes the amount of water intercepted by the foliage (e.g., under overhead sprinkler irrigation or rainfall) and evaporated without ever entering either the soil or the plant. In practice, direct evaporation is difficult to measure separately from transpiration, so the two terms are lumped together merely for the sake of convenience. Clearly, however, much of the water evaporated without entering the plant is consumed nonproductively. Therefore, any method of irrigation that minimizes evaporation is likely to increase the efficiency of water utilization by the crop. Some new irrigation methods are capable of doing just that: They introduce water directly into the root zone without sprinkling the foliage or wetting the entire soil surface.

Partial-area irrigation methods offer the additional benefit of keeping the part of the soil surface between the rows of crop plants (or fruit trees) dry. This discourages the growth of weeds that would otherwise not only compete with crop plants for nutrients and moisture in the root zone and for light above ground, but also hinder field operations and pest control.

All of the indexes of efficiency may be combined in a single concept, the *overall agronomic efficiency of water use*, WUE_{ag}:

$$WUE_{ag} = P/W \tag{21.1}$$

where P is crop production (in terms of either total dry matter or marketable product) and W is the volume of water applied.

The one component of the field water balance that generally cannot, and probably should not, be reduced is transpiration by the crop. In the open field, little can be done to limit transpiration if the conditions required for high yields are to be maintained. It appears that the greatest promise for increasing water-use efficiency lies in allowing the crop to transpire freely in response to the climatic demand by preventing or alleviating any possible water shortages while avoiding waste and obviating all other environmental constraints to attainment of the optimal productive potential of the crop. This is particularly important in the case of the new and superior varieties that have been developed in recent decades and that can provide high yields only if water stress is prevented and such other factors as soil fertility (availability of nutrients), aeration, salinity, and soil tilth are also optimized.

Sample Problem

In a given season, a field of 4 ha (hectares) was irrigated with a total volume of 48,000 m^3 of water. Evapotranspiration amounted to 900 mm. Assuming that the water content of the soil profile at the end of the season was the same as it was at the beginning, estimate the irrigation efficiency. How much water was drained? What would be the irrigation requirement for the field if an irrigation efficiency of 85% were to be attained? Assume negligible precipitation, runoff, capillary rise, and plant-water storage.

It is convenient to divide the volume of irrigation by the field's area (recalling that 1 ha = 10,000 m^2) to convert the irrigation quantity into depth units:

$$I = 48{,}000 \text{ m}^3/4 \times 10{,}000 \text{ m}^2 = 1.2 \text{ m} = 1200 \text{ mm}$$

Irrigation efficiency F_i has been defined as the volume of water used "consumptively" (i.e., in evapotranspiration) divided by the total volume applied to the field. Thus

$$F_i = (E + T)/I = 900 \text{ mm}/1200 \text{ mm} = 0.75 = 75\%$$

We now recall the overall water balance [see Eq. (21.2)]:

$$\Delta S + \Delta V = (P + I + U) - (R + D + E + T)$$

where ΔS is the soil storage increment, ΔV is the vegetational storage increment, P is precipitation, I is irrigation, U is upward capillary rise, R is runoff, D is drainage, E is evaporation, and T is transpiration. According to the problem statement, the terms ΔS, P, U, R, and ΔV are assumed to be negligible. Hence

$$I = D + (E + T)$$

To obtain the amount of drainage, we set

$$D = I - (E + T) = 1200 - 900 = 300 \text{ mm}$$

If irrigation efficiency were 85% and evapotranspiration were unchanged, our irrigation would be

$$I = (E + T)/F_i = 900/0.85 = 1059 \text{ mm} = 1.059 \text{ m}$$

which, over a field of 4 ha, would amount to approximately

$$1.059 \text{ m} \times 40{,}000 \text{ m}^2 = 42{,}360 \text{ m}^3$$

A net savings of 48,000 – 42,360 = 5460 m^3 would thus be realized.

Note: This calculation of "efficiency" includes evaporation as well as transpiration. In an "open" crop, such as a young orchard or row crop, direct evaporation from the exposed soil surface may be a partly avoidable loss. Hence it would stand to reason that evaporation should be counted as a contribution not to efficiency but to inefficiency. To the extent that transpiration by the crop (a largely inevitable consequence of its growth) can be separated from the direct evaporation of soil moisture, an alternative expression of irrigation efficiency can be attempted. However, the two components of evapotranspiration cannot easily be separated, and in any case there usually is a feedback effect such that a reduction of the one would tend to enhance the other, though not necessarily to a commensurate degree.

TRANSPIRATION IN RELATION TO PRODUCTION

Transpiration may be limited either by the supply of water or by the supply of energy needed for vaporization, the latter determined mainly by such climatic factors as radiation, temperature, wind, and humidity. When the supply of water is plentiful, it is the climate, and particularly the net radiation, which determines water use. When water is limiting, the assimilation rate, plant growth, and consequently crop yield are all related quantitatively to the water supply. The classical analysis of the relation between transpiration and yield was published by de Wit (1958). He reviewed the data then available and found that in climates with a large percentage of bright sunshine duration (i.e., arid regions), yield (Y_d) tends to be linearly related to the ratio of actual transpiration (E_a) to free-water evaporation (E_0):

$$Y_d = mE_a/E_0 \tag{21.4}$$

In climates with a limited duration of bright sunshine duration (i.e., cloudy regions), the relation

$$Y_d = nE_a \tag{21.5}$$

was found (that is to say, dry-matter production is proportional to transpiration and hence "water-use efficiency" is constant). These relationships evidently hold for both container-grown and field-grown plants, and the values of constants m and n were reported to be characteristic for each crop. The assumption of linearity between yield and transpiration has served as the basis for subsequent efforts toward comprehensive modeling of crop growth as influenced by soil water (e.g., Hanks, 1974).

An empirically based equation to predict dry-matter yield from known values of evapotranspiration was suggested by J. I. Stewart et al. (1977):

$$Y/Y_m = 1 - \beta E_{td} = 1 - \beta + \beta E_t/E_{tm} \quad (21.6)$$

where Y is yield, Y_m is maximum attainable yield, E_t is evapotranspiration, E_{tm} is maximum (potential) evapotranspiration, and β is the slope of relative yield (Y/Y_m) versus the evapotranspirational deficit ($E_{td} = 1 - E_t/E_{tm}$). To predict yield from this equation, one must know E_t, E_{tm}, Y_m, and β.

Determination of the fraction of evapotranspiration due to evaporation from the soil (rather than transpiration from the plants) can be important, since evaporation does not relate to plant function and growth; hence it can be considered a loss. However, evaporation and transpiration are not mutually independent, and the reduction of one may entail an increase of the other, though not necessarily to a commensurate degree.

Doorenbos and Kassam (1979) proposed a method for evaluating the yield response of crops to applied water in terms of the following relationship:

$$(1 - Y_a/Y_m) = f_y(1 - E_{ta}/E_{tp}) \quad (21.7)$$

where Y_a is actual yield, Y_m is maximum attainable yield when water requirements are fully met, f_y is a yield response factor, E_{ta} is actual evapotranspiration, and E_{tp} is potential evapotranspiration.

The relationship between the yield of a crop and the amount of water used is known as the *crop water production function* (Vaux and Pruitt, 1983) or *crop-response function* (Yaron and Bresler (1983). In many cases, the reported data pertain to the above-ground dry-matter yield. If the yield of interest is grain, fruit, or fiber, its relation to water use by the crop plants may be quite different.

The relationships found between yield and water use under limited water supply may not hold as potential evapotranspiration is attained and water ceases to be a limiting factor in plant growth. Beyond the point where transpiration reaches its climatic limit, the promise of increasing production may lie in identifying and obviating any other possible environmental constraints, such as light distribution in the canopy, carbon dioxide concentration in the air, and nutrients in the soil. Finally, we come up against genetic constraints. This is why environmental scientists (including soil physicists) must cooperate with plant geneticists in the effort to improve the productive potential of agricultural crops.

Sample Problem

A forage crop raised under two different irrigation treatments and two fertilization subtreatments gave the following yields (seasonal potential evapotranspiration was 1000 mm):

(a) *In the "wet" treatment,* water was applied in several large irrigations totaling 900 mm. The high-fertilization subtreatment produced a drainage of 100 mm and a dry-matter yield of 12 metric ton/ha. The low-fertilization subtreatment drained

200 mm and yielded 7 ton/ha. In neither subtreatment was the end-of-season soil-moisture content different from the start-of-season value.
(b) *In the "dry" treatment*, the amount of water applied was 500 mm. High fertilization resulted in a net depletion of soil moisture amounting to 100 mm and a dry-matter yield of 9 ton/ha. The low-fertilization subtreatment depleted soil moisture 50 mm and yielded 5.5 ton/ha.

Calculate the evapotranspiration and water-use efficiency values. Assuming direct evaporation from the soil surface to be 10% of evapotranspiration for the wet treatment and 15% for the dry treatment, calculate the transpiration ratio for each treatment.

To calculate evapotranspiration E_t we subtract the amount of drainage from the amount of irrigation given to the wet treatment and add the amount of soil-water depletion to the amount of irrigation given the dry treatment:

Wet treatment

High-fertilization subtreatment:

$$E_t = 900 \text{ mm} - 100 \text{ mm} = 800 \text{ mm}$$

Low-fertilization subtreatment

$$E_t = 900 \text{ mm} - 200 \text{ mm} = 700 \text{ mm}$$

Dry treatment

High-fertilization subtreatment:

$$E_t = 500 \text{ mm} + 100 \text{ mm} = 600 \text{ mm}$$

Low-fertilization subtreatment:

$$E_t = 500 \text{ mm} + 50 \text{ mm} = 550 \text{ mm}$$

To calculate water-use efficiency (WUE), we divide the dry-matter yield by the corresponding per-area mass of water used in evapotranspiration:

Wet treatment

High-fertilization subtreatment:

$$\begin{aligned} \text{WUE} &= (12 \text{ ton/ha})/(0.8 \text{ m} \times 10{,}000 \text{ m}^2\text{/ha} \times 1 \text{ ton/m}^3) \\ &= 12 \text{ ton}/8000 \text{ ton} = 1.5 \times 10^{-3} \text{ ton dry matter/ton water} \\ &= 1.5 \times 10^{-3} \text{ kg dry matter/kg water} \end{aligned}$$

Low-fertilization subtreatment:

$$\text{WUE} = (7 \times 1000/10{,}000)/\ (0.7 \times 1 \times 1 \times 1000) = 1.0 \times 10^{-3}$$

Dry treatment

High-fertilization subtreatment:

$$\text{WUE} = (9 \times 1 \times 1 \times 1000)/(0.6 \times 1 \times 1 \times 1000) = 1.5 \times 10^{-3}$$

Low-fertilization subtreatment:

$$\text{WUE} = (5.5 \times 1000 \times 10{,}000)/(0.55 \times 1 \times 1 \times 1000) = 1.0 \times 10^{-3}$$

To calculate the transpiration ratio (T_R) we can subtract direct soil-moisture evaporation from evapotranspiration (an arguable procedure, since evaporation and transpiration are not completely independent) to estimate the amount of transpiration and then divide the mass of water transpired by the mass of dry matter produced:

Wet treatment

High-fertilization subtreatment:

$$\begin{aligned} T_R &= [0.8(1 - 0.1)\text{m}^3/\text{m}^2(1\ \text{ton/m}^3)(1000\ \text{kg/ton})] \\ &\div [(12\ \text{ton/ha})(1000\ \text{kg/ton})/\ (10{,}000\ \text{m}^2/\text{ha})] \\ &= 600\ \text{kg water/kg dry matter} \end{aligned}$$

Low-fertilization subtreatment:

$$T_R = [0.7(1 - 0.1) \times 1 \times 1000]/(7 \times 1000/10{,}000) = 900\ \text{kg water/kg dry matter}$$

Dry treatment

High-fertilization subtreatment:

$$T_R = [0.6(1 - 0.15) \times 1 \times 1000]/(9 \times 1000/10{,}000) = 567\ \text{kg water/kg dry matter}$$

Low-fertilization subtreatment:

$$T_R = [0.55(1 - 0.15) \times 1 \times 1000]/(5.5 \times 1000/10{,}000) = 850\ \text{kg water/kg dry matter}$$

ESTIMATION OF CROP WATER REQUIREMENTS

Irrigation scheduling is the term used to describe the procedure by which an irrigator determines the timing and quantity of water application. Accordingly, the two classical questions of irrigation scheduling are when to irrigate and how much water to apply. To the first of these questions (when — that is, how frequently — to irrigate) the answer is, in principle, to irrigate as frequently as is feasible economically. To the second question, the answer is to water enough to meet the current evaporative demand and to prevent salination of the root zone.

The evaporative demand can be assessed by monitoring relevant weather variables (e.g., temperature, wind, atmospheric humidity, and solar radiation) and then applying any of several functional equations or formulae to calculate the potential evapotranspiration. Alternatively, and more simply, the evaporative demand can be estimated from standard evaporimeters. One of the simplest and most useful of such devices is the *evaporation pan*. It consists of a shallow water-filled container placed on the ground within the irrigated area. The amount evaporated daily can be obtained by measuring the volume of water per unit area of the pan that must be added to bring the water surface back up to a marked level. The pan evaporimeter is subject to the integrated effects of radiation, wind, temperature, and humidity, hence it can be correlated with evapotranspiration from the field in which it is placed.

Of the various standardized pans, the one used most widely is the Class A pan, introduced by the U.S. Weather Bureau. It is a circular container, 121 cm across and 25.5 cm deep, placed on a slatted wooden frame resting over the ground. The pan is filled with water to a height about 5 cm below the rim. This standard design is relatively easy to follow. However, while inexpensive and easy to install, maintain, and monitor, evaporation pans in general have several shortcomings.

Although in principle the water-filled pan is subjected to the same general climate as the crop, it does not necessarily respond in the same way. A vegetated surface generally differs from a free-water surface in reflectivity, thermal properties (heat storage), day–night temperature fluctuation, water transmissivity, and aerodynamic roughness. Such factors as the color of the pan, the depth and turbidity of the water, and shading from nearby plants can all affect the measurement to some degree. Pan evaporation depends on the exact placement of the pan relative to wind exposure. Pans surrounded by tall grass may evaporate some 20% less than pans placed in a fallow area. Rainfall may occur during the irrigation season and may add water to the pan, or thirsty animals wandering in the area may drink from the pan, thus detracting from its usefulness. To avoid water loss to drinking animals (especially birds), pans are often covered with screens, which may reduce the evaporation rate by 10–20%, thus requiring the use of a correction factor. All these shortcomings notwithstanding, pan evaporimeters, if properly sited and maintained, can provide valuable data on weather conditions affecting the crop. The problem is how to translate pan evaporation into an estimate of crop-water requirements.

The first step is to apply a correction factor to account for the fact that free water generally evaporates more than does a crop stand. Many experiments have shown that the appropriate correction factor varies from 0.5 to 0.85, with a mean value of about 0.66:

$$\mathrm{PET}_{\text{full cover}} \approx 0.66 E_{\text{pan}} \tag{21.8}$$

The second step is to account for the stage of crop growth, as indicated by the fractional ground cover. That can be estimated from ground observations of the area shaded by the crop. However, potential evapotranspiration, which is a function of the crop's coverage, is not simply proportional to it. Hillel (1997) proposed the following relationship:

$$\mathrm{PET}_{\text{partial cover}} \approx 0.33(1 + C)E_{\text{pan}} \tag{21.9}$$

where C is the crop's fractional ground cover, varying from 0 when the crop is sown to 1 when the crop stand is full. In the latter case, Eq. (21.9) becomes Eq. (21.8).

The third step is to estimate the irrigation requirements (I), including the actual crop water requirement (W), plus a leaching fraction (L), minus an effective rainfall that might have occurred since the previous irrigation (R). Assuming that the actual crop-water requirement is about 80% of PET and that the desirable leaching fraction is some 10% of PET (i.e., $W = 0.8$ PET, and $L = 0.1$ PET), the estimate is:

$$\begin{aligned} I &= [0.33\ (W + L)]E_{\text{pan}}\ (1 + C) - R \\ &= (0.33 \times 0.9)E_{\text{pan}}(1 + C) - R = 0.3E_{\text{pan}}(1 + C) - R \end{aligned} \tag{21.10}$$

These relationships are empirical and should only be regarded as estimates, to be tested in each case under local conditions of soil, weather, and crop (Hillel, 1997). Moreover, the estimates refer only to a crop's active growth stage. As some crops mature, their foliage senesces and transpiration (hence also water requirements) naturally diminishes. Irrigation is to be discontinued when its further contribution to crop yield no longer justifies its added cost.

BOX 21.2 The Sigmoid Response Curve

The typical yield-response curve, relating crop yield to the amount of water applied, is sigmoid in shape (Figure 21.4). The slope of the curve at each point represents the yield response per unit quantity of irrigation.

Along that curve, we can recognize four critical points: (1) a threshold quantity of water, designated *T*, below which no yield is obtained. (2) Beyond that threshold, the yield response is a rising curve with a steepening slope, up to a point of inflection designated *I*. That is the point at which the slope is greatest, indicating the highest incremental yield response to the addition of irrigation. (3) Beyond that point, the yield continues to rise as more water is added, but with diminishing returns per unit volume of water. At some point, an economic optimum may be reached beyond which any additional supply of water will no longer produce an incremental yield commensurate with the additional cost it involves. We designate that hypothetical optimal point *O*. (4) If still more water is supplied regardless of economic cost–benefit considerations, an absolute maximum yield is reached, designated *M*, after which soil moisture become excessive and the yield declines (owing to the restriction of aeration and the leaching of nutrients).

The aim of irrigation research is not to ascertain the amount of water that would result in maximal absolute yield, or necessarily the amount needed to attain maximal yield per unit amount of water, but the optimal irrigation that can produce the greatest net income. It is this elusive *O* point that is most difficult to define, because it depends on so many additional biophysical factors, including vagaries of climate (cold spells or hot spells during the growing season), diseases and pests, quality (as well as quantity) of the irrigation water, and nutrient (fertility) status. It also depends on such economic factors as the cost of the water itself, the cost of delivering it to the crop, the cost of other essential inputs, and — of course — the price commanded by the product.

Still another relevant factor is the possible contribution of rainfall to the water balance of the crop. In some areas, rainfall occurs during the growing season, and its amount reduces the amount of irrigation needed to attain optimum economic yield.

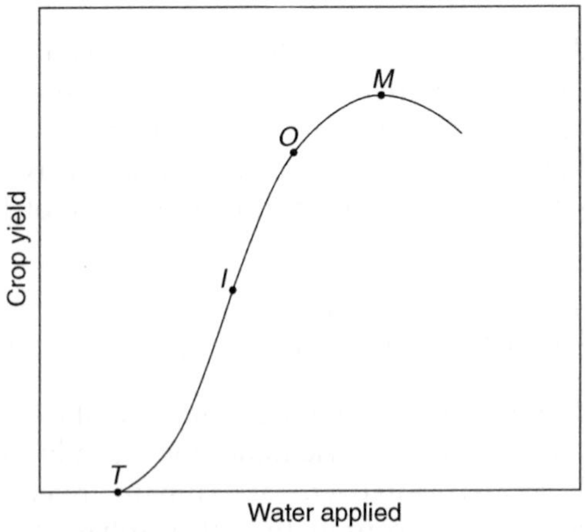

Fig. 21.4. The typical yield-response curve.

Even where there is no rainfall during the growing season, there may be a reserve of soil moisture remaining from prior-season rains, which could be utilized by the crop and hence should be taken into consideration. From all the foregoing, one may justifiably conclude that irrigation and water-use efficiency research is a challenging task indeed.

ENVIRONMENTAL ASPECTS OF IRRIGATION DEVELOPMENT

Irrigation development in semiarid and arid areas may have both positive and negative impacts on the environment. To be sustainable, irrigation must maximize the positive impacts and minimize the negative impacts (Letey, 1994). The positive aspect of irrigation is that intensifying food production in the favorable lands can allow a country to reduce pressure on marginal lands now under rain-fed cultivation or grazing. In many areas, such lands are already undergoing a process of degradation (including denuding of vegetation and erosion) known as *desertification* (Dregne, 1992, 1994; Hillel and Rosenzweig, 2002). Where the opportunity exists for irrigation development, it can serve as a constructive alternative to worsening degradation.

The potentially negative environmental impacts of irrigation development may occur off-site as well as on-site. The off-site effects may take place upstream of the land to be developed, such as where a river is to be dammed for the purpose of supplying irrigation water. Another set of problems may be generated downstream of the irrigated area by the disposal of excess water that may contain harmful concentrations of salts, organic wastes, pathogenic organisms, as well as residues of pesticides and fertilizers.

Of most direct concern are the on-site impacts. Irrigated lands, especially in river valleys prone to high water-table conditions, usually require drainage. Because groundwater drainage is a complex, exacting, and expensive operation (often more expensive than the initial development of irrigation itself), there is a temptation to start new irrigation projects while ignoring the need for drainage or delaying its installation "until it is actually needed." The trouble is that by the time the need for drainage becomes inescapable, much damage has already been done and the cost of rectifying it may be prohibitive.

One of the most serious problems in irrigation projects in developing countries is the potential health hazard resulting from the use of open irrigation channels for drinking, washing, and the disposal of human and animal wastes. All too often, "where water goes, disease follows." Unfortunately, water-storage and -conveyance structures present favorable breeding grounds for disease vectors (e.g., mosquitoes and snails) and for pathogens of some of the most debilitating illnesses rampant in the developing world. Among these are schistosomiasis (bilharzia), onchocerciasis (river blindness), malaria, cholera, dysentery, and other intestinal diseases. Measures to control water-borne diseases include the following: (1) concrete lining of channels (or conveyance in pipes) to prevent scouring, clogging, and stagnation of the water as well as animal wading and the proliferation of riparian weeds; (2) avoidance of waste disposal in channels and reservoirs; (3) treatment of the water used directly for human needs.

Food supply and security in the future must depend more on intensification than on continued expansion of cultivation at the expense of the remaining natural ecosystems. Only thus can the global environment and its biodiversity be saved. Irrigation already plays an important role in increasing and stabilizing agricultural production. Given rising populations and living standards, it is expected to play an even more important role in the future. This expectation will be met only if irrigation is based on thorough knowledge and sound sustainable management of the soil–plant–atmosphere system.

USE OF WASTEWATER FOR IRRIGATION

Decomposition in the soil is nature's way to recycle organic waste products of biological activity. As long as human habitation remained sparse, the waste products of human activity could be similarly accommodated. The growth of cities and of industries, however, produced quantities of solid and liquid wastes in excess of the ability of the soil to reprocess in the immediate environs of human habitation. Some societies (most notably, in China and other parts of Asia) continued nonetheless to transport human wastes from cities to agricultural land for the purpose of fertilizing crops and replenishing depleted soil nutrients. In many other places, on the other hand, uncontrolled disposal of raw garbage and sewage ended up polluting streams and wells, to the detriment of public health.

With the advent of sewerage systems in Europe at the beginning of the modern age, interest in wastewater farming or land application increased (Shuval et al., 1986). A report issued in 1865 by the Royal Commission on Sewage Disposal in England stated: "The right way to dispose of town sewage is to apply it continuously to the land and it is by such application that the pollution of rivers can be avoided." In 1868, Victor Hugo expressed a stronger opinion: "All the human and animal manure which the world loses . . . by discharge of sewage to rivers . . . . If returned to the land instead of being thrown into the sea, would suffice to nourish the world."

In the late 19th and early 20th centuries, sewage-irrigated farms were widespread in the environs of many cities. In addition to the aims of preventing pollution of water supplies and of conserving nutrients for crops, there arose, especially in arid areas, the impetus to utilize sewage as an additional water resource.

Subsequently, however, urban development encroached on the sewage farm areas, and many of the early wastewater irrigation projects in Europe and America were abandoned. Concerns over the odor problem and over the transmission of disease from crops irrigated with raw sewage also contributed to the decline of sewage farming. Another disadvantage in some areas was that, following heavy rainstorms, runoff from sewage-irrigated fields periodically conveyed pollutants into rivers and reservoirs. Many public health officials therefore came to believe that sewage farming was an unsanitary and hence undesirable practice of the past. The alternative was a sewage-treatment technology based on intensive centralized civil engineering systems in which the organic components of sewage could be digested and from which the "purified" fluid could be discharged more or less harmlessly to streams.

Still more recently, the pendulum has swung once again. Interest has revived in the use of sewage as a resource rather than merely as a problem, based on more rational scientific principles than before. Research on the fate of pathogens and of various organic and inorganic components of wastewaters applied to the soil has shown that in many cases treated or even partially treated sewage can indeed be used safely and advantageously for irrigation, provided that necessary precautions are taken to prevent deleterious effects on public health as well as on soil quality. Guidelines have been formulated to allow wastewater irrigation to become a socially acceptable and sanitary practice. The trend has been led by the State of California and by the State of Israel, in both of which growing urbanization and industrialization have increased the competitive demand (and hence the price) for scarce fresh water while simultaneously increasing the volume of generated wastewater.

Wastewater-treatment standards such as those established by the California State Health Department apply strict criteria of bacterial counts (e.g., no more than 2.2 coliform/100 mL) and require the application of chemical disinfection (chlorination) for treated sewage water to be allowed for general use. Partially treated sewage should be used only for the irrigation of industrial crops, such as cotton, or of tree-grown fruits that do not come in contact with the irrigation water, or — at most — for vegetable crops that are to be cooked before being consumed. Similar guidelines have been adopted elsewhere.

In Israel, well over two-thirds of domestic sewage is currently being recycled for use in agriculture, and the projection is for reuse of some 80% by the year 2000. In fact, recycled sewage is expected to become the main source of water for irrigation. Wastewater irrigation has been shown to contribute significantly to soil fertility (especially to nitrogen, phosphorus, and organic matter augmentation) in many areas and to produce greater crop yields (Muzafi et al., 1997).

Wastewater irrigation is not always a blessing. In places, the concentration of nitrates can become excessive and contribute to groundwater and surface-water pollution. Heavy metals and other toxic materials may tend to accumulate in the soil and thence to enter the biological food chain. Especially hazardous are agricultural pesticide residues (some of which are highly persistent) and industrial waste products, which may be carcinogenic. Wastewaters typically contain increased concentrations of soluble salts and therefore pose the danger soil salination. Where salination is prevented by leaching, the salts may accumulate in the underlying aquifer and contribute to the salinity of well waters. The variable pH of the water applied to the soil may cause either acidification or alkalination, with consequent effects on soil structure and fertility. Finally, the materials suspended in sewage, if not readily decomposed, may cause clogging of soil pores and thereby reduce soil permeability to water and air. For all these reasons, the practice of wastewater irrigation — promising though it is in principle — must be examined carefully in each case to ensure that short-term benefits are not negated by long-term damage.

GLOSSARY

Adhesion: The attraction between molecules of dissimilar substances, acting at surfaces of contact between the substances.

Adsorption: A phenomenon occurring at the boundary between phases, where cohesive and adhesive forces cause the concentration or density of a substance to be greater or smaller than in the interior of the separate phases.

Advection: The exchange of energy and matter between a field and its surroundings. More specifically, it is the transport of heat and vapor into or out of a field by winds sweeping over or through it. See: "Clothesline effect" and "Oasis effect."

Aeration of soil: The exchange of gases between the soil's air phase and the external atmosphere, supplying oxygen to growing roots and microorganisms and removing carbon dioxide (the product of aerobic respiration).

Aggregate stability: The ability of soil aggregates to withstand disruptive forces, whether such forces are imposed mechanically (as during tillage or trampling) or by the action of water (causing swelling, slaking, and dispersion of clay).

Aggregates: Soil structural units consisting of assemblages of primary particles that are often bonded together by flocculated clay and may be stabilized by organic (humus) or inorganic cementing agents.

Air encapsulation: The entrapment of air in bubbles or in bypassed pores within a nearly saturated soil.

Albedo: The reflectivity coefficient of a surface to short-wave radiation, i.e., the fraction of incoming solar radiation that is reflected rather than absorbed. For soils, it varies generally between 0.1 and 0.4, depending on color (whether light-hued or dark), surface roughness and wetness, and inclination of the incident radiation.

Alkalinity of soil: A high concentration of sodium ions in the soil's exchange complex, which affects soil properties and behavior. It typically results in an increase of soil pH, dispersion of clay, and breakdown of soil aggregates, which in turn reduces soil permeability and hence soil productivity.

Alumino-silicate clay minerals: Clay minerals in which the crystal lattice typically consists of alternating layers of alumina and silica ions in association with oxygen atoms or hydroxide ions. Particles of such minerals typically exhibit negative surface charges, which are usually countered by adsorbed (exchangeable) cations.

Anaerobic soil conditions: A consequence of restricted soil aeration, resulting in chemical and biochemical reduction reactions, including denitrification (from nitrate to the nitrite, thence to nitrous oxide), manganese reduction (from manganic to manganous state), iron reduction (from ferric to ferrous state), sulfate reduction (to hydrogen sulfide). Among the organic compounds formed under anaerobic conditions are methane, ethylene, and various organic acids.

Aquifer: A porous geological formation that contains and transmits groundwater to natural springs or to wells that supply water for human purposes.

Arid regions: Regions in which the natural supply of water by rainfall is below the potential evapotranspiration rate during most of the year. In such regions, crop plants are apt to suffer from a deficit of soil moisture much of the time, unless provided with artificial irrigation.

Bernoulli's law: An equation relating the pressure of a moving fluid to its velocity. It indicates that an increase in velocity entails a decrease in pressure.

Blackbody: A body whose surface absorbs rather than reflects incoming shortwave radiation, and that emits long-wave radiation at maximum efficiency.

Bound water: The thin layer of water tightly adsorbed onto the surface of clay particles, presumed by some investigators to be more rigid than the bulk water that resides in the wider pore spaces between the particles.

Boussinesq equation: An equation describing pressure distribution in uniform elastic materials. It is used in soil mechanics to estimate the stresses at any point within the soil profile due to a concentrated load applied to the surface.

Bowen ratio: An index that is proportional to the ratio of temperature gradient to vapor pressure gradient in the atmosphere above a field. It is relatively low when the field is wet and the evaporation rate is high, i.e., when the temperature gradient is small relative to the vapor pressure gradient. When the field is dry, the humidity gradients tend to be low and much of the received solar energy goes to warming the soil, hence the temperature gradients tend to be steep and the Bowen ratio becomes large.

Breakthrough curve: A plot of the relative solute concentration at the outflow boundary of a soil column, following a step-change in the concentration of the entering solution at the inflow boundary, as a function of the volume of effluent.

Bulk density of soil: Mass of soil solids per unit bulk volume of the soil (including the volume of the voids).

Caliche: A layer of soil cemented by the accumulation and re-precipitation of calcium and magnesium carbonate. It generally forms in the B-horizon of lime-rich soils in arid regions. In some places, erosion of the A horizon exposes the caliche so that it appears as a hardened layer capping the looser soil material beneath.

Capillarity: The condition of water drawn into the narrow ("capillary") pores of the soil, where it is generally at a sub-atmospheric pressure called tension or suction.

Capillarity: The tendency of a liquid to enter into the narrow pores within a porous body, due to the combination of the cohesive forces within the liquid (expressed in its surface tension) and the adhesive forces between the liquid and the solid (expressed in their contact angle).

Capillary fringe: The thin zone just above the water table that is still saturated, though under sub-atmospheric pressure (tension). The thickness of this zone (typically a few centimeters or decimeters) represents the suction of air entry for the particular soil.

Carbon sequestration by soils: The accumulation of carbonaceous (organic) compounds in the soil, brought about by the addition of plant and animal residues to the soil and their partial decomposition within it to form more-or-less stable humus.

Cementation: The process by which calcareous, siliceous, or ferruginous compounds tend to dissolve and then re-precipitate in certain horizons within the soil profile, thus binding the particles into a hardened mass.

Characteristic curve of soil moisture: A graph of the water content in a unit mass or unit volume of soil (mass wetness or volume wetness) as a function of matric suction. If measured in desorption, it is also called the "moisture retention" or "release" curve.

Clay: The colloidal fraction of mineral soils, consisting of particles smaller than 2 micrometers in diameter.

Clay minerals: A group of minerals found in the soil's clay fraction, generally formed within the soil by the decomposition of primary minerals in the parent rocks and their recomposition into secondary minerals. The most prevalent minerals in the clay fraction of temperate-region soils are the alumino-silicates.

Clothesline effect: The passage of warm air through a stand of plants growing in a field, especially where the density of vegetation is sparse, thus affecting the rate of evapotranspiration.

Cohesion: The internal mutual bonding of like molecules or particles of a particular substance, imparting strength to a body composed of that substance.

Compaction of soil: Densification of an unsaturated soil by the reduction of fractional air volume. Compaction can take place either under a static load or transient vibration or trampling by animals and machines.

Conditioners of soil: Polymeric substances that, when added to the soil, can promote aggregate stability by gluing particles together within the aggregates as well as by coating the surfaces of aggregates.

Consolidation of soil: The process by which a saturated body of soil is compressed, resulting in reduction of pore volume by expulsion of water. Tests of consolidation are intended to help predict the gradual settlement, or subsidence, of soil under prolonged loading such as that due to a heavy structure.

Contact angle of a liquid on a solid: A measure of the attraction or affinity (adhesion) between a liquid (e.g., water) and a solid surface (e.g., soil particles). A contact angle of zero implies that a drop of the liquid placed on the solid would spread over it (in which case the solid is considered "hydrophilic"). At the opposite extreme, a contact angle of 180 degrees implies a non-wetting of the surface (i.e., it is "hydrophobic").

Continuity equation: A statement, in mathematical form, that for a conserved substance (i.e., one, such as water, that is neither created nor destroyed in the soil), the time rate of change of content must equal the negative rate of the change of flux with distance (i.e., the amount per unit time entering minus the amount exiting a volume element of soil).

Convection: The mass flow of a fluid in response to a driving force acting on the entire body of fluid (such as a pressure or gravity gradient or their combined force).

Coulomb's law: The frictional resistance toward a tangential stress tending to slide one planar body (or surface) against another is proportional to the normal stress pressing the bodies together.

Crop-response function to irrigation: The functional relation between the yield of a crop and the amount of water applied to, or used by, the crop. A typical crop-response curve is sigmoid in shape, its slope at each point representing the incremental yield response to a unit quantity of added water.

Crust formation: The development of a dense, hard, and relatively impermeable layer at the soil surface, due to the breakdown of soil aggregates. Such a crust may inhibit aeration, infiltration, and the emergence of germinating seedlings.

Cylindrical (triaxial) shearing test: A method for measuring the shear strength of soil samples, in which the failure surface is not predetermined but allowed to form within the specimen as successive combinations of lateral and axial stresses are applied to a series of samples of the same soil.

Damping depth (thermal): The depth in the soil at which the temperature amplitude decreases to a fraction 1/e (1/2.718 = 0.37) of the amplitude at the soil surface. It is related to the thermal properties of the soil (volumetric heat capacity and thermal conductivity) as well as to the frequency of the temperature fluctuation considered.

Darcy's law: The rate of flow of a fluid (e.g., water) in a porous body (volume of fluid per unit time per unit area perpendicular to the flow direction) is proportional to the hydraulic gradient. The constant of proportionality is called the hydraulic conductivity, which depends on the properties of the fluid as well as of the porous medium.

Deflation: A process of wind erosion, by which the loose top layer of the soil is blown away, generally following the denudation and pulverization of the soil in arid regions.

Degradation of soil: The deterioration of soil productivity by such processes as erosion, organic matter depletion, leaching of nutrients, compaction, breakdown of aggregates, waterlogging, and/or salination.

Density of soil solids: Mass of soil solids per unit volume of those solids (excluding the volume of the voids).

Desertification: The degradation of land in semiarid regions, caused by human action including the destruction of the vegetative cover by overgrazing and tillage, depletion of soil organic matter and nutrients, deterioration of the soil's aggregated structure, compaction of the soil by trampling animals and machines, and baring the surface to erosion by water and wind. An area that was originally a rich and biodiverse ecosystem may thus come to resemble a desert. Where irrigation is practiced injudiciously, waterlogging and salination may further exacerbate the degradation of the land.

Dew point: The temperature at which the vapor pressure of the atmosphere reaches a point of saturation and the vapor begins to condense into droplets of liquid water.

Diffuse double layer: A phenomenon occurring at and near the surfaces of hydrated clay particles, consisting of the negative charges of the clay surfaces and the counter charges of the swarm of positive ions (cations) in the surrounding solution phase.

Diffusion: The movement of molecules of different substances relative to one another (due to their random thermal agitation) in non-homogeneous gaseous or liquid medium, tending toward equalizing the concentrations of all components throughout the medium.

Dilatancy: The tendency of a body under shearing stress to expand as it deforms. This property is typical of sandy soils, as a result of the sliding and rolling of particles over one another along the shearing plane.

Disperse system: A system in which at least one of the phases is subdivided into numerous minute particles, which together present a very large interfacial area per unit volume. Examples are colloidal sols, gels, emulsions, aerosols, and soils.

Drainage: Outflow of water from the soil, either naturally or artificially. Surface drainage refers to the downslope flow of excess water from the soil surface. Subsurface (groundwater) drainage refers to the removal of water from within or below the soil, generally involving the lowering of the water table.

Dryland farming: The production of crops in semiarid regions on the basis of natural rainfall alone, without irrigation. Its proper practice calls for measures to conserve both soil and scarce water by promoting infiltration and minimizing evaporation through the use of contour planting, mulching, and weed control.

Ecosystem: A community of plants, animals, and microorganisms that interact within a shared habitat and that are influenced by, and influence, its cycles of energy and materials.

Elasticity: The property of a body to deform instantaneously under an applied stress and in proportion to it, to retain the new form as long as the stress is maintained, then to regain the original dimensions when the stress is released.

Electrical resistance blocks: Porous bodies containing electrodes, placed in the soil to equilibrate with soil moisture. The electrical resistance measured within such blocks can be calibrated against independent measurements of soil wetness or soil water potential.

Emissivity coefficient: The property of a surface to emit radiation. Emissivity is unity for a perfect emitter, called a "blackbody," and smaller than unity for most surfaces.

Energy balance of a field: An expression of the classical law of energy conservation, stating that the sum of the energy inputs minus the energy outputs of a field (including radiant energy, sensible heat, latent heat) over a specified period must equal the change in energy content (both thermal and chemical).

Erodibility of soil: The vulnerability of a soil to erosion, which depends on texture, structure (size and stability of aggregates), looseness and roughness of the soil surface.

Erosion: The detachment of particles from the soil surface and their transport by running water or by wind.

Erosivity: The erosive power of rainfall, running water, or wind. In the case of rainfall, it depends on the kinetic energy of raindrops per unit area and time (i.e., on rainfall intensity as well as duration). In the case of running water, it depends on the velocity and turbulence, as well as on the particles carried in suspension. In the case of wind, it depends on wind speed and direction.

Eutrophication: The enrichment of water bodies with nutrients. In surface waters, it may lead to the proliferation of algae and to the oxygen-deprivation of fish species. In shallow aquifers, it may render the water unsuitable for drinking.

Evaporativity: The power of the atmosphere to vaporize and remove water from a moist surface, at a rate determined by such meteorological variables as intensity of solar radiation, ambient temperature and humidity, and wind speed. As an approximation, evaporativity is considered to be practically independent of the properties of the surface itself. In fact, however, the latter properties (including color and reflectivity, as well as roughness) always interact with the externally imposed meteorological conditions in determining the actual rate of evaporation.

Evapotranspiration: The transfer of water from its liquid state in the soil-plant system to the atmosphere as vapor. It includes evaporation from the soil surface and transpiration from the canopies of plants. The two processes are simultaneous and interactive and are difficult to quantify separately. Their relative rates depend on the density of plant cover.

Exchangeable sodium percentage (ESP): The percentage of the soil's exchange capacity that is occupied by sodium ions. When ESP exceeds 15%, the soil is considered to be an alkali soil.

Failure stress: The value of stress at which a stressed body collapses or fractures, i.e. loses its structural integrity.

Failure: The reaction of a body to stresses that exceed its strength, generally leading to loss of cohesion or structural integrity by such modes as fracturing, slumping, plastic yielding or apparent liquefaction.

Fertigation: The injection of soluble fertilizers into the water supply so as to fertilize a crop while irrigating it. Fertigation is often practiced with techniques of microirrigation.

Field capacity: An empirical measurement supposed to represent the soil profile's ability to retain water after the process of internal drainage has ceased. It is usually measured about two days after an infiltration event. The measured value depends on the initial depth of wetting and on the texture and layering of the profile.

Fingering: The appearance of protrusions in the normally smooth wetting front during infiltration. Such "fingers" may propagate downward into the subsoil and carry plumes of water and solutes while bypassing the greater volume of the soil matrix. The phenomenon is also called "wetting-front instability" and "unstable flow."

Flocculation and dispersion: The tendency of clay particles in an aqueous suspension to either clump together into flocs or to separate from one another and thus disperse in the fluid medium, depending on the composition and concentration of the electrolytes in the ambient solution.

Fluidity: The ratio between the density and dynamic viscosity of a fluid. In the case of flow in a porous medium, fluidity is the ratio of the hydraulic conductivity to the intrinsic permeability of the medium. It depends on the composition and temperature of the fluid.

Fourier's law: The law governing heat conduction, stating that the flux of heat in a homogeneous body is in the direction of, and proportional to, the temperature gradient, with a coefficient of proportionality known as "thermal conductivity."

Greenhouse effect: Inhibition of the atmospheric transmission of outgoing thermal radiation from the earth, due to the presence of certain gases in the atmosphere. The principal "greenhouse gas" is water vapor. Another important one is carbon dioxide, the concentration of which has been increasing due to forest clearing, cultivation of formerly virgin soils, and – especially – the burning of fossil fuels (coal, petroleum, natural gas).

Greenhouse gas emissions: The emission from the soil into the atmosphere of gases (such as carbon dioxide, methane, and nitrous oxide) that inhibit the release into outer space of heat from the earth, thus causing global warming. Agricultural sources may constitute as much as 20% of the total anthropogenic emissions of those greenhouse gases. Such emissions can be reversed by judicious soil management, with the soil as a reservoir capable of sequestering (rather than emitting) carbon.

Groundwater: The water contained in the saturated portion of the soil or the underlying porous formations, generally at a pressure greater than atmospheric.

Heat capacity, volumetric: The change in heat content of a unit bulk volume of a body per unit change in temperature. In the soil, it can be estimated from the sum of the heat capacities of the various constituents (solids, water, and air), each weighted according to its respective volume fraction.

Heat conduction: The propagation of heat within a body by internal molecular motion, driven by a temperature gradient. The quantitative expression of heat conduction is the equation known as Fourier's Law.

Heat flux plates: Thin, flat plates of constant thermal conductivity that can be placed in the soil at various depths to allow precise measurement of heat flux through them.

Heat of wetting: The heat released by a unit mass of initially dry soil when immersed in water. It is related to the soil's specific surface (i.e., the content and composition of the clay fraction).

Homogeneity: Uniformity of properties in all parts of a system. Such a condition may be safely assumed only in the case of a small volume or sample of soil. In the open field, the soil tends to be inhomogeneous, i.e., spatially variable.

Hooghoudt equation: An equation relating the height of the water-table between drains to depth of the drains and the distance between them, the hydraulic conductivity of the soil, and the rate of recharge by percolating water.

Hooke's law: The deformation (strain) of a body under stress is proportional to the stress applied to it. This law pertains to elastic bodies. The constant of proportionality between stress and strain is known as "Young's modulus."

Humid regions: Regions in which the amount of rainfall equals or exceeds the potential evapotranspiration throughout the growing season. In such regions, plants may suffer from excess water in the soil and impeded aeration, but seldom from shortage of water.

Humus: The fraction of the organic matter in the soil that remains relatively stable after the initial decomposition of the more labile components of plant and animal residues. It is usually dark-colored, found mostly in the surface zone (the A horizon), and it contributes to the stabilization of soil aggregates.

Hydraulic conductivity: The ratio between the flux of water through a porous medium and the hydraulic gradient (i.e., the flux per unit hydraulic gradient).

Hydraulic diffusivity: The ratio between the flux of water and the gradient of soil wetness. This term is somewhat misleading, since it does not refer to diffusion as such but to convection. The term is taken from the analogy to the diffusion equation (Fick's law), stating that the rate of diffusion is proportional to the concentration gradient.

Hydraulic head: The sum of the pressure head (hydrostatic pressure relative to atmospheric pressure) and the gravitational head (elevation relative to a reference level). The gradient of the hydraulic head is the driving force for water flow in porous media.

Hydrodynamic dispersion: The tendency of a flowing solution in a porous medium that is initially permeated with a solution of different composition to disperse, due to the non-uniformity of the flow velocity in the conducting pores. The process is somewhat analogous to diffusion, though it is a consequence of convection.

Hydrogen bonding: The electrostatic attraction between molecules due to the presence of hydrogen atoms bound to negatively charged atoms. A primary example is the mutual attraction of water molecules.

Hydrology: The study of the dynamic state, cyclic movements, transformations, spatial and temporal distribution, and interactions of the Earth's water in the various domains in which it occurs – oceans, continents, and the atmosphere, particularly as it affects living systems in general and human life in particular.

Hydrophobicity: The tendency of a surface or of an assemblage of particles to resist wetting by water, caused by the obtuse contact angle between water and the surface. Most mineral soils have a natural affinity for water (i.e., are hydrophilic). The surface zone of some soils, however, may be hydrophobic (water-repellent), especially when coated with oily resins, and tends to shed rather than absorb water applied to it.

Hysteresis: The dependence of an equilibrium state on the direction of the process leading up to it. A prime example is the difference between the soil moisture characteristic curves measured in sorption (i.e., as an initially dry soil is gradually wetted) and in desorption (i.e., as an initially saturated soil is gradually dried).

Infiltrability: The flux of infiltration resulting when water at atmospheric pressure is applied and maintained at the soil surface. Infiltrability is relatively high when water is first applied to a dry soil, but diminishes in time and, if the soil is deep, it eventually tends asymptotically toward a constant rate called the soil's "steady infiltrability."

Infiltration: Water entry into the soil, generally by downward flow through all or part of the soil surface.

Infiltrometer: A device to measure the infiltration rate and hence the infiltrability of a soil in situ, generally by applying a shallow head of water to the soil surface inside a ring. A special type is the "disc infiltrometer," which introduces water to the soil surface at subatmospheric pressure. Another type is the "sprinkling infiltrometer," which applies water to the surface by a controlled rate of sprinkling.

Internal drainage: The post-infiltration movement of soil moisture in an initially deeply wetted or saturated profile, or in the presence of a high water table.

Ion exchange: The adsorption of ions (particularly cations, but in some cases anions as well) by the surfaces of clay particles, and their exchange with cations in the surrounding aqueous solution.

Irrigation: The practice of supplying water to the root zone of a crop so as to permit farming in arid regions and to offset drought in semiarid or semihumid regions. The potential productivity of irrigated land generally exceeds that of unirrigated ("rainfed") land, due to increased yields per crop and to the possibility of multiple cropping.

Isotropy: Each point in the system having equal permeability in all directions. In contrast, an anisotropic system may have a higher permeability or conductivity in one direction than in another. Anisotropy is typically due to the layering of the soil or to a pattern of micropores or macropore with a distinct directional bias.

Kriging: A method of estimating the unmeasured value of a physical property for any location by means of a geostatistical interpolation procedure in which the known values of surrounding locations are analyzed spatially so as to minimize the overall variance.

Laminar flow: The flow of a fluid such that adjacent laminae (layers), moving at different velocities, slide over one another smoothly, without creating eddies.

Laterites: Soils typically occurring in the humid tropics, in the genesis of which silica dissolves and is leached while iron and aluminum oxides accumulate and impart a typical red color to the soil. Chunks dug up from laterites and dried in the sun tend to harden and form bricks, hence the name derived from the Latin word *later*, meaning "brick."

Leaching fraction: The additional fractional volume of water that must be applied per unit area (in excess of seasonal evapotranspiration) in order to leach the residual salts and to prevent their accumulation in the root zone.

Leaching: The removal of solutes from the soil by the downward percolation of water. This process is desirable to the extent that it removes excess salts, but may be harmful to soil productivity where it removes essential nutrients.

Loam: A soil of intermediate texture that contains a balanced proportion of the textural fractions. As such, a loam is often (but not always) considered to be the optimal soil for plant growth and for agriculture.

Macropores: Relatively large pores in the soil, generally several millimeters in width, occurring between aggregates or as cracks or fissures in the soil matrix. Some macropores are biogenic (e.g., channels formed by decaying roots or by earthworms). When open to the surface, macropores serve as pathways for the rapid infiltration of water. They may also allow evaporation to dry the soil to considerable depth.

Matrix of the soil: The assemblage and arrangement of the solid phase (mineral and organic components) constituting the body of the soil, and the interstices (pores) contained within it.

Mechanical analysis: A procedure in which the array of particle sizes of a soil is determined quantitatively, by means of sieving, sedimentation, and/or microscopy.

Meniscus: The curved surface of a liquid inside a capillary pore. The curvature of the meniscus depends on the liquid-solid contact angle as well as on the liquid-to-gas surface tension and the atmospheric pressure.

Microirrigation: The low-volume, high-frequency (or continuous) application of water to the soil at selected points (rather than over the entire surface), generally via narrow-orifice emitters set in tubes that convey water at relatively low pressure. Microirrigation includes such techniques as drip, trickle, bubbler, and microsprayer irrigation.

Miscible displacement: The displacement of one solution by another inside the soil, when the two solutions are miscible and tend to mix along their contact front, due to both diffusion and hydrodynamic dispersion.

Modulus of rupture test: A procedure to characterize the cohesive strength of dry soil briquettes. This test is applied mainly to the characterization of soil crusts.

Mohr circle: A graphic representation of the values of normal and tangential (shearing) stresses acting within a body at any angle of inclination.

Mohr envelope: If a series of stress states just sufficient to cause failure is imposed on the same material and these are plotted as a set of Mohr circles, the tangent line (envelope) to these circles can be used as a criterion of shearing strength. In soils, this envelope often forms a straight line that is seen to obey Coulomb's law.

Mulch: A layer of loose material on the soil, serving to protect its surface against the direct impact of raindrops and against rapid desiccation by sun and wind. A virgin soil is generally covered with a natural mulch consisting of the remains of dead plants and animal droppings. Special practices are needed, however, to maintain mulch over a cultivated soil. Materials used for the purpose may include straw, wood shavings, paper, plastic sheeting, sprayable asphalt, sand, and even gravel.

Neutron moisture meter: A device to measure the wetness of a soil profile, consisting of a probe that includes a radioactive source of fast neutrons and a detector of slow (thermalized) neutrons. The fast neturons are emitted radially into the soil and are slowed by repeated collisions with the hydrogen nuclei in soil water. The density of the cloud of slowed neutrons around the probe is proportional to the volumetric wetness of the soil.

Oasis effect: The extraction of sensible heat from a warm mass of air flowing over the top of a field and the conversion of this heat to latent heat of evaporation.

Osmosis: Diffusion of molecules or ions of a substance in solution through a barrier having pores of molecular size.

Osmotic Pressure: The pressure that must be applied to a solution that is in contact with pure water via a semipermeable membrane (i.e., permeable to the molecules of water but not to those of the solute) so as to prevent the net diffusion of pure water into the solution through that membrane.

Overland flow: The flow of uninfiltrated rainwater over the surface of the ground. The term is sometimes used synonymously with "runoff," though some soil physicists tend to distinguish between the two terms, using the former to describe unchanneled "sheet" flow toward a stream and the latter term for channeled flow in streams.

Particle size distribution: The continuous array of particle sizes in the soil, typically represented in graphic form as a "particle-size distribution curve."

Ped: A term used by pedologists to describe aggregates formed by natural processes in the course of soil formation, in contrast with the aggregates occurring in cultivated soils, which are referred to as "clods."

Penetrometer: An instrument for assessing soil strength in the field. It consists of a narrow probe driven into the soil. The soil's resistance to penetration is measured in terms of the energy expended per unit depth.

Penman equation: An equation based on the simultaneous solution of the heat and mass balance equations, designed to estimate the potential evapotranspiration from a smooth, well-watered field with a dense, low crop.

Penman-Monteith equation: An extension of the Penman equation, estimating the rates of heat and vapor transfers from less-than-saturated surfaces. Specifically, this form of the equation accounts for the existence of "canopy resistance," due to the physiological restriction of transpiration by the crop (the so-called "stomatal control mechanisms").

Perched groundwater: An accumulation of water that is under positive pressure, at some depth in or under the soil profile, resting on a relatively impermeable layer that lies above the general (regional) water table.

Percolation: The downward flow of water through saturated or nearly saturated layers of the soil profile, generally due to gravity (i.e., where suction gradients are negligible).

Permeability (intrinsic): A property of a porous medium affecting fluid flow. It is presumed to be independent of the properties of the permeating fluid. Permeability is proportional to the hydraulic conductivity, with the proportionality constant being the ratio of viscosity to the product of density and gravitational acceleration. This definition assumes that the fluid and the medium do not interact is such a way as to change the properties of either.

Planck's law: The intensity distribution of radiation (energy flux in a given wavelength range) emitted from a body as a function of absolute temperature.

Plow pan: A compact layer formed in cultivated soils, just below the tilled topsoil, as a result of the pressures exerted by tillage implements and tractor wheels.

Poiseuille's law: The volume-rate of laminar flow through a narrow (capillary) tube is inversely proportional to the viscosity and directly proportional to the pressure drop per unit distance and to the fourth power of the tube radius.

Porosity: Fractional volume of voids (pores) in the bulk volume of the soil.

Potential evapotranspiration: The rate of evapotranspiration from a dense stand of an actively growing herbaceous crop (e.g., grass) that is well endowed with water. As such, potential evapotranspiration represents the maximal, weather-determined, rate of evaporation from a vegetated surface. It depends on such meteorological variables as solar radiation, air temperature and humidity, and wind speed.

Pressure-plate apparatus: An instrument to measure the soil moisture characteristic in the range of 1 to 15 or more bars of tension. Pressurized air is used to extract water incrementally from initially saturated soil samples resting on a porous plate.

Profile of the soil: A vertical section of the soil, from the surface through the soil's various horizons, down to the underlying parent material.

Puddling: The mechanical manipulation (e.g., kneading) of a wet soil, especially of a clayey soil that tends to be plastic when very wet, resulting in the destruction of the soil's aggregates. After drying, a puddled soil typically forms massive, hardened clods.

Radiation balance: The sum of all incoming minus outgoing radiant energy fluxes for a given surface. The radiation balance for a typical soil surface is generally positive during daytime (as the incoming shortwave solar radiation flux exceeds the outgoing longwave radiation flux) and negative during nighttime.

Radiation: The emission of energy in the form of electromagnetic waves from all bodies above 0 degrees Kelvin. The intensity of the emitted radiation is proportional to the fourth power of the body's surface temperature. It also depends on a property of the body called emissivity, according to the Stefan-Boltzmann law.

Redistribution of soil moisture: The post-infiltration movement of soil moisture in a partially wetted (unsaturated) profile in the absence of a shallow water table, during which water moves progressively (albeit at a diminishing rate) from the initially wetted upper part of the profile to the unwetted lower part of the profile and the initially sharp wetting front becomes indistinct.

Redox potential: A quantitative measure of the reducing or oxidizing power (i.e. of the electron activity) of a medium or solution. The redox potential of anaerobic soils is usually determined by measuring the difference in potential between an inert platinum electrode and a reference electrode (such as a saturated calumel or Ag:AgCl solution).

Representative elementary volume: Minimum volume of a soil sample needed to obtain a consistent value of a measured parameter.

Respiration (aerobic): The release of chemical energy from organic matter by living organisms via the process of oxidation. This process is the reversal of photosynthesis, by which green plants transform radiant energy into chemical energy by synthesizing carbohydrates while releasing oxygen.

Reynolds number: A quantitative criterion for the transition from laminar to turbulent flow, equal to the product of mean flow velocity, tube diameter, and fluid density divided by dynamic viscosity. In straight tubes, the critical value is of the order of 1000. That value is reduced greatly when the tube is curved and its diameter varies. In porous media, it is safe to assume that flow remains linear with hydraulic gradient only if the Reynolds number is smaller than unity.

Rheology: A branch of the science of mechanics, describing the deformation of bodies under applied stresses, and more specifically the stress-strain-time relations of deformable bodies. When applies to soil bodies, the science dealing with that set of phenomena is often designated "soil dynamics."

Richards equation: A flow equation combining Darcy's law with the continuity equation, along with the proviso that in an unsaturated soil the hydraulic conductivity is a function of matric potential (or of soil wetness).

Richardson number: A criterion for estimating the relative importance of buoyancy versus frictional forces in producing atmospheric turbulence, hence for assessing the stability or instability of the air over a field. The air temperature profile tends to be stable when the Richardson number is positive, and unstable when it is negative.

Root zone: The part of the soil profile that is penetrated and permeated by plant roots, and that provides them with their water, nutrient, and oxygen requirements.

Runoff inducement: The practice of treating the soil surface on sloping ground so as to reduce infiltrability and increase runoff; as well as of collecting, directing, and utilizing the runoff thus obtained. Also called "water harvesting," this practice has been applied since ancient times by societies in arid regions, to obtain drinking water and irrigation.

Runoff: The volume of water that runs off the soil surface and flows into streams during episodes of intense rainfall or during an irrigation that exceeds the soil's infiltrability.

Salinity of soil: The excessive accumulation of salts in the soil, such that restricts plant growth and causes a decline of soil productivity (in extreme cases, it may even result in soil sterility). Salinity affects plants directly through the reduced osmotic potential of the soil solution and by the toxicity of specific ions such as boron, chloride, and sodium.

Saltation: The bouncy movement of sand particles caused by wind blowing over the surface. Gusts of wind pick up protruding particles and carry them downwind in short spurts. As they fall to the ground at an angle, they strike other particles and bounce them up in sequence.

Sedimentation: The deposition of eroded particles, which takes place where the flowing water or blowing wind that had detached and transported the particles becomes quiescent.

Seepage: The loss of water from a pond or reservoir by percolation into and through the soil. An alternative use of the term is to describe the emergence of water from an exposed stratum of a saturated soil.

Semipermeable membrane: A membrane that permits the passage of solvent molecules but not the molecules or ions of certain solutes. When such a membrane is placed between solutions of different concentration, osmotic effects occur. Naturally occurring semipermeable membranes play a vital role in the physiology of plants and animals. Artificial semipermeable membranes serve in many industrial processes, including the desalination process of reverse osmosis.

Sodium adsorption ratio (SAR): The ratio between the concentration of sodium ions in an aqueous solution and the square root of the combined concentration of calcium and magnesium. This ratio serves to predict the equilibrium exchangeable sodium percentage (ESP) of a soil that is to be irrigated with the given solution.

Soil physics: The branch of soil science that deals with the state and transport of matter and with the state and transformations of energy within the soil, as well as between the soil and the adjacent domains (namely, the atmosphere above and the substrata below).

Soil: The weathered and fragmented outer layer of the earth's terrestrial surface, formed initially through the disintegration and decomposition of rocks by physical and chemical processes and influenced subsequently by the activity and accumulated residues of numerous species of microscopic and macroscopic biota.

Soil-Plant-Atmosphere Continuum: A concept recognizing that the field with all its components – soil, plant, and ambient atmosphere taken together – constitutes a physically integrated, dynamic system in which the various flow processes involving energy and matter occur simultaneously and interdependently like links in a chain.

Spatial variability of properties and processes: The non-uniform distribution of specifiable attributes of an area, whether random or systematic (structured), and the range of their variation expressed by means of statistical (probabilistic) criteria.

Specific heat: The change in heat content of a unit mass of a substance or body per unit change in temperature. Of the various substances constituting the soil, water exhibits the highest specific heat.

Specific surface: The total surface area of particles per unit mass of the particles or per unit volume of the soil, usually expressed as square meters per gram or per cubic centimeter.

Splash erosion: The action of raindrops on bare soil, in which the impacting drops detach soil particles and splash them in all directions but with a net tendency toward a downslope direction. This lashing action initiates the process of water erosion, which is then continued by running water (i.e., particle-laden overland flow and runoff).

Stefan-Boltzmann law: The total energy flux emitted by a radiating body, integrated over all wavelengths, is proportional to the fourth power of the absolute temperature of the body's surface.

Stokes law: An equation relating the velocity of a spherical particle settling in an aqueous suspension, under the influence of gravity, to the particle radius and density, as well as to the viscosity and density of the fluid.

Strain: The ratio of the deformation of a body that is subject to stress (compressive, tensile, or shear) to the body's original dimensions.

Strength: The capacity of a body to withstand stresses without experiencing failure, whether by rupture, fragmentation, collapse, or flow. In quantitative terms, it is the maximal stress a given body can bear without undergoing failure, or the minimal stress that will cause the body to fail.

Stress: A force per unit area acting on a body and tending to deform it by compression, tension, or shear.

Structural stability: The ability of the soil's aggregated structure to resist the forces tending to cause its disruption, such as slaking by water or grinding and compaction by machinery.

Structure of soil: The arrangement and organization of the particles in the soil (i.e., the internal configuration of the soil matrix, or fabric), whether single-grained (unattached particles variously stacked) or associated in more or less stable aggregates. As such, the structure of the soil determines the geometric configuration of the pores and the permeability of the soil to fluids.

Surface storage capacity: The maximal volume of water per unit area that can accumulate over the surface during periods of intense rainfall (i.e., rainfall exceeding the soil's infiltrability). When surface storage capacity is exceeded, overland flow begins.

Surface tension: The resistance of a liquid body to an increase in its surface area. It is due to the cohesion between the liquid molecules and is expressed quantitatively as the force per unit length or the energy per unit area needed to overcome that resistance and to increase the surface of the liquid.

Surface-water excess: The excess of rainfall intensity over the infiltration rate. That excess tends to accumulate in pockets or depressions over the surface, or – if the surface is smooth and sloping – to run off the surface in the form of overland flow.

System: A group of interrelated, interdependent elements constituting an integrated entity; e.g., the soil is a heterogeneous, polyphasic, particulate, and porous system.

Tensiometer: A device to measure the matric potential of soil moisture in situ, consisting of a porous (ceramic) cup filled with water, with a manometer to monitor the pressure of the water in the cup at equilibrium with soil moisture. As soil moisture diminishes, its matric tension increases, hence it draws water from the cup, which in turn registers a subatmospheric pressure, called tension.

Tension-plate assembly: A device for equilibrating an initially saturated soil sample with a known matric tension value, applicable to the tension range of 0 to 1 bar.

Terracing: The practice of building horizontal barriers on the contour, aimed at checking surface runoff and preventing it from accelerating downslope.

Texture of soil: The range of particle sizes in a soil, expressed in terms of the proportions by mass of the fractions known as sand (the coarsest particles), silt (intermediate-size particles), and clay (the finest particles).

Thermal conductivity: The flux of heat (i.e., the amount of heat conducted across a unit crosssectional area in unit time) per unit temperature gradient.

Thermal regime of a soil profile: The variation of soil temperature with time for various depths in the profile.

Thermocouple psychrometer: A device to measure relative humidity of the air phase in the soil. Soil-moisture potential (sum of the matric and osmotic potentials) is related to the logarithm of the relative humidity. The measurement is made by a thermocouple (two junctions of dissimilar metals), one junction in a hollow porous cup embedded in the soil, the other kept in an insulated medium.

Thixotropy: The tendency of bodies composed of certain materials to change their mechanical properties when subjected to abrupt stresses; e.g., the tendency of an apparently solid body of moist clay to liquefy and slump when jarred suddenly.

Threshold hydraulic gradient: The minimal value of hydraulic gradient above which Darcy's law is valid. Below that value the flow rate may be either zero (the water being effectively immobile) or lower than predicted by Darcy's law. The possible existence of a threshold hydraulic gradient in soils is, howver, a controversial issue.

Tillage: Mechanical manipulation of the soil's upper layer by means of implements designed to slice, pulverize, loosen, invert, and/or mix the soil. The aims of tillage are to prepare a seedbed, eradicate weeds, enhance infiltration, and/or shape the surface.

Time-domain reflectometry: A technique for measuring soil wetness, consisting of parallel metal rods inserted into the soil. When a step pulse of electricity is sent through the rods, the received signal is related to the dielectric properties of the medium. Since the dielectric constant of water is about 81 while that of soil solids is 4 to 8 and that of air is about 1, the reading depends mainly on the fractional volume of water present.

Tortuosity: The ratio of the average "roundabout" path of flow in a porous medium to the apparent, or straight, flow path. Tortuosity is thus a dimensionless geometric parameter of porous media, which is always greater than 1 and may exceed 2. The "tortuosity factor" is sometimes defined as the reciprocal of the above ratio.

Transpiration: Evaporation from plants, principally through the open stomates in the leaves. When plants experience water stress, they tend to close their stomates, thereby restricting transpiration. By so doing, however, they also curtail the process of photosynthesis, i.e., the absorption of atmospheric carbon dioxide by the stomates.

Turbulent flow: The internally disordered flow of a fluid such that "packets" of the fluid swirl about in eddies, and more energy is dissipated in internal friction than is the case in laminar flow.

Two-phase flow: The simultaneous flow in the soil of two fluids, either immiscible (e.g., air and water, or petroleum and water) or miscible (e.g., saline and fresh water).

Universal soil loss equation: An empirical equation that considers the rate of erosion as a combined product of rainfall erosivity, soil erodibility, slope steepness or length, surface cover, and management.

Vadose zone: The generally unsaturated soil and the underlying porous strata that overlie the permanent water table. It is also called the "unsaturated zone" or the "aerated zone," in contrast with the saturated zone that lies underneath the water table.

Vapor pressure: The partial pressure of water vapor in the atmosphere. It depends on the pressure and temperature of the atmosphere, as well as on the state of water in a water-containing body (such as soil) at equilibrium with the atmosphere.

Vertisols: Soils rich in expansive clay (e.g., montmorillonite). When subject to alternating cycles of wetting and drying, as in a semiarid region, such soils tend to heave and then to settle and form wide, deep cracks, as well as slanted sheer planes extending deep into the soil profile.

Viscosity: The resistance of a fluid to shear (i.e., to a force causing the adjacent layers of the fluid to slide over each other). That resistance is proportional to the velocity of the shearing. As such, viscosity can be visualized as the fluid's internal friction.

Void ratio: Fractional volume of voids (pores) per volume of soil solids.

Water application efficiency: The net amount of water added to the root zone as a fraction of the amount applied to the field. Surface irrigation methods such as flooding or furrowirrigation may result in runoff and/or in percolation beyond the root zone, thus reducing the application efficiency. Sprinkling irrigation may involve losses due to wind-drift. Microirrigation methods, if well managed, offer the potential for relatively high application efficiency.

Water balance of a field: an equation embodying the law of mass conservation by summing all quantities of water added to, subtracted from, and stored within a given volume of soil (e.g., the root zone of a cropped field or of a natural plant habitat) during a specified period of time.

Water conveyance efficiency: the volume of water arriving at the field as a fraction of the volume taken from some source (say, a reservoir). The difference between those volumes represents the loss of water incurred in conveyance and distribution, such as seepage and evaporation from open (often unlined) canals or leakage from pipes.

Water potential in the soil: A measure of the energy state of water in the soil relative to that of pure water at atmospheric pressure and a standard elevation. At each point, water potential depends on temperature, hydrostatic pressure (being "negative" in unsaturated soil, hence called tension or suction), solute concentration, and relative elevation.

Water Table: The top surface of a body of groundwater, that surface being at atmospheric pressure. All groundwater beneath the water table is at a pressure greater than atmospheric, whereas soil moisture above it is at a pressure smaller than atmospheric (i.e., at a state of tension, or suction).

Water use efficiency: The measure of crop produced (in terms of total vegetative growth or – preferably – in terms of the weight or value of the marketable product), either per unit amount of water applied or per unit amount of water consumed (evaporated and transpired). Unlike other measures of efficiency, this index is not expressed in dimensionless terms as percentage, but as crop mass or value per unit volume of water.

Waterlogged soil: A soil that is nearly saturated with water much of the time, such that its air phase is restricted and anaerobic conditions prevail. In extreme cases of prolonged waterlogging, anaerobiosis occurs, the roots of mesophytes suffer, and chemical reduction processes occur (including denitrification, methangenesis, and the reduction of iron and manganese oxides).

Watershed (catchment): A section of the land surface that drains its surface-water excess (i.e., the excess of rainfall over infiltration) to a specified point along a stream.

Watershed divide: The line delineating the upper edge of a watershed and separating between it and adjacent watersheds.

Wetness: Water content per unit mass of soil solids (mass-wetness); or per unit bulk volume of the soil (volume-wetness).

Wettability: The tendency of a soil to accept water (in contrast with the opposite tendency of some soils to repel water, called "hydrophobicity"). Water applied to a wettable soil typically forms an acute contact angle, and is drawn into the soil's pores.

Wien's law: A statement that the wavelength of maximal radiation intensity emitted by a blackbody is inversely proportional to the absolute temperature of the body's surface.

Windbreaks: Strips of trees or fences built of slats or brush, arranged perpendicularly to the direction of the prevailing wind, aimed at reducing its speed so as to prevent wind erosion and desiccation, as well as mechanical damage to crops and roads by drifting sand and snow. Vegetated windbreaks are also called shelterbelts.

Zonal soil: A mature, residual soil formed under local climatic conditions and reflecting the cumulative and lasting influence of its formative conditions.

BIBLIOGRAPHY

Abdalla, A. M., Hetteriaratchi, D. R. P., and Reece, A. R. (1969). The mechanics of root growth in granular media. *J. Agr. Eng. Res.* 14, 236–248.

Abramopoulos, F., Rosenzweig, C., and Choudhury, B. (1988). Improved ground hydrology calculations for global climate models (GCMs): Soil water movement and evapotranspiration. *J. Climate* 1, 921–91.

Adam, K. M., Bloomsburg, G. L., and Corey, A. T. (1969). Diffusion of trapped gas from porous media, *Water Resour. Res.* 5, 840–849.

Adams, J. E. (1970). Effect of mulches and bed configuration. *Agron. J.* 62, 785–790.

Adams, J. E., and Hanks, R. J. (1964). Evaporation from soil shrinkage cracks. *Soil Sci. Soc. Am. Proc.* 28, 281–284.

Adams, R. McC. (1981). "Heartland of Cities: Surveys of Ancient Settlements and Land Use on the Central Floodplain of the Euphrates." University of Chicago Press, Chicago.

Addiscott, T. M., Whitmore, A. P., and Powlson, D. S. (1991). "Farming, Fertilizers and the Nitrate Problem." CAB International, Wallingford, UK.

Adrian, D. D., and Franzini, J. B. (1966). Impedance to infiltration by pressure build-up ahead of the wetting front. *J. Geophys. Res.* 71, 5857–5862.

Agassi, M., Bloem, D., and Ben-Hur, M. (1994). Effect of drop energy and soil and water chemistry on infiltration and erosion. *Water Resour. Res.* 30, 1187–1193.

Agren, G. I., and Bosatta, E. (1997). "Theoretical Ecosystem Ecology: Understanding Element Cycles." Cambridge University Press, Cambridge, UK.

Ahlfeld, D. P., Dahmani, A., and Ji, W. (1994). A conceptual model of field behavior of air sparging and its implications for application. *Ground Water Monitor. Rev.* Fall, 132–139.

Ahuja, L. R., and Swartzendruber, D. (1992). Flow through crusted soils: analytical and numerical approaches. Adv. Soil Sci. pp. 93–122.

Alexander, M. (1977). "Introduction to Soil Microbiology." Wiley, New York.

Alharthi, A., and Lange, J. (1987). Soil water saturation: Dielectric determination. *Water Resour. Res.* 23, 591–595.

Ambus, P., and Christensen, S. (1994). Measurement of N_2O emission from a fertilized grassland: An analysis of spatial variability. *J. Geophys. Res.* 99, 16549–16555.

American Society for Testing Materials (ASTM). (1956). "Symp. Vane Shear Testing of Soils," Spec. Tech. Publ. 193. Am. Soc. Testing Mater., Philadelphia, PA.

American Society for Testing Materials (ASTM). (1958). "Book of Standards," Part 11, pp. 217–224. Am. Soc. Testing Mater., Philadelphia, PA.

American Society for Testing Materials (ASTM). (1984). "Procedures for Soil Testing." Am. Soc. Testing Mater., Philadelphia, PA.

Amerman, C. R., Hillel, D. I., and Peterson, A. E. (1970). A variable-intensity sprinkling infiltrometer. *Soil Sci. Soc. Am. Proc.* 34, 830–832.

Amoozegar, A., and Warrick, A. W. (1986). Hydraulic conductivity of saturated soils: Field methods. In: Klute, A., ed., "Methods of Soil Analysis, Part 1: Physical and Mineralogical Methods." Monograph No. 9. Am. Soc. Agron., Madison, WI.

Anat, A., Duke, H. R., and Corey, A. T. (1965). Steady upward flow from water tables. Colorado State Univ. Hydrol. Paper No. 7.

Anderson, D. M., and Tice, A. R. (1971). Low-temperature phases of interstitial water in clay water systems. *Soil Sci. Soc. Am. Proc.* 35, 47–54.

Anderson, J. P. E. (1982). Soil respiration. In: Page, A. L., ed., "Methods of Soil Analysis, Part 2." Monograph No. 9. Amer. Soc. Agron., Madison, WI.

Andraski, B. J., and Scanlon, B. R. (2002). Thermocouple psychrometry. In: Dane, J. H., and Topp, G. C., eds., "Methods of Soil Analysis, Part 4: Physical Methods." Soil Science Society of America, Madison, WI.

Andraski, B. J., Mueller, D. H., and Daniel, T. C. (1985). Effects of tillage and rainfall simulation date on water and soil losses. *Soil Sci. Soc. Am. J.* 49, 1512–1517.

Andreae, M. O., and Schimel, D. S. (1989). "Exchange of Trace Gases Between Terrestrial Ecosystems and the Atmosphere." Wiley, New York.

Angus, D. E., and Watts, P. J. (1984). Evapotranspiration — How good is the Bowen-ratio method? In: Sharma, M. L., ed., "Evapotranspiration from Plant Communities." Elsevier, Amsterdam.

Artzy, M., and Hillel, D. (1988). A defense of the theory of progressive soil salinization in ancient southern Mesopotamia. *Geoarchaeology* 3, 235–238.

Aslyng, H. C. (1963). Soil physics terminology. *Int. Soc. Soil Sci. Bull.* 23, 7.

Awadhwal, N. K., and Singh, C. P. (1985). Mechanical and viscoelastic characteristics of a puddled soil. In: "Proc. Int. Conf. on Soil Dynamics," Vol. 3, pp. 471–480, National Soil Dynamics Laboratory, Auburn, AL.

Aylmore, L. A. G. (1994). Application of computer assisted tomography to soil-plant-water studies: An overview. In: Anderson, S. H., and Hopmans, J. W., Soil Sci. Soc. Am., Madison, WI.

Bachmat, Y., and Elrick, D. E. (1970). Hydrodynamic instability of miscible fluids in a vertical porous column. *Water Resour. Res.* 6, 156–171.

Bagnold, R. A. (1941, 1965). "The Physics of Blown Sand and Desert Dunes." Methuen, London.

Bailey, A. C., and Vanden Berg, G. E. (1968). Yielding by compaction and shear in unsaturated soils. *Trans. Am. Soc. Agric. Eng.* 11, 307–311, 317.

Baker, J. L., and Johnson, H. P. (1979). The effect of tillage systems on pesticides in runoff from small watersheds. *Trans. Am. Soc. Agric. Eng.* 22, 554–559.

Baker, J. L., and Johnson, H. P. (1981). Nitrate-nitrogen in tile drainage as affected by fertilization. *J. Environ. Qual.* 10, 519–522.

Baker, J. L., and Laflen, J. M. (1982). Effects of corn residue and fertilizer management on soluble nutrient runoff losses. *Trans. Am. Soc. Agric. Eng.* 25, 344–348.

Baker, J. M., and Spaans, E. J. A. (1994). Measuring water exchange between soil and atmosphere with TDR-microlysimetry. *Soil Sci.* 158, 22–30.

Baker, J. M., and Norman, J. M. (2002). Evaporation from natural surfaces. In: Dane, J. H., and Topp, G. C., eds., "Methods of Soil Analysis, Part 4: Physical Methods." Soil Science Society of America, Madison, WI.

Baker, R. S. (1998). Bioventing systems: a critical review. In Adriano, D. C., Bollag, J.-M., Frankenberger, W., and Sims, R., eds., "Bioremediation of Contaminated Soils." Soil Sci. Soc. Am./Am. Soc. Agron., Madison, WI (in press).

Baker, R. S., and Becker, D. J. (1998). Soil vapor extraction. In: Meyers, R. A., ed., "The Encyclopedia of Environmental Analysis and Remediation." John Wiley & Sons, New York, (in press).

Baker, R. S., and Hillel, D. (1990). Laboratory tests of a theory of fingering during infiltration into layered soils. *Soil Sci. Soc. Am. J.* 54, 20–30.

Baker, R. S., and Hillel, D. (1991). Observations of fingering behavior during infiltration into layered soils. In: Gish, T. J., and Shirmohammadi, A., eds., "Proc. National Symp. on Preferential Flow," Dec. 16–17, 1991, Chicago, IL, pp. 87–99. Amer. Soc. Agri. Eng., St. Joseph, MI.

Baker, R. S., Hayes, M. E., and Frisbie, S. H. (1995). Evidence of preferential vapor flow during in situ air sparging. In: Hinchee, R. E., Miller, R. N., and Johnson, P. C., eds., "*In Situ* Aeration: Air Sparging, Bioventing, and Related Remediation Processes," pp. 63–73. Battelle Press, Columbus, OH.

Ball, B. C., O'Sullivan, M. F., and Hunter, R. (1988). Gas diffusion, fluid flow and derived pore continuity indices in relation to vehicle traffic and tillage. *J. Soil Sci.* 39, 327–339.

Ball, B. C., Horgan, G. W., Clayton, H., and Parker, J. P. (1997). Spatial variability of nitrous oxide fluxes and controlling soil topographic properties. *J. Env. Qual.* 26, 1399–1408.

Ball, B. C., and Schjønning, P. (2002). Air permeability. In: Dane, J. H., and Topp, G. C., eds., "Methods of Soil Analysis, Part 4: Physical Methods." Soil Science Society of America, Madison, WI.

Baluska, F., Ciamporova, M., and Barlow, P. W., eds. (1995). "Structure and Function of Roots." Kluwer Academic Publishers, Dordrecht, The Netherlands.

Bamford, S. J., Parker, C. J., and Carr, M. K. J. (1991). Effects of soil physical conditions on root growth and water use of barley grown in containers. *Soil Tillage Res.* 21, 309–323.

Barber, E. S. (1965). Stress distribution. In: "Methods of Soil Analysis, Part 1," Monograph No. 9. Am. Soc. Agron., Madison, WI.

Barley, K. P. (1962). The effect of mechanical stress on the growth of roots. *J. Exp. Bot.* 13, 95–110.

Barrs, H. D. (1968). Determination of water deficits in plant tissues. In: T. T. Kozlowski, ed., "Water Deficits and Plant Growth," pp. 235–368. Academic Press, New York.

Bear, J. (1969). "Dynamics of Fluids in Porous Media." Elsevier, Amsterdam.

Bear, J., Zaslavsky, D., and Irmay, S. (1968). "Physical Principles of Water Percolation and Seepage." UNESCO, Paris.

Bedaiwy, M. N., and Rolston, D. E. (1993). Soil surface densification under simulated high intensity rainfall. *Soil Technol.* 8, 365–376.

Bekker, M. G. (1960). "Off-the-Road Locomotion: Research and Development in Terramechanics." Univ. Michigan Press, Ann Arbor, MI.

Bekker, M. G. (1961). Mechanical properties of soil and problems of compaction. *Trans Am. Soc. Agr. Eng.* 4, 231–234.

Belmans, C., Feyen, J., and Hillel, D. (1979). An attempt at experimental validation of macroscopicscale models of soil moisture extraction by roots. *Soil Sci.* 127, 174–186.

Ben Porath, A., and Baker, D. N. (1990). Taproot restriction effects on growth, earliness, and dry weight partitioning of cotton. *Crop Sci.* 30, 809–814.

Ben-Hur, M., and Agassi, M. (1997). Predicting interrill erodibility factor from measured infiltration rate. *Water Resour. Res.* 33, 2409–2416.

Ben-Hur, M., Stern, R., van der Merve, A. J., and Shainberg, I. (1992). Slope and gypsum effects on infiltration and erodibility of dispersive and nondispersive soils. *Soil Sci. Soc. Am. J.* 56, 1571–1576.

Bennie, A. T. P. (1991). Growth and mechanical impedance. In: Waisel, Y., Asher, A., and Katkafi, U., eds., "Plant Roots: The Hidden Half." Marcel Dekker, New York.

Berkowitz, B., Bear, J., and Braester, C. (1988). Continuum models for contaminant transport in fractured porous formations. *Water Resour. Res.* 24, 1225–1236.

Betrand, A. R. (1967). Water conservation through improved practices. In: "Plant Environment and Efficient Water Use." Am. Soc. Agron., Madison, WI.

Beven, K. J., and Clarke, R. T. (1986). On the variation of infiltration into a homogeneous soil matrix containing a population of macropores. *Water Resour. Res.* 22, 383–388.

Beven, K. J., and Germann, P. (1982). Macropores and water flow in soils. *Water Resour. Res.* 18, 1311–1325.

Biggar, J. W., and Nielsen, D. R. (1976). Spatial variability of the leaching characteristics of a field soil. *Water Resour. Res.* 12, 78–84.

Binley, A., Elgy, J., and Beven, K. J. (1989). A physically based model of heterogeneous hillslopes. 1. Runoff prediction. *Water Resour. Res.* 25, 1219–1226.

Bird, R. B., Stewart, W., and Lightfoot, E. (1960). "Transport Phenomena." Wiley, New York.

Birkeland, P. W. (1974). "Pedology, Weathering, and Geomorphological Research." Oxford Univ. Press, London and New York.

Birkholzer, J., and Tsang, C. F. (1997). Solute channeling in unsaturated heterogeneous porous media. *Water Resour. Res.* 23, 2221–2238.

Birkle, D. E., Letey, J., Stolzy, L. H., and Szuszkiewicz, T. E. (1964). Measurement of oxygen diffusion rates with the platinum microelectrode. *Hilgardia* 35, 555–556.

Bishop, A. W., and Henkel, D. J. (1964). "The Measurement of Soil Properties in the Triaxial Test." Arnold, London.

Black, C. A., ed. (1965). "Methods of Soil Analysis," Part 1. Am. Soc. Agron., Madison, WI.

Black, C. A. (1993). "Soil Fertility Evaluation and Control." Lewis Publishers, Boca Raton, FL.

Black, T. A., Thurtell, G. W., and Tanner, C. B. (1968). Hydraulic load-cell lysimeter, construction, calibration, and tests. *Soil Sci. Soc. Am. Proc.* 32, 632–639.

Blackwell, P. S. (1983). Measurement of aeration in waterlogged soils. J. Soil Sci. 34: 271–285.

Blanchart, E. (1992). Restoration by earthworms of the macroaggregate structure of a destructed savanna soil under field conditions. Soil Biol. Biochem. 24: 1587–1594.

Blaney, H. F., and Criddle, W. D. (1950). Determining water requirements in irrigated areas from climatological and irrigation data. USDA Soil Conservation Service Tech. Paper No. 96.

Blevins, R. L., and Fry, W. W. (1992). Conservation tillage: An ecological approach to soil management. *Adv. Agron.* 51, 33–78.

Boeker, E., and van Grondelle, R. (1995). "Environmental Physics." John Wiley & Sons, Chichester, England.

Boggs, S. (1995). "Principles of Sedimentation and Stratigraphy." Prentice Hall, Englewood Cliffs, N. J.

Bohn, H. L., McNeal, B. L., and O'Connor, G. A. (1979). "Soil Chemistry." John Wiley and Sons, New York.

Bolt, G. H. (1976). Soil physics terminology. *Bull. Int. Soc. Soil Sci.* 49, 26–36.

Bolt, G. H., and Bruggenwert, M. G. M., eds. (1976). "Soil Chemistry." Elsevier, Amsterdam.

Bolt, G. H., and Frissel, M. J. (1960). Thermodynamics of soil moisture. *Neth. J. Agr. Sci.* 8, 57–78.

Boltzmann, L. (1894). Zur integration des diffusiongleichung bei variabeln diffusions coefficienten. *Ann. Phys.* 53, 959–964.

Bond, J. J., and Willis, W. O. (1969). Soil water evaporation: Surface residue rate and placement effects. *Soil Sci. Soc. Am. Proc.* 33, 445–448.

Bonnifield, P. (1979). "The Dust Bowl: Men, Dirt, and Depression." University of New Mexico Press, Albuquerque.

Bootlink, H. W. G., and Bouma, J. (2002). Suction crust infiltrometer. In: Dane, J. H., and Topp, G. C., eds., "Methods of Soil Analysis, Part 4: Physical Methods." Soil Science Society of America, Madison, WI.

Bossert, I. D., and Compeau, G. C. (1995). Cleanup of petroleum hydrocarbon contamination in soil. In: Young L. Y., and Cerniglia, C. E., eds., "Microbial Transformation of Toxic Organic Chemicals." Wiley, New York.

Bouma, J. (1981). Soil morphology and preferential flow along macropores. *Agric. Water Manage.* 3, 235–250.

Bouma, J., Hillel, D. I., Hole, F. D. and Amerman. C. R. (1971). Field measurement of unsaturated hydraulic conductivity by infiltration through artificial crust. *Soil Sci. Soc. Am. Proc.* 32, 362–364.

Bouwer, H. (1961). A double tube method for measuring hydraulic conductivity of soil in situ above a water table. *Soil Sci. Soc. Am. Proc.* 25, 334–342.

Bouwer, H. (1962). Field determination of hydraulic conductivity above a water table with the double tube method. *Soil Sci. Soc. Am. Proc.* 26, 330–335.

Bouwer, H. (1978, 1986). "Groundwater Hydrology." McGraw-Hill, New York.

Bouwer, H. (1986). Intake rate: Cylinder infiltrometer. In: Klute, A., ed., "Methods of Soil Analysis," Monograph No. 9. Am. Soc. Agron., Madison, WI.

Bouwer, H. (1987). Effect of irrigated agriculture on groundwater. *J. Irrig. Drain. Eng. ASCE* 113, 4–15.

Bouwman, A. F., ed. (1990). "Soils and the Greenhouse Effect." Wiley, Chichester, England.

Bowen, H. D. (1981). Alleviating mechanical impedance. In: Arkin, G., and Taylor, H. M., eds., "Modifying the Root Environment." Amer. Soc. Agric. Engineers, St. Joseph, MI.

Boyer, J. S. (1995). "Measuring the Water Status of Plants and Soils." Academic Press, San Diego.

Boyle, M. (1988). Radon testing of soils. *Environ. Sci. Tech.* 22, 1397–1399.

Bradford, J. M. (1986). Penetrability. In: Klute, A., ed., "Methods of Soil Analysis, Part 1," Monograph No. 9. Am. Soc. Agron., Madison, WI.

Bradford, J. M., and Huang, C. (1993). Comparison of interrill soil loss for laboratory and field procedures. *Soil Technol.* 6, 145–156.

Bradford, J. M., and Peterson, G. A. (2000). Conservation tillage. In: Summer, M. E., ed., "Handbook of Soil Science." CRC Press, Boca Raton, FL.

Bradford, J. M., Farrell, D. A., and Larson, W. E. (1973). Mathematical evaluation of factors affecting gully stability. *Soil Sci. Soc. Am. Proc.* 37, 103–107.

Bradford, J. M., Remley, P. A., Ferris, J. E., and Santini, J. B. (1986). Effect of soil surface sealing on splash from a single waterdrop. *Soil Sci. Soc. Am. J.* 50, 1547–1552.

Bradford, J. M., Ferris, J. E., and Remley, P. A. (1987). Interrill soil erosion processes, I. Effect of surface sealing on infiltration, runoff, and soil loss. *Soil Sci. Soc. Am. J.* 51, 1566–1571.

Brady, N. C. (1990). "The Nature and Properties of Soils." Macmillan, New York.

Brandon, J. R., Wertzel, T. T., Perumpral, J. V., and Woeste, F. E. (1986). Shear rate effects on strength parameters of agricultural soils, Paper 86–1044. Amer. Soc. Agr. Engineering, St. Joseph, MI.

Bras, R. L. (1990). "Hydrology: An Introduction to Hydrologic Science." Addison Wesley, Reading, MA.

Bremner, J. M., and Blackmer, A. M. (1982). Composition of soil atmosphere. In: Page, A. L., ed., "Methods of Soil Analysis, Part 2," Monograph No. 9. Am. Soc. Agron., Madison, WI.

Bresler, E. (1972). Control of soil salinity. In: Hillel, D., ed., "Optimizing the Soil Physical Environment Toward Greater Crop Yields," pp. 102–132. Academic Press, New York.

Bresler, E. (1973). Simultaneous transport of solutes and water under transient unsaturated flow conditions. *Water Resour. Res.* 9, 975–986.

Bresler, E. (1978). Theoretical modeling of mixed-electrolyte solution flows for unsaturated soils. *Soil Sci.* 125, 196–203.

Bresler, E. (1991). Soil spatial variability. In: Hanks, J., and Ritchie, J. T., eds., "Modeling Plant and Soil Systems," Monograph No. 31. Am. Soc. Agron., Madison, WI.

Bresler, E., and Dagan, G. (1979). Solute dispersion in unsaturated heterogeneous soil at field scale. II. Applications. *Soil Sci. Soc. Am. J.* 43, 467–472.

Bresler, E., and Dagan, G. (1983). Unsaturated flow in spatially variable fields. *Water Resour. Res.* 19, 421–428.

Bresler, E., and Hanks, R. J. (1969). Numerical method for estimating simultaneous flow of water and salt in unsaturated soils. *Soil Sci. Soc. Am. Proc.* 33, 827–832.

Bresler, E., Kemper, W. D., and Hanks, R. J. (1969). Infiltration, redistribution, and subsequent evaporation of water from soil as affected by wetting rate and hysteresis. *Soil Sci. Soc. Am. Proc.* 33, 832–840.

Bresler, E., McNeal, B. L., and Carter, D. L. (1982). "Saline and Sodic Soils: Principles-Dynamics-Modeling." Springer-Verlag, Berlin.

Bridge, B. J., and Collis-George, N. (1973). An experimental study of vertical infiltration into a structurally unstable swelling soil with particular reference to the infiltration throttle. *Aust. J. Soil Res.* 11, 121–132.

Briggs, L. J., and Shantz, H. L. (1921). The relative wilting coefficient for different plants. *Bot. Gaz. (Chicago)* 53, 229–235.

Bristow, K. L. (2002). Thermal conductivity. In: Dane, J. H., and Topp, G. C., eds., "Methods of Soil Analysis, Part 4: Physical Methods," Soil Science Society of America, Madison, WI.

Bristow, K. L., Campbell, G. S., and Calissendorff, K. (1993). Test of a heat-pulse probe for measuring changes in soil water content. *Soil Sci. Soc. Am. J.* 57, 930–934.

Brooks, R. H., and Corey, A. T. (1966). Properties of porous media affecting fluid flow. *Proc. Am. Soc. Civ. Eng., J. Irrigation Drainage Div.* IR2, 61–68.

Brown, R. (1991). "Fluid Mechanics of the Atmosphere." Academic Press, San Diego.

Browning, G. M. (1950). Principles of soil physics in relation to tillage. *Agr. Eng.* 31, 341–344.

Bruce, R. R., and Luxmoore, R. J. (1986). Water retention: Field methods. In: Klute, A., ed., "Methods of Soil Analysis, Part 1: Physical and Mineralogical Methods," Monograph No. 9. Am. Soc. Agron., Madison, WI.

Brunauer, S., Emmett, P. H., and Teller, E. (1938). Adsorption of gases in multimolecular layers. *J. Am. Chem. Soc.* 60, 309–319.

Brustkern, R. L., and Morel-Seytoux, H. J. (1970). Analytical treatment of two-phase infiltration. *J. Hydraul. Div. ASCE* 96, 2535–2548.

Bryan, R. B., and Luk, S. H. (1981). Laboratory experiments on the variation of soil loss under simulated rainfall. *Geoderma* 26, 245–265.

Bryan, R. B., ed. (1987). "Rill Erosion: Processes and Significance," Supplement No. 8. Catena, Cremlingen, Germany.

Buckingham, E. (1904). Contributions to our knowledge of the aeration of soils, U.S. Bur. Soils Bulletin 25.

Bucks, D. A., Nakayama, F. S., and Warrick, A. W. (1982). Principles, practices, and potentialities of trickle (drip) irrigation. In Hillel, D., ed., "Advances in Irrigation," Vol. 1. Academic Press, New York.

Buol, S. W., Sanchez, P. A. Weed, S. B., and Kimble, J. M. (1990). Predicted impact of climatic warming on soil properties and use. In: Kimball, B. A., Rosenberg, N. J., and Jones, L. H., eds., "Impact of Carbon Dioxide, Trace Gases, and Climate Change on Global Agriculture," Amer. Soc. Agron. Special Publication No. 53, Madison, WI.

Buol, S. W., Southard, R. J., Graham, R. C., and McDaniel, P. A. (2003). "Soil Genesis and Classification." Iowa State Press, Ames, Iowa.

Buras, N. (1974). Water management systems. In: van Schilfgaarde, J., ed., "Drainage for Agriculture," Monograph No. 17. Am. Soc. Agron., Madison, WI.

Burkart, M. R., and Kolpin, D. W. (1993). Hydrologic and land-use factors associated with herbicide and nitrate in near-surface aquifers. *J. Environ. Qual.* 22, 646–656.

Burke, W., Gabriels, D., and Bouma, J. (1986). "Soil Structure Assessment." Balkema, Rotterdam.

Busscher, W. J. (1979). Simulation of infiltration from a continuous and intermittent subsurface source. *Soil Sci.* 128.

Campbell, G. S. (1977). "An Introduction to Environmental Biophysics." Springer-Verlag, New York.

Campbell, G. S. (1985). "Soil Physics with Basic: Transport Models for Soil-Plant Systems." Elsevier, Amsterdam.

Campbell, G. S. (1991). Simulation of water uptake by plant roots. In: Hanks, R. J., and Ritchie, J. T., eds., "Modeling Plant and Soil Systems," Monograph No. 31. Am. Soc. Agron., Madison, WI.

Campbell, G. S., and Campbell, M. D. (1982). Irrigation scheduling using soil moisture measurement: Theory and practice. In: Hillel, D., ed., "Advances in Irrigation," Vol. 1. Academic Press, New York.

Campbell, G. S., and Norman, J. M. (1998). "A Introduction to Environmental Biophysics." Springer-Verlag, New York.

Campbell, G. S., Calissendorff, K., and Williams, J. H. (1991). Probe for measuring soil specific heat using a heat-pulse method. *Soil Sci. Soc. Am. J.* 55, 291–293.

Cannell, R. Q. (1977). Soil aeration and compaction in relation to root growth and soil management. *Adv. Appl. Biol.* 2, 1–86.

Capriel, P., Harter, P., and Stephenson, D. (1992). Influence of management on the organic matter of a mineral soil. *Soil Sci.* 153, 122–128.

Capriel, P., Beck, T., Borchert, H., Gronholz, J., and Zachmann, G. (1995). Hydrophobicity of the organic matter in arable soils. *Soil Biol. Biochem.* 27, 1453–1458.

Carslaw, H. S., and Jaeger, J. C. (1959). "Conduction of Heat in Solids." Oxford Univ. Press, London and New York.

Carson, J. E. (1961). Soil Temperature and Weather Conditions, Report 6470, Argonne National Laboratories, Argonne, IL.

Carter, D. L., Mortland, M. M., and Kemper, W. D. (1986). Specific surface. In: Klute, A., ed., "Methods of Soil Analysis, Part 1: Physical and Mineralogical Methods," Monograph No. 9. Am. Soc. Agron., Madison, WI.

Cary, J. W. (1963). Onsager's relations and the non-isothermal diffusion of water vapor. *J. Phys. Chem.* 67, 126–129.

Cary, J. W. (1964). An evaporation experiment and its irreversible thermodynamics. *Int. J. Heat Mass Transfer* 7, 531–538.

Cassell, D. K., and Klute, A. (1986). Water potential: tensiometry. In: Klute, A., ed., "Methods of Soil Analysis, Part 1: Physical and Mineralogical Methods," Monograph No. 9. Am. Soc. Agron., Madison, WI.

Cedergren, H. R. (1967). "Seepage, Drainage and Flow Nets." Wiley, New York.

Celia, M. A., Bouloutas, E. T., and Zarba, R. L. (1990). A general mass-conservative solution for the unsaturated flow equation. *Water Resour. Res.* 26, 1483–1496.

Chancellor, W. J., ed. (1994). "Advances in Soil Dynamics, Volume 1," Monograph No. 12, Amer. Soc. Agric. Engineers, St. Joseph, MI.

Chancellor, W. J., and Schmidt, R. H. (1962). Soil deformation beneath surface loads. *Trans. Am. Soc. Agr. Eng.* 5, 204–246, 249.

Chepil, W. S. (1962). A compact rotary sieve and the importance of dry sieving in physical soil analysis. *Soil Sci. Soc. Am. Proc.* 26, 4–6.

Childs, E. C. (1969). "An Introduction to the Physical Basis of Soil Water Phenomena." Wiley, New York.

Childs, E. C., and Collis-George, N. (1950). The permeability of porous materials. *Proc. R. Soc. London Ser. A.* 201, 392–405.

Chudnovskii, A. F. (1966). "Fundamentals of Agrophysics." Israel Program for Scientific Translations, Jerusalem.

Cicerone, R. J., and Oremland, R. S. (1988). Biogeochemical aspects of atmospheric methane. *Global Biogeochem. Cycles* 2, 299–327.

Clapp, R. B., and Hornberger, G. M. (1977). Empirical equations for some soil hydraulic properties. *Water Resour. Res.* 14, 601–604.

Clarke, R. T., and Newson, M. D. (1978). Some detailed water balance studies of research catchments. *Proc. R. Soc. London Ser. A.* 363, 21–42.

Clifford, S. M., and Hillel, D. (1986). Knudsen diffusion: gaseous transport in small soil pores. *Soil Sci.* 141, 289–297.

Clothier, B., and Scotter, D. (2002). Unsaturated water transmission parameters. In: Dane, J. H., and Topp, G. C., eds., "Methods of Soil Analysis, Part 4: Physical Methods." Soil Science Society of America, Madison, WI.

Coleman, D. C., and Crossley, J. R., Jr. (1996). "Fundamentals of Soil Ecology." Academic Press, San Diego.

Collins, J. R., Leow, G. H., Luke, B., and White, H. D. (1988). Theoretical investigations of the role of clay edges in brebiotic peptide bond formation. *Origins Life Evol. Biospheres* 18, 107–119.

Collis-George, N., and Yates, D. B. (1990). The first stage of drainage from ponded soils with encapsulated air. *Soil Sci.* 149, 103–111.

Conca, J. L., and Wright, J. V. (1992). Diffusion and flow in gravel, soil, and whole rock. *Appl. Hydrogeol.* 1, 5–24.

Cooper, A. W., Trouse, A. C., and Dumas, W. T. (1969). Controlled traffic in row crop production. In: "Proc. Int. 7th Cong. Agric. Eng. (CIGR)," Baden-Baden, Germany, Section III, pp. 1–6.

Corapciouglu, M. Y., and Baehr, A. L. (1987). A compositional multiphase model for groundwater contamination by petroleum products: I. Theoretical considerations. *Water Resour. Res.* 23, 191–200.

Corey, A. T. (1977). "Mechanics of Heterogeneous Fluids in Porous Media." Water Resources Publications, Fort Collins, CO.

Corey, A. T. and Brooks, R. H. (1975). Drainage characteristics of soils. *Soil Sci. Soc. Am. Proc.* 39, 251–255.

Corey, A. T., and Klute, A. (1985). Application of the potential concept to soil water equilibrium and transport. *Soil Sci. Soc. Am. J.* 49, 3–11.

Cowan, I. R. (1965). Transport of water in the soil-plant-atmosphere system. *J. Appl. Ecol.* 2, 221–229.

Cowan, I. R. and Milthorpe, F. L. (1968). Plant factors influencing the water status of plant tissues. In: Kozlowski, T. T., ed., "Water Deficits and Plant Growth," pp. 137–193. Academic Press, New York.

Crank, J. (1975). "The Mathematics of Diffusion." Oxford Univ. Press, London and New York.

Cuenca, R. H., and Amegee, K. Y. (1987). Analysis of evapotranspiration as a regionalized variable. In Hillel, D., ed., "Advances in Irrigation," Vol. 4. Academic Press, Orlando, FL.

Currie, J. A. (1961). Gaseous diffusion in porous media, Part 3: Wet granular material. *Br. J. Appl. Phys.* 12, 275–281.

Currie, J. A. (1970). In: "Sorption and Transport Processes in Soils," Chem. Ind. Monograph No. 37, 152–171. London.

Currie, J. A. (1975). Soil respiration. In: "Soil Physical Conditions and Crop Production," Tech. Bulletin No. 29. Min. of Ag., Fish., and Food, London.

Dagan, G. (1986). Statistical theory of groundwater flow and transport: Pore to laboratory, laboratory to formation, and formation to regional scale. *Water Resour. Res.* 22, 102–135.

Dagan, G., and Bresler, E. (1979). Solute dispersion in unsaturated heterogeneous soil at field scale: I. Theory. *Soil Sci. Soc. Am. J.* 43, 461–467.

Dalton, F. N., and Rawlins, S. L. (1968). Design criteria for Peltier effect thermocouple psychrometers. *Soil Sci.* 105, 12–17

Dalton, F. N., Herkelrath, W. N., Rawlins, D. S., and Rhoades, J. D. (1984). Time–domain reflectometry: Simultaneous measurement of soil water content and electrical conductivity with a single probe. *Science* 224, 989–990.

Dane, J. H. (1978). Calculation of hydraulic conductivity decreases in the presence of mixed NaCa solutions. *Can. J. Soil Sci.* 58, 145–152.

Dane, J. H., and Klute, A. (1977). Salt effects on the hydraulic properties of a swelling soil. *Soil Sci. Soc. Am. J.* 41, 1043–1049.

Dane, J. H., and Hopmans, J. W. (2002). Water retention and storage. In: Dane, J. H., and Topp, G. C., eds., "Methods of Soil Analysis, Part 4: Physical Methods." Soil Science Society of America, Madison, WI.

Darcy, H. (1856). "Les Fontaines Publique de la Ville de Dijon." Dalmont, Paris.

Dasberg, S., Hillel, D., and Arnon, I. (1966). Response of grain sorghum to seedbed compaction. *Agron. J.* S8, 199–201.

Davies, D. B., Finney, J. B., and Richardson, S. J. (1973). Relative effects of tractor weight and wheel slip in causing soil compaction. *J. Soil Sci.* 24, 399–408.

Davis, J. L., Singh, R., Waller, M. J., and Gower, P. (1983). Time-domain reflectometry and acousticemission monitoring techniques for locating linear failures. Submitted to Office of Research and Development, U. S. Environmental Protection Agency, Cincinnati, OH, Contract No. 6803-3030.

de Boer, J. H. (1953). "The Dynamical Character of Adsorption." Oxford Univ. Press, London and New York.

de Pastrovich, T. L., Baradat, Y., Barthel, R., Chiarelli, A., and Fussell, D. R. (1979). Protection of groundwater from oil pollution. Report 3/79, CONCAWE. The Hague, Netherlands.

de Ploey, J., ed. (1983). "Rainfall Simulation, Runoff, and Soil Erosion." Catena Verlag, Cremlingen, Germany.

de Vries, D. A. (1950). Some remarks on gaseous diffusion in soils. *Trans. 4th Int. Cong. Soil Sci.* 2, 41–43.

de Vries, D. A. (1963). Thermal properties of soils. In: van Wijk, W. R., ed., "Physics of the Plant Environment." North-Holland, Amsterdam.

de Vries, D. A. (1975a). "The Thermal Conductivity of Soil." Med. Landbouwhogeschool, Wageningen, Netherlands.

de Vries, D. A. (1975b). Heat transfer in soils. In: de Vries, D. A. and Afgan, N. H., eds., "Heat and Mass Transfer in the Biosphere," pp. 5–28. Scripta Book Co., Washington, DC.

de Wiest, R. J. M., ed. (1969). "Flow Through Porous Media." Academic Press, New York.

de Wit, C. T. (1958). Transpiration and plant yields. *Versl. Landbouwk. Onderz.* 646, 59–84.

de Wit, C. T. (1965). Photosynthesis of leaf canopies, Agr. Res. Rep. 663, pp. 1–57. PUDOC, Wageningen, The Netherlands.

de Wit, C. T., and Goudriaan, J. (1978). "Simulation of Ecological Processes," 2nd ed. PUDOC, Wageningen, The Netherlands.

de Wit, C. T., and van Keulen, H. (1972). "Simulation of Transport Processes in Soils." PUDOC, Wageningen, Netherlands.

Deacon, E. L., Priestley, C. H. B., and Swinbank, W. C. (1958). Evaporation and the water balance. *Climatol. Rev. Res. Arid Zone Res.* 9–34. UNESCO, Paris.

Decker, W. L. (1959). Variations in the net exchange of radiation from vegetation of different heights. *J. Geophys. Res.* 64, 1617–1619.

Delwiche, C. C., ed. (1981). "Denitrification, Nitrification, and Atmospheric Nitrous Oxide." Wiley, New York.

Demond, A. H. (1988). Capillarity in Two-Phase Liquid Flow of Organic Contaminants in Groundwater. Ph. D. Diss., Dept. of Civil Eng., Stanford Univ., Menlo Park, CA.

Denmead, O. T., and Shaw, R. H. (1962). Availability of soil water to plants as affected by soil moisture content and meteorological conditions. *Agron. J.* 54, 385–390.

Denmead, O. T., Freney, J. R., and Simpson, J. R. (1979). Nitrous oxide emission during denitrification in a flooded field. *Soil Sci. Soc. Am. J.* 43, 716–718.

Deresiewicz, H. (1958). Mechanics of granular matter. *Adv. Appl. Mech.* 5, 233–306.

Dexter, A. R. (1987). Mechanics of root growth. *Plant Soil* 98, 303–312.

d' Hollander, E., and Impens, I. (1975). Hybrid simulation of a dynamic model for water movement in a soil-plant-atmosphere continuum. In: Vansteenkiste, G. C., ed., "Computer Simulation of Water Resources Systems," pp. 349–360. North-Holland, Amsterdam.

Diamond, S. (1970). Pore size distribution in clays. *Clays Clay Mineral.* 18, 7–24.

Dickinson, R. E. (1984). Modeling evapotranspiration for three-dimensional global climate models. In: Hansen, J. E., and Takahashi, T., eds., "Climate Processes and Climate Sensitivity," pp. 58–72. American Geophysical Union.

Diment, G. A., and Watson, K. K. (1983). Stability analysis of water movement in unsaturated porous materials. 2. Numerical studies. *Water Resour. Res.* 21, 979–984.

Diment, G. A., Watson, K. K., and Blennerhassett, P. J. (1982). Stability analysis of water movement in unsaturated porous materials. 1. Theoretical considerations. *Water Resour. Res.* 18, 1248–1254.

Dixon, R., and Linden, M. (1972). Soil air pressure and water infiltration under border irrigation. *Soil Sci. Soc. Am. Proc.* 36, 948–953.

Dobson, M. C., and Ulaby, F. T. (1986). Active microwave soil moisture research. *IEEE Trans. Geosci. Remote Sens.* GE-24, 23–36.

Domenico, P. A. (1972). "Concepts and Models in Groundwater Hydrology." McGraw-Hill, New York.

Donahue, R. L., Shickluna, J. C., and Robertson, L. S. (1971). "Soils: An Introduction to Soils and Plant Growth," 3rd ed. Prentice-Hall, Englewood Cliffs, NJ.

Doolette, J. B., and Smyle, J. W. (1990). Soil and moisture conservation technologies. In: "Watershed Development in Asia: Strategies and Technologies," Tech. Paper 127. The World Bank, Washington, DC.

Doorenbos, J., and Kassam, A. H. (1979). Yield response to water, Irrig. Drain. Paper 33. U. N. Food and Agriculture Organization, Rome.

Dracos, T. (1987). Immiscible transport of hydrocarbons infiltrating in unconfined aquifers. In: Vandermeulen, J. H., and Hrudney, S. E., eds., "Oil in Freshwater: Chemistry, Biology, Countermeasure Technology." Pergamon, New York.

Dregne, H. E., ed. (1992). "Degradation and Restoration of Arid Lands." Texas Tech University Press, Lubbock.

Dregne, H. E. (1994). Land degradation in the world's arid zones. In: Baker, R. S., Gee, G. W., and Rosenzweig, C., eds., "Soil and Water Science: Key to Understanding Our Global Environment," Special Publication No. 41. Soil Sci. Soc. Am., Madison, WI.

Dullien, F. A. L. (1992). "Porous Media; Fluid Transport and Pore Structure." Academic Press, San Diego.

Dumas, W. T., Trouse, A. C., Smith, L. A., Kummer, F. A., and Gill, W. R. (1973). Development and evaluation of tillage and other cultural practices in a controlled traffic system for cotton in the Southern Coastal Plain. *Trans. Am. Soc. Agr. Eng.* 16, 872–875, 880.

Dumas, W. T., Trouse, A. C., Smith, L. A., Kummer, F. A., and Gill, W. R. (1975). Traffic control as a means of increasing cotton yields by reducing soil compaction, Paper 75–1050. Presented at the Annual Meeting of Am. Soc. Agric. Eng., Davis, CA.

Eagleson, P. S. (1970). "Dynamic Hydrology." McGraw-Hill, New York.
Eching, S. O., Hopmans, J. W., and Wendroth, O. (1994). Unsaturated hydraulic conductivity from transient multistep outflow and soil water pressure data. *Soil Sci. Soc. Am. J.* 58, 687–695.
Edwards, W. M., and Larson, W. E. (1970). Infiltration of water into soils as influenced by surface seal development. *Soil Sci. Soc. Am. Proc.* 34, 101.
Edwards, W. M., Norton, L. D., and Redmond, C. E. (1988). Characterizing macropores that affect infiltration in nontilled soil. *Soil Sci. Soc. Am. J.* 52, 483–487.
Egashira, K., and Matsumoto, J. (1981). Evaluation of the axial ratio of soil clays from gray lowland soils based on viscosity measurements. Soil Sci. Plant Nutr. 27: 273–279.
Ehlers, W., Popke, V., Hesse, F., and Bohm, W. (1983). Penetration resistance and root growth of oats in tilled and untilled loam soil. *Soil Tillage Res.* 3, 261–275.
Eisenbud, M. (1987). "Environmental Radioactivity." Academic Press, Orlando, FL.
El Swaify, S. A., Moldenhauer, W. C., and Lo, A., eds. (1985). "Soil Erosion and Conservation." Soil Conservation Society of America, Ankeny, OH.
El-Haris, M. K. (1987). Soil spatial variability: Areal interpolation of physical and chemical properties. Ph. D. dissertation. Univ. of Arizona, Tucson.
El-Kadi, A. I. (1987). Variability of infiltration under uncertainty in unsaturated zone parameters. *J. Hydrol.* 90, 61–80.
Elrick, D. E., and Laryea, K. B. (1979). Sorption of water in soils: A comparison of techniques for solving the diffusion equation. *Soil Sci.* 128, 210.
Elrick, D. E., and Reynolds, W. D. (1992). Infiltration from constant head well permeameters and infiltrometers. In: Topp, G. C., Reynolds, W. D., and Green, R. E., eds., "Advances in the Measurement of Soil Physical Properties: Bringing Theory into Practice," Special Publ. No. 30. Soil Sci. Soc. Am., Madison, WI.
Elrick, D. E., Laryea, K. B., and Groenevelt, P. H. (1979). Hydrodynamic dispersion during infiltration of water into soil. *Soil Sci. Soc. Am. J.* 43, 856–865.
Eltahir, E. A. B. and Bras, R. L. (1993). Estimation of the fractional coverage of rainfall in climate models. *J. Climate* 6, 639–644.
Elwell, H. A. (1986). Determination of erodibility of a subtropical clay soil: A laboratory rainfall simulator experiment. *J. Soil Sci.* 37, 345–350.
Emerson, E. W. (1959). The structure of soil crumbs. *J. Soil Sci.* 10, 235.
Engman, E. T. (1991). Applications of microwave remote sensing of soil moisture for water resources and agriculture. *Remote Sens. Environ.* 35, 213–226.
Engman, E. T. (1995). Recent advances in remote sensing in hydrology. In: "Reviews of Geophysics," pp. 967–975, American Geophysical Union, Washington, DC.
Engman, E. T., and Chauhan, B, (1995) Status of microwave soil moisture measurements with remote sensing. *Remote Sens. Eviron.* 51, 189–198.
Engman, E. T., Angus, G., Kustas, W. P., and Wang, J. R. (1989). Relationships between the hydrologic balance of a small watershed and remotely sensed soil moisture, *IAHS Publ.* 186, 76–84.
Entekhabi, D. and Eagleson, P. S. (1989). Land surface hydrology parameterization for atmospheric general circulation models including subgrid scale spatial variability. *J. Climate* 2, 816–829.
Environmental Technology Transfer Committee. (1994). "Remediation Technologies Screening Matrix and Reference Guide." 2nd ed. EPA/542/B-94/013. Federal Remediation Technologies Roundtable, Washington, DC.
Epstein, E. (1973). Roots. *Sci. Am.* 228, 48–58.
Epstein, E. (1977). The role of roots in the chemical economy of life on earth. *Bioscience* 27, 783–787.
Erickson, A. E. (1972). Improving the water properties of sand soil. In: Hillel, D. ed., "Optimizing the Soil Physical Environment Toward Greater Crop Yields," pp. 35–42. Academic Press, New York.

Evett, S. R., Matthias, A. D., and Warrick, A. W. (1994). Energy balance model of spatially variable evaporation from bare soil. Soil Sci. Soc. Am. J. 58: 1604–1611.

Evett, S. R. (2002). Water and energy balances at soil-plant-atmosphere interfaces. In: Warnik, A. W., ed., "Soil Physics Companion." CRC Press, Boca Raton, FL.

Fairbourn, M. L., and Cluff, C. B. (1974, April–May). Use of gravel mulch to save water for crops. *Crops Soil Mag.*

Fairbourn, M. L., and Kemper, W. D. (1971). Microwatersheds and ground color for sugarbeet production. *Agron. J.* 63, 101–104.

Famiglietti, J. S. and Wood, E. F. (1994). Multiscale modeling of spatially variable water and energy balance processes. *Water Resources Res.*, 30, 3079–3093.

Fanning, D. S., and Fanning, M. C. B. (1989). "Soil Morphology, Genesis, and Classification," John Wiley and Sons, New York.

FAO. (1965). "Soil Erosion by Water." Food and Agriculture Organization of the United Nations, Rome, Italy.

FAO. (1989). Guidelines for designing and evaluating surface irrigation systems, Technical Paper 45. U. N. Food and Agriculture Organization, Rome, Italy.

FAO. (1990). "Soil Map of the World." World Soil Resources Report No. 60, U. N. Food and Agriculture Organization, Rome, Italy.

FAO. (1996). Control of water pollution from agriculture, Irrig. Drain. Paper 55. U. N. Food and Agriculture Organization, Rome, Italy.

FAO-UNESCO. (1971–1981). Soil Map of the World, 1, 5,000,000. Vols. I-X, UNESCO, Paris.

FAO-UNESCO. (1988). "Soil Map of the World: Revised Legend." World Resources Report No. 60. 138 pp. (Available from FAO, Via delle Terme di Caracalla. 0100 Rome, Italy.)

Farr, A. M., Houghtalen, R. J., and McWhorter, D. B. (1990). Volume estimation of light nonaqueous phase liquids in porous media. *Ground Water* 28(1), 48–56.

Farr, E., and Henderson, W. C. (1986). "Land Drainage." Longmans, London and New York.

Farrell, D. A., Greacen, E. L., and Gurr, C. G. (1966). Vapor transfer in soil due to air movement. *Soil Sci.* 102, 305–313.

Farrell, R. E., DeJong, E., and Elliott, J. A. (2002). Gas sampling and analysis. In: Dane, J. H., and Topp, G. C., eds., "Methods of Soil Analysis, Part 4: physical Methods." Soil Science Society of America, Madison, WI.

Faybisheuks, B. (1995). Hydraulic behavior of quasi–saturated soils in the presence of entrapped air: Laboratory experiments. *Water Resour. Res.* 31, 2421–2435.

Fayer, M. J., and Hillel, D. (1986a). Air encapsulation: 1. Measurement in a field soil. *Soil Sci. Soc. Am. J.* 50, 568–572.

Fayer, M. J., and Hillel, D. (1986b). Air encapsulation: 2. Profile water storage and shallow water table fluctuations. *Soil Sci. Soc. Am. J.* 50, 572–577.

Feddes, R. A., Bresler, E., and Neuman, S. P. (1974). Field test of a modified numerical model for water uptake by root systems. *Water Resour. Res.* 10, 1199–1206.

Feddes, R. A., Neuman, S. P., and Bresler, E. (1976). Finite element analysis of two-dimensional flow in soils considering water uptake by roots: II. Field applications. *Soil Sci. Soc. Am. Proc.* 39, 231–237.

Ferguson, H., and Gardner, W. H. (1962). Water content measurement in soil columns by gamma ray absorption. *Soil Sci. Soc. Am. Proc.* 26, 11–14.

Feyen, J., Belmans, C., and Hillel, D. (1980). Comparison between measured and simulated plant water potential during soil water extraction by potted ryegrass. *Soil Sci.* 129, 180.

Fiscus, E. L., and Kaufman, M. R. (1990). The nature and movement of water in plants. In: Stewart, B. A., and Nielsen, D. R., eds., "Irrigation of Agricultural Crops," Monograph No. 30. Am. Soc. Agron., Madison WI.

Fleming, G. (1972). "Computer Simulation Techniques in Hydrology." American Elsevier, New York.

Flint, A. L., and Flint, L. E. (2002). Porosity. In: Dane, J., and Topp, G. C., eds., "Methods of Soil Analysis, Part 4: Physical Methods." Soil Science Society of America, Madison, WI.

Flury, M., Fluhler, W., Jury, W. A., and Leuenberger, J. (1994). Susceptibility of soils to preferential flow of water: A field study. *Water Resour. Res.* 30, 1945–1954.

Fok, Y. S. (1970). One-dimensional infiltration into layered soils. *J. Irrig. Drainage Div. ASCE* 96, 121–129.

Foley, J. A. (1995). An equilibrium model of the terrestrial carbon budget. *Tellus* 371, 52–54.

Foster, G. R. (1988). "User Requirements, USDA Water Erosion Prediction Project (WEPP)." USDA-ARS National Soil Erosion Lab., West Lafayette, IN.

Foster, G. R., and Highfill, R. E. (1983). Effect of terraces on soil loss: USLE P-factor values for terraces. *J. Soil Water Conserv.* 38, 48–51.

Foster, S. S. D., Cripps, A. C., and Smith-Carington, A. (1982). Nitrate leaching to groundwater. *Phil. Trans. R. Soc. (London)* 296, 477–489.

Foth, H. (1978). "Fundamental of Soil Science." Wiley, New York.

Fountain, J. C., Starr, R. C., Middleton, T., Beikirch, M., Taylor, C., and Hodge, D. (1996). A controlled field test of surfactant-enhanced aquifer remediation. *Ground Water* 34, 910–916.

Fox, S. W. (1988). "The Emergence of Life." Basic Books, New York.

Fredlund, D. G., and Vanapalli, S. K. (2002). Shear strength of unsaturated soils. In: Dane, J. H., and Topp, G. C., eds., "Methods of Soil Analysis, Part 4: Physical Methods." Soil Science Society of America, Madison, WI.

Freeze, R. A. (1980). A stochastic-conceptual analysis of rainfall-runoff processes on a hillslope. *Water Resour. Res.* 16, 391–408.

Freeze, R. A., and Cherry, J. A. (1979). "Groundwater." Prentice-Hall, Englewood Cliffs, NJ.

Freney, J. R., Denmead, O. T., Watanabe, I., and Crasswell, E. T. (1981). Ammonia and nitrous oxide losses following applications of ammonium sulfate to flooded rice. *Aust. J. Agric. Res.* 32, 37–445.

Frenkel, H., Goertzen, J. O., and Rhoades, J. D. (1978). Effects of clay type and content, exchangeable sodium percentage, and electrolyte concentration on clay dispersion and soil hydraulic conductivity. *Soil Sci. Soc. Am. J.* 42, 32–39.

Fried, J. J. (1976). "Ground Water Pollution." Elsevier, Amsterdam.

Fritton, D. D., Kirkham, D., and Shaw, R. H. (1970). Soil water evaporation, isothermal diffusion, and heat and water transfer. *Soil Sci. Soc. Am. Proc.* 34, 183–189.

Fuchs, M. (1986). Heat flux. In: Klute, A., ed., "Methods of Soil Analysis," Monograph No. 9. Am. Soc. Agron., Madison, WI.

Fuchs, M., and Tanner, C. B. (1967). Evaporation from a drying soil. *J. Appl. Meteorol.* 6, 852–857.

Fuchs, M., and Tanner, C. B. (1968). Calibration and field test of soil heat flux plates. *Soil Sci. Soc. Am. Proc.* 32, 326–328.

Gajem, Y. M., Warrick, A. W., and Myers, D. E. (1981). Spatial structure of physical properties of a typic torrifluvent. *Soil Sci. Soc. Am. J.* 45, 709–715.

Gao, Z., Fan, X., and Bian, L. (2003). An analytical solution to one-dimensional thermal conduction-convection in soil. Soil Sci. 168: 99–107.

Gardner, W. R. (1956). Representation of soil aggregate-size distribution by a logarithmic-normal distribution. *Soil Sci. Soc. Am. Proc.* 20, 151–153.

Gardner, W. R. (1958). Some steady state solutions of the unsaturated moisture flow equation with application to evaporation from a water table. *Soil Sci.* 85(4), 228–232.

Gardner, W. R. (1959). Solutions of the flow equation for the drying of soils and other porous media. *Soil Sci. Soc. Am. Proc.* 23, 183–187.

Gardner, W. R. (1960a). Dynamic aspects of water availability to plants. *Soil Sci.* 89, 63–73.

Gardner, W. R. (1962a). Approximate solution of a non-steady state drainage problem. *Soil Sci. Soc. Am. Proc.* 26, 129–132.

Gardner, W. R. (1962b). Note on the separation and solution of diffusion type equations. *Soil Sci. Soc. Am. Proc.* 26, 404.

Gardner, W. R. (1964). Relation of root distribution to water uptake and availability. *Agron. J.* 56, 35–41.

Gardner, W. R., ed. (1996). "A New Era for Irrigation." National Academy Press, Washington, DC.

Gardner, W. R., and Hillel, D. (1962). The relation of external evaporative conditions to the drying of soils. *J. Geophys. Res.* 67, 4319–4325.

Gardner, W. R., and Mayhugh, M. S. (1958). Solutions and tests of the diffusion equation for the movement of water in soil. *Soil Sci. Soc. Am. Proc.* 22, 197–201.

Gardner, W. R., Hillel, D., and Benyamini, Y. (1970a). Post irrigation movement of soil water: I. Redistribution. *Water Resour. Res.* 6, 851–861.

Gardner, W. R., Hillel, D., and Benyamini, Y. (1970b). Post irrigation movement of soil water: II. Simultaneous redistribution and evaporation. *Water Resour. Res.* 6, 1148–1153.

Gash, J. H. C., Nobre, C. A., Roberts, J. M., and Victoria, R. L., eds. (1996). *Amazonian Deforestation and Climate.* John Wiley, New York. 430 pp.

Gates, D. M. (1980). "Biophysical Ecology." Springer-Verlag, New York.

Gates, D. M. (1993). "Climate Change and Its Biological Consequences." Sinauer, Sunderland, MA.

Gee, G. W., and Hillel, D. (1988). Groundwater recharge in arid regions. *Hydrol. Proc.* 2, 255–266.

Gee, G. W., and Or, D. (2002). Particle size analysis. In: Dane, J., and Topp, G. C., eds., "Methods of Soil Analysis, Part 4: Physical Methods." Soil Science Society of America, Madison, WI.

Gee, G. W., Stiver, J. F., and Borchert, H. R. (1976). Radiation hazard from americium-beryllium neutron moisture probes. *Soil Sci. Soc. Am. J.* 40, 492–494.

Gerba, C. P. (1996). Municipal waste and drinking water treatment. In: Pepper, I. L., Gerba, C. P., and Brusseau, M. L., eds., "Pollution Science." Academic Press, San Diego.

Germann, P. F., and Beven, K. J. (1985). Kinematic wave approximation in infiltration into soils with sorting macropores. *Water Resour. Res.* 21, 990–996.

Germann, P. F., Edwards, W. M., and Owens, L. B. (1984). Profiles of bromide and increased moisture after infiltration into soils with macropores. *Soil Sci. Soc. Am. J.* 48, 237–244.

Ghassemi, F., Jakeman, A. J., and Nix, H. A. (1995). "Salinization of Land and Water Resources." Australian National University, Canberra.

Gill, W. R. (1979). Tillage. In: Fairbridge, R. W., and Finkl, C. W., Jr., eds., "The Encyclopedia of Soil Science, Part I," pp. 566–571. Dowden, Hutchinson & Ross, Stroudsburg, PA.

Gill, W. R., and Cooper, A. (1972). Soil compaction — its causes and remedies. In: "Proc. Western Cotton Production Conf.," Bakersfield, CA, pp. 11–13.

Gill, W. R., and Vanden Berg, G. E. (1967). "Soil Dynamics in Tillage and Traction," Handbook 316. Agr. Res. Service, U.S. Dept. of Agriculture, Washington, DC.

Gish, T. J., and Jury, W. A. (1983). Effect of plant roots and root channels on solute transport. *Trans. Am. Soc. Agric. Eng.* 26, 440–444, 451.

Gish, T. J., and Shirmohammadi, A., eds. (1991). "Preferential Flow." Proc. Natl. Symp. Am. Soc. Agric. Eng., St. Joseph, MI.

Glass, R. J., Steenhuis, T. S., and Parlange, J. Y. (1988). Wetting front instability as a rapid and far-reaching hydrological process in the vadose zone. *J. Contam. Hydrol.* 3, 207–226.

Glass, R. J., Steenhuis, T. S., and Parlange, J. Y. (1989). Wetting front instability. *Water Resour. Res.* 25, 1187–1194.

Glinski, J., and Stepniewski, W. (1985). "Soil Aeration and its role for Plants." CRC Press, Boca Raton, FL.

Goldberg, D., and Shmueli, M. (1970). Drip irrigation: A method used under arid and desert conditions of high water and soil salinity. *Trans. ASAE* 13, 38–41.

Goss, M. J. (1970). Further studies on the effect of mechanical resistance on the growth of plant roots, Rept. Agr. Res. Coun. Lethcombe Lab. ARCRL 20, 43–45.

Grable, A. R., and Siemer, E. G. (1968). Effects of bulk density, aggregate size, and soil water suction on oxygen diffusion, redox potentials and elongation of corn roots. *Soil Sci. Soc. Am. Proc.* 32, 180–186.

Graham, W. G., and King, K. M. (1961). Fraction of net radiation utilized in evapotranspiration from a corn crop. *Soil Sci. Soc. Am. Proc.* 25, 158–160.

Grant, R. F., Nyborg, M., and Laidlaw, J. W. (1993). Evolution of nitrous oxide from soil: I. Model development. *Soil Sci.* 156, 259–265.

Greacen, E. L., Barley, K. P., and Farrell, D. A. (1969). The mechanics of root growth in soils with particular reference to the implications for root distribution. In: Whittington, W. J., ed., "Root Growth," pp. 256–269. Butterworths, London.

Greacen, E. T., Ponsana, P., and Barley, K. P. (1976). Resistance to water flow in the roots of cereals. In: "Water and Plant Life, Ecological Studies 19," pp. 86–100. Springer-Verlag, Berlin and New York.

Greb, B. W. (1966). Effect of surface-applied wheat straw on soil water losses by solar distillation. *Soil Sci. Soc. Am. Proc.* 30, 786–788.

Green, R. E., and Corey, J. C. (1971). Calculation of hydraulic conductivity: A further evaluation of predictive methods. *Soil Sci. Soc. Am. Proc.* 35, 3–8.

Green, R. E., Alwja, L. R., and Chang, S. K. (1986). Hydraulic conductivity, diffusivity and sorptivity of unsaturated soils: Field methods. In: Klute, A., ed., "Methods of Soil Analysis, Part 1: Physical and Mineralogical Methods," Monograph No. 9. Am. Soc. Agron., Madison, WI.

Green, W. H., and Ampt, G. A. (1911). Studies on soil physics: I. Flow of air and water through soils. *J. Agr. Sci.* 4, 1–24.

Greenland, D. J. (1965). Interaction between clays and organic compounds in soils. Part I. Mechanisms of interaction between clays and defined organic compounds. *Soil Fert.* 28, 415–425.

Greenland, D. J., Lindstrom, G. R., and Quirk, J. P. (1962). Organic materials which stabilize natural soil aggregates. *Soil Sci.* Am. Proc. 26, 236–371.

Greenwood, D. J. (1971). Soil aeration and plant growth. *Rep. Prog. Appl. Chem.* 55, 423–431.

Gregorich, E. G., Turchenek, L. W., Carter, M. R., and Angers, D. A., eds. (2002). "Soil and Environmental Science Dictionary." CRC Press, Boca Raton, FL.

Groenevelt, P. H., and Bolt, G. H. (1969). Non-equilibrium thermodynamics of the soil-water system. *J. Hydrol.* 7, 358–388.

Grossman, R. B., and Reinsch, T. G. (2002). Bulk density and linear extensibility. In: Dane, J., and Topp, G. C., eds., "Methods of Soil Analysis, Part 4: Physical Methods." Soil Science Society of America, Madison, WI.

Grover, B. L. (1956). Simplified air permeameter for soil in place. *Soil Sci. Soc. Am. Proc.* 19, 414–418.

Gumaa, G. S. (1978). Spatial variability of *in situ* available water. Ph.D. Diss., Univ. of Arizona, Tucson, AZ (Available as 78-24365 from Xerox, University Microfilms, Ann Arbor, MI.)

Gurr, C. G. (1962). Use of gamma rays in measuring water content and permeability in unsaturated columns of soil. *Soil Sci.* 94, 224–449.

Hadas, A., and Fuchs, M. (1973). Prediction of the thermal regime of bare Soils. In: Hadas, A., Swartzendruber, D., Rijtema, P. E., Fuchs, M., and Yaron, B., eds., "Physical Aspects of Soil Water and Salts in Ecosystems," Springer-Verlag, Berlin and New York.

Hadas, A., and Hillel, D. (1968). An experimental study of evaporation from uniform columns in the presence of a water table. In: "Trans. 9th Int. Congr. Soil Sci.," Adelaide, Vol. I, pp. 67–74.

Hadas, A., Schmulevich, I., Hadas, O., and Wolf, D. (1990). Forage wheat yields as affected by compaction and conventional vs. wide-frame tractor traffic patterns. *Trans. Am. Soc. Agric. Eng.* 23, 836–839.

Hagen, L. J. (1991). A wind erosion prediction system to meet user needs. *J. Soil Water Conserv.* 46, 106–111.

Haines, W. B. (1930). Studies in the physical properties of soils. V. The hysteresis effect in capillary properties and the modes of moisture distribution associated therewith. *J. Agr. Sci.* 20, 97–116.

Hanks, R. J. (1974). Model for predicting plant yield as influenced by water use. *Agron. J.* 66, 660–665.

Hanks, R. J. (1983). Yield and water use relationships: An overview. In: Taylor, H. M., Jordan, W. R., and Sinclair, T. R., eds., "Limitations to Efficient Water Use in Crop Production." Am. Soc. Agron., Madison, WI.

Hanks, R. J. (1991a). Infiltration and redistribution. In: Hanks, R. J., and Ritchie, J. T., eds. "Modeling Plant and Soil Systems," Monograph No. 31. Am. Soc. Agron., Madison, WI.

Hanks, R. J. (1991b). Soil evaporation and transpiration. In: Hanks, R. J., and Ritchie, J. T., eds. "Modeling Plant and Soil Systems," Monograph No. 31. Am. Soc. Agron., Madison, WI.

Hanks, R. J. (1992). "Applied Soil Physics: Soil Water and Temperature Applications." Springer-Verlag, New York.

Hanks, R. J., and Bowers, S. A. (1962). Numerical solution of the moisture flow equation for infiltration into layered soils. *Soil Sci. Soc. Am. Proc.* 26, 530–534.

Hanks, R. J., and Gardner, H. R. (1965). Influence of different diffusivity-water content relations on evaporation of water from soils. *Soil Sci. Soc. Am. Proc.* 29, 495–498.

Hanks, R. J., and Ritchie, J. T., eds. (1991). "Modeling Plant and Soil Systems," Monograph No. 31. Am. Soc. Agron., Madison, WI.

Hanks, R. J., Bowers, S. B., and Boyd, L. D. (1961). Influence of soil surface conditions on net radiation, soil temperature, and evaporation. *Soil Sci.* 91, 233–239.

Hanks, R. J., Gardner, H. R., and Fairbourn, M. L. (1967). Evaporation of water from soils as influenced by drying with wind or radiation. *Soil Sci. Soc. Am. Proc.* 31, 593–598.

Hansen, G. K. (1975). A dynamic continuous simultation model of water state and transportation in the soil-plant-atmosphere system. *Acta Agr. Scand.* 25, 129–149.

Harris, R. F., Chester, G., and Allen, O. N. (1965). Dynamics of soil aggregation. *Adv. Agron.* 18, 107–160.

Harriss, R. C., Sebacher, D. I., Barttlett, K. B., and Grill, P. M. (1988). Sources of atmospheric methane in the south Florida environment. *Global Biogeochem. Cycles* 2, 231–243.

Hatfield, J. L. (1983). Evapotranspiration obtained from remote sensing methods. In: Hillel, D., ed., "Advances in Irrigation," Vol. 2. Academic Press, New York.

Haverkamp, R., and Reggiani, P. (2002). Physically based water retention prediction models. In: Dane, J., and Topp, G. C., eds., "Methods of Soil Analysis, Part 4: Physical methods." Soil Science Society of America, Madison, WI.

Haverkamp, R., Vauclin, M., Touma, J., Wierenga, P. J., and Vachaud, G. (1977). A comparison of numerical simulation models for one-dimensional infiltration. *Soil Sci. Soc. Am. J.* 41, 285–294.

Haverkamp, R., Kutilek, M., Parlange, J. Y., Rendon, L., and Krejca, M. (1988). Infiltration under ponded conditions: 2. Infiltration equations tested for parameter time-dependence and predictive use. *Soil Sci.* 145, 317–329.

Haverkamp, R., Parlange, J. Y., Starr, J. L., Schmitz, G., and Fuentes, C. (1990). Infiltration under ponded conditions: 3. A predictive equation based on physical parameters. *Soil Sci.* 149, 292–300.

Hemond, F., and Fechner, E. J. (1994). "Chemical Fate and Transport in the Environment." Academic Press, San Diego.

Hendrickx, J. M. H., Dekker, L. W., and Raats, P. A. C. (1988a). Formation of sand columns caused by unstable wetting fronts. *Grondboor en Hamer* 6, 173–175 (in Dutch).

Hendrickx, J. M. H., Dekker, L. W., van Zuilen, E. J., and Boersma, O. H. (1988b). Water and solute movement through a water repellent soil with grass cover. In: Wierenga, P. J., and Bachelet, D., eds., "Proc. Conf. on Validation of Flow and Transport Models for the Unsaturated Zone," New Mexico State University, Las Cruces, NM.

Hendrickx, J. M. H., Wraith, J. M., Corwin, D. L., and Kachanoski, R. G. (2002). Solute content and concentration, In: Dane, J. H., and Topp, G. C., eds. "Methods of Soil Analysis, Part 4: Physical Methods." Soil Science Society of America, Madison, WI.

Herkelrath, W. N. (1975). Water uptake by plant roots. Ph.D. Diss., Univ. of Wisconsin, Madison, WI.

Herkelrath, W. N., Hamburg, S. P., and Murphy, F. (1991). Automatic, real-time monitoring of soil moisture in a remote field area with time domain reflectometry. *Water Resour. Res.* 27, 857–864.

Heywood, H. (1947). Syposium on particle size analysis. Trans. Inst. Chem. Eng. 22: 214.

Hiemenez, P. C. (1986). "Principles of Colloid and Surface Chemistry." Marcel Dekker, New York.

Hignett, C., and Evett, S. R. (2002). Neutron thermalization. In: Dane, J., and Topp, G. C., eds., "Methods of Soil Analysis, Part 4: Physical Methods." Soil Science Society of America, Madison, WI.

Hill, D. (1984). Diffusion coefficients of nitrate, chloride, sulphate, and water in cracked and uncracked chalk. *J. Soil Sci.* 35, 27–33.

Hill, E. D., and Parlange, J. Y. (1972). Wetting front instability in layered soils. *Soil Sci. Soc. Am. Proc.* 36, 697–702.

Hill, R. L. (1990). Long-term conventional and no-tillage effects on selected soil physical properties. *Soil Sci. Soc. Am. J.* 54, 161–166.

Hill, R. L. (1991). Irrigation scheduling. In: Hanks, R. J., and Ritchie, J. T., eds. "Modeling Plant and Soil Systems," Monograph No. 31. Am. Soc. Agron., Madison, WI.

Hill, R. L., and Meza-Montalvo, M. (1990). Long-term wheel traffic effects on soil physical properties under different tillage systems. *Soil Sci. Soc. Am. J.* 54, 865–870.

Hill, R. L., Gross, C. M., and Angle, J. S. (1991). Rainfall simulation for evaluating agrochemical surface loss. In: Nash, R. G., and Leslie, A. R., eds. "Groundwater Residue Sampling Design." Am. Chem. Soc., New York.

Hillel, D. (1959). "Studies of Loessial Crusts," Bulletin 63. Israel Agr. Res. Inst., Beit Dagan. Israel.

Hillel, D. (1960). Crust formation in loessial soils. In: "Trans. 7th Int. Soil Sci. Congr.," Madison, WI. Vol. 1, pp. 330–339.

Hillel, D. (1964). Infiltration and rainfall runoff as affected by surface crusts. In: "Trans. 8th Int. Soil Sci. Congr.," Bucharest, Vol. 2, pp. 53–62.

Hillel, D. (1967). "Runoff Inducement in Arid Lands," Special Publ. Hebrew University of Jerusalem, Israel.

Hillel, D. (1968). Soil water evaporation and means of minimizing it, Report to U.S. Dept. Agr., Hebrew Univ. of Jerusalem, Israel.

Hillel, D. (1971). "Soil and Water: Physical Principles and Processes." Academic Press, New York.

Hillel, D., ed. (1972a). "Optimizing the Soil Physical Environment Toward Greater Crop Yields." Academic Press, New York.

Hillel, D. (1972b). Soil moisture and seed germination. In: Koziowski, T. T., ed., "Water Deficits and Plant Growth," pp. 65–89. Academic Press, New York.

Hillel, D. (1974). Methods of laboratory and field investigation of physical properties of soils. In: "Trans. 10th Int. Soil Sci. Congr.," Moscow, Vol. 1, pp. 301–308.

Hillel, D. (1975). Evaporation from bare soil under steady and diurnally fluctuating evaporativity. *Soil Sci.* 120, 230–237.

Hillel, D. (1976a). On the role of soil moisture hysteresis in the suppression of evaporation from bare soil. *Soil Sci.* 122, 309–314.

Hillel, D. (1976b). Soil management. In: "McGraw-Hill Yearbook of Science and Technology." McGraw-Hill, New York.

Hillel, D. (1977). "Computer Simulation of Soil-Water Dynamics." Int. Dev. Res. Centre, Ottawa, Canada.

Hillel, D. (1980a). "Fundamentals of Soil Physics." Academic Press, New York.

Hillel, D. (1980b). "Application of Soil Physics." Academic Press, New York.

Hillel, D. (1982). "Negev: Land, Water and Life in a Desert Environment." Praeger, New York.

Hillel, D. (1987a). "The Efficient Use of Water in Irrigation: Principles and Practices for Improving Irrigation in Arid and Semi-Arid Regions," Technical Publication No. 64. The World Bank, Washington.

Hillel, D. (1987b). On the tortuous path of research. *Soil Sci.* 143, 304–305.

Hillel, D. (1991a). SPACE: A modified soil-plant-atmosphere continuum electroanalog. *Soil Sci.* 151, 399–404.

Hillel, D. (1991b). "Out of the Earth: Civilization and the Life of Soil." Free Press, New York; University of California Press, Berkeley.

Hillel, D. (1991c). The problem of desertification: A critical reconsideration. Arid Lands Research Centre (Tottori, Japan) Annual Report 1990–91, 33–35.

Hillel, D. (1993a). Science and the crisis of the environment. *Geoderma* 60, 377–382.

Hillel, D. (1993b). Unstable flow: A potentially significant mechanism of water and solute transport to groundwater. In: Russo, D., and Dagan, G., eds., "Water Flow and Solute Transport in Soils." Springer-Verlag, Berlin.

Hillel, D. (1994). "Rivers of Eden: The Struggle for Water and the Quest for Peace in the Middle East." Oxford University Press, New York.

Hillel, D. (1997). "Small-Scale Irrigation for Arid Zones: Principles and Options," Development Monograph No. 2. United Nations Food and Agriculture Organization, Rome.

Hillel, D. (1998). "Environmental Soil Physics." Academic Press, San Diego, CA.

Hillel, D. (2000). "Salinity Management for Sustainable Irrigation: Integrating Science, Environment, and Economics," The World Bank, Washington, DC.

Hillel, D., and Baker, R. S. (1988). A descriptive theory of fingering during infiltration into layered soils. *Soil Sci.* 146, 51–56.

Hillel, D., and Benyamini, Y. (1974). Experimental comparison of infiltration and drainage methods for determining unsaturated hydraulic conductivity of a soil profile in situ. In: "Proc. FAD/IAEA Symp. Isotopes and Radiation Techniques in Studies of Soil Physics," Vienna, pp. 271–275.

Hillel, D., and Berliner, p. (1974). Water-proofing surface zone soil aggregates for water conservation. *Soil Sci.* 118, 131–135.

Hillel, D., and Elrick, D. E., eds. (1990). "Scaling and Similitude in Soil Physics: Principles and Applications," Soil Sci. Soc. Amer., Madison, WI.

Hillel, D., and Gardner, W. R. (1969). Steady infiltration into crust-topped profiles. *Soil Sci.* 108, 137–142.

Hillel, D., and Gardner, W. R. (1970a). Transient infiltration into crust-topped profiles. *Soil Sci.* 109, 69–76.

Hillel, D., and Gardner, W. R. (1970b). Measurement of unsaturated conductivity and diffusivity by infiltration through an impeding layer. *Soil Sci.* 109, 149–154.

Hillel, D., and Guron, Y. (1973). Relation between evapotranspiration rate and maize yield. *Water Resour. Res.* 9, 743–748.

Hillel, D., and Hornberger, G. M. (1979). Physical model of the hydrology of sloping heterogeneous fields. *Soil Sci. Soc. Am. J.* 43, 434–439.

Hillel, D., and Mottes, J. (1966). Effect of plate impedance, wetting method and aging on soil moisture retention. *Soil Sci.* 102, 135–140.

Hillel, D., and Rawitz, E. (1972). Soil moisture conservation. In: Koziowski, T. T., ed., "Water Deficits and Plant Growth," pp. 307–337. Academic Press, New York.

Hillel, D., and Talpaz, H. (1976). Simulation of root growth and its effect on the pattern of soil water uptake by a nonuniform root system. *Soil Sci.* 121, 307–312.

Hillel, D., and Talpaz, H. (1977). Simulation of soil water dynamics in layered soils. *Soil Sci.* 123, 54–62.

Hillel, D., and Rosenzweig, C. (2002). Desertification in relation to climate variability and change. In: Sparks, D. L., ed., "Advances in Agronomy, Vol. 77."

Hillel, D., and van Bavel, C. H. M. (1976). Dependence of profile water storage on soil hydraulic properties: A simulation model. *Soil Sci. Soc. Am. J.* 40, 807–815.

Hillel, D., Gairon, S., Falkenflug, V., and Rawitz, E. (1969a). New design of a low-cost hydraulic lysimeter for field measurement of evapo-transpiration. *Israel J. Agr. Res.* 3(19), 57–63.

Hillel, D., Ariel, D., Orlowski, S., Stibbe, E., Wolf, D., and Yavnai, A. (1969b). Soil-crop-tillage interactions in dryland and irrigated farming, Research report submitted to the U.S. Dept. Agric. Hebrew University of Jerusalem, Jerusalem, Israel.

Hillel, D., Krentos, V., and Stylianou, Y. (1972). Procedure and test of an internal drainage method for measuring soil hydraulic characteristics in situ. *Soil Sci.* 114, 395–400.

Hillel, D., van Beek, C., and Talpaz, H. (1975). A microscopic-scale model of soil water uptake and salt movement to plant roots. *Soil Sci.* 120, 385–399.

Hillel, D., Talpaz, H., and van Keulen, H. (1976). A macroscopic-scale model of water uptake by a nonuniform root system and of water and salt movement in the soil profile. *Soil Sci.* 121, 242–255.

Hills, R. G., Porro, I., Hudson, D. B., and Wierenga, P. J. (1989). Modeling one-dimensional infiltration into very dry soils 1. Model development and evaluation. *Water Resour. Res.* 25, 1259–1269.

Holy, Milos. (1980). "Erosion and Environment." Pergamon Press, Oxford.

Hoogenboom, G., Huck, M. G., and Hillel, D. (1987). Modification and testing of a model simulating root and shoot growth as related to soil water dynamics. In: Hillel, D., ed., "Advances in Irrigation," Vol. 4. Academic Press, Orlando, FL.

Hooghoudt, S. B. (1937). Bijdregen tot de kennis van eenige natuurkundige grootheden van de grond, 6. *Versl. Landb. Ond.* 43, 461–676.

Hoogmoed, W. D., and Bouma, J. (1980). A simulation model for predicting infiltration into cracked soil. *Soil Sci. Soc. Am. J.* 44, 458–461.

Hornung, U. (1983). Identification of nonlinear soil physical parameters from an input-output experiment. In: Deuflhardt, O., and Hairer, E., eds., "Workshop on Numerical Treatment of Inverse Problems in Differential Integral Equations." Birkhauser, Boston.

Horowitz, J., and Hillel, D. (1983). A critique of recent attempts at characterizing the spatial heterogeneity of soil physical properties in the field. Soil Sci. Soc. Amer. J. 47: 614–616.

Horton, R. E. (1940). An approach toward a physical interpretation of infiltration-capacity. *Soil Sci. Soc. Am. Proc.* 5, 399–417.

Horton, R. (2002). Soil thermal diffusivity. In: Dane, J. H., and Topp, G. C., eds., "Methods of Soil Analysis, Part 4: Physical Methods," Soil Science Society of America, Madison, WI.

Horton, R., and Chung, S. O. (1991). Soil heat flow. In: Hanks, R. J., and Ritchie, J. T., eds. "Modeling Plant and Soil Systems," Monograph No. 31. Am. Soc. Agron., Madison, WI.

Huang, P. M., and Schnitzer, M., eds. (1986). "Interactions of Soil Minerals with Natural Organics and Microbes," Special Publication No. 17. Soil Sci. Soc. Am., Madison, WI.

Hubbert, M. K. (1956). Darcy's law and the field equations of the flow of underground fluids. *Am. Inst. Min. Met. Petl. Eng. Trans.* 207, 222–239.

Huck, M. G. (1977). Root distribution and water uptake patterns. In Marshall, J. K., ed., "The Belowground Ecosystem: A Synthesis of Plant-Associated Processes," Range Science Series No. 26. Colorado State Univ., Fort Collins, CO.

Huck, M. G., and Hillel, D. (1983). A model of root growth and water uptake accounting for photosynthesis, respiration, transpiration, and soil hydraulics. In: Hillel, D., ed., "Advances in Irrigation," Vol. 2. Academic Press, New York.

Hudson, N. (1981). "Soil Conservation." Cornell University Press, Ithaca, NY.

Huyakorn, P. S., Lester, B., and Mercer, J. W. (1983). An efficient finite element technique for modeling transport in fractured porous media. 1. Single species transport. *Water Resour. Res.* 19, 841–854.

Huyakorn, P. S., Mercer, J. W., and Ward, D. S. (1985). Finite element matrix and mass balance computational schemes for transport in variably saturated porous media. *Water Resour. Res.* 21, 346–358.

Incropera, F. P., and DeWitt, D. P. (1990). "Fundamentals of Heat and Mass Transfer." Wiley, New York.

Intergovernmental Panel on Climate Change (IPCC). (1990). "Climate Change: The Scientific Assessment." Cambridge University Press, Cambridge.

Intergovernmental Panel on Climate Change (IPCC). (1995). "Climate Change 1994: Radiative Forcing of Climate Change and an Evaluation of the IPCC 1992 Emission Scenarios." Intergovernmental Panel on Climate Change. Cambridge University Press, Cambridge.

Iwata, S., Tabuchi, T., and Warkentin, B. P. (1995). "Soil-Water Interactions: Mechanisms and Applications." Marcel Dekker, New York.

Jackson, R. D. (1964a). Water vapor diffusion in relatively dry soil: I. Theoretical considerations and sorption experiments. *Soil Sci. Soc. Am. Proc.* 28, 172–176.

Jackson, R. D. (1964b). Water vapor diffusion in relatively dry soil: II. Desorption experiments. *Soil Sci. Soc. Am. Proc.* 28, 464–466.

Jackson, R. D. (1964c). Water vapor diffusion in relatively dry soil: III. Steady state experiments. *Soil Sci. Soc. Am. Proc.* 28, 466–470.

Jackson, R. D. (1972). On the calculation of hydraulic conductivity. *Soil Sci. Soc. Am. Proc.* 36, 350–383.

Jackson, R. D. (1973). Diurnal changes in soil water content during drying. In "Field Soil Water Regimes," pp. 37–55. Soil Sci. Soc. Am., Madison, WI.

Jackson, R. D., and Whisler, F. D. (1970). Approximate equations for vertical nonsteady-state drainage: I. Theoretical approach. *Soil Sci. Soc. Am. Proc.* 34, 715–718.

Jackson, R. D., and Taylor, S. A. (1986). Thermal conductivity and diffusivity. In: Klute, A., ed., "Methods of Soil Analysis, Part 1: Physical and Mineralogical Methods." Monograph No. 9. Am. Soc. Agron., Madison, WI.

Jackson, R. D., Kimball, B. A., Reginato, R. J., and Nakayama, S. F. (1973). Diurnal soil water evaporation: Time-depth-flux patterns. *Soil Sci. Soc. Am. Proc.* 37, 505–509.

Jackson, R. D., Reginato, R. J., Kimball, B. A., and Nakayama, F. S. (1974). Diurnal soil-water evaporation: Comparison of measured and calculated soil-water fluxes. *Soil Sci. Soc. Am. Proc.* 38, 861–866.

Jackson, T. J. (1993). Measuring surface soil moistures using passive microwave remote sensing. *Hydrol. Proc.* 7, 139–152.

Jackson, T. J., and Schmugge, T. J. (1989). Passive microwave remote sensing system for soil moisture: Some supporting research. *IEEE Trans. Geosci. Remote Sens.* GE-27, 225–235.

Jackson, T. J., and Schmugge, T. J. (1991). Correction for the effects of vegetation on the microwave emission of soils. *IEEE Intl. Geosci. Remote Sensing Symp. (IGARSS) Digest* 753–756.

Jacobsen, T. (1982). "Salinity and Irrigation Agriculture in Antiquity." Undena Publications, Malibu, CA.

Jenkinson, D. S., and Smith, K. A., eds. (1998). "Nitrogen Efficiency in Agricultural Soils." Elsevier, London.

Jenny, H. F. (1941). "Factors of Soil Formation." McGraw-Hill, New York.

Jensen, K. H., and Mantoglou, A. (1992). Application of stochastic unsaturated flow theory, numerical simulations, and comparisons to field observations. *Water Resour. Res.* 28, 296–284.

Jensen, M. E., ed. (1973). "Consumptive Use of Water and Irrigation Water Requirements." Am. Soc. Civil Eng., New York.

Johnson, P. C., and Ettinger, R. A. (1994). Considerations for the design of *in situ* vapor extraction systems: Radius of influence vs. zone of remediation. *Ground Water Monitor. Rev.* Summer, 123–128.

Johnson, P. C., Stanley, C. C., Kemblowski, M. W., Byers, D. L., and Colthart, J. D. (1990). A practical approach to the design, operation, and monitoring of *in situ* soil-venting systems. *Ground Water Monitor Rev.* 10(2), 159–178.

Johnson, P. C., Johnson, R. L., Neaville, C., Hansen, E. E., Stearns, S. M., and Dortch, I. J. (1995). Do conventional monitoring practices indicate *in situ* air sparging performance? In: Hinchee, R. E., Miller, R. N., and Johnson, P. C., eds., "*In Situ* Aeration: Air Sparging, Bioventing, and Related Remediation Processes," pp. 1–20. Battelle Press, Columbus, OH.

Jones, C. A., Bland, W. L., Ritchie, J. T., and Williams, J. R. (1991). Simulation of root growth. In: Hanks, R. J., and Ritchie, J. T., eds. "Modeling Plant and Soil Systems," Monograph No. 31. Am. Soc. Agron., Madison, WI.

Jones, F. E. (1991). "Evaporation of Water, with Emphasis on Applications and Measurements." Lewis Publishers, Chelsea, MI.

Journel, A. G., and Huijbregts, Ch. J. (1978). "Mining Geostatistics." Academic Press, New York.

Jurinak, J. J. (1990). The chemistry of salt-affected soils and waters. In: Tanji, K. K., ed. "Agricultural Salinity Assessment and Management." Am. Soc. Civil Eng., New York.

Jury, W. A. (1973). Simultaneous transport of heat and moisture through a medium sand. Ph.D. Thesis, Univ. of Wisconsin, Madison, WI.

Jury, W. A. (1982a). Simulation of solute transport with a transfer function model. *Water Resour. Res.* 18, 363–368.

Jury, W. A. (1982b). Use of solute transport models to estimate salt balance below irrigated cropland. In: Hillel, D., ed. "Advances in Irrigation," Vol. I. Academic Press, New York.

Jury, W. A. (1983). Chemical transport modeling: Current approaches and unresolved problems. In: Nelson, D. W., Elrick, D. E., and Tanji, K. K., eds., "Chemical Mobility and Reactivity in Soil Systems," Special Publication No. 42. Am. Soc. Agron., Madison, WI.

Jury, W. A., and Bellantuoni, B. (1976). Heat and water movement under surface rocks in a field soil. *Soil Sci. Soc. Am. J.* 40, 505–513.

Jury, W. A., and Sposito, G. (1985). Field calibration and validation of solute transport models for the unsaturated zone. *Soil Sci. Soc. Am. J.* 49, 1331–1341.

Kadlec, R. H., and Knight, R. L. (1996). "Treatment Wetlands." Lewis Publishers, Boca Raton, FL.

Kanemasu, E. T., Stone, L. R., and Powers, W. L. (1976). Evapotranspiration model tested for soybean and sorghum. *Agron. J.* 68, 569–572.

Katchalsy, A., and Curran. P. F. (1965). "Nonequilibrium Thermodynamics in Biophysics." Harvard Univ. Press, Cambridge, MA.

Kavanau, J. L. (1965). Water. In "Structure and Function in Biological Membranes," pp. 170–248. Holden-Day, San Francisco, CA.

Kay, B. D., and Angers, D. A. (2002). Soil structure. In: Warrick, A. W., ed., "Soil Physics Companion." CRC Press, Boca Raton, FL.

Kemp, D. D. (1994). "Global Environmental Issues: A Climatological Approach." Routledge, London.

Kemper, W. D., and Rosenau, R. C. (1986). Aggregate stability and size distribution. In: Klute, A., ed., "Methods of Soil Analysis, Part 1: Physical and Mineralogical Methods," Monograph No. 9. Am. Soc. Agron., Madison, WI.

Kezdi, A. (1974). "Handbook of Soil Mechanics, Vol. I. Soil Physics." Elsevier, Amsterdam.

Kincaid, DC. (1986). Intake rate: border and furrow. In: Klute, A., ed., "Methods of Soil Analysis," Monograph No. 9. Am. Soc. Agron., Madison, WI.

Kinnell, P. J. A. (1993a). Interrill erodibilities based on the rainfall intensity — flow discharge erosivity factor. *Aust. J. Soil Res.* 31, 319–332.

Kinnell, P. J. A. (1993b). Runoff as a factor influencing experimentally determined interrill erodibilities. *Aust. J. Soil Res.* 31, 333–342.

Kinniburgh, D. G., and Miles, D. L. (1983). Extraction and chemical analysis of interstitial water from soils and rocks. *Environ. Sci. Technol.* 17, 362–368.

Kirby, M. J., ed. (1978). "Hillslope Hydrology." Wiley, New York.

Kirkham, D. (1967). Explanation of paradoxes in Dupuit–Forchheimer seepage theory. *Water Resour. Res.* 3, 609–622.

Kirkham, D., and Powers, W. L. (1972). "Advanced Soil Physics." Wiley-Interscience, New York.

Kirkham, D., DeBoodt, M. F., and DeLeenheer, L. (1959). Modulus of rupture determination on undisturbed soil core samples. *Soil Sci.* 87, 141–144.

Kittrick, J. A., ed. (1986). "Soil Mineral Weathering." Van Nostrand Reinhold, New York.

Klotz, I. M. (1974). Water. In: Kasha, M., and Pullman, B., eds., "Horizons in Biochemistry," pp. 253–550. Academic Press, New York.

Kluitenberg, G. J. (2002). Heat capacity and specific heat. In: Dane, J. H., and Topp, G. C., eds. "Methods of Soil Analysis, Part 4: Physical Methods," Soil Science Society of America, Madison, WI.

Klute, A., ed. (1986a). "Methods of Soil Analysis, Part 1: Physical and Mineralogical Methods," Monograph No. 9. Am. Soc. Agron., Madison, WI.

Klute, A. (1986b). Water retention: Laboratory methods. In: Klute, A., ed., "Methods of Soil Analysis, Part 1: Physical and Mineralogical Methods," Monograph No. 9. Am. Soc. Agron., Madison, WI.

Klute, A., and Dirksen, C. (1986). Hydraulic conductivity and diffusivity: Laboratory methods. In: Klute, A., ed., "Methods of Soil Analysis, Part 1: Physical and Mineralogical Methods," Monograph No. 9. Am. Soc. Agron., Madison, WI.

Klute, A., and Peters, D. B. (1969). Water uptake and root growth. In: Whittington, W. J., ed., "Root Growth," pp. 105–133. Butterworths, London.

Knisel, W. G. (1980). CREAMS: A field-scale model for chemicals, runoff, and erosion from agricultural management systems, Conservation Research Report 26. USDA, Washington, DC.

Knowles, R. (1993). Methane: Processes of production and consumption. In: Harper, L. A., Mosier, A. R., Duxbury, J. M., and Rolston, D., eds., "Agricultural

Ecosystem Effects on Trace Gases and Global Climate Change." Special Publ. 55. Am. Soc. Agron., Madison, WI.

Komar, P. D., and Reimers, C. E. (1978). Grain shape effects on settling rates. J. Geol. 86, 193–209.

Kool, J. B., and Parker, J. C. (1988). Analysis of the inverse problem for transient unsaturated flow. *Water Resour. Res.* 24, 817–830.

Koolen, A. J., and Kuipers, H. (1983). "Agricultural Soil Mechanics." Springer-Verlag, Berlin.

Kramer, P. J., and Boyer, J. S. (1995). "Water Relations of Plants and Soils." Academic Press, San Diego., CA.

Kravtchenko, J., and Sirieys, P. M. (eds.) (1966). "Rheology and Soil Mechanics." Springer-Verlag, Berlin and New York.

Kristenson, K. J., and Lemon, E. R. (1964). Soil aeration and plant root relations: 111. Physical aspects of oxygen diffusion in the liquid phase of the soil. *Agron. J.* 56, 295–301.

Kueper, B. H., and Frind, E. O. (1988). An overview of immiscible fingering in porous media. *J. Contam. Hydrol.* 2, 95–110.

Kung, K. J. S. (1990). Preferential solute transport in a sandy vadose zone. *Geoderma* 46, 12–19.

Kussow, W., El-Swaify, S. A., and Mannering, J., eds. (1982). "Soil Erosion and Conservation in the Tropics," Special Publ. 43. Am. Soc. Agron., Madison, WI.

Kutilek, M. (1984). Some theoretical and practical aspects of infiltration in clays. In: Bouma, J., and Raats, P. A. C., eds., "Water and Solute Movement in Heavy Clay Soils," Int. Inst. Land Reclam. and Improve. Publ. 37. ILRI, Wageninger, Netherlands.

Kutilek, M., and Nielsen, D. R. (1994). "Soil Hydrology." Catena Verlag, Cremlingen, Germany.

Lagerwerff, J. V., Nakayama, F. S., and Frere, M. H. (1969). Hydraulic conductivity related to porosity and swelling of soil. *Soil Sci. Soc. Am. Proc.* 33, 3–11.

Lahav, N. (1994). Minerals and the origin of life: Hypotheses and experiments in heterogeneous chemistry. *Heterogeneous Chem. Rev.* 1, 159–179.

Lal, R. (1994). Global overview of soil erosion. In: Baker, R. S., ed., "Soil and Water Science: Key to Understanding Our Global Environment," Special Publ. 41. Soil Sci. Soc. Am., Madison, WI.

Lal, R., Kimble, J., Levine, E., and Stewart, B. A., eds. (1995b). "Soil Management and Greenhouse Effect." Lewis Publishers, Boca Raton, FL.

Lal, R., Kimble, J. M., Follet, R. F., and Stewart, B. A., eds. (1997). "Soil Processes and the Carbon Cycle." CRC Press, Boca Raton, FL.

Lal, R., Kimble, J. M., Follett, R. F., and Stewart, B. A., eds. (1998). "Management of Carbon Sequestration in Soil." CRC Press, Boca Raton, FL.

Laliberte, G. E. (1969). A mathematical function for describing capillary pressure-desaturation data. *Bull. Int. Assoc. Sci. Hydrol.* 142, 131–149.

Lane, L. J., Foster, G. R., and Nicks, A. D. (1987). Use of fundamental erosion mechanics in erosion prediction, Paper 87–2540. Am. Soc. Agr. Engineers, St. Joseph, MI.

Langmuir, I. (1918). The adsorption of gases on plane surfaces of glass, mica, and platinum. *J. Am. Chem. Soc.* 40, 1361–1402.

Langmuir, I. (1938). Distribution of cations between two charged plates. *Science* 811, 430–433.

Larson, W. E., Pierce, F. J., and Dowdy, R. H. (1983). The threat of soil erosion to long-term crop productivity. *Science* 219, 458–465.

Law, J. P. (1964). Effect of fatty alcohol and a nonionic surfactant on soil moisture evaporation in a controlled environment. *Soil Sci. Sci. Am. Proc.* 28, 695–699.

Leahy, J. G., and Colwell, R. R. (1990). Microbiol degradation of hydrocarbons in the environment. *Microbiol. Rev.* 54, 305–315.

Leeson, A., and Hinchee, R. E. (1995). "Principles and Practices of Bioventing," 2 vol. Prepared for Tyndall AFB, FL, USEPA, Cincinnati, OH, and U.S. Air Force Center for Environmental Excellence, Brooks AFB, TX, by Battelle Memorial Institute, Columbus, OH.

Legates, D. R. and C. J. Willmott. (1990a). Mean seasonal and spatial variability in global surface air temperature. *J. Theor. Appl. Climatol.* 41: 11–21.

Legates, D. R. and C. J. Willmott. (1990b). Mean seasonal and spatial variability in gauge-corrected, global precipitation. *Int. J. Climatol.* 10:111–127.

Leggett, J., ed. (1990). "Global Warming: The Greenpeace Resport." Oxford University Press, New York.

Leij, F. J., and van Genuchten, M. Th. (2002). Solute transport. In: Warrick, A. W., ed., "Soil Physics Companion" CRC Press, Boca Raton, FL.

Lemon, E. R. (1960). Photosynthesis under field conditions. II. An aerodynamic method for determining the turbulent carbon dioxide exchange between the atmosphere and a corn field. *Agron. J.* 52, 697–703.

Lemon, E. R. (1962). Soil aeration and plant root relations. 1. Theory. *Agron. J.* 54, 167–170.

Lemon, E. R., and Erickson, A. E. (1952). The measurement of oxygen diffusion in the soil with a platinum microelectrode. *Soil Sci. Soc. Am. Proc.* 16, 160–163.

Lenhard, R. J., and Parker, J. C. (1990). Estimation of free hydrocarbon volume from fluid levels in monitoring wells. *Ground Water.* 28(1), 57–67.

Leopold, L. B. (1974). "Water: A Primer." Freeman, San Francisco, CA.

Lesaffre, B., ed. (1990). "Land Drainage," Proc. 4th Int. Drainage Workshop, Cairo, Egypt, CEMAGREF-DICOVA, Antony Cedex, France.

Letey, J. (1968). Movement of water through soil as influenced by osmotic pressure and temperature gradients. *Hilgardia* 39, 405–418.

Letey, J. (1994). Is irrigated agriculture sustainable? In: Baker, R. S., Gee, G. W., and Rosenzweig, C., eds., "Soil and Water Science: Key to Understanding Our Global Environment," Special Publ. 41. Soil Sci. Soc. Am., Madison, WI.

Letey, J., and Stolzy, L. H. (1964). Measurement of oxygen diffusion rates with the platinum microelectrode: 1. Theory and equipment. *Hilgardia* 55, 545–554.

Letey, J., and Stolzy, L. H., and Kemper, W. D. (1967). Soil aeration. In: "Irrigation of Agricultural Lands," pp. 943–948. Am. Soc. Agron., Madison, WI.

Levy, G. J., Levin, J., and Shainberg, I. (1994). Seal formation and interrill soil erosion. *Soil Sci. Soc. Am. J.* 58, 203–209.

Levy, R., and Hillel, D. (1968). Thermodynamic equilibrium constants of Na-Ca exchange in some Israeli soils. *Soil Sci.* 106, 393–398.

Liebenow, A. M., Elliot, W. J., Laflen, J. M., and Kohl, K. D. (1990). Interrill erodibility: Collection and analysis of data from cropland soils. *Trans. ASAE* 33, 1882–1888.

Lin, C. C. (1955). "The Theory of Hydrodynamic Stability." Cambridge Univ. Press, London and New York.

Lin, D. S., Mancini, M., Troch, P., Wood, E. F., and Jackson, T. J. (1994). Comparisons of remotely sensed and model simulated soil moisture over a heterogeneous watershed. *Remote Sensing of Environment* 48, 159–171.

Lowdermilk, W. C. (1953). "Conquest of the Land Through Seven Thousand Years," Bulletin 99. U.S. Department of Agriculture, Washington, DC.

Lowery, B., and Morrison, J. E. (2002). Soil penetrometers and penetrability. In: Dane, J. H., and Topp, G. C., eds., "Methods of Soil Analysis, Part 4: Physical Methods." Soil Science Society of America, Madison, WI.

Luk, S. H. (1979). Effect of soil properties on erosion by wash and splash. *Earth Surf. Proc. Landforms* 4, 241–255.

Lundegard, P. D., and LaBrecque, D. (1996). Integrated geophysical and hydrologic monitoring of air sparging flow behavior. In: "Proc. of the 1st Int. Symp. *In Situ* Air

Sparging for Site Remediation," Oct. 24–25, (1996, Las Vegas, NV. INET, Potomac, MD.

Luth, H. A., and Wismer, R. D. (1971). Performance of plane soil cutting blade in sand. *Trans. ASAE* 14, 225–259, 262.

Luthin, J. N. (1966). "Drainage Engineering." Wiley, New York.

Luthin, J. N. (1974). Drainage analogues. In: "Drainage for Agriculture," Monograph No. 17. Am. Soc. Agron., Madison, WI.

Luxmoore, R. J., and Ferrand, L. A. (1993). Towards pore-scale analysis of preferential flow and chemical transport. In: Russo, D., and Dagan, G., eds., "Water Flow and Solute Transport in Soils: Development and Applications." Springer-Verlag, Berlin.

Mack, W. N., and Leistikow, E. A. (1996). Sands of the world. *Sci. Am.* 275, 44–49.

Mallants, D., Mohanty, B. P., Vervoort, A., and Feyen, J. (1997). Spatial analysis of saturated hydraulic conductivity in a soil with macropores. *Soil Technol.* 10, 115–132.

Marshall, C. E. (1964). "The Physical Chemistry and Mineralogy of Soils." Wiley, New York.

Marshall, T. J. (1959). The diffusion of gases through porous media. *J. Soil Sci.* 10, 79–82.

Marshall, T. J., and Holmes, J. W. (1979). "Soil Physics." Cambridge University Press, Cambridge.

Matheron, G. (1971). The theory of regionalized variables and its applications. Ecole des Mines, Fountainebleu, France.

Matheron, G. (1973). The intrinsic random functions and their applications. *Adv. Appl. Probability* 5, 439–468.

Matias, P., Correia, F. N., and Pereira, L. S. (1989). Influence of spatial variability of saturated hydraulic conductivity on the infiltration process. In: Morel-Seytoux, H. J., ed., "Unsaturated Flow in Hydraulic Modeling: Theory and Practice." Kluwer Academic Publishers, London.

Matthews, E. (1983). Global vegetation and land use: New high-resolution data bases for climate studies. *J. Climate and Appl. Meteorol.* 22, 474–487.

Matthews, E. (1984). Prescription of Land-surface Boundary Conditions in GISS GCM II: A Simple Method Based on High Resolution Vegetation Databases. NASA Tech. Memo. 86096. 20 pp. National Aeronautics and Space Administration. Greenbelt, MD. 20 pp. (Available from NTIS, Springfield, VA 22161.)

McBratney, A. B. and Odeh, I. O. A. (1997). Application of fuzzy sets in soil science: Fuzzy logic, fuzzy measurements and fuzzy decisions. *Geoderma* 77, 85–113.

McBride, M. B., (1994). "Environmental Chemistry of Soils." Oxford University Press, New York.

McCool, D. K., Foster, G. R., Mutchler, C. K., and Meyer, L. D. (1989). Revised slope length factor for the universal soil loss equation. *Trans. ASAE* 32, 1571–1576.

McCray, J. E., and Falta, R. W. (1996). Defining the air sparging radius of influence for ground-water remediation. *J. Contam. Hydrol.* 24, 25–52.

McDonald, J. E. (1954). The shape of raindrops. In: "The Physics of Everyday Phenomena." Freeman, San Francisco, CA.

McGregor, K. C., Greer, J. D., and Gurley, G. E. (1975). Erosion control with no-till cropping practices. *Trans. Am. Soc. Agr. Eng.* 18, 918–920.

McInnes, K. (2002). Soil heat: Temperature. In: Dane, J. H. and Topp, G. C., eds., "Methods of Soil Analysis, Part 4: Physical Methods." Soil Science Society of America, Madison, WI.

McIntyre, D. C. (1958). Permeability measurements of soil crusts formed by raindrop impact. *Soil Sci.* 85, 185–189.

McIntyre, D. S. (1970). The platinum microelectrode method for soil aeration measurement. In: "Advances in Agronomy," Vol. 22, pp. 235–283. Academic Press, New York.

McIntyre, D. S., and Philip, J. R. (1964). A field method for measurement of gas diffusion into soils. *Aust. J. Soil Res.* 2, 133–145.

McKyes, E. (1985). "Soil Cutting and Tillage." Elsevier, Amsterdam.

McLane, M. (1995). "Sedimentology." Oxford University Press, New York.

McNaughton, K. (1997). *WISPAS 67 (Water in the Soil-Plant-Atmosphere System).* Hort Research, Palmerston North, New Zealand.

McNeal, B. L. (1974). Soil salts and their effects on water movement. In: van Schilfgaarde, J., ed., "Drainage for Agriculture," Monograph No. 17. Am. Soc. Agron., Madison, WI.

McNeal, B. L., and Coleman, N. T. (1966). Effect of solution composition on soil hydraulic conductivity. *Soil Sci. Soc. Am. Proc.* 30, 308–312.

McWhorter, D. B. (1990). Unsteady radial flow of gas in the vadose zone. *J. Contam. Hydrol.* 5, 297–314.

McWhorter, D. B. (1995). Relevant processes concerning hydrocarbon contamination in low permeability soils. In: Walden, T., ed., "Petroleum Contaminated Low Permeability Soil: Hydrocarbon Distribution Processes, Exposure Pathways and *In Situ* Remediation Technologies." API Pub. No. 4631, pp. A-1–A-34. American Petroleum Institute, Washington, DC.

Meidner, H., and Sheriff, D. W. (1976). "Water and Plants." Halsted Press, New York; Wiley, New York.

Mein, R. G., and Larson, C. L. (1973). Modeling infiltration during a steady rain. *Water Resour. Res.* 9, 384–394.

Melillo, J. M., McGuire, A. D., Kicklighter, D. W., Moore III, B., Vorosmarty, C. J., and Schloss, A. L. (1993). Global climate change and terrestrial net primary production. *Nature* 363, 234–240.

Mercer, J. W., and Cohen, R. M. (1990). A review of immiscible fluids in the subsurface: Properties, models, characterization, and remediation. *J. Contam. Hydrol.* 6, 107–163.

Metting, F. B., Jr., ed. (1993). "Soil Microbial Ecology: Applications in Agricultural and Environmental Management." Marcel Dekker, New York.

Meyer, L. D. (1981). How rain intensity affects interrill erosion. *Trans. ASAE* 24, 1472–1475.

Meyer, L. D., and Harmon, W. C. (1984). Susceptibility of agricultural soils to interrill erosion. *Soil Sci. Soc. Am. J.* 48, 1152–1157.

Mihara, Y. (1951). Raindrops and soil erosion. Bulletin of Natural Institute of Agricultural Sciences Series A, 1.

Miller, B., and Buchan, G. (1996). TDR vs. neutron probe — how do they compare? *WISPAS (Water in Soil-plant-Atmosphere System)* 65, 8–9.

Miller, D. E., and Gardner, W. H. (1962). Water infiltration into stratified soil. *Soil Sci. Soc. Am. Proc.* 26, 115–118.

Miller, E. E. (1975). Physics of swelling and cracking soils. *J. Colloid Interface Sci.* 52, 434–443.

Miller, E. E. (1980). Similitude and scaling of soil-water phenomena. In: Hillel, D., "Applications of Soil Physics." Academic Press, New York.

Miller, E. E., and Klute, A. (1967). Dynamics of soil water. Part I — Mechanical forces. In: "Irrigation of Agricultural Lands," Monograph No. 11, pp. 209–244. Am. Soc. Agron., Madison, WI.

Miller, E. E., and Miller, R. D. (1955a). Theory of capillary flow: 1. Practical implications. *Soil Sci. Soc. Am. Proc.* 19, 267–271.

Miller, E. E., and Miller, R. D. (1955b). Theory of capillary flow: 2. Experimental information. *Soil Sci. Soc. Am. Proc.* 19, 271–275.

Miller, E. E., and Miller, R. D. (1956). Physical theory for capillary flow phenomena. *J. Appl. Phys.* 27, 324–332.

Miller, R. D. (1980). Freezing phenomena in soils. In: Hillel, D., "Applications of Soil Physics." Academic Press, New York.

Miller, R. J., and Low, P. F. (1963). Threshold gradient for water flow in clay systems. *Soil Sci. Soc. Am. Proc.* 27, 605–609.

Miller, R. M. (1996). Biological processes affecting contaminant fate and transport. In: Pepper, I. L., Gerba, C. P., and Brusseau, M. L., eds., "Pollution Science." Academic Press, San Diego.

Milligan, G. W. E., and Houlsby, G. T. (1986). "BASIC Soil Mechanics." Butterworths, London.

Millington, R. J. (1959). Gas diffusion in porous media. *Science* 130, 100–102.

Millot, G. (1979). Clay. *Sci. Am.* 240, 108–118.

Milly, P. C. D. (1985). A mass-conservative procedure for time-stepping in models of unsaturated flow. *Adv. Water Resour.* 8, 32–36.

Mishra, S., Parker, J. C., and Singhal, N. (1988). Estimation of soil hydraulic properties and their uncertainty from particle size distribution data. *J. Hydrol.* 108, 1–18.

Mitchell, A. R., and van Genuchten, M. T. (1991). Deterministic modeling of preferential flow in a cracked soil during flood irrigation. In: Gish, T. J., and Shirmohammadi, A., eds., "Preferential Flow." Am. Soc. Agric. Eng., St. Joseph, MI.

Moldenhauer, W. C., and Long, DC. (1964). Influence of rainfall energy on soil loss and infiltration rates: I. Effect over a range of texture. *Soil Sci. Soc. Am. J.* 28, 813–817.

Molz, F. J. (1976). Water transport in the soil-root system: Transient analysis. *Water Resour. Res.* 12, 805–807.

Monteith, J. L. (1973). "Principles of Environmental Physics." American Elsevier, New York.

Monteith, J. L. (1978). Models and measurement in crop climatology. In: "Proc. 11th Int. Soil Sci. Cong.," Edmonton,

Monteith, J. L. (1980). The development and extension of Penman's evaporation formula. In: Hillel, D., "Applications of Soil Physics." Academic Press, New York.

Monteith, J. L. (1997). *WISPAS 66 (Water in the Soil-Plant-Atmosphere System).* Hort Research, Palmerston North, New Zealand.

Monteith, J. L., Szeicz, G., and Yabuki, K. (1964). Crop photosynthesis and the flux of carbon dioxide below the canopy. *J. Appl. Ecol.* 6, 321–337.

Monteith, J. L., Szeicz, G., and Waggoner, P. E. (1965). The measurement and control of stomatal resistance in the field. *J. Appl. Ecol.* 2, 345–357.

Moore, D. C., and Singer, M. J. (1990). Crust formation effect on soil erosion processes. *Soil Sci. Soc. Am. J.* 54, 1117–1123.

Morel-Seytoux, H. J. (1983). Infiltration affected by air, seal, crust, ice, and various sources of heterogeneity. In: "Advances in Infiltration," Publ. 11–83. Am. Soc. Agric. Eng., St. Joseph, MI.

Morel-Seytoux, H. J. (1993). Capillary barrier at the interface of two layers. In: Russo, D., and Dagan, G., eds., "Water Flow and Solute Transport in Soils: Developments and Applications." Springer-Verlag, Berlin.

Morel-Seytoux, H. J., and Billica, J. A. (1985). A two-phase numerical model for prediction of infiltration: Application to a semi-infinite column. Water Resour. Res. 21: 607–615.

Morgan, R. P. C. (1986). "Soil Erosion and Conservation." Longman, Harlow, Essex, England.

Morin, J., and Benyamini, Y. (1977). Rainfall infiltration into bare soils. *Water Resour. Res.* 14, 813–837.

Morin, J., Goldberg, D., and Seginer, 1. (1967). A rainfall simulator with a rotating disk. *ASAE Trans.* 10, 74–77.

Morin, J., Benyamini, Y., and Michaeli, A. (1981). The effect of raindrop impact on the dynamics of soil surface crusting and water movement in the profile. *J. Hydrol.* 52, 321–335.

Mualem, Y. (1976). A new model for predicting the hydraulic conductivity of unsaturated porous media. *Water Resour. Res.* 12, 513–522.

Mualem, Y. (1984). A modified dependent domain theory of hysteresis. *Water Resour. Res.* 12, 513–522.

Mualem, Y., Assouline, L., and Rohdenburg, H. (1990). Rainfall induced soil seal. A. A critical review of observations and models. B. Applications of a new model to saturated soils. C. A dynamic model with kinetic energy instead of cumulative rainfall as independent variable. *Catena* 17, 185–203, 205–218, 289–303.

Mulla, D. J., and McBratney, A. B. (2000). Soil spatial variability. In: Sumner, M. E., ed., "Handbook of Soil Science." CRC Press, Boca Raton, FL.

Munn, J. R., Jr., and Huntington, G. L. (1976). A portable rainfall simulator for erodibility and infiltration measurements on rugged terrain. *Soil Sci. Soc. Am. J.* 59, 727–734.

Muzafi, S., Mandelbaum, R., and Kautski, L. (1997). Influence of long-term irrigation with wastewater on biological activity in the soil and on decomposition of pesticides. *Mayim ve Hashkaya* 368, 39–45 (In Hebrew).

Nalven, G. F., ed. (1997). "The Environment: Air, Water, and Soil." Am. Inst. Chem. Eng., New York.

Nash, H., and McCall, G. J. H., eds. (1995). "Groundwater Quality." Chapman and Hall, London.

National Research Council (1996). "A New Era for Irrigation." National Academy Press, Washington, DC.

Neilson, R. P., King, G. A., and Koerper, G. (1992). Toward a rule-based biome model. *Landscape Ecology* 7, 27–43.

Nemethy, G., and Scheraga, H. A. (1962). Structure of water and hydraulic bonding in proteins. 1. A model for the thermodynamic properties of liquid water. *J. Chem. Phys.* 36, 3382–3400.

Nerpin, S., Pashkina, S., and Bondarenko, N. (1966). The evaporation from bare soil and the way of its reduction. In: "Symp. Water Unsaturated Zone." Wageningen, The Netherlands.

Neuman, S. P. (1975). Galerkin approach to saturated-unsaturated flow in porous media. In: Gallagher, R. H., Oden, J. T., Taylor, C., and Zienkiewicz, O. C., eds., "Finite Elements in Fluids, Vol. 1: Viscous Flow and Hydrodynamics." John Wiley, London.

Newman, E. I. (1974). Root and soil water relations. In: Carson, E. W., ed., "The Plant Root and Its Environment," pp. 363–440. Univ. Press of Virginia, Charlottesville, VA.

Nielsen, D. R., and Biggar, J. W. (1961). Miscible displacement in soils: I. Experimental information. *Soil Sci. Soc. Am. Proc.* 25, 1–5.

Nielsen, D. R., and Biggar, J. W. (1962). Miscible displacement: III. Theoretical consideration. *Soil Sci. Soc. Am. Proc.* 26, 216–221.

Nielsen, D. R., and Bouma, J., eds. (1985) "Soil Spatial Variability." PUDOC, Wageningan, The Netherlands.

Nielsen, D. R., Jackson, R. D., Cary, J. W., and Evans, D. D., eds. (1972). "Soil Water." Am. Soc. Agron., Madison, WI.

Nielsen, D. R., Biggar, J. W., and Erh, K. T. (1973). Spatial variability of field-measured soil-water properties. *Hilgardia* 42, 215–259.

Nimah, M. N., and Hanks, R. J. (1973). Model for estimating soil, water, plant, and atmosphere interactions: 1. Description and sensitivity. *Soil Sci. Soc. Am. Proc.* 37, 522–527.

Nimmo, J. R. (2002). Property transfer from particle and aggregate size to water retention. In: Dane, J., and Topp, G. C., eds., "Methods of Soil Analysis, Part 4: Physical Methods." Soil Science Society of America, Madison, WI.

Nimmo, J. R., and Perkins, K. S. (2002). Aggregate stability and size distribution. In: Dane, J., and Topp, G. C., eds., "Methods of Soil Analysis, Part 4: Physical Methods." Soil Science Society of America, Madison, WI.

Nimmo, J. R., Rubin, J., and Hammermeister, D. P. (1987). Unsaturated flow in a centrifugal field: Measurement of hydraulic conductivity and testing of Darcy's law. *Water Resour. Res.* 23, 124–134.

Nobel, P. S. (1974). "Introduction to Biophysical Plant Physiology." Freeman, San Francisco, CA.

Noborio, K., McInnes, K. J., and Heilman, J. L. (1996). Measurements of soil water content, heat capacity, and thermal conductivity with a single TDR probe. *Soil Sci.* 161, 22–28.

Nori, F., Sholtz, P., and Bretz, M. (1997). Booming sand. *Sci. Am.* 277, 64–69.

Nyhan, J. W., Schofield, T. G., and Starmer, R. H. (1997). A water balance study of four landfill cover designs varying in slope for semi arid regions. *J. Env. Qual.* 26, 1385–1392.

Ogden, F. L., and Saghafian, B. (1997). Green and Ampt infiltration with redistribution. *J. Irrig. Drain. Eng.* 123, 386–393.

O'Neill, P. E., Chauhan, N. S., and Jackson, T. J. (1996). Use of active and passive microwave remote sensing for soil moisture estimation through corn. Internat. J. Remote Sens. 17, 1851–1865.

Osawa, S., and Honjo, T., eds. (1991). "Evolution of Life." Springer-Verlag, Tokyo.

Oster, J. D. (1982). Gypsum usage in irrigated agriculture: A review. *Fertilizer Res.* 3, 73–89.

Oster, J. D., and Schroer, F. W. (1979). Infiltration as influenced by irrigation water quality. *Soil Sci. Soc. Am. J.* 43: 611–615.

Oster, J. D., and Willardson, L. S. (1971). Reliability of salinity sensors for the management of soil salinity. *Agron. J.* 63, 695–698.

Pachepsky, Y. A., Timlin, D., and Varallyay, G. (1996). Artificial neural networks to estimate soil water retention from easily measurable data. *Soil Sci. Soc. Ma. J.* 60, 727–733.

Pan, L., and Wierenga, P. J. (1995). A transformed pressure-head based approach to solve Richards' equation for variably saturated soils. *Water Resour. Res.* 31, 925–931.

Papendick, R. I. and Runkles, J. R. (1965). Transient-state oxygen diffusion in soil. I. The case when rate of oxygen consumption is constant. *Soil Sci.* 1W, 251–261.

Parker, J. C. (1989). Multiphase flow and transport in porous media. *Rev. Geophys.* 27(3), 311–328.

Parker, J. J., and Taylor, H. M. (1965). Soil strength and seedling emergence relations. *Agron. J.* 57, 289–291.

Parlange, J. Y. (1971). Theory of water movement in soils. I. One dimensional absorption. *Soil Sci.* 111, 134–137.

Parlange, J. Y., and Hill, D. E. (1976). Theoretical analysis of wetting front instability in soils. *Soil Sci.* 122, 236–239.

Parlange, J. Y., Lisle, I., Braddock, R. D., and Smith, R. E. (1982). The three parameter infiltration equation. *Soil Sci.* 133, 337–341.

Payne, D., and Gregory, P. J. (1988). The soil atmosphere. In: Wild, A., ed., "Soil Conditions and Plant Growth." Longman, Harlow, Essex, England.

Pearse, J. F., Oliver, T. R., and Newitt, D. M. (1949). The mechanism of the drying of solids: Part I. The forces giving rise to movement of water in granular beds during drying. *Trans. Inst. Chem. Eng. (London)* 27, 1–8.

Peck, A. J. (1965). Moisture profile development and air compression during water uptake by bounded porous bodies: 3. Vertical columns. *Soils Sci.* 100, 44–51.

Peck, A. J. (1969). Entrapment, stability, and persistence of air bubbles in soil water. *Aust. J. Soil Res.* 7, 79–90.

Peck, A. J. (1970). Redistribution of soil water after infiltration. *Aust. J. Soil Res.* 7.

Peixoto, J. P. and Oort, A. H. (1992). "Physics of Climate." American Institute of Physics, New York, 520 pp.

Pelton, W. L. (1961). The use of lysimetric methods to measure evapotranspiration. *Proc. Hydrol. Symp.* 2, 106–134 (Queen's Printer, Ottawa, Canada. Cat. No. R32361/2).

Penman, H. L. (1940). Gas and vapor movements in the soil: 1. The diffusion of vapors through porous solids. *J. Agr. Sci.* 30, 437–461.

Penman, H. L. (1948). Natural evaporation from open water, bare soil and grass. *Proc. R. Soc. London ser. A* 193, 120–146.

Penman, H. L. (1949). The dependence of transpiration on weather and soil conditions. *J. Soil Sci.* 1, 74–89.

Penman, H. L. (1956). Evaporation: An introductory survey. *Neth J. Agr. Sci.* 4, 9–29.

Pennell, K. D. (2002). Specific surface area. In: Dane, J., and Topp, G. C., eds. "Methods of Soil Analysis, Part 4: Physical Methods," Soil Science Society of America, Madison, WI.

Penning de Vries, F. W. T. (1975). The cost of maintenance process in plant cells. *Amm. Botany* 39, 77–92.

Penin, s., Livingston, N. J., and Hook, W. R. (1995). Temperature-dependent measurement errors in time-domain reflectometry determinations of soil water. *Soil Sci. Soc. Am. J.* 59, 38–43.

Pepper, I. L., Gerba, C. P., and Brusseau, M. L., eds. (1996) "Pollution Science." Academic Press, San Diego.

Perfect, E., McLaughlin, N. B., Kay, B. D., and Topp, G. C. (1996). An improved fractal equation for the soil water retention curve. *Water Resour. Res.* 32, 281–287.

Perrier, E., Rieu, M., Spositio, G., and deMarsily, G. (1996). Models of the water retention curve for soils with a fractal pore size distribution. *Water Resour. Res.* 32, 3025–3031.

Perroux, K. M., and White, I. (1988). Design for disc permeameter. *Soil Sci. Soc. Am. J.* 52, 1205–1215.

Pessarakli, M. (1996). "Handbook of Plant and Crop Physiology." Marcel Dekker, New York.

Peters, D. B. (1965). Water availability. In: Black, C. A., ed., "Methods of Soil Analysis," pp. 279–285. Monograph No. 9. Am. Soc. Agron., Madison, WI.

Petersen, L. W., Thomsen, A., Moldrug, P., Jacobsen, D. H., and Rolston, D. E. (1995). High-resolution time domain reflectometry: Sensitivity dependency on probe-design. *Soil Sci.* 159, 149–154.

Peterson, L. W., Rolston, D. E., Moldrup, P., and Yamaguchi, T. (1994). Volatile organic vapor diffusion and adsorption in soils. *J. Env. Qual.* 23, 799–805.

Phene, C. J., (1986). Oxygen electrode measurement. In: Klute, A. ed., "Methods of Soil Analysis, Part 1: Physical and Mineralogical Methods." Amer. Soc. Agron., Madison, WI.

Phene, C. J., McCormick, R. L., Davis, K. R., and Oierro, J. (1989). A lysimeter system for precise evapotranspiration measurements and irrigation control. *Trans. Am. Soc. Agric. Eng.* 32, 477–484.

Phene, C. J., Clark, D. A., Cardon, G. E., and Mead, R. M. (1992). Soil matric potential sensor research and applications. In: "Advances in Measurement of Soil Physical Properties: Bringing Theory into Practice," Special Publ. 30, pp. 263–280. Soil Sci. Soc. Am., Madison, WI.

Philip, J. R. (1955a). Numerical solution of equations of the diffusion type with diffusivity concentration dependent. *Trans. Faraday Soc.* 51, 885–892.

Philip, J. R. (1995b). The concept of diffusion applied to soil water. *Proc. Natl. Acad. Sci. India* 24A, 93–104.

Philip, J. R. (1957a). Numerical solution of equations of the diffusion type with diffusivity concentration-dependent II. *Aust. J. Phys.* 10, 29–42.

Philip, J. R. (1957b). The theory of infiltration: 2. The profile at infinity. *Soil Sci.* 83, 435–448.
Philip, J. R. (1957c). The theory of infiltration: 3. Moisture profiles and relation to experiment. *Soil Sci.* 84, 163–178.
Philip, J. R. (1957d). The theory of infiltration: 4. Sorptivity and algebraic infiltration equations. *Soil Sci.* 84, 257–264.
Philip, J. R. (1957e). Evaporation, moisture and heat fields in the soil. *J. Meteorol.* 14, 354–366.
Philip, J. R. (1960). Absolute thermodynamic functions in soil-water studies. *Soil Sci.* 89, 111.
Philip, J. R. (1966a). Absorption and infiltration in two– and three–dimensional systems. In: Rijtema, R. E., and Wassink, H. eds., "Water in the Unsaturated Zone," Vol. 2, pp. 503–525. IASH/UNESCO Symp., Wageningen, The Netherlands.
Philip, J. R. (1966b). Plant water relations: Some physical aspects. *Ann Rev. Plant Physiol.* 17, 245–268.
Philip, J. R. (1968). Absorption and infiltration in two and three dimensional systems. In: "Proc. UNESCO Symp. Water in the Unsaturated Zone," Wageningen, The Netherlands.
Philip, J. R. (1969a). Theory of infiltration. *Adv. Hydrosci.* 5, 215–290.
Philip, J. R. (1969b). Hydrostatics and hydrodynamics in swelling soils. *Water Resour. Res.* 5, 1070–1077.
Philip, J. R. (1972). Hydrology of swelling soils. In: "Salinity and Water Use" (a national symposium on hydrology sponsored by the Australian Academy of Science). Macmillan, New York.
Philip, J. R. (1974). Water movement in soil. In: de Vries, D. A., and Afgan, N. H., eds., "Heat and Mass Transfer in the Biosphere," pp. 29–47. Halsted Press–Wiley, New York.
Philip, J. R. (1975). Stability analysis of infiltration. *Soil Sci. Soc. Am. Proc.* 39, 1042–1049.
Philip, J. R. (1993) Constant-rainfall infiltration on hillslopes and slopecrests. In: Russo, D., and Dagan, G. eds., "Water Flow and Solute Transport in Soils: Developments and Applications." Springer-Verlag, Berlin.
Philip, J. R. (1995). Desperately seeking Darcy in Dijon. *Soil Sci. Soc. Am. J.* 59, 319–324.
Philip, J. R., and de Vries, D. A. (1957). Moisture movement in porous materials under temperature gradients. *Trans. Am. Geophys. Un.* 38, 222–228.
Phillips, R. E., Blevius, R. L., Thomas, G. W., Frye, W. W., and Phillips, S. H. (1980). No tillage agriculture. *Science* 208, 1108–1113.
Pimentel, D. (1997). Pest management in agriculture. In: "Techniques for Reducing Pesticide Use." John Wiley, New York.
Pinder, F. G., and Gray, W. G. (1977). "Finite Element Simulation in Surface and Subsurface Hydrology." Academic Press, New York.
Polmann, D. J., McLaughlin, D., Luis, S., Gelhar, L. W., and Abadou, R. (1991). Stochastic modeling of large-scale flow in heterogeneous unsaturated soils. *Water Resour. Res.* 27, 1447–1458.
Post, D. F. (1996). Sediment (soil erosion) as a source of pollution. In: Pepper, I. L., Gerba, C. P., and Brusseau, M. L., eds., "Pollution Science." Academic Press, San Diego, CA.
Post, W. M., Emanuel, W. R., Zinke, P. J., and Stangenberger, A. G. (1982). Soil carbon pools and world life zones. *Nature* 298, 156–159.
Potter, C. S., Randerson, J. T., Field, C. B., Matson, P. A., Vitousek, P. M., Mooney, H. A., and Klooster, S. A. (1993). Terrestrial ecosystem production: A process model based on global satellite and surface data. *Global Biogeochemical Cycles* 7, 811–841.
Poulovassilis, A. (1962). Hysteresis of pore water, an application of the concept of independent domains. *Soil Sci.* 93, 405–412.

Power, J. F., and Schepers, J. S. (1989). Nitrate contamination of groundwater in North America. *Agric. Ecosyst. Environ.* 26, 165–187.

Prentice, I. C., Cramer, W., Harrison, S. P., Leemans, R., Monserud, R. A., and Solomon, A. M. (1992). A global biome model based on plant physiology and dominance, soil properties and climate. *Journal of Biogeography* 19, 117–134.

Price, J. C. (1982). Estimation of regional scale evapotranspiration through analysis of satellite thermal-infrared data. IEEE Trans. Geosci. and Rem. Sensing. GE 20, 286–292.

Prigogine, I. (1961). "Introduction to Thermodynamics of Irreversible Processes." Wiley, New York.

Pringle, J. (1975). The assessment and significance of aggregate stability in soil. In "Soil Physical Conditions and Crop Productions," Tech. Bulletin 29, pp. 249–260. Min. Agr., Fisheries and Food, London.

Quirk, J. P. (1986). Soil permeability in relation to sodicity and salinity. *Phil. Trans. R. Soc. (London)* A316, 297–317.

Quirk, J. P., and Schofield, R. K. (1955). The effect of electrolyte concentration on soil permeability. *J. Soil Sci.* 6, 163–178.

Raats, P. A. C. (1973). Unstable wetting fronts in uniform and nonuniform soils. *Soil Sci. Soc. Am. Proc.* 37, 681–685.

Rawitz, E., and Hillel, D. (1974). Progress and problems of drip irrigation in Israel. In: "Proc. Int. Conf. Drip Irrig.," San Diego, CA.

Rawitz, E., Margolin, M., and Hillel, D. (1972). An improved variable-intensity sprinkling infiltrometer. *Soil Sci. Soc. Am. Proc.* 36, 533–535.

Rawlins, S. L., and Campbell, G. S. (1986). Water potential: Thermocouple psychrometry. In: Klute, A., ed., "Methods of Soil Analysis, Part 1: Physical and Mineralogical Methods," Monograph No. 9. Am. Soc. Agron., Madison, WI.

Rawlins, S. L., and Raats, P. A. C. (1975). Prospects for high frequency irrigation. *Science* 188, 604–610.

Rechcigl, J. E., ed. (1995). "Soil Amendments: Impacts on Biotic Systems." CRC Press, Boca Raton, FL.

Reeve, R. C. (1965b). Modulus of rupture. In: "Methods of Soil Analysis," Part I. Monograph No. 9. Am. Soc. Agron., Madison, WI.

Reeve, R. C., and Bower, C. A. (1960). Use of high salt water as a flocculant and source of divalent cations for reclaiming sodic soils. *Soil Sci.* 90, 139–144.

Reicosky, D. C., and Ritchie, J. T. (1976). Relative importance of soil resistance and plant resistance in root water absorption. *Soil Sci. Soc. Am. J.* 40, 293–297.

Reicosky, D. C., Cassel, D. K., Blevius, R. L., Gill, W. R., and Naderman, G. C. (1977). Conservation tillage in the Southeast. *J. Soil Water Conserv.* 32, 13–19.

Remson, I., Drake, R. L., McNeary, S. S., and Walls, E. M. (1965). Vertical drainage of an unsaturated soil. *Am. Soc. Civil Eng. Proc. J. Hyd. Div.* 9, 55–74.

Remson, I., Fungaroli, A. A., and Hornberger, G. M. (1967). Numerical analysis of soil moisture systems. *Am. Soc. Civil Eng. Proc. J. Irrig. Drain. Div.* 3, 153–166.

Remson, I., Hornberger, G. M., and Molz, F. (1971). "Numerical Methods in Subsurface Hydrology." Wiley (Interscience), New York.

Renard, K. G., Foster, G. R., Weesies, G. A., and Porter, J. P. (1991). RUSLE revised universal soil loss equation. *J. Soil Water Conserv.* 46, 30–33.

Renard, K. G., Foster, G., Yoder, D. and McCool, D. (1994). RUSLE revisited: Status, questions, answers, and the future. *J. Soil Water Conserv.* 49, 213–220.

Reynolds, W. D., Elrick, D. E., Youngs, E. G., Amoozegar, A., Bootlink, H. W. G., and Bouma, J. (2002). Saturated and unsaturated water flow parameters. In: Dane, J. H., and Topp, G. C., eds., "Methods of Soil Analysis, Part 4: Physical Methods." Soil Science Society of America, Madison, WI.

Rhoades, J. D. (1974). Drainage for salinily control. In: van Schilfgaarde, J., ed., "Drainage for Agriculture," Monograph No. 17. Am. Soc. Agron., Madison, WI.

Rhoades, J. D., Chanduvi, F., and Lesch, S. M. (1999). Soil Salinity Assessment Methods and Interpretation of Electrical Conductivity Measurements. Irrigation and Drainage Paper 57, FAO, Rome.

Richards, L. A. (1931). Capillary conduction of liquids in porous mediums. *Physics* 1, 318–333.

Richards, L. A., ed. (1954). "Diagnosis and Improvement of Saline and Alkali Soils." U. S. Dept. Agr. Handbook No. 60.

Richards, L. A. (1960). Advances in soil physics. "*Trans. 7th Int. Congr. Soil Sci.*," Madison, WI, Vol. 1, pp. 67–69.

Richards, L. A., and Wadleigh, C. H. (1952). Soil water and plant growth. In: "Soil Physical Conditions and Plant Growth," Monograph No. 2. Am. Soc. Agron., Madison, WI.

Richards, L. A., Gardner, W. R., and Ogata, G. (1956). Physical processes determining water loss from soil. *Soil Sci. Soc. Am. Proc.* 20, 310–314.

Richards, S. J. (1965). Soil suction measurements with tensiometers. In: "Methods of Soil Analysis," Monograph No. 9, pp. 153–163. Am. Soc. Agron., Madison, WI.

Ripple, C. D., Rubin, J., and van Hylkama, T. E. A. (1972). Estimating steady-state evaporation rates from bare soils under conditions of high water table. U. S. Geol. Survey, Water Supp. Paper 2019-A.

Ritchie, J. T., and Adams, J. E. (1974). Field measurement of evaporation from soil shrinkage cracks. *Soil Sci. Soc. Am. Proc.* 38, 131–134.

Ritchie, J. T., and Nesmith, D. S. (1991). Temperature and crop development. In: Hanks, R. J., and Ritchie, J. T., eds., "Modeling Plant and Soil Systems," Monograph No. 31. Am. Soc. Agron., Madison, WI.

Robbins, C. W. (1991). Solute transport and reactions in salt-affected soils. In: Hanks, R. J., and Ritchie, J. T., eds., "Modeling Plant and Soil Systems," Monograph No. 31. Am. Soc. Agron., Madison, WI.

Rolston, D. E. (1986). Gas Flux. In: Klute, A., ed., "Methods of Soil Analysis, Part 1: Physical and Mineralogical Methods," Monograph No. 9. Am. Soc. Agron., Madison, WI.

Rolston, D. E., and Moldrup, P. (2002). Gas diffusivity. In: Dane, J. H., and Topp, G. C., eds., "Methods of Soil Analysis, Part 4: Physical Methods." Soil Science Society of America, Madison, WI.

Romano, N., and Santini, A. (2002). Water retention and storage: Field. In: Dane, J. H., and Topp, G. C., eds., "Methods of Soil Analysis, Part 4: Physical Methods." Soil Science Society of America, Madison, WI.

Romkens, M. J. M., Dabney, S. M., Govers, G., and Bradford, J. M. (2002). Soil erosion by water and tillage. In: Dane, J. H., and Topp, G. C., eds. "Methods of Soil Analysis, Part 4: Physical Methods." Soil Science Society of America, Madison, WI.

Rose, C. W. (1966). "Agricultural Physics." Pergamon, Oxford.

Rose, C. W. (1968). Evaporation from bare soil under high radiation conditions. In: "Trans. 9th Int. Congr. Soil Sci.," Adelaide, Vol. I, pp. 57–66.

Rose, C. W., and Stern, W. R. (1967a). Determination of withdrawal of water from soil by crop roots as a function of depth and time. *Aust. J. Soil Res.* 5, 11–19.

Rose, C. W., and Stern, W. R. (1967b). The drainage component of the water balance equation. *Aust. J. Soil Res.* 3, 95–100.

Rose, C. W., and Stern, W. R., and Drummond, J. E. (1965). Determination of hydraulic conductivity as a function of depth and water content for soil in situ. *Aust. J. Soil Res.* 3, 19.

Rose, C. W., Byrne, G. F. and Begg, J. E. (1966). An accurate hydraulic lysimeter with remote weight recording. CSIRO Div. Land Res. and Reg. Survey Tech. Paper 27. Canberra, Australia.

Rose, C. W., Byrne, G. F., and Hansen, G. K. (1976). Water transport from soil through plant to atmosphere: A lumped–parameter model. *Agric. Meteorol.* 16, 171–184.

Rosenberg, N. J. (1974). "Microclimate: The Biological Environment." Wiley, New York.

Rosenzweig, C. and Abramopoulos, F. (1997). Land-surface model development for the GISS GCM. *J. Climate* 10, 2040–2054.

Rosenzweig, C., and Hillel, D. (1998). "Climate Change and the Global Harvest: Potential Impacts of the Greenhouse Effect on Agriculture." Oxford University Press, New York.

Ross, P. J. (1990). Efficient numerical methods for infiltration using Richard's equation. Water Resour. Res. 26: 279–290.

Rowell, D. L. (1981). Oxidation and reduction. In: Greenland, D. J., and Hayes, M. H. D., eds., "The Chemistry of Soil Processes." Wiley, Chichester, UK.

Rubin, J. (1966). Theory of rainfall uptake by soils initially drier than their field capacity and its applications. *Water Resour. Res.* 2, 739–749.

Rubin, J. (1967). Numerical method for analyzing hysteresis-affected, post-infiltration redistribution of soil moisture. *Soil Sci. Soc. Am. Proc.* 31, 13–20.

Russell, C. S., and Shogren, J. F., eds. (1993). "Theory, Modelling, and Experience in the Management of Nonpoint–Source Pollution." Kluwer Academic Publishers, Boston.

Russell, E. W. (1973). "Soil Conditions and Plant Growth," 10th ed. Longman, London.

Russell, G. (1980). Crop evaporation, surface resistance, and soil water status. *Agric. Meteorol.* 21, 213–226.

Russell, R. S., and Goss, M. J. (1974). Physical aspects of soil fertility: The response of roots to mechanical impedance. *Neth. J. Agr. Sci.* 22, 305–318.

Russo, D. (1993). Analysis of solute transport in partially saturated heterogeneous soils. In: Russo, D., and Dagan, G. eds., "Water Flow and Solute Transport in Soils: Developments and Applications." Springer-Verlag, Berlin.

Russo, D. and Bresler, E. (1977). Analysis of saturated-unsaturated hydraulic conductivity in a mixed sodium-calcium soil system. *Soil Sci. Soc. Am. J.* 41, 706–710.

Russo, D., and Dagan, G. eds. (1993). "Water Flow and Solute Transport in Soils." Springer-Verlag, Berlin.

Russo, D., Zaidel, J., and Lanfer, A. (1994). Stochastic analysis of solute transport in partially saturated heterogeneous soil, 1. Numerical experiments. *Water Resour. Res.* 30, 769–779.

Sauer, T. J. (2002). Heat flux density. In: Dane, J. H. and Topp, G. C., eds., "Methods of Soil Analysis, Part 4: Physical Methods," Soil Science Society of America, Madison, WI.

Sauer, T. J., and Daniel, T. C. (1987). Effect of tillage system on runoff losses of surface-applied pesticides. *Soil Sci. Soc. Am. J.* 51, 410–415.

Scanlon, B. R., Nicot, J. P., and Massmann, J. W. (2002). Soil gas movement in unsaturated systems. In: Warrick, A. W., ed., "Soil Physics Companion." CRC Press, Boca Raton, FL.

Schaap, M. G. and Bouten, W. (1996). Modeling water retention curves of sandy soils using neural networks. *Water Resourc. Res.* 32, 3033–3040.

Scheinost, A. C., Sinowski, W., and Auerswalt, K. (1997). Regionalization of soil water retention curves in a highly variable soilscape. I. Developing a new pedotransfer function. *Geoderma* 78, 129–143.

Schlesinger, W. H. (1991). "Biogeochemistry: An Analysis of Global Change." Academic Press, San Diego.

Schmugge, T. (1990). Measurements of surface soil moisture and temperature. In: Hobbs, R. J., and Mooney, H. A., eds., "Remote Sensing of Biosphere Functioning." Springer-Verlag, New York.

Schnitzer, M. (1982). Organic matter characterization. In: Page, A. L. ed., "Methods of Soil Analysis, Part 2." Monograph No. 9, Amer. Soc. Agron., Madison, WI.

Schofield, C. S. (1940). Salt balance in irrigated areas. *J. Agr. Res.* 61, 17–39.

Schofield, R. K. (1946). Ionic forces in thick films between charged surfaces. *Trans. Faraday Soc.* 42B, 219–225.

Schwab, G. O., Fangmeier, D. D., Elliot, W. J., and Frevert, R. K. (1993). "Soil and Water Conservation Engineering." John Wiley & Sons, New York.

Schwertmann, U., and Fitzpatrick, R. W. (1992). Iron minerals in surface environments. *Catena Suppl.* 129, 7–29.

Scott Russell, R. (1977). "Plant Root Systems: Their Function and Interaction with the Soil." McGraw-Hill Book Co., London.

Selim, H. M., and Kirkham, D. (1970). Soil temperature and water content changes during drying as influenced by cracks: A laboratory experiment. *Soil Sci. Soc. Am. Proc.* 34, 565–569.

Sellers, W. D. (1965). "Physical Climatology." Univ. of Chicago Press, Chicago, IL.

Sellin, R. H. J. (1969). "Flow in Channels." Macmillan, New York.

Sextone, A. J., Revsbech, N. P., Parkin, T. B., and Tiedje, J. M. (1985). Direct measurement of oxygen profiles and denitrification rates in soil aggregates. Soil Sci. Soc. Am. J. 49: 645–651.

Shainberg, I. (1973). Ion exchange properties in irrigated soils. In: "Arid Zone Irrigation (B. Yaron, et al., eds.). Springer Verlag, Berlin."

Shainberg, I., and Letey, J. (1984). Response of soils to sodic and saline conditions. *Hilgardia* 52, 1–57.

Shainberg, I., and Shalhevet, J., eds. (1984). "Soil Salinity Under Irrigation." Springer-Verlag, New York.

Shanan, L., and Tadmor, N. H. (1976). "Micro-Catchment Systems for Arid Zone Development," Special Publ. Hebrew University of Jerusalem and the Center for International Agricultural Cooperation, Rehovot, Israel.

Shanan, L., Tadmor, N. H., Evenari, M., and Reiniger, P. (1970). Runoff farming in the desert. III. Microcatchments for improvement of desert range. *Agron. J.* 62, 445–449.

Sharma, M. L. (1985). Estimating evapotranspiration. In Hillel, D., ed., "Advances in Irrigation," Vol. 3. Academic Press, Orlando, FL.

Sheppard, M. I., Kay, B. D. and Loch, J. P. G. (1978). Development and testing of a computer model for heat and mass flow in freezing soils. *Soil Sci. Soc. Am. J.* 42, 38.

Shuttleworth, W. J. (1976). A one-dimensional theoretical description of the vegetation atmosphere interaction. *Boundary Layer Meteorol.* 10, 273–302.

Shuval, H. I., Adin, A., Fattal, B., Rawitz, E., and Yekutiel, P. (1986). "Wastewater Irrigation in Developing Countries." Tech. Paper No. 51, The World Bank, Washington, DC.

Silebi, C. A., and Schiesser, W. E. (1992). "Dynamic Modeling of Transport Process Systems." Academic Press, San Diego, CA.

Sinai, G., Jain, P. K., and Hillel, D. (1987). On the drainage of irrigated lands under sequential water application. In: Hillel, D., ed., "Advances in Irrigation," Vol. 4. Academic Press, Orlando, FL.

Sinclair, T. R. (1990). Theoretical considerations in the description of evaporation and transpiration. In Stewart, B. A., and Nielsen, D. R., eds., "Irrigation of Agricultural Crops," Monograph No. 30. Am. Soc. Agron., Madison, WI.

Singh, M. J., Janitzky, P., and Blackard, J. (1982). The influence of exchangeable sodium percentage on soil erodibility. *Soil Sci. Soc. Am. J.* 46, 117–121.

Skaggs, R. W. (1991). Drainage. In: Hanks, R. J., and Ritchie, S. T., eds. "Modeling Plant and Soil Systems," Monograph No. 31. Am. Soc. Agron., Madison, WI.

Skaggs, T. H., and Leij, F. J. (2002). Solute transport. In: Dane, J. H., and Topp, G. C., eds. "Methods of Soil Analysis, Part 4: Physical Methods." Soil Science Society of America, Madison, WI.

Skidmore, E. L., and Williams, J. R. (1991). Modified EPIC wind erosion model. In: Hanks, R. J., and Ritchie, S. T., eds. "Modeling Plant and Soil Systems," Monograph No. 31. Am. Soc. Agron., Madison, WI.

Skopp, J. M. (2002). Physical properties of primary particles. In: Warrick, A. W., ed., "Soil Physics Companion." CRC Press, Boca Raton, FL.

Slack, D. C., ed. (1983). "Advances in Infiltration." Am. Soc. Agr. Engineers, St. Joseph, MI.

Slatyer, R. O. (1967). "Plant Water Relationships." Academic Press, New York.

Slatyer, R. O., and McIlroy, I. C. (1961). "Practical Microclimatology." CSIRO, Australia.

Slichter, C. S. (1899). U. S. Geol. Sur. Ann. Rep. 19–II, pp. 295–384.

Smedema, L. K., and Rycroft, D. W. (1983). "Land Drainage: Planning and Design of Agricultural Drainage Systems." Cornell University Press, Ithaca, NY.

Smiles, D. E. (1974). Infiltration into swelling material. *Soil Sci.* 117, 110–116.

Smiles, D. E. (1976). On the validity of the theory of flow in saturated swelling materials. *Aust. J. Soil Res.* 14, 389–395.

Smiles, D. E., Philip, J. R., Knight, J. H., and Elrick, D. E. (1978). Hydrodynamic dispersion during sorption of water by soil. *Soil Sci. Soc. Am. J.* 42, 229–234.

Smith, R. E., and Woolhiser, D. A. (1971). Overland flow on an infiltrating surface. *Water Resources Res.* 5, 144–152.

So, H. B., Aylmore, L. A. C., and Quirk, J. P. (1976). The resistance of intact maize roots to water flow. *Soil Sci. Soc. Am. J.* 40, 222–225.

Soane, B. D. (1975). Studies on some soil physical properties in relation to cultivations and traffic. In "Soil Physical Conditions and Crop Production," Tech. Bulletin 29, pp. 249–260. Min. Agr., Fisheries and Food, London.

Söhne, W. H. (1966). Characterization of tillage tools. *Sairtryck ur Grundforbattring* 1, 31–48.

Soil Conservation Service (1988). Wind erosion. In: "National Agronomy Manual." Am. Soc. Agron., Madison, WI.

Soil Science Society of America (SSSA). (1996). "Glossary of Soil Science Terms." Soil Science Soc. Am., Madison, WI.

Soil Survey Division Staff (1992). "Keys to Soil Taxonomy," Technical Monograph No. 19. Soil Management Support Services, U. S. Department of Agriculture, Pocahontas Press, Blacksburg, VA.

Soil Survey Division Staff (1993). "Soil Survey Manual," Handbook 18. U. S. Department of Agriculture, U. S. Govt. Printing Office, Washington, DC.

Sorensen, V. M., Hanks, R. J., and Cartee, R. L. (1979). Cultivation during early season and irrigation influences on corn production. *Agron. J*, 67, 30–34.

Spalding, R. F., and Exner, M. E. (1993). Occurrence of nitrate in groundwater — a review. *J. Environ. Qual.* 22, 392–402.

Spangler, M. G., and Handy, R. L. (1982). "Soil Engineering." Harper and Row, New York.

Sparks, D. L. (1995). "Environmental Soil Chemistry." Academic Press, San Diego.

Sposito, G. (1984). "The Surface Chemistry of Soils." Oxford University Press, New York.

Sposito, G. (1989). "The Chemistry of Soils." Oxford University Press, New York.

Spreat, J. I. (1987). "The Ecology of the Nitrogen Cycle." Cambridge University Press, Cambridge.

Stanhill, G. (1965). Observation on the reduction of soil temperature. *Agr. Meteorol.* 2, 197–203.

Staple, W. J. (1969). Comparison of computed and measured moisture redistribution following infiltration. *Soil Sci. Soc. Am. Proc.* 33, 206.

Starr, J. L., DeRoo, H. C., Frink, C. R., and Parlange, J. Y. (1978). Leaching characteristics of a layered field soil. *Soil Sci. Soc. Am. J.* 42, 386–391.

Steenhuis, T. S., and Muck, R. E. (1988). Preferred movement of nonadsorbed chemicals on wet, shallow, sloping soils. *J. Environ. Qual.* 17, 367–384.

Steenhuis, T. S., Stanbitz, W. Andreini, M. S., Surface, J., Richard, T. L., Paulsen, R., Pickering, N. B., Hagerman, J. R., and Goehring, L. D. (1990). Preferential movement of pesticides and tracers in agricultural soils. *J. Irrig. Drain. Eng.* 116, 50–66.

Steinhardt, R., and Hillel, D. (1966). A rainfall simulator for laboratory and field use. *Soil Sci. Soc. Am. Proc.* 30, 680–682.

Stephens, D. B. (1995). "Vadose Zone Hydrology." Lewis Publishers, Boca Raton, FL.

Stern, R., Ben-Hur, M., and Shainberg, I. (1991). Clay mineralogy effect on rain infiltration, seal formation, and soil losses. *Soil Sci.* 152, 455–461.

Stevenson, F. J. (1982). "Humus Chemistry." Wiley, New York.

Stevenson, F. J. (1986). "Cycles of Soil." Wiley, New York.

Stewart, J. I., Danielson, R. E., Hanks, R. J., Jackson, E. B., Hagan, R. M., Pruitt, W. O., Franklin, W. T., and Riley, J. P. (1977). Optimizing crop production through control of water and salinity levels in the soil, PRWG 151-1, p. 191. Utah Water Lab, Logan, Utah.

Strebel, O., Duynisveld, W. H. M., and Bottcher, J. (1989). Nitrate pollution of groundwater in Western Europe. *Agric. Ecosyst. Environ.* 26, 189–214.

Suklje, L. (1969). "Rheological Aspects of Soil Mechanics." Wiley (Interscience). New York.

Sutcliffe, J. (1968). "Plants and Water." Edward Arnold, London.

Swain, R. W. (1975). Subsoiling. In: "Soil Physical Conditions and Crop Production," Tech. Bulletin 29, pp. 189–204. Min. Agr. Fish Food. HMSO, London.

Swarzendruber, D. (1962). Non Darcy behavior in liquid saturated porous media. *J. Geophys. Res.* 67, 5205–5213.

Swartzendruber, D. (1997). Exact mathematical solution of a two-term infiltration equation. *Water Resour. Res*, 33, 2502–2507.

Swartzendruber, D., and Hillel, D. (1973). The physics of infiltration. In: Hadas, A., *et al.*, eds., "Physics of Soil, Water and Salts in Ecosystems." Springer-Verlag, Berlin and New York.

Swartzendruber, D., and Hillel, D. (1975). Infiltration and runoff for small field plots under constant intensity rainfall. *Water Resour. Res.* 11, 445–451.

Swoboda-Colberg, N. G. (1995). Chemical contamination of the environment, types and fate of synthetic organic chemicals. In: Young, L. Y., and Cerniglia, C. E., eds., "Microbial Transformation and Degradation of Toxic Organic Chemicals." Wiley, New York.

Szeicz, G., van Bavel, C. H. M., and Takami, S. (1973). Stomatal factor in water use and dry matter production by sorghum. *Agr. Meteorol.* 12, 361–389.

Tackett, J. L., and Pearson, R. W. (1965). Some characteristics of soil crusts formed by simulated rainfall. *Soil Sci.* 99, 407–413.

Tadmor, N. H., Evenari, M., Shanan, L., and Hillel, D. (1957). The ancient desert agriculture of the Negev. *Isr. J. Agric. Res.* 8, 127–151.

Talsma, T. (1969). In site measurement of sorpivity. *Aust. J. Soil Res.* 7, 269–276.

Tamari, S., Wosten, J. H. M., and Ruiz-Suarez, J. C. (1996). Testing an artificial neural network for prediction soil hydraulic conductivity. *Soil Sci. Soc. Am. J.* 60, 1732–1741.

Tan, K. H. (1993). "Principles of Soil Chemistry." Marcel Dekker, New York.

Tanji, K. K., ed. (1990). "Agricultural Salinity Assessment and Management," Manual No. 71. Am. Soc. Civil Eng., New York.

Tanji, K. K. and Nour El Din, M. (1991). Nitrogen solute transport. In: Hanks, R. J., and Ritchie, J. T., eds., "Modeling Plant and Soil Systems," Monograph No. 31. Am. Soc. Agron., Madison, WI.

Tanner, C. B. (1957). Factors affecting evaporation from plants and soils. *J. Soil Water Conserv.* 12, 221–227.

Tanner, C. B. (1960). Energy balance approach to evapotranspiration from crops. *Soil Sci. Soc. Am. Proc.* 24, 1–9.

Tanner, C. B. (1968). Evaporation of water from plants and soil. In: "Water Deficits and Plant Growth." Academic Press, New York.

Tanner, C. B., and Lemon, E. R. (1962). Radiant energy utilized in evaporation. *Agron. J.* 54, 207–212.

Tanner, C. B., and Pelton, W. L. (1960). Potential evapotranspiration estimates by the approximate energy balance method of Penman. *J. Geophys. Res.* 65, 3391–3413.

Tarara, J. M., and Ham, J. M. (1997). Measuring soil water content in the laboratory and field with dual-probe heat-capacity sensors. *Agron. J.* 89, 535–542.

Taylor, S. A., and Cary, J. W. (1960). Analysis of the simultaneous flow of water and heat or electricity with the thermodynamics of irreversible processes. In: "Trans. 7th Int. Congr. Soil Sci.," Madison, WI, Vol. I, pp. 80–90.

Taylor, S. A., and Jackson, R. D. (1986). Heat capacity and specific heat. In: Klute, A., ed., "Methods of Soil Analysis, Part 1: Physical and Mineralogical Methods," Monograph No. 9. Am. Soc. Agron., Madison, WI.

Terzaghi, K. (1953). "Theoretical Soil Mechanics." Wiley, New York.

Tetzlaff, D. M., and Harbaugh, J. W. (1989). "Simulating Clastic Sedimentation." Van Nostrand, New York.

Thornthwaite, C. W. (1948). An approach toward a rational classification of climate. *Geograph. Rev.* 38, 55–94.

Tietje, O., and Hennings, V. (1996). Accuracy of the saturated hydraulic conductivity prediction by pedo-transfer functions compared to the variability within FAO textural classes. *Geoderma* 69, 71–84.

Tietje, O., and Tapkenhinrichs, M. (1993). Evaluation of pedo-transfer functions. *Soil Sci. Soc. Am. J.* 57, 1088–1095.

Tindall, J. A., Petrusak, R. L., and McMahon, P. B. (1995). Nitrate transport and transformation processes in unsaturated porous media. *J. Hydrol.* 169, 51–94.

Todd, D. K. (1967). "Ground Water Hydrology," 6th printing. Wiley, New York.

Topp, E., and Pattey, E. (1997). Soils as sources and sinks for atmospheric methane. *Can. J. Soil Sci.* 77, 167–178.

Topp, G. C., and Miller, E. E. (1966). Hysteresis moisture characteristics and hydraulic conductivities for glass–bead media. *Soil Sci. Soc. Am. Proc.* 30, 156–162.

Topp, G. C., and Davis, J. L. (1985). Time–domain reflectometry (TDR) and its application to irrigation scheduling. In: Hillel, D., ed., "Advances in Irrigation," Vol. 3. Academic Press, San Diego.

Topp, G. C., and Sattlecker, S. (1983). A rapid measurement of horizontal and vertical components of saturated hydraulic conductivity. *Can. Agric. Eng.* 25, 193–197.

Topp, G. C., and Ferré, P. A. (2002). Water content. In: Dane, J., and Topp, G. C., eds., "Methods of Soil Analysis, Part 4: Physical Methods." Soil Science Society of America, Madison, WI.

Touma, J., and Vauclin, M. (1986). Experimental and numerical analysis of two-phase infiltration in a partially saturated soil. *Trans. Porous Media* 1, 27–55.

Tseng, P. H., and Jury, W. A. (1994). Comparison of transfer function and deterministic modeling of area-averaged solute transport in a heterogeneous field. *Water Resour. Res.* 30, 2051–2063.

Tyler, S. W., and Wheatcraft, S. W. (1990). The consequences of fractal scaling in heterogeneous soils and porous media. In: D. Hillel and D. E. Elrick, eds., "Scaling in Soil Physics: Principles and Applications." *Soil Sci. Soc. Am. Special Pub.* 25, pp. 109–122.

Ulaby, F. T., Moore, R. K., and Fung, T. J. (1986). Microwave remote sensing: active and passive, Vol. III, "From Theory to Application," Artech House, Dedham, MA.

Unger, P. W. (1990). Conservation tillage systems. In Singh, R. P., ed., "Dryland Agriculture: Strategies for Sustainability," *Adv. Soil Sci.*, Vol. 13, Springer-Verlag, New York.

Unger, P. W., and Van Doren, D. M., eds. (1982). "Predicting Tillage Effects on Soil Physical Properties and Processes." Special Publ. 44, Am. Soc. Agron., Madison, WI.

Uri, N. D. (1999). "Conservation Tillage in U.S. Agriculture: Environmental, Economic, and Policy Issues." Food Products Press (Hawthorne Press), New York.

Vachaud, G., and Thony, J. L. (1971). Hysteresis during infiltration and redistribution in a soil column at different initial water contents. *Water Resour. Res.* 7, 111–127.

Vachaud, G., and Dane, J. H. (2002). Instantaneous profile. In: Dane, J. H., and Topp, G. C., eds. "Methods of Soil Analysis, Part 4: Physical Methods." Soil Science Society of America, Madison, WI.

van Bavel, C. H. M. (1952). Gaseous diffusion and porosity in porous media. *Soil Sci.* 73, 91–104.

van Bavel, C. H. M. (1963). Neutron scattering measurement of soil moisture: Development and current status. In: "Proc. Int. Symp. Humidity Moisture," Washington, DC, pp. 171–184.

van Bavel, C. H. M. (1966). Potential evaporation: The combination concept and its experimental verification. *Water Resour. Res.* 2, 455–467.

van Bavel, C. H. M. (1972). Soil temperature and crop growth. In: Hillel, D., ed., "Optimizing the Soil Physical Environment Toward Greater Crop Yields," pp. 23–33. Academic Press, New York.

van Bavel, C. H. M., and Hillel, D. (1975). A simulation study of soil heat and moisture dynamics as affected by a dry mulch. In: "Proc. Summer Comput. Simulat. Conf.," San Francisco, CA. Simulation Councils, LaJolla, CA.

van Bavel, C. H. M., and Hillel, D. (1976). Calculating potential and actual evaporation from a bare soil surface by simulation of concurrent flow of water and heat. *Agr. Meteorol.* 17, 453–476.

van Bavel, C. H. M., and Lascano, R. (1980). A numerical method to compute soil water content and temperature profiles under a bare surface. Texas A&M University, College Station, Texas.

van Bavel, C. H. M., Lascano, R. J., and Baker, J. M. (1985). Calibrating two-probe, gamma-gauge densitometers. *Soil Sci.* 140, 393–395.

van de Griend, A., Camillo, P. J., and Gurney, R. J. (1985). Discrimination of soil physical purameters, thermal inertia, and soil moisture from diurnal surface temperature fluctuations. *Water Resour. Res.* 21, 997–1009.

van der Ploeg, R. R., Ringe, H., Machulla, G., and Hermsmeyer, D. (1997). Postwar nitrogen use efficiency in West German agriculture and groundwater quality. *J. Environ. Qual.* 26, 1203–1211.

van Es, H. M. (2002). Soil variability. In: Dane, J. H., and Topp, G. C., eds., "Methods of Soil Analysis, Part 4: Physical Methods." Soil Science Society of America, Madison, WI.

van Genuchten, M. Th. (1978a, September). Calculating the unsaturated hydraulic conductivity with a new closed-form analytical model. Publication of the Water Resour. Prog. Dept. Civ. Eng., Princeton Univ., Princeton, NJ.

van Genuchten, M. Th. (1978b). Numerical solutions of the one-dimensional saturated-unsaturated flow equation. Res. Report 78-WR-9. Water Resour. Prog., Dept. Civil Eng., Princeton Univ., Princeton, NJ.

van Genuchten, M. Th. (1980). A closed-form equation for predicting the hydraulic conductivity of unsaturated soils. *Soil Sci. Soc. Am. J.* 44, 892–898.

van Genuchten, M. Th., and Shouse, P. J. (1989). Solute transport in heterogeneous field soils. In: Allen, D. T., Cohen, Y., and Kaplan, I. R., eds., "Intermedia Pollutant Transport: Modeling and Field Measurement." Plenum Publishing Corp., New York.

van Genuchten, M. Th., and Wierenga, P. J. (1974). Simulation of one-dimensional solute transfer in porous media. Agric. Exp. Stn. Bulletin 628, New Mexico State Univ., Las Cruces, NM.

van Genuchten, M. Th., and Wierenga, P. J. (1976). Mass transfer studies in sorbing porous media: I. Analytical solutions. Soil Sci. Soc. Am. J. 40: 473–479.

van Genuchten, M. Th., and Alves, W. J. (1982). Analytical solutions of the one-dimensional convective – dispersive solute-transport equation. USDA Tech. Bull. 1661.

van Genuchten, M. Th., Tang, D. H., and Guennelon, R. (1984). Some exact solutions for solute transport through soils containing large cylindrical macropores. *Soil Sci. Soc. Am. J.* 20, 1303–1310.

van Keulen, H., and Hillel, D. (1974). A simulation study of the drying-front phenomenon. *Soil Sci.* 118, 270–273.

van Olphen, H. (1963). "An Introduction to Clay Colloid Chemistry." Wiley (Interscience), New York.

van Wijk, W. R., ed. (1963). "Physics of the Plant Environment." North-Holland, Amsterdam.

van Wijk, W. R., and de Vries, D. A. (1963). Periodic temperature variations in homogeneous soil. In: van Wijk, W. R., ed., "Physics of Plant Environment." North-Holland, Amsterdam.

Vanoni, V. A. (1975). "Sedimentation Engineering." Am. Soc. Civil Eng., New York.

Vaux, H. J., and Pruitt, W. O. (1983). Crop-water production functions. In Hillel, D., ed., "Advances in Irrigation," Vol. 2. Academic Press, New York.

Veihmeyer, F. J., and Hendrickson, A. H. (1950). Soil moisture in relation to plant growth. *Ann. Rev. Plant Physiol.* 1, 285–304.

Veihmeyer, F. J., and Hendrickson, A. H. (1955). Does transpiration decrease as the soil moisture decreases? *Trans. Am. Geophys. Un.* 36, 425–448.

Vernadsky, Vladimir (1986). "The Biosphere." Synergistic Press, Oracle, AZ. (Abridged version based on 1929 French edition.)

Vieira, S. R., Nielsen, D. R., and Biggar, J. W. (1981). Spatial variability of field-measured infiltration rate. *Soil Sci. Soc. Am. J.* 45, 1040–1048.

Viets, F. G., Jr. (1962). Fertilizers and the efficient use of water. *Adv. Agron.* 14, 228–261.

Visser, W. C. (1966). Progress in the knowledge about the effect of soil moisture content on plant production, Tech. Bulletin 45. Inst. Land Water Management, Wageningen, Netherlands.

Vomocil, J. A., and Flocker, W. J. (1961). Effect of soil compaction on storage and movement of soil air and water. *Trans. Am. Soc. Agr. Eng.* 4, 242–246.

Wagenet, R. J., and Hutson, J. L. (1987). "LEACHM: A Finite-Difference Model for Simulating Water, Salt, and Pesticide Movement in the Plant Root Zone." New York State Resources Institute, Cornell Univ., Ithaca, NY.

Wallace, A., and Terry, R. E., eds. (1997). "Handbook of Soil Conditioners: Substances that Enhance the Physical Properties of Soil." Marcel Dekker, New York.

Wang, D., Yates, S. R., Simuneck, J., and van Genuchten, M. T. (1997). Solute transport in simulated conductivity fields under different irrigations. *J. Irrig. Drain. Eng.* 123, 336–343.

Wang, F. C., and Lakshminarayana, V. (1968). Mathematical simulation of water movement through unsaturated nonhomogeneous soil. *Soil Sci. Soc. Am. Proc.* 32, 329–334.

Ward, R. C., and Robinson, M. (1990). "Principles of Hydrology." McGraw-Hill, London.

Warnant, P., Francois, L., Strivay, D., and Gerard, J. C. (1994). CARAIB: A global model of terrestrial biological productivity. *Global Biogeochemical Cycles* 8, 225–270.

Warrick, A. W. (1988). Additional solutions for steady-state evaporation from a shallow water table. *Soil Sci.* 146, 63–66.

Warrick, A. W. (1990a). Application of scaling to the characterization of spatial variability of soils. In: D. Hillel, ed., "Scaling in soil physics: Principles and applications." *Soil Sci. Soc. Amer. Special Pub.* 25, pp. 39–51.

Warrick, A. W. (1990b). Nature and dynamics of soil water. In: Stewart, B. A., and Nielsen, D. R., eds., "Irrigation of Agricultural Crops," Monograph No. 30. Am. Soc. Agron., Madison, WI.

Warrick, A. W. (1992). Models for disc infiltrometers. *Water Resour. Res.* 28, 1319–1327.

Warrick, A. W. (1993). Inverse estimation of soil hydraulic properties with scaling: One-dimensional infiltration. *Soil Sci. Soc. Am. J.* 57, 631–636.

Warrick, A. W. (1998). Spatial variability. In: Hillel, D., "Environmental Soil Physics." Academic Press, San Diego, CA.

Warrick, A. W., and Nielsen, D. R. (1980). Spatial variability of soil physical properties in the field. In: Hillel, D., ed., "Applications of Soil Physics." Academic Press, New York.

Watson, D. A., and Laflen, J. M. (1986). Soil strength, slope, and rainfall intensity effects on interrill erosion. *Trans. Am. Soc. Agr. Eng.* 29, 98–102.

Watson, J. E. (1996). Pesticides as a source of pollution. In: Pepper, I. L., *et al.*, eds., "Pollution Science." Academic Press, San Diego.

Watson, K. K. (1996). An instantaneous profile method for determining the hydraulic conductivity of unsaturated porous materials. *Water Resour. Res.* 2, 709–715.

Webster, R., and Oliver, M. A. (1990). "Statistical Methods in Soil and Land Resource Survey." Oxford University Press, New York.

Wendroth, O., Katul, G. G., Parlange, M. B., Puente, C. E., and Nielsen, D. R. (1993). A nonlinear filtering approach for determining hydraulic conductivity functions. *Soil Sci.* 156, 293–301.

Wesseling, J. (1962). Some solutions of the steady-state diffusion of carbon dioxide through soils. *Neth. J. Agr. Sci.* 10, 109–117.

Wesseling, J. (1974). Crop growth and wet soils. In: van Schilfgaarde, J., ed., "Drainage for Agriculture," Monograph No. 17, pp. 7–38. Am. Soc. Agron., Madison, WI.

West, C. C., and Harwell, J. H. (1992). Surfactants and sub-surface remediation. *Environ. Sci. Technol.* 26, 2324–2330.

Whalley, W. R. (1993). Considerations on the use of time-domain reflectometry (TDR) for measuring soil water content. *J. Soil Sci.* 44, 1–9.

White, I., and Broadbridge, P. (1988). Constant rate rainfall infiltration: A versatile nonlinear model. *Water Resour. Res.* 24, 155–162.

White, I., and Perroux, K. M. (1987). Use of sorptivity to determine field soil hydraulic properties. *Soil Sci. Soc. Am. J.* 51, 1093–1101.

White, N. F., Duke, H. R., Sunada, D. K., and Corey, A. T. (1970). Physics of desaturation in porous materials. *J. Irr. Div., ASCE Proc.* JR-2, 165–191.

White, R. E. (1985). The influence of macropores on the transport of dissolved and suspended matter through soil. In: "Advances in Soil Science," Springer-Verlag, New York.

Whittig, L. D., and Allardice, W. R. (1986). X-ray diffraction techniques. In: Klute, A., ed., "Methods of Soil Analysis, Part 1: Physical and Mineralogical Methods," Monograph No. 9. Am. Soc. Agron., Madison, WI.

Wickramanayake, G. B., Gupta, N., Hinchee, R. E., and Nielsen, B. J. (1991). Free petroleum hydrocarbon volume estimates from monitoring well data. *J. Environ. Eng.* 117(5), 686–691.

Wiedenroth, M. (1981). Relations between photosynthesis and root metabolism of cereal seedlings influenced by root anaerobiosis. *Photosynthetica* 15, 575–591.

Wierenga, P. J., and de Wit, C. T. (1970). Simulation of heat transfer in soils. *Soil Sci. Soc. Am. Proc.* 32, 326–328.

Wild, A. (1993). "Soils and the Environment." Cambridge University Press, Cambridge.

Wild, J. R., Varfolomeyer, S. D., and Scozzafava, A., eds. (1996). "Perspectives in Bioremediation: Technologies for Environmental Improvement." Kluwer Academic Publishers, Dordrecht, The Netherlands.

Williams, J. R. (1991). Runoff and water erosion. In: Hanks, R. J., and Ritchie, J. T., eds., "Modeling Plant and Soil Systems," Monograph No. 31. Am. Soc. Agron., Madison, WI.

Willmott, C., M. Rowe, and Y. Mintz. (1985). Climatology of the terrestrial seasonal water cycle. *J. Climatol.* 5: 589–606.

Winger, R. J. (1960). In-place permeability tests and their use in subsurface drainage. Off. of Drainage and Ground Water Eng., Burl of Reclamation, Denver, CO.

Winter, E. J. (1974). "Water, Soil and the Plant." Macmillan, London.

Wischmeier, W. H., and Smith, D. D. (1978). Predicting rainfall erosion losses — a guide to conservation planning. Handbook 537. U.S. Department of Agriculture, U.S. Govt. Print. Off., Washington, DC.

Wittmus, H. D., Triplett, G. B., Jr., and Greb, B. W. (1973). Concepts of conservation tillage using surface mulches. In: "Conservation Tillage," pp. 5–12. Soil Conserv. Soc. Am., Ames, IA.

Wittmus, H., Olson, L., and Delbert, L. (1975). Energy requirements for conventional versus minimum tillage. *J. Soil Water Conserv.* 30, 72–75.

Wood, J. A., and Anthony, D. H. J. (1997). Herbicide contamination of prairie springs at ultratrace levels of detection. *J. Environ. Qual.* 26, 1308–1317.

Wooding, R. A. (1969). Growth of fingers at an unstable diffusing interface in a porous medium or Hele-Shaw cell. *J. Fluid Mech.* 39, 477–495.

Woodruff, N. P., and Siddoway, F. H. (1965). A wind erosion equation. *Soil Sci. Soc. Am. Proc.* 29, 602–608.

Yaron, B., and Bresler, E. (1983). Economic analysis of on-farm irrigation using response functions of crops. In Hillel, D., ed., "Advances in Irrigation," Vol. 2. Academic Press, New York.

Yaron, B., Calvet, R., and Prost, R. (1996). "Soil Pollution: Processes and Dynamics." Springer-Verlag, Berlin.

Yaron, B., Gerstl, Z., and Spencer, W. F. (1985). Behavior of herbicides in irrigated soils. *Adv. Soil Sci.* 3, 121–211.

Yavorsky, B., and Detlaf, A. (1972). "Handbook of Physics." Mir, Moscow.

Yiridoe, E. K., Voroney, R. P. and Weersink, A. (1997). Impact of alternative farm management practices on nitrogen pollution of groundwater: Evaluation and application of CENTURY model. *J. Environ. Qual.* 26, 1255–1263.

Yong, R. N., and Osler, J. C. (1966). On the analysis of soil deformation under a moving rigid wheel. In: "Proc. Int. Conf. Soc. Terrain-Vehicle Sys.," p. 341.

Yong, R. N., and Warkentin, B. P. (1975). "Soil Properties and Behaviour." Elsevier, Amsterdam.

Young, M. H., and Sisson, J. B. (2002). Tensiometry. In: Dane, J. H., and Topp, G. C., eds., "Methods of Soil Analysis, Part 4: Physical Methods." Soil Science Society of America, Madison, WI.

Young, M. H. (2002). Water potential: Piezometry. In: Dane, J. H., and Topp, G. C., eds. "Methods of Soil Analysis, Part 4: Physical Methods." Soil Science Society of America, Madison, WI.

Young, M. H., Fleming, J. B., Wierenga, P. J., and Warrick, A. W. (1997). Rapid laboratory calibration of time domain reflectometry using upward infiltration. *Soil Sci. Soc. Am. J.* 61, 707–712.

Youngs, E. G. (1958a). Redistribution of moisture in porous materials after infiltration. *Soil Sci.* 86, 117–125.

Youngs, E. G. (1958b). Redistribution of moisture in porous materials after infiltration. *Soil Sci.* 86, 202–207.

Youngs, E. G. (1960a). The drainage of liquids from porous materials. *J. Geophys. Res.* 65, 4025–4030.

Youngs, E. G. (1960b). The hysteresis effect in soil moisture studies. In: "Trans. 7th Int. Soil Sci. Congr.," Madison, WI, Vol. 1, pp. 107–113.

Youngs, E. G. (1964). An infiltration method of measuring the hydraulic conductivity of porous materials. *Soil Sci.* 97, 307–311.

Youngs, E. G. (1983). The use of similar media theory in the consideration of soil water redistribution in infiltrated soils. In: "Advances in Infiltration," Publ. 11–83, Am. Soc. Agr. Eng., St. Joseph, MI.

Yu, C., Warrick, A. W., Conklin, M. H., Young, M. H., and Zreda, M. (1997). Two-and-three-parameter calibrations of time domain reflectometry for soil moisture measurement. *Water Resour. Res.* 33, 2417–2422.

Zaradny, H., and Feddes, R. A. (1979). Calculation of non-steady flow towards a drain in saturated-unsaturated soil by finite elements. *Agric. Water Manage.* 2, 37–53.

Zimmerman, M. H. (1983). "Xylem Structure and the Ascent of Sap." Springer-Verlag, Berlin.

INDEX

B

C

F

G

H

I

K

L

M

N

O

P

T

X

Y

Z

This book is but a clearing at the edge of the woods
where students might observe a few of the trees
as they prepare to set forth independently
to explore the great forest that yet lies beyond, –
and to care for it.

Printed and bound by CPI Group (UK) Ltd, Croydon, CR0 4YY
03/10/2024
01040309-0018